# WIMAX NETWORK PLANNING AND OPTIMIZATION

# WIRELESS NETWORKS AND MOBILE COMMUNICATIONS

Dr. Yan Zhang, Series Editor
Simula Research Laboratory, Norway
E-mail: yanzhang@ieee.org

**Broadband Mobile Multimedia:**
**Techniques and Applications**
Yan Zhang, Shiwen Mao,
Laurence T. Yang, and Thomas M. Chen
ISBN: 978-1-4200-5184-1

**Cooperative Wireless**
**Communications**
Yan Zhang, Hsiao-Hwa Chen,
and Mohsen Guizani
ISBN: 978-1-4200-6469-8

**Distributed Antenna Systems:**
**Open Architecture for Future**
**Wireless Communications**
Honglin Hu, Yan Zhang, and Jijun Luo
ISBN: 978-1-4200-4288-7

**The Internet of Things:**
**From RFID to the Next-Generation**
**Pervasive Networked Systems**
Lu Yan, Yan Zhang, Laurence T. Yang,
and Huansheng Ning
ISBN: 978-1-4200-5281-7

**Millimeter Wave Technology in**
**Wireless PAN, LAN and MAN**
Shao-Qiu Xiao, Ming-Tuo Zhou
and Yan Zhang
ISBN: 978-0-8493-8227-7

**Mobile WiMAX:**
**Toward Broadband Wireless**
**Metropolitan Area Networks**
Yan Zhang and Hsiao-Hwa Chen
ISBN: 978-0-8493-2624-0

**Resource, Mobility, and Security**
**Management in Wireless Networks**
**and Mobile Communications**
Yan Zhang, Honglin Hu, and Masayuki Fujise
ISBN: 978-0-8493-8036-5

**Security in RFID and Sensor Networks**
Yan Zhang and Paris Kitsos
ISBN: 978-1-4200-6839-9

**Security in Wireless Mesh**
**Networks**
Yan Zhang, Jun Zheng and Honglin Hu
ISBN: 978-0-8493-8250-5

**Unlicensed Mobile Access**
**Technology: Protocols, Architectures,**
**Security, Standards, and Applications**
Yan Zhang, Laurence T. Yang,
and Jianhua Ma
ISBN: 978-1-4200-5537-5

**WiMAX Network Planning and**
**Optimization**
Yan Zhang
ISBN: 978-1-4200-6662-3

**Wireless Mesh Networking:**
**Architectures, Protocols, and**
**Standards**
Yan Zhang, Jijun Luo, and Honglin Hu
ISBN: 978-0-8493-7399-2

**Wireless Quality-of-Service:**
**Techniques, Standards, and**
**Applications**
Maode Ma, Mieso K. Denko, and Yan Zhang
ISBN: 978-1-4200-5130-8

## AUERBACH PUBLICATIONS

www.auerbach-publications.com
To Order Call: 1-800-272-7737 • Fax: 1-800-374-3401
E-mail: orders@crcpress.com

# WIMAX NETWORK PLANNING AND OPTIMIZATION

Edited by Yan Zhang

CRC Press
Taylor & Francis Group
Boca Raton London New York

CRC Press is an imprint of the
Taylor & Francis Group, an **informa** business

AN AUERBACH BOOK

Auerbach Publications
Taylor & Francis Group
6000 Broken Sound Parkway NW, Suite 300
Boca Raton, FL 33487-2742

© 2009 by Taylor & Francis Group, LLC
Auerbach is an imprint of Taylor & Francis Group, an Informa business

**Library of Congress Cataloging-in-Publication Data**

Zhang, Yan, 1977-
    WiMAX network planning and optimization / Yan Zhang.
        p. cm.
    Includes bibliographical references and index.
    ISBN 978-1-4200-6662-3 (alk. paper)
    1. Wireless communication systems. 2. Broadband communication systems. 3. IEEE 802.16
(Standard) I. Title.

TK5103.2.Z55 2009
621.382--dc22                                                                    2008040983

**Visit the Taylor & Francis Web site at**
**http://www.taylorandfrancis.com**

**and the Auerbach Web site at**
**http://www.auerbach-publications.com**

# Contents

## PART I  WiMAX Fundamentals

## PART II     WiMAX Planning and Optimization

# Editor

**Yan Zhang** received his BS in communication engineering from Nanjing University of Post and Telecommunications, China; his MS in electrical engineering from Beijing University of Aeronautics and Astronautics, China; and his PhD in the School of Electrical & Electronics Engineering, Nanyang Technological University, Singapore. He is an associate editor on the editorial board of *Wiley Wireless Communications and Mobile Computing (WCMC)*; *Security and Communication Networks* (Wiley); *International Journal of Network Security*; *International Journal of Ubiquitous Computing*; *Transactions on Internet and Information Systems (TIIS)*; *International Journal of Autonomous and Adaptive Communications Systems (IJAACS)*; *International Journal of Ultra Wideband Communications and Systems (IJUWBCS)*; and *International Journal of Smart Home (IJSH)*. He is currently serving as the book series editor for the Wireless Networks and Mobile Communications book series (Auerbach Publications, CRC Press, Taylor & Francis Group). He serves as guest coeditor for the following: *IEEE Intelligent Systems*, special issue on "Context-Aware Middleware and Intelligent Agents for Smart Environments"; *Wiley Security and Communication Networks* special issue on "Secure Multimedia Communication"; *Springer Wireless Personal Communications* special issue on selected papers from ISWCS 2007; *Elsevier Computer Communications* special issue on "Adaptive Multicarrier Communications and Networks"; *International Journal of Autonomous and Adaptive Communications Systems (IJAACS)* special issue on "Cognitive Radio Systems"; *The Journal of Universal Computer Science (JUCS)* special issue on "Multimedia Security in Communication"; *Springer Journal of Cluster Computing* special issue on "Algorithm and Distributed Computing in Wireless Sensor Networks"; *EURASIP Journal on Wireless Communications and Networking (JWCN)* special issue on "OFDMA Architectures, Protocols, and Applications"; and *Springer Journal of Wireless Personal Communications* special issue on "Security and Multimodality in Pervasive Environments."

He is serving as coeditor for several books: *Resource, Mobility and Security Management in Wireless Networks and Mobile Communications*; *Wireless Mesh Networking: Architectures, Protocols and Standards*; *Millimeter-Wave Technology in Wireless PAN, LAN and MAN*; *Distributed Antenna Systems: Open Architecture for Future Wireless Communications*; *Security in Wireless Mesh Networks*; *Mobile WiMAX: Toward Broadband Wireless Metropolitan Area Networks*; *Wireless Quality-of-Service: Techniques, Standards and Applications*; *Broadband Mobile Multimedia: Techniques and Applications*; *Internet of Things: From RFID to the Next-Generation Pervasive Networked Systems*; *Unlicensed Mobile Access Technology: Protocols, Architectures, Security, Standards and Applications*; *Cooperative Wireless Communications*; *WiMAX Network Planning and Optimization*; *RFID Security: Techniques, Protocols and System-On-Chip Design*; *Autonomic Computing and Networking*; *Security in RFID and Sensor Networks*; *Handbook of Research on Wireless Security*; *Handbook of Research on Secure Multimedia Distribution*; *RFID and Sensor Networks*; *Cognitive Radio Networks*; *Wireless Technologies for Intelligent Transportation Systems*; *Vehicular Networks: Techniques, Standards and Applications*; *Orthogonal Frequency Division Multiple Access (OFDMA)*; *Game Theory for Wireless Communications and Networking*; and *Delay Tolerant Networks: Protocols and Applications*.

Dr. Zhang serves as symposium cochair for the following: ChinaCom 2009; program cochair for BROADNETS 2009; program cochair for IWCMC 2009; workshop cochair for ADHOCNETS 2009; general cochair for COGCOM 2009; program cochair for UC-Sec 2009; journal liasion chair for IEEE BWA 2009; track cochair for ITNG 2009; publicity cochair for SMPE 2009; publicity cochair for COMSWARE 2009; and publicity cochair for ISA 2009. He has also served as general cochair for WAMSNet 2008; publicity cochair for TrustCom 2008; general cochair for COGCOM 2008; workshop cochair for IEEE APSCC 2008; general cochair for WITS-08; program cochair for PCAC 2008; general cochair for CONET 2008; workshop chair for SecTech 2008; workshop chair for

SEA 2008; workshop co-organizer for MUSIC'08; workshop co-organizer for 4G-WiMAX 2008; publicity cochair for SMPE-08; international journals coordinating cochair for FGCN-08; publicity cochair for ICCCAS 2008; workshop chair for ISA 2008; symposium cochair for ChinaCom 2008; industrial cochair for MobiHoc 2008; program cochair for UIC-08; general cochair for CoNET 2007; general cochair for WAMSNet 2007; workshop cochair FGCN 2007; program vice-cochair for IEEE ISM 2007; publicity cochair for UIC-07; publication chair for IEEE ISWCS 2007; program cochair for IEEE PCAC'07; special track cochair for "Mobility and Resource Management in Wireless/Mobile Networks" in ITNG 2007; special session co-organizer for "Wireless Mesh Networks" in PDCS 2006; and a member of technical program committee for numerous international conferences, including ICC, GLOBECOM, WCNC, PIMRC, VTC, CCNC, AINA, and ISWCS. He received the Best Paper Award in the IEEE 21st International Conference on Advanced Information Networking and Applications (AINA-07).

Since August 2006, he has been working with Simula Research Laboratory, Norway (http://www.simula.no/). His research interests include resource, mobility, spectrum, data, energy, and security management in wireless networks and mobile computing. He is a member of IEEE and IEEE ComSoc.

# Contributors

**Yousry Abdel-Hamid**
Department of Electrical and Computer
 Engineering
University of Victoria
Victoria, British Columbia, Canada

**Iftekhar Ahmad**
School of Engineering
Edith Cowan University
Joondalup, Western Australia, Australia

**Mohamad Khattar Awad**
Electrical and Computer Engineering
University of Waterloo
Waterloo, Ontario, Canada

**Bruno Baynat**
Information Laboratory
Université Pierre et Marie Curie
Paris, France

**Yan Q. Bian**
ProVision Communications
Bristol, United Kingdom

**Thomas Michael Bohnert**
SAP Research CEC
Zurich, Switzerland

**Noureddine Boudriga**
Communication Networks and Security
 Research Laboratory
University of Carthage
Ariana, Tunisia

**Hsiao-Hwa Chen**
Department of Engineering Science
National Cheng Kung University
Taiwan, Taiwan

**Marceau Coupechoux**
TELECOM ParisTech
Paris, France

**Fabio D'Andreagiovanni**
Dipartimento di Informatica e Sistemistica
Sapienza—Università di Roma
Roma, Italy

**Xiaoming Fu**
Institute for Computer Science
University of Goettingen
Goettingen, Germany

**Fayez Gebali**
Department of Electrical and Computer
 Engineering
University of Victoria
Victoria, British Columbia, Canada

**Vasken Genc**
School of Computer Science
 and Informatics
University College Dublin
Dublin, Ireland

**André Girard**
Energy, Materials and Telecommunications
National Institute of Scientific Research
Montréal, Quebec, Canada

**T. Aaron Gulliver**
Department of Electrical and Computer
 Engineering
University of Victoria
Victoria, British Columbia, Canada

**Daryoush Habibi**
School of Engineering
Edith Cowan University
Joondalup, Western Australia, Australia

**Aun Haider**
National Institute of Information and
 Communications Technology
Tokyo, Japan

**Richard J. Harris**
School of Engineering and Advanced
 Technology
Massey University
Palmerston North, New Zealand

**Hossam S. Hassanein**
School of Computing
Queen's University
Kingston, Ontario, Canada

**Jianhua He**
Institute of Advanced Telecommunications
Swansea University
Swansea, United Kingdom

**Roberto Carlos Hincapié**
Department of Informatics and
  Telecommunication Engineering
Universidad Pontificia Bolivariana
Medellín, Colombia

**Xuemin Huang**
Nokia Siemens Networks
Munich, Germany

**Mohamed Ibnkahla**
Department of Electrical and
  Computer Engineering
Queen's University
Kingston, Ontario, Canada

**Antonio Iera**
Department of Computer Science,
  Mathematics, Electronics,
  and Transportation
University Mediterranea
  of Reggio Calabria
Calabria, Italy

**Neila Krichene**
Communication Networks and Security
  Research Laboratory
University of Carthage
Ariana, Tunisia

**Jijun Luo**
Nokia Siemens Networks
Munich, Germany

**Carlo Mannino**
Dipartimento di Informatica e Sistemistica
Sapienza—Università di Roma
Roma, Italy

**Masood Maqbool**
TELECOM ParisTech
Paris, France

**Mehri Mehrjoo**
Electrical and Computer Engineering
University of Waterloo
Waterloo, Ontario, Canada

**Roberto Bustamante Miller**
Department of Electrical Engineering
Universidad de los Andes
Bogotá, Colombia

**Melody Moh**
Department of Computer Science
San Jose State University
San Jose, California

**Teng-Sheng Moh**
Department of Computer Science
San Jose State University
San Jose, California

**Antonella Molinaro**
Department of Computer Science,
  Mathematics, Electronics,
  and Transportation
University Mediterranea
  of Reggio Calabria
Calabria, Italy

**Liam Murphy**
School of Computer Science
  and Informatics
University College Dublin
Dublin, Ireland

**Seán Murphy**
School of Computer Science
  and Informatics
University College Dublin
Dublin, Ireland

**Andrew R. Nix**
Centre for Communications Research
University of Bristol
Bristol, United Kingdom

**Georges Nogueira**
Information Laboratory
Université Pierre et Marie Curie
Paris, France

**Konstantinos Ntagkounakis**
School of Electrical, Electronic and
  Computer Engineering
Newcastle University
Newcastle upon Tyne, United Kingdom

**Ioannis Papapanagiotou**
Electrical and Computer Engineering
North Carolina State University
Raleigh, North Carolina

**Georgios S. Paschos**
Centre of Research and Technology Hellas
The Informatics and Telematics Institute
Thessaloniki, Greece

**Dongming Peng**
Department of Computer and Electronics
  Engineering
University of Nebraska–Lincoln
Omaha, Nebraska

**Sara Pizzi**
Department of Computer Science,
  Mathematics, Electronics,
  and Transportation
University Mediterranea
  of Reggio Calabria
Calabria, Italy

**Saifur Rahman**
Advanced Research Institute
Virginia Tech
Arlington, Virginia

**Brunilde Sansò**
Department of Electrical Engineering
École Polytechnique de Montréal
Montreal, Quebec, Canada

**Bayan Sharif**
School of Electrical, Electronic and
  Computer Engineering
Newcastle University
Newcastle upon Tyne, United Kingdom

**Hamid Sharif**
Department of Computer and Electronics
  Engineering
University of Nebraska–Lincoln
Omaha, Nebraska

**Xuemin (Sherman) Shen**
Electrical and Computer Engineering
University of Waterloo
Waterloo, Ontario, Canada

**Yucheng Shih**
Department of Computer Science
San Jose State University
San Jose, California

**Shekhar Srivastava**
Wireless Research
Huawei Technologies
Plano, Texas

**Ahmed Iyanda Sulyman**
Department of Electrical and Computer
  Engineering
Royal Military College of Canada
Kingston, Ontario, Canada

**Yong Sun**
Toshiba Research Europe Limited
Bristol, United Kingdom

**Zuoyin Tang**
Department of Electronic and Electrical
  Engineering
University of Strathclyde
Glasgow, Scotland, United Kingdom

**Honggang Wang**
Department of Computer and Electronics
  Engineering
University of Nebraska–Lincoln
Omaha, Nebraska

**Wei Wang**
Department of Computer and Electronics
  Engineering
University of Nebraska–Lincoln
Omaha, Nebraska

**Matthew W. Webb**
Centre for Communications Research
University of Bristol
Bristol, United Kingdom

**Jie Xiang**
Simula Research Laboratory
Lysaker, Norway

**Abdulrahman Yarali**
Murray State University
Murray, Kentucky

**Yang Yu**
School of Computer Science
  and Informatics
University College Dublin
Dublin, Ireland

**Yan Zhang**
Simula Research Laboratory
Oslo, Norway

# *Part I*

---

## *WiMAX Fundamentals*

# 1 Quality of Service in WiMAX

*Iftekhar Ahmad and Daryoush Habibi*

## CONTENTS

Wireless networks are generally less efficient and unpredictable compared to wired networks, which make quality of service (QoS) provisioning a bigger challenge for wireless communications. The wireless medium has limited bandwidth, higher packet error rate, and higher packet overheads that altogether limit the capacity of the network to offer guaranteed QoS. In response to the increasing QoS challenge in wireless networks, researchers have made significant modifications in the legacy IEEE 802.11 standards to facilitate QoS to end users. The design constraints at several layers of the IEEE 802.11, however, restrict its capacity to deliver guaranteed QoS. Recently, the IEEE 802.16 standard, also known as worldwide interoperability for microwave access (WiMAX), has emerged as the strongest contender for broadband wireless technology with promises to offer guaranteed QoS to wireless application users. In this chapter, the details of QoS framework in WiMAX technology are presented. The definition of QoS as well as the legacy wireless QoS framework is presented first. The key technological strengths of WiMAX to address QoS challenges are identified. We present the details of WiMAX QoS mechanisms and finally conclude the chapter by identifying some open research issues.

## 1.1  INTRODUCTION

Although QoS in communication networks has been a topic of great interest for a long time, interestingly there is no common or formal definition of the term QoS. However, there are a number of definitions for QoS put forward by different standards.

The International Telecommunication Union standard X.902 [1,2] refers to QoS as "a set of quality requirements on the collective behavior of one or more objects." A number of QoS parameters are used to describe the speed and reliability of data transmission.

The ATM Lexicon [1,2] defines QoS as "a term which refers to the set of ATM performance parameters that characterize the traffic over a given virtual connection." QoS parameters apply mostly to the lower level protocol layers and are not meant to be directly observable or verifiable by the application. These parameters include cell loss ratio, error rate, misinsertion rate, delay variation, transfer delay, and average cell transfer delay.

The Internet Engineering Task Force (IETF) defines QoS [1,2] "as the demand for networked real time services grows, so does the need for shared networks to provide deterministic delivery services. Such deterministic delivery services demand that both the source application and the network infrastructure have capabilities to request, setup, and enforce the delivery of the data. Collectively these services are referred to as bandwidth reservation and Quality of Service (QoS)."

In general, the term QoS is used for different strategies and techniques designed to provide end users with a predefined and predictable level of service from the network and other components associated with the network. Although the key issue behind the definition of QoS is to achieve and guarantee the users' perceived needs, it is important to quantify the perceived needs in terms of a list of quantitative parameters so that different strategies and techniques can target these parameters. A widespread survey and analysis [3–5] have led researchers to identify the following major parameters for broad wireless networks: (1) maximum sustained rate (MSR), (2) minimum reserved rate (MRR), (3) maximum latency, (4) maximum jitter, and (5) priority. MSR, according to the 802.16 standard, is the peak information rate of the service expressed in bits per second. The service must conform to this parameter on the average over time on the wireless link and that no additional policing is needed for the base station (BS) in the downlink direction. MRR defines the minimum data rate in bits per second reserved for service flow. MRR ensures that the service does not experience starvation and satisfies the minimum level of perceived QoS. Maximum latency is defined as the time required in the worst case by a packet to reach the destination. Maximum jitter measures the worst case variation in the end-to-end latency and is mainly a function of queuing delay along the source to destination path. Priority defines the differentiation in treatment of services at each level of communication according to the importance of various services.

For a wireless communication network, the first and foremost QoS challenge remains providing end users with a predefined and predictable level of QoS parameters in the form of MSR, MRR, maximum latency, maximum jitter, and priority for both fixed and mobile broadband wireless communications. Additional challenges due to reliability of the wireless channel and mobility of end users arise. Enabling QoS in such a network implies the exercise of various radio link resource management mechanisms including Medium Access Control (MAC), resource scheduling, policy control, call admission control (CAC), and error control that altogether satisfy and maintain the QoS. Researchers have been working on these issues for a long time and have incorporated changes in the legacy IEEE 802.11 standard. The improvement, however, is not sufficient for guaranteed QoS and

WiMAX finally appears as the most promising candidate to deliver guaranteed QoS in a wireless network.

## 1.2  LEGACY QoS SOLUTIONS FOR WIRELESS COMMUNICATIONS

Over the past decade, researchers have introduced various wireless QoS mechanisms and almost all of them are incorporated at the MAC or physical (PHY) layer [6,7]. Considering the wireless channel reliability issue, various error control mechanisms have also been introduced as part of the transmission protocols to improve jitter, loss rate, and overall throughput. The legacy QoS mechanisms include the following.

### 1.2.1  MAC QoS MECHANISMS

The MAC, also known as the medium access control, is a part of the data link layer that acts as an interface between the logical link control sublayer and the network's physical layer. The MAC layer is responsible for controlling which node of the wireless network is allowed to access the shared channel and how nodes communicate with each other, hence it has significant impacts on the MSR, MRR, latency, jitter, and priority characteristics observed at each node. Many different wireless MAC schemes have been developed to support a wide variety of services while trying to ensure QoS. Figure 1.1 depicts the QoS schemes in the legacy 802.11. Each scheme has been optimized to support a particular application or set of applications. The optimization of a particular scheme leads to its inherent strengths and weaknesses. These strengths and weaknesses determine how effectively the scheme functions in real life for a particular mix of applications.

The MAC, designed for the IEEE 802.11b [8–12], was originally intended to allow quick, easy, and robust access to a wireless channel without complicated addressing or queuing techniques. Differentiation of services is usually all that it achieves, and latency and jitter are still unpredictable due to the random nature of the waiting time at each client for the channel access. The average throughput in a saturated network running 802.11 distributed coordination function (DCF) MAC is equal for all nodes if they all have the same traffic pattern. The IEEE 802.11e standard [13–16,38] implements an enhanced version of DCF. This is still a contention-based MAC using carrier sense multiple access with collision avoidance (CSMA/CA). Traffic at each node is differentiated into up to eight queues, each having a different arbitrary interframe space (AIFS) and a different minimum contention window time. Traffic classes with a shorter AIFS and window size will have a higher probability of getting access to the medium. This scheme guarantees bandwidth for high priority traffic very well while still maintaining connectivity for low priority traffic. The enhanced distributed channel access (EDCA) also achieves reasonably good latency performance. However, each queue essentially works like its own DCF, meaning that as the number of users rises, the collision rate increases quite rapidly limiting the throughput. SpectaLink is one of the world's largest provider of voice over IP (VoIP) telephony products and as such have developed their own scheme

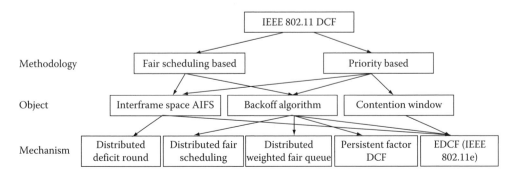

**FIGURE 1.1**  QoS framework in IEEE 802.11.

SpectraLink voice priority (SVP) for providing QoS in 802.11 networks, in the absence of a suitable standard. SVP [17] is a modification of 802.11 which specifies that the back-off time for higher priority packets should be set to zero. In the original specification of SVP, setting the contention window to zero for high priority traffic is only done at the access point. SVP also specifies that higher priority packets should either be put at the head of the queue or put in a separate queue completely. Both these methods are designed to give priority access to packets that contain higher priority data and allow them to access the network in a timely manner at the expense of more collisions. Collisions, however, often reduce the total throughput of data in the system. Because SVP is based on the concept of DCF, many of DCF's shortcomings are also evident in SVP. The wireless token network (WTN) [18] is another MAC design that incorporates the overheads that are absolutely necessary to provide good throughput and QoS. All decisions during the design phase leaned toward lower transmission overhead and hence WTN is more efficient with the bit rate compared to the 802.11. The WTN, however, cannot offer guaranteed QoS when the network is overloaded and also suffers from the problem of higher jitter because of its design issues.

### 1.2.2   PHY Layer QoS Mechanisms

The 802.11 standard specifies multiple transmission rates [15] that can be achieved by different modulation techniques at the PHY layer. The philosophy behind it is to adapt the modulation techniques according to the channel conditions so that the received error remains within a limit and QoS does not degrade substantially. The standard, however, leaves the rate adaptation and signaling mechanisms open. Because transmission rates depend on the channel conditions, an optimized link adaptation mechanism is desirable to maximize the throughput under different channel conditions. Most of the existing link adaptation mechanisms focus on algorithms to switch among transmission rates specified in the physical layer convergence procedure (PLCP), without the need to modify existing standards. The 802.11b, however, incorporates a novel method to adjust the length of direct sequence spread spectrum systems (DSSS) pseudo-noise (PN) code with slight modifications of its DCF. Metrics that are also commonly used in existing link adaptation algorithms include channel signal-to-noise ratio/carrier-to-interference ratio (SNR/CIR), average payload length, received power level, or transmission acknowledgments. Received signal strength (RSS)[19] is a metric used in the adaptation algorithm with the assumption that transmission power is fixed. The RSS metric also assumes that there is a linear relationship between the average RSS and SNR. Based on the measured RSS, the station dynamically switches to an appropriate transmission rate.

Packet error rate (PER) prediction [20] is another link adaptation scheme in which decisions are made based on PER prediction that not only depends on SNR/CIR but also on the momentary channel transfer function. MAC protocol data unit (MPDU)-based link adaptation [21] is another link adaptation scheme that uses a combination of SNR, average payload length, and frame retry count as the metric for the link adaptation algorithm. The proposed algorithm pre-established a table of best transmission rate for decision making. Link adaptation with success/fail (S/F) thresholds [22] uses the ACKs of transmitted frames as a measurement of channel condition and adjusts the transmission rate depending on the subsequent successful transfer of frames. Code Adapts To Enhance Reliability (CATER) [23] is an adaptive PN code algorithm for DSSS used in 802.11b and it is designed to improve the throughput under high bit error rate (BER) channel conditions.

### 1.2.3   Error Control Mechanisms

A wireless network is not as reliable as a wired network and error in transmitted packets is common in wireless communication. The error is more evident when the nodes have the mobility that causes error in the received packets due to slow and fast fading. An error control mechanism attempts to address the problems caused by error in received signals and thereby maintains QoS by improving loss rate and jitter performances and overall throughput. The Transmission Control Protocol (TCP) [24] is a popular protocol designed to provide reliable and orderly delivery of a stream of bytes and is a key part of the TCP/IP protocol suit. The TCP provides a simpler interface to applications by hiding most of the underlying packet structures, rearranging out-of-order packets, minimizing network congestion, and retransmitting corrupted packets. Forward error correction (FEC) [25] is another error control mechanism for data transmission. In FEC, the sender incorporates additional redundant data to its messages, which allows the receiver to detect and correct errors within a certain limit without the need for retransmission. FEC block codes are applied to a sequence of packets, and in case of a loss/error in packets, a receiver reconstructs the missing packets from the redundant information carried in error-correcting codes. Naturally, error-correcting capability in FEC comes at some costs because the FEC codes represent redundant information that increases the overall transmission rate. FEC is highly effective where the communication media is unreliable and retransmissions of too many packets prove costly in context of available bandwidth.

Although considerable effort has gone into improving the QoS in the 802.11 standard, the most it can achieve is to differentiate traffic and treat them with their corresponding priority and also to adapt the transmission rates at various environments to offer graceful degradation of throughput. Due to its design limitations at different layers, the 802.11 standard cannot offer guaranteed QoS, which is one of the key motivations behind introducing another standard, the IEEE 802.16, also known as WiMAX.

## 1.3 TECHNOLOGICAL STRENGTHS OF WiMAX TO ADDRESS QoS

WiMAX is designed with QoS in mind and it has some underlying technological strengths that help it offer improved QoS. Some of these strengths are outlined in this section.

### 1.3.1 WiMAX PHY Layer

In WiMAX, the upstream PHY layer consists of time division multiple access (TDMA) and demand assigned multiple access (DAMA) [3,26]. For TDMA, the channel for upstream communication is divided into multiple time slots and the access of time slots for various clients is governed by the MAC layer at the receiver end. The time slots allocated for various clients can be varied depending on demands. The downstream traffic can be continuous time division multiplexing (TDM) or burst mode transfer. In continuous TDM, data for various clients is multiplexed onto the same stream and is received by all clients at the same coverage sector. For bursty data, bursts are sent to the receiver in a similar fashion to the TDMA upstream burst. With time slots-based communication, overheads due to contentions and collisions can be reduced significantly, which can improve the QoS.

The modulation used in WiMAX is the orthogonal frequency division multiplexing (OFDM). WiMAX OFDM features multiple subcarriers ranging from a minimum of 256 up to 2048, each modulated with either BPSK, QPSK, 16 QAM, or 64 QAM modulation [27]. The advantage of orthogonality is that it minimizes self-interference, a major source of error in received signals in wireless communications. WiMAX supports different signal bandwidths ranging from 1.25 to 20 MHz to facilitate transmission over longer ranges in different multipath environments. Multipath signals [3,28], another limiting factor for higher sustained throughput in wireless communications, specially when the terminal nodes have the mobility, are caused by reflections between a transmitter and receiver whereby the reflections arrive at the receiver at different times. Interference caused by multipath tends to be highly problematic when the delay spread, the time span separating the reflection, is on the order of the transmitted symbol time. For WiMAX, due to its OFDMA, symbol times tend to be in the order of $100\,\mu s$, which makes multipath less of a problem. Moreover, in WiMAX, a guardband of about $10\,\mu s$, called the cyclic prefix, is inserted after each symbol to mitigate the effect of multipath. Another feature of WiMAX PHY is the use of advanced multiantenna signal processing techniques, mainly in the form of multiple input multiple output (MIMO) processing and beamforming [3,29]. For MIMO, the received signal from one transmitting antenna can be quite different to the received signal from a second antenna, a common scenario in indoor or dense metropolitan areas where there are many reflections and multipaths between the transmitter and the receiver. In such cases, a different signal can be transmitted from each antenna at the same frequency and still be recovered at the receiver by signal processing. Beamforming, on the other hand, attempts to form a coherent construction of the multiple transmitters at the receiver, which can ultimately offer a higher SNR at the receiver resulting in higher bandwidth or longer range communication. In WiMAX, it is also possible to combine both MIMO and beamforming in cases like 4-antenna systems.

All these features in the WiMAX PHY layer contribute to higher throughput and stability at the receiver end, which makes WiMAX an excellent platform to deliver a predefined level of QoS. With improved throughput and stability, management of QoS is considerably easier in WiMAX compared to other similar wireless standards. Increased throughput, however, does not ensure guaranteed QoS, and bandwidth management is another crucial part that plays a big role for maintaining QoS. This is where WiMAX MAC comes into action.

### 1.3.2 WiMAX MAC

WiMAX MAC [3,30,31] is designed for the point-to-multipoint wireless communication with the capability to support higher-layer protocols including ATM, IP, and other future protocols. One of the design considerations of WiMAX MAC is to accommodate very high bit rates of the broadband PHY layer, while delivering ATM-compatible QoS at the same time. A connection oriented MAC architecture in WiMAX provides a platform for strong QoS control. MAC uses a scheduling algorithm that enables the subscriber station (SS) to only compete once for initial entry into the network and upon successful entry, the SS is allocated a time slot by the BS. The time slot can increase or decrease according to the needs and it remains assigned to the SS for the whole communication period. The time slot assigned to an SS cannot be used by other subscribers, which makes WiMAX MAC increasingly stable under overload and over-subscriptions. It also works as a key tool for the BS to control QoS by adjusting the time-slot assignments according to the applications' needs of the SSs.

### 1.4 QoS IN WiMAX

WiMAX is known for its capability to deliver ATM-like connection-oriented QoS guarantee in broadband wireless networks and in the following we present how WiMAX achieves it.

### 1.4.1 ATM LIKE QoS

ATM is considered as the pioneer technology for QoS and a bulk of its success is directly attributed by its capability to offer QoS guarantees [32,39]. ATM attempts to deliver QoS by defining different classes of service based on constant and variable bit rates. Traffic is sent at a constant rate in constant bit rate (CBR) applications with examples including telephony, interactive voice, audio, and early video conferencing standards like H.261. Since the traffic rate is constant, it is relatively easy for the network to maintain the required QoS for applications with CBR traffic. Bit rate is variable for applications with variable bit rate (VBR) traffic. Examples of traffic that requires this type of service include compressed video streams like MPEG video and MP3 audio. Provisioning of QoS for VBR applications is a complex process for the network, though it affords greater flexibility as well. Best effort services include available bit rate (ABR) and unspecified bit rate (UBR) and apply to local area network (LAN) traffic that is more tolerant of delays and cell loss. ATM features are built with QoS capabilities on target and have the capability to provide the equivalent of a leased line through CBR virtual circuits. ATM has to be deployed at every hop of the path to assure the required end-to-end QoS.

The 802.16 standard is designed to offer the ATM like QoS and a key aspect of the design is the polling-based MAC layer that is more deterministic than contention-based MAC used in other similar standards. The MAC layer employs a single scheduling data service for each connection and

**TABLE 1.1**
**WiMAX Service Classes**

| Class | Description | Minimum Rate | Maximum Rate | Latency | Jitter | Priority |
|---|---|---|---|---|---|---|
| Unsolicited grant service | VOIP, E1; fixed-size packets on periodic basis | | X | X | X | |
| Real-time polling service | Streaming audio/video | X | X | X | | X |
| Enhanced real-time polling service | VOIP with activity detection | X | X | X | X | X |
| Nonreal-time polling service | FTP | X | X | | | X |
| Best effort | Data transfer, Web browsing, etc. | | X | | | X |

X = QoS specified

each data service is associated with a set of QoS parameters that quantify its behavior. Like ATM, the 802.16 MAC also defines different service classes (Table 1.1) and specifies up to five separate service classes [30,31,33] to provide QoS for various types of applications. The service classes include:

*Unsolicited grant service (UGS):* It is designed to support real-time service flows generated at CBR. The UGS will be granted periodically without a polling-request procedure and thereby reducing the latency.

*Real-time polling service (rtPS):* It is designed to support real-time service flows where packets are generate at VBR. This service requires more request overheads and latency compared to UGS, but supports variable grant sizes. The rtPS is suitable for connections carrying services like VoIP or video streaming services.

*Extended real-time polling service (ertPS):* It is designed to support real-time service flows where packets are generated at variable-size rate on a periodic basis, like VoIP services. ertPS is intended to utilize the efficiency of both UGS and rtPS.

*Nonreal-time polling service (nrtPS):* It is designed to accommodate delay-tolerant data streams that consist of variable-size data packets. These services are capable of tolerating longer delays and are relatively insensitive to delay jitter. The nrtPS is appropriate for Internet services with a minimum guaranteed rate like File Transfer Protocol (FTP) and Hypertext Transfer Protocol (HTTP).

*Best Effort (BE) service:* The BE service is designed to facilitate data streams that have no minimum service requirement and therefore may be supported on a resource availability basis such as e-mail. For BE, throughput and delay guarantees are not required.

To achieve ATM like QoS for different service classes, WiMAX defines a QoS framework, which is presented in the following section.

### 1.4.2 WiMAX QoS Framework

The WiMAX Forum has specified a framework [3,26,32] for service management and QoS. Service flow and its management is a crucial part of that framework and Figure 1.2 depicts the steps involved in the management of service flows in WiMAX networks.

#### 1.4.2.1 Service Flows

The service flow management (SFM) entity governs the creation, admission, activation, modification, and deletion of 802.16 service flows. The service flow that associates packets traversing the MAC

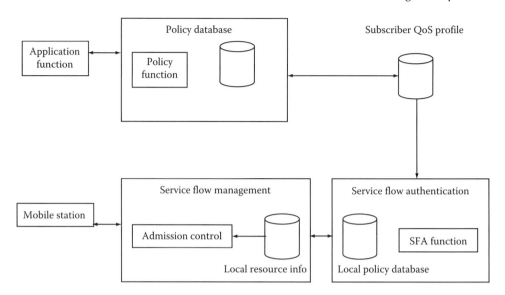

**FIGURE 1.2**    WiMAX QoS framework.

interface serves as a crucial mechanism for providing QoS in WiMAX. A service flow as identified by a connection ID is a bidirectional flow of packets that provides a particular QoS. This QoS is facilitated by the joint cooperation of SS and BS according to a set of parameters defined for the service flow. Service flows are maintained in both the uplink and downlink direction and may exist without actually being activated to carry traffic. In 802.16, all service flows have a 32-bit service flow identifier (SFID) and active service flows also have a 16-bit connection identifier (CID). SFID is assigned to all existing service flows and serves as the primary identifier in the SS and the BS for the service flow. An SFID and an associated direction are the key requirements for an existing service flow. Another key attribute of a service flow includes a QoS parameter set provisioned via the network management system. The parameter set is mapped to resources, mainly in terms of bandwidth, but may also offer require guarantees on latency and the BS (and possibly the SS) reserves resources accordingly. Service flow also includes a set of QoS parameters defining the service actually being provided to the service flow. Only an active service flow may forward packets. A service flow normally has its QoS parameter set specified in any of three ways: (1) by explicitly including all traffic parameters, (2) by indirectly referring to a set of traffic parameters by indicating a service class, or (3) by specifying a service class name along with modifying parameters.

Service flows can be either dynamic or static. Dynamic service flows can be created, changed, or deleted through a series of MAC management messages commonly known as dynamic service addition (DSA) for creating a new service flow, dynamic service change (DSC) for changing an existing flow, and dynamic service deletion (DSD) for deleting an existing service flow. These protocols allow providers to add new subscribers, modify traffic contracts, or reclaim resources on the fly without interfering with other existing subscribers. A DSA request (DSA-REQ) can be triggered by either the BS or SS. A DSA-REQ that is initiated by the SS (Figure 1.3), contains a service flow reference and a set of QoS parameters. The BS then responds to that request by generating a DSA-RSP which may indicate either acceptance or rejection. If the request is rejected due to any nonsupported parameter that specific parameter is notified to the BS through DSA-RSP. A DSA-REQ generated from BS (Figure 1.4) contains an SFID for either an uplink or a downlink service flow. An associated CID and a set of active or admitted QoS parameters are optional. The BS then waits for a DSA-RSP from the SS and makes an acceptance or rejection decision response with a DSA-ACK accordingly. If the request is rejected due to any nonsupported parameter that specific parameter may be indicated in the SC's DSA-RSP.

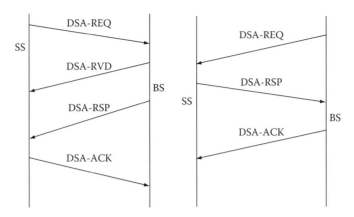

**FIGURE 1.3**    Service flow establishment in WiMAX.

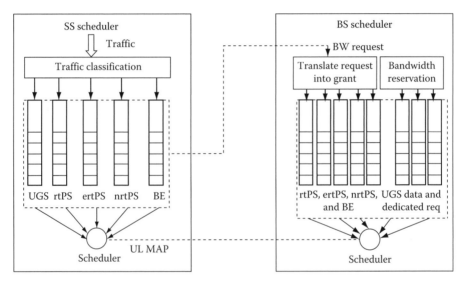

**FIGURE 1.4**    WiMAX MAC scheduler and service classification.

### 1.4.2.2   Policy Control

Policy functions reside in both home and the visited network comprising their respective databases that include general policy rules as well as application-dependent policy rules. The authentication, authorization, and accounting (AAA) servers contain the subscriber's QoS profile and the associated policy rules per subscriber. The policy function of the framework ensures that the service flow carrying data packets for an application follows the policy rules established for the corresponding subscriber.

### 1.4.2.3   Service Flow Authorization

Service flow authorization (SFA) is responsible for evaluating any service request against the subscriber's QoS profile. Initialization and authentication are an important part of the 802.16 standard and the standard defines the procedures for when a subscriber enters the network and the authentication process to allow carriers to establish the identity of the subscriber and the contracted QoS provisions. According to the WiMAX standard, every change to the service flow through messages like DSA, DSC, or DSD QoS parameters must be approved by an authorization module. The standard specifies identification through an x.509 certificate, a unique signature that is hard coded on every 802.16 equipment. This carries the subscriber's identity and allows mapping of the user's identification to the service level agreement (SLA) contract. 802.16 is the only wireless standard that provides carriers this ability to offer SLAs along with the mechanisms to deliver the required QoS parameters.

### 1.4.2.4   Connection Admission Control

The SFM consists of a Connection Admission Control (CAC) function and the associated local resource information. CAC is an important part of WiMAX QoS framework, which contributes to maintaining the QoS of new and existing connections by preventing oversubscription for a fixed amount of bandwidth. Without an efficient CAC, networks will not be able to maintain QoS of applications like voice and video under an increased load scenario. Bandwidth based CAC (BW-CAC) [34,35] makes a decision on connection admission/rejection based on the available bandwidth. The BW-CAC considers all the DSA/DSC/DSD requests and updates the available bandwidth after the processing of each request. There is another type of CAC known as QoS-based CAC (QoS-CAC) [34,35], where a new connection request is classified into a particular queue depending on the associated service class type. QoS-CAC treats the UGS connection queue with the highest priority, followed by rtPS and then by nrtPS queues. In QoS-CAC, there is no need for admission control of BE connections because they do not demand any guarantee. The new request queues of different service classes are scanned in the order as mentioned before at every scheduling interval and a decision is made whether a connection can be admitted while satisfying the requested QoS of the new connection as well as the existing connections. Call admission decision can be made depending on policies as well-known as policy-based admission control.

The above-mentioned mechanisms in WiMAX contribute to QoS management by ensuring that resources are not overcommitted; users closely follow the policy rules so that QoS for other users' applications can be maintained and only the authorized users can use the resources. The next challenge of QoS management is the distribution of resources according to the QoS requirements of various applications. Another key challenge is to maintain a sustained supply of resources at changing channel conditions so that committed resources can be supplied. To achieve these goals, WiMAX standard has introduced some modifications at its PHY, MAC, and transmission control layers, which are detailed in the following sections.

### 1.4.3   QoS Mechanisms in WiMAX PHY Layer

The time division duplex (TDD) radio in WiMAX provides some extra advantages with more efficient use of spectrum and higher, steady data rate efficiency by using multiple OFDM instead of a single carrier approach. This parallel carrier ability also known as multicarrier modulation (MCM) or discrete multitone (DMT) is ideal for addressing signal errors that may arise in indoor and outdoor wireless environments. By using multiple carriers to transmit the data, reliable communication is still maintained when one or more carriers are subject to propagation errors. Spectral efficiency, expressed in bits per hertz, is a measure of the number of bits that can be transferred over a channel. An 802.16 256 fast Fourier transform (FFT) OFDM system is capable of supporting 5 bits/Hz at 64 QAM modulation resulting a maximum of around 70 Mbit/s of usable data in a single 20 MHz channel. With higher and steady bandwidth, management of QoS is more stable in WiMAX. Adaptive modulation and coding (AMC) is another attractive feature of the 802.16 standard, which allows providers to maintain prescribed QoS levels by adjusting the modulation level and coding according to the wireless environment. Wireless links have BER that are often varying due to environmental conditions like multipath, fast fade, and refraction. The 802.16 standard addresses this problem by employing AMC to mitigate the environmental effects and maintain a constant BER.

In AMC, the carrier to interference and noise ratio (CINR) serves as an indicator for the wireless signal quality. The higher the value of the CINR, the higher will be the received signal quality and higher throughput can be achieved at the receiver end. When the CINR value decreases, which is an indication of poor signal quality, the 802.16 standard adapts to signal degradation by dropping to a lower modulation in steps, thereby maintaining low BER at the cost of throughput. Retransmission of packets is another solution when a received packet contains error in it. Retransmission of packets, however, is not suitable for many real-time applications that require rigid latency and jitter limits. Real-time services cannot tolerate the problem of high latency and jitter, therefore a sustained BER is critical to deliver acceptable QoS.

### 1.4.4 WiMAX MAC Scheduler and QoS

MAC scheduler plays a vital role in addressing the QoS issue in WiMAX. As shown in Figure 1.4, 802.16's MAC layer enables classification of QoS and non-QoS-dependent application flows and maps them to connections with distinct scheduling services enabling both guaranteed handling and traffic enforcement. Each connection is associated with a single scheduling data service and each data service is associated with a set of QoS parameters that quantify aspects of its behavior. The scheduling services in WiMAX are described below.

#### 1.4.4.1 UGS Algorithm

In UGS algorithm that is designed to support real-time service flows generating fixed-size data packets, the BS periodically assigns fixed size grants to the SS and these grants are sufficient to send data packets generated at the maximum data rate and thereby guarantee a predefined level of QoS. The UGS algorithm therefore minimizes MAC overhead and uplink access delay caused by the bandwidth request process of the SS to send data packets. However, since the BS always assigns fixed-size grants that are sufficient to send data packets generated at the maximum data rate, it causes a waste of many uplink resources in cases where the SS does not upload data at the maximum rate. In Figure 1.5, the dotted line and the solid line represent the amount of assigned uplink resources by the

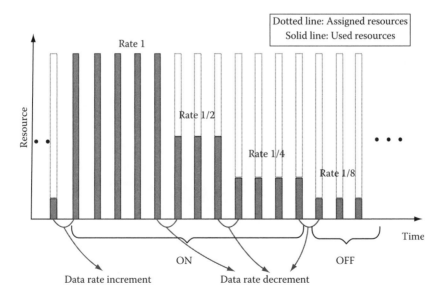

**FIGURE 1.5** UGS scheduler in WiMAX MAC. (Redrawn from Lee, H., Kwon, T., and Cho, D., *Proceedings of IEEE ICC06*, 2060, 2006.)

BS and the amount of used uplink resources by the user, respectively. The blank regions represent the uplink resources that are wasted.

### 1.4.4.2  rtPS Algorithm

The rtPS algorithm is designed to facilitate real-time service flows that generate variable size data packets. In this algorithm, the BS assigns uplink resources that are sufficient for unicast bandwidth requests to the user. This process is called a bandwidth request or polling process. The rtPS algorithm is more efficient than the UGS algorithm because the rtPS algorithm always uses a bandwidth request process for suitable size grants. However, this involves extra MAC overhead and additional access delay. The rtPS algorithm has larger MAC overhead and access delay compared to the UGS algorithm. As evident in Figure 1.6 for the rtPS algorithm, the dot line and the solid line are nearly the same because the user requests exact amount of uplink resources for transmitting its packets.

### 1.4.4.3  UGS-AD Algorithm

The UGS-AD algorithm which is basically a combined form of the UGS and rtPS algorithms is designed to support real-time service flows that generate fixed size data packets on a semiperiodic basis. As shown in Figure 1.7, there are two scheduling modes (UGS, rtPS) in UGS-AD and the modes switch according to the status of the users. This algorithm first starts at the rtPS mode when the services start. The BS maintains this mode as long as the user requested bandwidth size is zero byte. When the user requests another nonzero size bandwidth, the BS switches its mode to the UGS. The UGS-AD algorithm partially addresses the problems of uplink resource wastage caused by the UGS algorithm as well as MAC overhead and access delay problems caused by the rtPS algorithm.

### 1.4.4.4  ertPS Algorithm

In an ertPS algorithm (Figure 1.8), the BS changes its polling size following the user requests of bandwidth for sending the voice packets. The BS keeps the polling size unchanged until the user sends another request. The ertPS algorithm follows the data rates of the voice users, which helps the BS obtain improved efficiency compared to the UGS, rtPS, and UGS-AD algorithms.

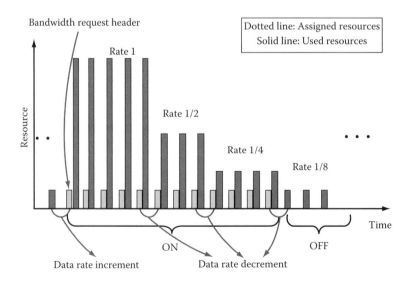

**FIGURE 1.6**  rtPS scheduler in WiMAX. (Redrawn from Lee, H., Kwon, T., and Cho, D., *Proceedings of IEEE ICC06*, 2060, 2006.)

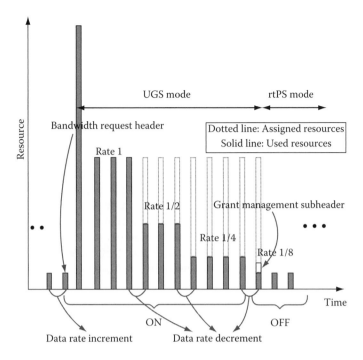

**FIGURE 1.7** UGS-AD scheduler in WiMAX MAC. (Redrawn from Lee, H., Kwon, T., and Cho, D., *Proceedings of IEEE ICC06*, 2060, 2006.)

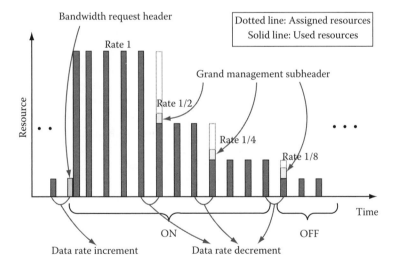

**FIGURE 1.8** ertPS scheduler in WiMAX MAC. (Redrawn from Lee, H., Kwon, T., and Cho, D., *Proceedings of IEEE ICC06*, 2060, 2006.)

### 1.4.4.5 nrtPS and BE Algorithms

The nrtPS scheduling algorithm is designed to facilitate nonreal-time data with variable size data grant burst. The BS offers unicast polls at regular intervals so that the flow receives request opportunities even at high network load. The BS polls nrtPS connection identifiers at a regular interval and the SS can use contention request opportunity or unicast request opportunity or unsolidated data burst types to request for service. The BE algorithms are designed to support nonreal-time service flows like HTTP and FTP that are tolerant of latency and jitter. The SS can request for service by using contention request opportunities as well as unicast request opportunities and unsolidated data grant burst types.

MAC scheduler in WiMAX ensures that connections are provided resources as per their QoS requirements and therefore, it is a key part of WiMAX QoS design.

### 1.4.4.6   Error Control Mechanisms in WiMAX

Error correction techniques have been incorporated into WiMAX to address the received bit error at the receiver end at various channel conditions. Strong Reed Solomon FEC, convolutional encoding, and interleaving are the recommended algorithms that can be used to detect and correct errors to maintain QoS by improving jitter and throughput. These robust error correction techniques help recover corrupted packets that may have been lost due to slow or fast fading or burst errors. Automatic repeat request (ARQ) is used to retransmit the packets once they cannot be fixed by the FEC. This significantly improves the BER performance that contributes to maintaining a predefined level of QoS.

### 1.4.5   QoS and Mobility in WiMAX

WiMAX (802.16e) [26,36,37] is designed to support mobility of the terminal nodes. The OFDMA technology used in WiMAX offers the advantages of bandwidth scalability, tolerance to multipath and self-interference, and advanced antenna support that offers WiMAX extra tools to achieve improved QoS for mobile nodes. Majority of the changes required to support mobile nodes are made at the MAC layer. Smooth handoff service is important to maintain the QoS of SS when the SS moves from one BS to another. Based on the signal quality, either the SS or the BS may initiate the handoff process and before executing the actual handoff the neighboring BS are notified through the backbone of the handoff request. Information such as QoS, security, registration that describe the SS are the sent to the neighboring BS and the BS in return sends feedback about its capacity to support the QoS and other parameters concerning the SS. During the actual handoff, the current BS notifies the SS that the handoff should be executed and may recommend a new BS based on the capacity of the new BS to support the QoS and other requirements concerning the SS. WiMAX supports soft handoff ahead of hard handoff, which ensures the continuity of QoS for the SS. In case the SS has stopped communication with the old BS before the handoff, the SS attempts to connect to the new BS and in most cases this will suffice, as neighbor BS will either be aware of the new SS or will be able to communicate to its old SS via the backbone to collect the required information. The BSs communicate to each other through backbone to perform handoff ensuring the interoperability between BSs from different manufacturers, which addresses the need to maintain continuous QoS as the mobile SS travels from one place to another. WiMAX uses fast scheduling on a per-frame basis and broadcasts the downlink/uplink scheduling in the media access protocol (MAP) messages at the beginning of each frame. For mobile SS, the scheduling can be adjusted very quickly as can the amount of resources, which makes WiMAX well suited to bursty data traffic and rapidly changing channel conditions evident in mobile wireless communication.

Mobile WiMAX supports adaptive modulation and coding in both downlink and uplink with variable packet size and this helps to combat signal noises and received errors at the receiver end that arises when the SS has the mobility. Other enhancements to support mobility at WiMAX PHY layer include extended OFDMA forward automatic power control (APC) that facilitates finer control of variations in the mobile channel and fast correction of uplink power, frequency, and timing that enables fast frequency and timing correction on the uplink and offers better tracking of the variations introduced by the mobile stations.

## 1.5  CONCLUSION

This chapter presented the QoS framework in WiMAX for broadband wireless communications. As shown, WiMAX is well equipped with tools to offer improved QoS to the end users. However, providing guaranteed QoS to the mobile terminals is still quite challenging, given the extremely high PER at high speeds. Seamless handoff at high speed requires some more investigations, specifically when the mobile terminal's speed and directions vary quite rapidly. The MAC scheduler in WiMAX is fundamentally well designed for QoS, but not completely free from overheads. The overheads can be significant when the network becomes highly loaded or the available bandwidth becomes very limited due to unfavorable channel conditions. Forward error control eliminates the needs to retransmit packets, but on the other hand, FEC incurs some costs in the form of extra overheads. More investigations are required to improve the error recovery capacity that will incur minimal costs. Further research is required to take the speed and direction information of the mobile terminals into account to predict the PER, which in turn can help design an efficient error control mechanism.

## REFERENCES

[1] ITU. [Online]. Available at: http://www.itu.int/rec/T-REC-X.902-199511-I/, last accessed December 2007.

[2] Quality of Service (QoS). [Online]. Available at: http://www.objs.com/ survey/QoS.htm, last accessed December 2007.

[3] T. Cooklev, *Wireless Communication Standards—A Study of IEEE 802.11, 802.15 and 802.16*, New York: IEEE Press, 2004.

[4] B. Hayat and R. Nasir, 802.16 2001 MAC layer QoS, *Ubiquity CM IT Magazine*, 7(17), May 2006.

[5] M. Pidutti, 802.16 Tackles Broadband Wireless QoS Issues. [Online]. Available at: http://www.comms-design.com/article/printableArticle.jhtml?articleID=54201623, last accessed December 2007.

[6] H. Zhu, M. Li, I. Chlamtac, and B. Prabhakaran, A survey of quality of service in IEEE 802.11 networks, *IEEE Wireless Communications*, 11(4), 6–14, August 2004.

[7] N. Ramos, D. Panigrahi, and S. Dey, Quality of service provisioning in 802.11e networks: Challenges, approaches, and future directions, *IEEE Network*, 19(4), 14–20, August 2005.

[8] W. Pattara-Atikom, P. Krishnamurthy, and S. Banerjee, Distributed mechanisms for quality of service in wireless lans, *IEEE Wireless Communications*, 10(3), 26–34, 2003.

[9] L. Zhao and C. Fan, Enhancement of QoS differentiation over IEEE 802.11 WLAN, *IEEE Communications Letters*, 8(8), 494–496, 2004.

[10] J. Zhao, Z. Guo, Q. Zhang, and W. Zhu, Distributed MAC adaptation for WLAN QoS differentiation, *Proceedings of IEEE Global Telecommunications Conference (GLOBECOM 2003)*, vol. 6, pp. 3442–3446, 2003.

[11] W. Liu, W. Lou, X. Chen, and Y. Fang, A QoS-enabled MAC architecture for prioritized service in IEEE 802.11 WLANS, *Proceedings of IEEE Global Telecommunications Conference (GLOBECOM 2003)*, vol. 7, pp. 3802–3807, 2003.

[12] K. Kim, A. Ahmad, and K. Kim, A wireless multimedia lan architecture using DCF with shortened contention window for QoS provisioning, *IEEE Communications Letters*, 7(2), 97–99, 2003.

[13] S. Mangold, S. Choi, G. R. Hiertz, O. Klein, and B. Walke, Analysis of IEEE 802.11e for QoS support in wireless LANS, *IEEE Wireless Communications*, 10(6), 40–50, 2003.

[14] Y. Xiao, Enhanced DCF of IEEE 802.11e to support QoS, *Wireless Communications and Networking*, 2, 1291–1296, 2003.

[15] Q. Ni, L. Romdhani, and T. Turletti, A survey of QoS enhancements for IEEE 802.11 wireless LAN, *Wireless Communications and Mobile Computing*, 4(5), 547–566, 2004.

[16] L. A. Grieco, G. Boggia, S. Mascolo, and P. Camarda, A control theoretic approach for supporting quality of service in IEEE 802.11e WLANS with HCF, *Proceedings of 42nd IEEE Conference on Decision and Control*, 2, 1586–1591, 2003.

[17] SpectraLink, Spectralink Voice Priority—Quality of Service for Voice Traffic on Wireless LANS. [online]. Available at: http://www.spectralink.com/resources/index.jsp, last accessed December 2007.

[18] J. Wyatt, D. Habibi, and I. Ahmad, Providing QoS for symmetrical voice/video traffic in wireless networks, *Proceedings of IEEE ICON'07*, Adelaide, Australia, 2007.

[19] J. Pavon and S. Choi, Link adaptation strategy for IEEE 802.11 WLAN via received signal strength measurement, *IEEE ICC03*, 2, 1108–1113, May 2003.

[20] M. Lampe, H. Rohling, and J. Eichinger, PER-Prediction for link adaptation in OFDM systems, OFDM Workshop, Hamburg, Germany, 2002.

[21] D. Qiao, S. Choi, and K. G. Shin, Goodput Analysis and Link Adaptation for IEEE 802.11a Wireless LANs, *IEEE Transactions on Mobile Computing*, 1(4), 27892, 2002.

[22] P. Chevillat et al., A dynamic link adaptation algorithm for IEEE 802.11a wireless LANs, *IEEE ICC03*, 2, 1141–1145, 2003.

[23] B. E. Mullins, N. J. Davis IV, and S. F. Midkiff, An adaptive wireless local area network protocol that improves throughput via adaptive control of direct sequence Spectrum Parameters, *ACM SIGMOBILE Mobile Computing and Communication Review*, 1(3), 9–20, September 1997.

[24] L. Ka-Cheong, V. Li, and Y. Daiqin, An overview of packet reordering in transmission control protocol (TCP): Problems, solutions, and challenges, *IEEE Transactions on Parallel and Distributed Systems*, 18(4), 522–535, April 2007.

[25] M. Smadi and B. Szabados, Error-recovery service for the IEEE 802.11b protocol, *IEEE Transactions on Instrumentation and Measurement*, 55(4), 1377–1382, August 2006.

[26] C. Huang, H. Juan, M. Lin, and C. Chang, Radio resource management of heterogeneous services in mobile WiMAX systems, *IEEE Wireless Communications*, 14(1), 20–26, February 2007.

[27] C. Eklund, R. B. Marks, K. L. Stanwood, and S. Wang, IEEE standard 802.16: A technical overview of the WirelessMAN air interface for broadband wireless access, *IEEE Communications Magazine*, 40(6), 98–107, June 2002.

[28] K. Leung, S. Mukherjee, and G. Rittenhouse, Mobility support for IEEE 802.1.6d wireless networks, *Proceedings of IEEE Wireless Communications and Networking Conference*, pp. 1446–1452, 2005.

[29] W. Fan, A. Ghosh, C. Sankaran, and S. Benes, WiMAX system performance with multiple transmit and multiple receive antennas, *Proceedings of IEEE VTC07*, Ireland, pp. 2807–2811, 2007.

[30] H. Lee, T. Kwon, and D. Cho, Extended-rtPS algorithm for VoIP services in IEEE 802.16 systems, *Proceedings of IEEE ICC06*, Istanbul, pp. 2060–2065, June 2006.

[31] C. Cicconetti, L. Lenzini, and E. Mingozzi, Quality of service support in IEEE 802.16 networks, *IEEE Network*, 20(2), 50–55, March–April 2006.

[32] S. Jha and M. Hasan, *Engineering Internet QoS*, London, U.K.: Artech House, 2002.

[33] V. Nair, Evolution of QoS and Charging Framework in WiMAX, WiMAX Forum. [online]. Available at: http://www.wimax.com/commentary/spotlight/evolution-of-qos-and-charging-framework-in-wimax, last accessed December 2007.

[34] J. Chen, W. Jiao, and H. Wang, A service flow management strategy for IEEE 802.16 broadband wireless access systems in TDD mode, *Proceedings of IEEE ICC05*, Korea, pp. 3422–3426, 2005.

[35] S. Chandra and A. Sahoo, An efficient call admission control for IEEE 802.16 networks, *Proceedings of IEEE Workshop on Local and Metropolitan Area Networks 2007*, pp. 188–193, 2007.

[36] Enabling Mobility in WiMAX Networks, White Paper, WiMAX Forum, http://www.wimax.com/research/whitepapers/WiMAX_07Nov06.pdf, last accessed December 2007.

[37] L. Doo, K. Kyamakya, and J. Umondi, Fast handover algorithm for IEEE 802.16e broadband wireless access system, *Proceedings of IEEE Symposium on Wireless Pervasive Computing*, 2006.

[38] H, Zen, D. Habibi, I. Ahmad, and J. Wyatt, A segregation based MAC protocol for real-time multimedia traffic in WLANs, *Proceedings of Wireless and Optical Communications Networks (WOCN)*, Singapore, 2007.

[39] I. Ahmad, J. Kamruzzaman, and S. Aswathanarayaniah, A dynamic approach to reduce preemption in book-ahead reservation in QoS-enabled networks, *Computer Communications*, 29(9), 1443–1457, 2006.

# 2 Scheduling in WiMAX

*Aun Haider and Richard J. Harris*

## CONTENTS

Scheduling is a critical component of worldwide interoperability of microwave access (WiMAX) that impacts significantly on its performance. Scheduling schemes help in providing service guarantees to heterogenous classes of traffic where there are a variety of different quality-of-service (QoS) requirements. In addition to scheduling, bandwidth requests, admission control, and bandwidth allocation mechanisms also play crucial roles in QoS provisioning for WiMAX. In general, a scheduler for WiMAX needs to be simple, efficient, fair, scalable, and have low computational complexity. It should also be able to protect against misbehaving flows and provide decoupling and necessary bounds on throughput and delay performance. The mesh and point-to-multi-point (PMP) modes of WiMAX require different scheduling architectures.

It has been confirmed in many earlier studies that most of the existing wireline and wireless schedulers do not perform very well with respect to the five traffic classes defined in WiMAX. In addition, each of these traffic classes has a different scheduling requirement and, consequently, it has become necessary to design appropriate heterogenous scheduling frameworks. In such setups, several modern techniques such as cross-layer, queue length aware, and multiuser diversity, are found to be very promising. Furthermore, the performance of newly designed schedulers must be compared and contrasted within both test-bed and real-life environments. In this chapter, the current major scheduling architectures are discussed, together with the need to improve their performance and adapt to widely varying traffic characteristics and the many different performance objectives in both static and mobile WiMAX networks.

A large part of the wireless broadband world is evolving toward the adoption of worldwide interoperability for microwave access (WiMAX) standards. It has a wide range of uses, such as cellular backhaul, wireless service provider backhaul, banking networks, education networks, public safety, offshore communications, campus connectivity, access networks, and rural connectivity. It will be a leading technology for providing scalable wireless broadband services in wide area networks. Further, it will have a deep and universal impact on the way that the current Internet and other data services are conceived, by providing cost-effective and easy to deploy alternative solutions to wired backhaul and last mile deployments at customer premises. WiMAX is ideally suited for providing voice, video, Internet, and other data transmission services to residential customers. Therefore, due to its flexibility and versatility in applications for both fixed and mobile wide area networks, it has a huge potential to achieve the objective of accessing global information at any time and at any place by any mobile wireless device, in a seamless manner.

The benefits promised by WiMAX will provide it with a significant edge over existing access technologies, e.g., data over cable service interface specification (DOCSIS) cable modems, Digital Subscriber Line technologies (xDSL), T-carrier and E-carrier (T-x/E-x) systems, and Optical Carrier Level (OC-X) technologies. WiMAX technology can reach theoretical data rates of up to 75 Mbps over a 30 mi radius and closer, with 1.5 Mbps at longer distances that deliver rates equivalent to T-1 line. Thus, similar to wired broadband services, a tiered pricing approach can be used, [37]. Also, it offers greater benefits over wireless fidelity (Wi-Fi) standards. Hence, it can be safely anticipated that in the future, WiMAX-based solutions will be widely preferred by operators.

Basically, WiMAX is a set of profiles based on IEEE 802.16 standards, where the choice of profiles depends upon various factors such as market demand, spectrum availability, regulatory constraints, type of service required, and vendor interests. Two standardized versions of WiMAX are 802.16-2004 and 802.16e-2005, whose essential features can be briefly described as follows [16,17]: 802.16-2004 WiMAX is based on the IEEE 802.16-2004 standard, which is a revised version of the IEEE Std 802.16-2001, and is also close to the European Telecommunications Standards Institute (ETSI) HiperMAN standards. It uses orthogonal frequency division multiplexing (OFDM) and supports fixed and nomadic access in both line-of-sight and nonline-of-sight environments. IEEE 802.16-2004 based WiMAX profiles are more applicable to fixed applications with directional antennae, as OFDM is less complex, has lower cost and is easy to deploy.

802.16e-2005 WiMAX has been optimized for dynamic radio channels and is based on the IEEE 802.16e standard. It is an enhanced version of 802.16-2004, employing scalable orthogonal frequency division multiplexing access (SOFDMA), which is a multi-carrier modulation technique using subchannelization. In addition, it has better QoS support than the older version; as the former version has four classes of service, whereas five classes of service have been defined for the latter. IEEE 802.16e has been standardized to minimize the interference for user devices with omnidirectional antennae, support for multiple input multiple output (MIMO), adaptive antenna systems, soft hand-offs, better security features, and improved power saving by introducing a sleep mode.

The Medium Access Control (MAC) layer of the WiMAX standard consists of three sublayers: Service-Specific Convergence Sublayer (CS), MAC Common Part Sublayer (MAC CPS), and the Security Sublayer [19]. The CS encapsulates wireline data from IEEE 802.3/Ethernet, ATM, and IPv4 onto the air interface. The core functionalities of system access, bandwidth allocation, connection establishment, and maintenance are provided by MAC CPS. The security sublayer provides authentication, secure key exchange, and encryption [19]. In the physical layer, WiMAX supports three modes: single carrier, OFDM, and OFDMA, details are presented in Refs. [16,17]. The primary focus of this chapter is scheduling in WiMAX, which is one of the key functions in the MAC CPS.

It should be noted that most of the work presented in this chapter has been derived by summarizing the existing literature on QoS provisioning and packet scheduling in wireline and wireless networks, with an emphasis on WiMAX-based networks. References to original work, for further study and research, have been provided at the end of this chapter.

## 2.1 INTRODUCTION TO QoS SCHEDULING IN WiMAX

One of the main objectives of WiMAX is to manage bandwidth resources at the radio interface in an efficient manner, while ensuring that QoS levels, negotiated at the time of connection setup, are met in an appropriate way. In the sequel, the provision of guaranteed levels of QoS in WiMAX is fundamentally dependent upon traffic policing, traffic shaping, connection admission control, and packet scheduling. To utilize the bandwidth most efficiently, the IEEE 802.16 standards employ operations of concatenation, fragmentation, and packing of MAC protocol data units (PDUs) and MAC SDUs [19,20].

Due to the highly variable nature of multimedia traffic, subscriber stations (SSs) of WiMAX can perform traffic shaping and policing for efficient utilization of resources and conformance to service level agreements. The nonconforming traffic can be penalized or rejected by an SS. A centralized connection admission control guarantees that newly admitted traffic will not cause congestion or degradation of services in the existing traffic. In WiMAX, admission control is implemented at the base station (BS). Despite the importance of the aforementioned mechanisms, the most important component for QoS provisioning is the Packet Scheduler. Thus, providing efficient scheduling mechanisms in WiMAX is the main focus of this chapter. However, several other related concepts are also described briefly.

In its broadest sense, scheduling refers to the mechanism for serving the enqueued resource requests of various users. A scheduling discipline has two orthogonal components: deciding the order of servicing the users' requests and management of the service queues [21]. A sketch of basic operations in a typical wireless scheduler is presented in Figure 2.1. Scheduling is important in both best effort and QoS networks. In the former case, the fair allocation of network bandwidth among a wide variety of network users is a prime objective. Whereas, in networks providing QoS guarantees (such as WiMAX), scheduling disciplines play a key role in ensuring that negotiated service level agreements are fully complied with. It should be noted that the requirements for scheduling algorithms in wired and wireless networks, such as WiMAX, are different from each other, and this will be explained in more detail in forthcoming sections.

To provide QoS provisioning, the following three methods have been devised for WiMAX: Service flow QoS scheduling, dynamic service establishment, and two-phase activation model. The details of these concepts can be found in Ref. [19]. A service flow in WiMAX has been defined as a

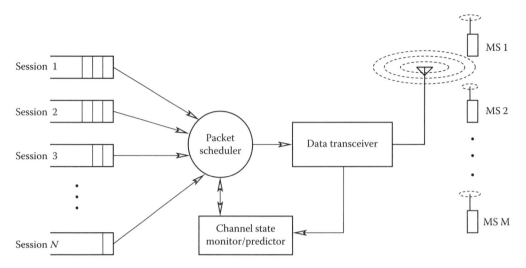

**FIGURE 2.1**    Basic operation of scheduler in a wireless network. (Adapted from Bhagwat, P., Bhattacharya, P., Krishna, A., and Tripathi, S.K., *IEEE INFOCOM*, 3, 1133, March 1996.)

MAC transport service that provides unidirectional transport of data packets either to uplink packets transmitted by the SS or to downlink packets transmitted by the BS. It is characterized by latency, jitter, and throughput assurances. It has the following major attributes [19,20]:

- Service flow ID (SFID): It is assigned to each existing service flow and serves as its principal identifier in the network
- Connection ID (CID): It is a mapping to the SFID that exists only when a connection has been admitted or it is an active service flow
- ProvisionedQoSParameterSet: It is a set of QoS parameters that is provisioned from outside the standard, such as a network management system belonging to the provider
- AdmittedQoSParameterSet: It defines a set of QoS parameters for which both the BS and the SS reserve resources (bandwidth, memory, and other time-based resources)
- ActiveQoSParameterSet: It is a set that defines the service actually being provided to active service flows
- Authorization module: It is a logical module within the BS that approves or denies every change to the QoS parameters and classifiers associated with a service flow

The relationship between the various sets of QoS parameters has been depicted in Figure 2.2. It can be noticed that the ActiveQoSParameterSet is always a subset of the AdmittedQoSParameterSet, which in turn is always a subset of the authorized envelope [19]. The scheduling algorithms to be used at the SSs of WiMAX will need to comply with values of the QoS parameters as indicated by the envelope.

Note that the automatic repeat request (ARQ) mechanism is optional in WiMAX. If implemented, it is done on a per connection basis and is specified and negotiated at the time of creation of the connection. Also, a connection cannot have a mixture of ARQ and non-ARQ traffic.

## 2.2   WiMAX TRAFFIC CLASSES

In the IEEE 802.16-2004 standard [19], four traffic classes or scheduling services are supported, viz., Unsolicited grant service (UGS), real-time polling service (rtPS), nonreal-time polling service (nrtPS), and best effort (BE).

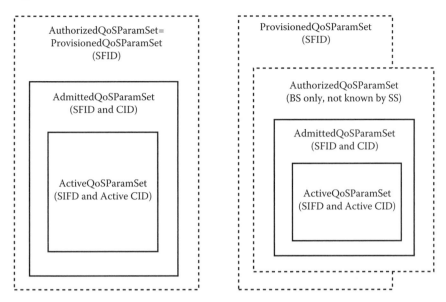

**FIGURE 2.2** Envelopes for provisioned and dynamic authorization models. (From IEEE 802.16e-2004, IEEE standard for local and metropolitan area networks—Part 16: An interface for fixed and mobile broadband wireless access systems, October 2004. With permission.)

The WiMAX standard dictates that the UGS traffic must only require fixed allocations for bandwidth and no bandwidth requests, thus eliminating the need to process UGS connections for scheduling purposes, which is true, both for BS to SS and SS to BS directions of data transmission in the wireless channel of a WiMAX network. It enables the BS to provide fixed sized bandwidth grants periodically to the registered UGS traffic flows. Thus, allowing the use of channel bandwidth by the SSs, without any additional signalling requirements, which helps to meet the stringent latency and jitter requirements of the UGS traffic class. Note also that the size of the bandwidth grant is based on the maximum sustainable data rate at the application interface and is negotiated at the time of connection setup.

In Ref. [20], a fifth service, called the extended rtPS (ertPS), has also been added to the original IEEE 802.16-2004 standard. In this traffic class, the BS will provide unsolicited unicast grants (as for the UGS service) that are not of a fixed size but are dynamic in nature. Thus, ertPS builds on the efficiency of both UGS and rtPS scheduling services. Each of the five traffic classes of the WiMAX standard have been designed for specific types of traffic characteristic, as summarized in Table 2.1:

The mandatory QoS parameters for various scheduling services of the various types of traffic over WiMAX, as given in Table 2.1, are listed as follows:

- Maximum sustained traffic rate ($\rho_{max}$): It defines a bound on the peak information rate of the service (bits/s). It does not include MAC overheads. It does not limit the instantaneous rate of the service which is governed by the channel state. At the SS in the uplink direction, the service is policed for conformity on the basis of a time average. However, no policing is done on the downlink at the BS. If this parameter is set to 0 or omitted, then there is no mandated maximum rate.
- Minimum reserved traffic rate ($\rho_{min}$): It specifies the minimum value of the averaged traffic rate (bits/s) for a service flow. Its value will be honored only when a sufficient amount of data is available for scheduling, otherwise all available traffic will be transmitted as soon as possible.

**TABLE 2.1**

**QoS Service Classes in IEEE 802.16e-Based WiMAX**

| Scheduling Service | Design Description |
|---|---|
| UGS | Supports real-time, fixed sized, and periodic data packets, such as in T1/E1 and VoIP without silence suppression |
| rtPS | Real-time, variable sized, and periodic data packets, such as in Moving Pictures Expert Group (MPEG) video |
| ertPS | Real-time, variable sized, and periodic data packets, such as in VoIP with silence suppression |
| nrtPS | Designed for delay-tolerant, variable sized, miniumum requred rate data rates, such as FTP |
| BE | No minimum service level is needed and can be handled on the basis of available space |

*Source:* IEEE 802.16e-2004, IEEE standard for local and metropolitan area networks—Part 16: An interface for fixed and mobile broadband wireless access systems, October 2004; IEEE 802.16e-2005, IEEE standard for local and metropolitan area networks—Part 16: An interface for fixed and mobile broadband wireless systems, February 2006; Hawa, M. and Petr, D.W., *10th IEEE International Workshop on QoS*, 247, May 2002. With permission.

- Maximum latency ($\tau_{max}$): It specifies the maximum latency (ms) between reception of a packet by BS or SS on its network interface and forwarding that packet to the RF interface. It represents a service commitment or an admission criteria at the BS or the SS and will be guaranteed. It is not mandatory for the service flows exceeding their minimum reserved rate.
- Tolerated jitter ($\xi_{max}$): It defines the maximum delay variation (ms).
- Traffic priority ($\phi$): It assigns priority to traffic flows with the same QoS parameters and with the same service class. A traffic flow with a higher priority will be given lower delay and preference in the queue buffer.
- Request/transmission policy (R/TP): It provides the capability to specify certain attributes (PDU formation, restriction on bandwidth request) for the associated service flows.

It should be noted that these QoS parameters must lie within the boundaries defined by the envelopes given in Figure 2.2. The set of QoS parameters to be complied with, for each of the five traffic classes of WiMAX, is given in Table 2.2.

The scheduling algorithms at the BS and SSs of WiMAX will need to work in concert with each other, so as to observe the values of mandatory QoS service flow parameters. It is evident that, because a different set of QoS parameters has been associated with each traffic class in Ref. [19], correspondingly it is clear that different types of schedulers will be necessary at the SSs.

A simple block diagram model of WiMAX's BS and SS indicating the necessity for schedulers and admission control, is presented in Figure 2.3. It is worth clarifying that the WiMAX standards define: (1) signaling mechanisms between BS and SSs and the uplink scheduling for UGS traffic classes; however, they do not define admission control mechanisms, traffic policing/shaping, and uplink scheduling for rtPS, ertPS, nrtPS, and BE traffic classes. Therefore, it is an important requirement for a service provider to design proper scheduling algorithms for the various traffic classes in WiMAX. Toward this end, this chapter focuses on describing efficient scheduling algorithms for various traffic classes in WiMAX-based wireless networks.

For the sake of completeness, we also present a brief introduction to bandwidth requests, admission control, and bandwidth allocation mechanisms in WiMAX networks.

**TABLE 2.2**
**List of QoS Parameters Associated with Various Scheduling Services**

| Scheduling Service | Mandatory QoS Service Flow Parameters |
| --- | --- |
| UGS | $\rho_{max}$, $\tau_{max}$, $\xi_{max}$, R/TP |
| rtPS | $\rho_{min}$, $\rho_{max}$, $\tau_{max}$, R/TP |
| ertPS | $\rho_{max}$, $\rho_{min}$, $\tau_{max}$, R/TP |
| nrtPS | $\rho_{min}$, $\rho_{max}$, $\phi$, R/TP |
| BE | $\rho_{max}$, $\phi$, R/TP |

*Source:* IEEE 802.16e-2004, IEEE standard for local and metropolitan area networks—Part 16: An interface for fixed and mobile broadband wireless access systems, October 2004; IEEE 802.16e-2005, IEEE standard for local and metropolitan area networks—Part 16: An interface for fixed and mobile broadband wireless systems, February 2006. With permission.

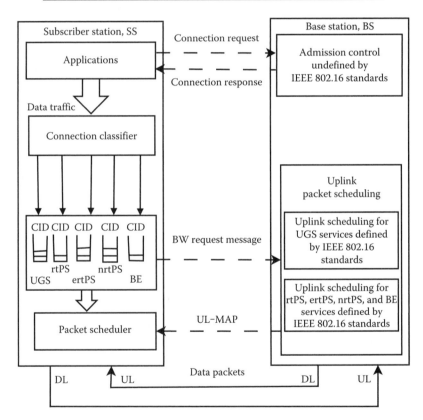

**FIGURE 2.3** A model of typical BS and SS for WiMAX. (Adapted from Wongthavarawat, K. and Ganz, A., *Inter. J. Commun. Syst.*, 16, 81, 2003.)

## 2.3 BANDWIDTH REQUEST MECHANISMS

During network entry and initialization processes, each SS is assigned up to three dedicated CIDs for the purpose of sending and receiving control messages [19]. They are used for allowing a differentiated level of QoS. In WiMAX, an SS can get a bandwidth request to the BS using several methods, these include requests, grants, UGS, Unicast Polling, Multicast/Broadcast Polling, Contention-based focused bandwidth requests, Contention-based code division multiple access (CDMA) bandwidth requests, and Optional Mesh topology support. Vendors are allowed to optimize the performance of their systems by employing different combinations of these schemes. Requests refer to mechanisms used by SSs to indicate to the BS that they require Uplink allocation of bandwidth. It can be as a stand-alone bandwidth request header or as a piggyback request. The use of piggyback is optional.

Another important concept in WiMAX is bandwidth stealing [19]. It refers to an optional strategy, adopted by an SS, in which a portion of bandwidth allocated (in response to a request by a connection) is used to send another bandwidth request rather than sending data. It has been modified in the mobile WiMAX standard [20], as referring to the use by a SS, of a portion of the bandwidth allocated in response to a bandwidth request for a connection to send a bandwidth request or data for any of its connections.

Polling refers to a process where the BS allocates some bandwidth to SSs specifically for making bandwidth requests. It can be for an individual (unicast) or a group (multicast/broadcast) of SSs. The contention-based bandwidth request mechanisms of WiMAX are only allowed for ertPS, nrtPS, and BE traffic classes. UGS and rtPS are not allowed to participate in the contention process. To resolve contentions, a mandatory, truncated binary exponential backoff scheme has been specified in the standard. It has an initial and a maximum backoff window controlled by the BS. Its value is specified by the uplink channel descriptor message and it follows the power-of-two rule, see Ref. [19] for more details. Readers interested in the modeling of polling and contention-based bandwidth requests are referred to Refs. [55,56].

To optimize the bandwidth request latency from an SS to a BS, and thus the response time, for elastic traffic generated by various sources (such as TCP or Push-to-Talk), Ref. [60] defined the Poll-Me (PM) bit in a generic MAC header. Currently, the PM bit is part of the WiMAX standard and it may be used to request to be polled for a different non-UGS connection. The SSs with currently active UGS connections can set the PM bit in the grant management subheader to indicate to the BS that they need to be polled to request bandwidth for non-UGS connections [19]. It is noted that except for UGS, piggyback and bandwidth stealing is allowed for all other traffic classes [19,20].

Recently, an adaptive polling mechanism has been presented in Ref. [61]. Its basic idea is to adapt polling intervals according to ON/OFF periods of traffic. During an ON period, polling intervals are short and of fixed length, whereas during an OFF period the polling intervals are lengthened exponentially, thus, reducing the signaling overhead.

## 2.4   ADMISSION CONTROL AND BANDWIDTH ALLOCATION

In general, admission control is a network's QoS mechanism that determines whether a new session (or connection), with given bandwidth and delay requirements, can be established or not. For providing QoS, this procedure has been applied to both wireline and wireless networks. In the case of WiMAX, whenever a new session wants to make use of the wireless network, an admission control request is sent by the SS to the BS. This admission control request will be accepted by the BS if there is enough available bandwidth, QoS guarantees for bandwidth and delay can be met and the QoS of existing connections is not disturbed. In Ref. [11], an admission control scheme for WiMAX has been proposed together with the derivation of rules for each of the four classes of WiMAX [19]. In addition, a token bucket based admission control for rtPS flows has been proposed in Ref. [50]. Omitting any further discussion involving admission control, we now present a brief overview of bandwidth allocation mechanisms in WiMAX.

The BS allocates bandwidth on a per SS basis, known as the grant per subscriber station (GPSS); further, each SS distributes this bandwidth among all of its active connections. The SS can efficiently distribute the allocated bandwidth as it has up-to-date information about the queue status of each connection. Thus, GPSS requires a packet scheduler at each SS, which may increase the complexity

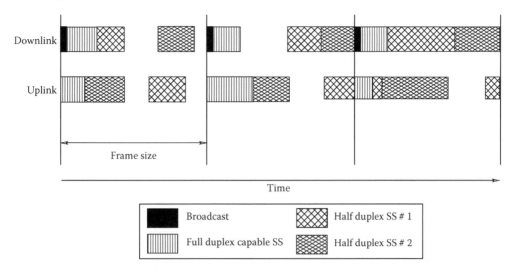

Downlink

Uplink

Frame size

Time

| ■ | Broadcast | ⧆ | Half duplex SS # 1 |
| ▥ | Full duplex capable SS | ⧆ | Half duplex SS # 2 |

**FIGURE 2.4** An example of Burst FDD bandwidth allocation. (From IEEE-802.16-2004, IEEE standard for local and metropolitan area networks—Part 16: An interface for fixed and mobile broadband wireless access systems, October 2004. With permission.)

and the cost of an SS. However, GPSS is scalable to a large number of SSs and is, therefore, the only bandwidth allocation mechanism being employed in the current WiMAX.*

With GPSS, each SS treats various connections separately at its own level and these are then pooled together as one entity for bandwidth allocation at the BS. Thus, the scheduler at the BS will only need a small amount of information about the overall bandwidth required by a particular SS. This approach has the additional advantage of avoiding the time lag in receiving updated information about individual connections at the SS. Once a lump of bandwidth has been granted to a particular SS, then it is responsible for the appropriate scheduling, according to priorities and the QoS for each active connection. This process greatly reduces the workload on the BS. For instance, suppose that an urgent packet arrives at the SS, then the BS does not need to have information about it and it is the duty of the scheduler at the SS to provide the required throughput and delay.

BS and SS communicate with each other by using a bidirectional path, viz., Uplink (UL: SS to BS) and Downlink (DL: BS to SS); whereas the bandwidth requirements are made by the UL and grants are made by the DL. WiMAX supports both Frequency Division Duplex (FDD) and Time Division Duplex (TDD) modes as shown in Figures 2.4 and 2.5, respectively.

In FDD mode, both the UL and DL are operating at separate frequencies and DL data can be sent in bursts. To facilitate various types of modulation, a fixed duration frame is used for both the DL and UL transmissions [19]. Also, it allows the simultaneous use of both full and half duplex SSs; a full duplex SS can transmit and receive data at the same time whereas a half duplex SS can either transmit or receive data at any given time. If half duplex SSs are used, then bandwidth controller will not allocate UL bandwidth at the same time that it is expecting to receive data on DL channel. It should also take into account the allowance for propagation delay, SS transmit/receive transition gap, and SS receive/transmit transition gap [19]. In FDD mode, the use of a fixed duration frame, for both the DL and UL channels, also helps in simplifying the design of algorithms for bandwidth allocation. It can be noted that a full duplex SS can listen to a DL channel continuously, whereas the half duplex SS can only listen to a DL when it is not transmitting on the UL channel.

---

* In the IEEE Std 802.16-2001 an alternative mechanism, known as grant per connection (GPC), for bandwidth allocation has also been defined. However, it is not present in the recent IEEE Std 802.16-2004 and IEEE Std 802.16e-2005 versions. In GPC mode, the BS allocates bandwidth to each connection on an individual basis. It is a purely centralized mechanism with all intelligence placed in the BS, while SSs work as passive entities.

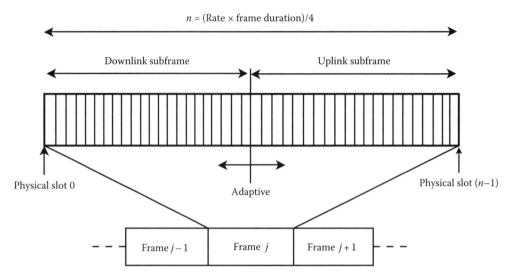

**FIGURE 2.5**   Frame structure of TDD. (From IEEE-802.16-2004, IEEE standard for local and metropolitan area networks—Part 16: An interface for fixed and mobile broadband wireless access systems, October 2004. With permission.)

In the case of TDD, the UL and DL transmissions occur at different time intervals, while usually employing the same frequency. A TDD frame is also of fixed duration and is composed of one DL and one UL subframe. For easy partitioning of bandwidth, a TDD frame is divided into an integer number of physical slots. Also, TDD framing is adaptive and the bandwidth allocated to the UL and DL parts can vary and is controlled by the higher layers. The DL-MAP and UL-MAP messages define the usage of the corresponding transmission intervals. The BS also regularly transmits DL and UL channel descriptors, DCD and UCD, for the physical description of the corresponding channel. The complete list of MAC management messages can be found in Refs. [19,20]. It should be noted that a WiMAX network can be planned with either FDD or TDD, but the former mode has been discussed more frequently in the literature; for a simple introduction to WiMAX design see Refs. [72] and [73].

## 2.5   GENERALIZED WIRELESS PACKET SCHEDULING

In this section, we consider the general problem of packet scheduling in wireless networks. For later sections it will set the stage for defining the classification/framework for scheduling algorithms in WiMAX. The fundamental characteristics of packet-networks operating over wireless channels include (1) time-varying wireless channel capacity, (2) location-dependent channel errors and traffic that is bursty in nature, (3) contention among mobile hosts, (4) mobile hosts do not have a global channel state (5) proper type of scheduler required for both UL and DL flows, and (6) mobile hosts often have limited battery and processing power. The above-mentioned factors need to be considered very carefully, while designing schedulers for wireless networks. Otherwise, the performance will not be optimal.

Two basic performance measures used in the literature for packet schedulers are throughput and fairness. Throughput usually refers to the amount of data transferred from the BS to the SS, in its own traffic class; whereas fairness refers to, ideally, equal allocation of allotted bandwidth to all SSs for a particular traffic class. If an SS lies in a bad channel state, then there should not be bandwidth allocation to that particular SS. The concept of fairness is discussed in more detail in the latter part of this section.

In wired networks, the retransmitted packets can be excluded from throughput computation to give another performance measure, known as goodput. Similarly, in WiMAX networks, either throughput or goodput can be used as measures of data transferred from BS to SS and vice versa. It should be noted that customers in various traffic classes will try to maximize their throughputs, which may lead to classical selfish game-theoretic behavior, which needs to be policed by mechanisms at the SSs or at the BSs. In such a competitive environment, each user will be trying to maximize its own utility function.

While taking into account the temporal characteristics of channels, a wireless packet scheduler should have the following essential features [1–3,10]:

- Efficient utilization of wireless channel bandwidth: The wireless scheduling algorithm should utilize the channel efficiently and should avoid wasting resources on links operating in bad state. An efficient service discipline will be able to meet the end-to-end performance guarantees for various service classes under all load conditions.
- Throughput bound: For each service class, the scheduler should be able to provide a short-term throughput bound for flow with a clean channel and a long-term throughput bound for all flows including those in an error state in the channel.
- Short-term and long-term fairness: The scheduler should be able to provide fair allocation of bandwidth to all flows, from various traffic classes, within a good channel state as well as to those lying within a bad state.
- Delay bound: The scheduling algorithm should be able to provide a guaranteed delay bound on various traffic classes.
- Implementational complexity: The scheduling algorithm should be simple and have a low time complexity to select and forward a packet from the queues of various classes. Generally, fairness and delay bound requirements collide with the complexity of the scheduling algorithm. Schedulers having good fairness and strict delay bounds are harder to implement, whereas $\mathcal{O}(1)$ algorithms are simplest but provide poor fairness and delay bounds.
- Graceful degradation of service: The scheduler should be able to compensate for back-logged flows, which have not received service due to bad channel conditions, at the expense of those flows which have received extra service due to good channel conditions. This corrective reduction in the service allocation of certain flows belonging to wireless channel in a good state should be smooth and gradual.
- Protection against misbehaving flows: The scheduling algorithm should be able to protect service guarantees for various classes and eliminate the effects of misbehaving flows, network load fluctuations, and best effort uncontrolled traffic flows (such as in the case of denial-of-service (DOS) attacks).
- Decoupling between delay and bandwidth: The scheduler should be able to decouple bandwidth and delay and thus should support both delay sensitive and error sensitive flows. Usually, the classes having higher reserved data rate also have low delay requirements; however, some high bandwidth applications can work well even with larger delays, such as web browsing.
- Flexibility and scalability: The scheduling algorithm should be flexible enough to cope with the vast number of different current types of IP traffic, as well for future traffic characteristics. Also, it should perform well when there is an increase in the number of connections for each traffic class.
- Power efficiency: The scheduling algorithm should be power efficient. It is especially important for the mobile subscriber's equipment, wherein currently available batteries have a limited (charged) life.

It can be noticed that designing a packet scheduler for WiMAX which satisfies all of the above-mentioned salient features is not a trivial task. Various existing schedulers can usually satisfy a limited subset of the above properties but not the whole set. To make the scheduling problem more difficult, the negotiated limits on all of QoS parameters for various traffic classes, Table 2.2, and associated envelopes as defined in Figure 2.2 need to be completely satisfied.

Next, we briefly discuss some relevant fairness principles. Speaking informally, fairness refers to the equal allocation of network resources (bandwidth in wireless channels or link capacity in wireline networks) among the various users. It is one of the fundamental requirements in both wireline and wireless networks. A major part of a wireless scheduler's performance is determined by how to provide good fairness among diverse users operating in both good and bad channel states. In general, fairness can be classified into one of the following two types [35]:

- Max-Min Fairness: For a total link capacity $C$, a set of throughputs for $i$ users is said to be max-min fair if it is feasible (i.e., $r_i > 0$ and $\sum_{i=1}^{n} r_i < C$) and the throughput of each user, $r_i$, cannot be increased without maintaining feasibility and decreasing the throughput of any other user, $r_j$, for which $r_j < r_i$.
- Proportional Fairness: A set of throughputs $r_i$ is proportionally fair if it is feasible and for any other feasible vector $r_i^*$, the sum of proportional changes is 0 or negative, i.e.,

$$\sum_{i=1}^{n} \frac{r_i^* - r_i}{r_i} \leq 0 \tag{2.1}$$

It should be noted that the above two types of fairness were originally defined in the context of wireline networks. Their application to wireless networks, in general, and WiMAX, in particular, will require modification by considering the effects of the channel state as well; otherwise precious bandwidth will get wasted by SS lying in a bad channel state. Fairness can quantified using Jain's fairness index (JFI) [35]:

$$\text{JFI} = \frac{\left(\sum_{i=1}^{n} r_i\right)^2}{n \cdot \sum_{i=1}^{n} r_i^2}, \tag{2.2}$$

where its value ranges from $1/n$ to 1 (best fairness). Also, borrowing from the field of Economics, the Gini fairness index has been defined as [65,74]

$$I = \frac{1}{2n^2 \bar{u}} \sum_{k=1}^{n} \sum_{l=1}^{n} |u_k - u_l|, \tag{2.3}$$

whereas for an SS with a long-term average transmission rate of $r_k$ and fair share weight of $w_k$, we have $u_k = r_k/w_k$, $u = (u_1, ..., u_n)$ and $\bar{u} = \sum_{k=1}^{n} u_k/n$; $I$ varies between 0 (best fairness) and 1. Now, that the basics of wireless packet scheduling have been defined, further we can focus on the scheduling setup for WiMAX.

## 2.6   SCHEDULING SETUP IN WiMAX

IEEE 802.16e has been developed to serve mobile SSs through a centralized BS by employing point-to-multi point (PMP) as well as through the optional mesh mode architecture of a wireless network topology. In the former operating mode, the downlink from a BS to an SS operates on the basis of PMP. However, in the mesh mode, there are no separate DL and UL subframes and there can be SSs that are not directly connected to the BS but only through intermediary SSs, which is in contrast to the PMP mode. Hence, a larger number of SS can be supported in the mesh mode than in the PMP mode or equivalently, mesh mode can offer the least number of BSs for economy. Furthermore, in the mesh mode of WiMAX the SSs can consume less power, thereby efficiently using battery life, as it is not mandatory to be always connected to the BS. The intermediate SSs will greatly reduce the power consumption of far off users [40].

Three types of scheduling are supported in the mesh mode of WiMAX; they are coordinated distributed, uncoordinated distributed, and centralized scheduling. In coordinated scheduling, all nodes coordinate in their two-hop neighborhood and broadcast their schedules (available resources, requests and grants) to all of their neighbors; whereas, in uncoordinated scheduling there are direct uncoordinated requests and grants between two nodes [19]. The main difference between the two types of distributed scheduling methods is in the use of the control subframe: transmitting collision-free scheduling messages in the coordinated type and with possible collision in the uncoordinated type. In the centralized method, the resources are distributed centrally and this is similar to the PMP case.

The performance of coordinated distributed scheduling has been investigated in Ref. [75]. It has been reported that this mechanism has a scalability problem that leads to poor performance in dense networks and aggravates QoS provisioning. To overcome these problems, the *XmtHoldoffTime* [19] has been made adaptive at every node, which has been shown to improve contention, and thus enhance the throughput in dense meshes. In Ref. [68] a combined distributed and centralized scheduling scheme has been proposed for mesh networks in WiMAX; wherein, through simulation studies, it has been shown that the *minislot** utilization can be significantly improved with the proposed scheme.

For synchronization of distributed and centralized control mesh networks, the WiMAX standard provides network configuration (MSH-NCFG) and network entry (MSH-NENT) packets as a basic level of communication between various nodes. The scheduling of transmission for the next MSH-NCFG is done by a mesh-election procedure. It is carried out among all eligible competing and local nodes. The NetEntry scheduling protocol provides slots for transmission of MSH-NENT packets by new nodes that are not yet fully functional members of the mesh [19].

In contrast to mesh mode, a detailed QoS architecture has been defined for the PMP mode in Refs. [19,20]. Scheduling services refer to data-handling mechanisms supported by the MAC scheduler for data transport on each connection. A single scheduling service will be associated with each data connection. Each of the data services will be characterized by a set of parameters that will quantify the QoS aspects of its behavior. These QoS parameters are managed by dynamic service addition (DSA), dynamic service change (DSC), and dynamic service deletion (DSD) message dialogues, where each of these signaling schemes can be initiated by either a BS or an SS [19].

It can be seen that scheduling mechanisms for a PMP mode are also applicable to a mesh mode; however, since all transmission between two nodes is managed by a link, PMP scheduling is not directly applicable to the mesh mode [40]. By default, at the time of connection establishment, each mesh SS is assigned a unique node identifier; whereas in Ref. [40] a Service Adaptive QoS has been proposed for mesh mode, which assigns five node IDs to each SS instead of a single ID. These five virtual nodes correspond to five traffic classes, as shown in Table 2.1, and each node requests bandwidth individually and the mesh mode BS handles these requests on the basis of their scheduling services [40]. Hence, mesh mode WiMAX can be treated by scheduling the services of the PMP mode. Therefore, subsequently in this chapter, we shall only consider scheduling in the PMP mode for WiMAX networks.

In Section 2.7, we present a classification and an overview of the important types of packet schedulers, originally designed for wireline and wireless networks, that can be tailored for use in WiMAX networks.

## 2.7  CLASSIFICATION AND OVERVIEW OF PACKET SCHEDULERS

Broadly speaking, packet schedulers can be classified into the following two types: work conserving and nonwork conserving types [3]. The former type of scheduler is never idle if there is a backlogged

---

* A unit of UL bandwidth allocation equivalent to $n$ time units for bandwidth allocation (physical slots); where $n = 2^m$, $m$ is an integer between 0 and 7 [19].

packet in any queue; whereas, the latter type of scheduler can be idle even if there are packets already in the queue and it is waiting for the arrival of some higher priority packet. Examples of work conserving scheduling algorithms include Generalized Processor Sharing (GPS), Weighted Fair Queueing (WFQ) (also known as Packet by Packet GPS), Virtual clock, Weighted Round Robin (WRR), Deficit Round Robin (DRR), Self Clocked Fair Queueing (SCFQ); whereas Hierarchical Round Robin (HRR), Stop-and-Go, and Jitter-Earliest-Due-Date are some examples of nonwork conserving schedulers [3]. In this chapter we have only considered the work conserving schedulers.

The evolution of scheduler design can also be identified under two major categories: (1) time-stamped based and (2) round robin based. The time-stamped or deadline-based schedulers have better fairness and delay properties, but need to sort the packet time stamps and thus suffer from complexity which can be logarithmic with the number of flows in a service class [18]. On the other hand, round robin based schedulers have $\mathcal{O}(1)$ worst-case per packet complexity * and fair bandwidth allocation to all flows (irrespective of whether they are in a good or bad channel state, therefore reducing throughput), but cannot provide guaranteed bounds on delay.

Thus, it appears that time-stamped based schedulers will be more widely used for WiMAX traffic classes that have strict delay and throughput guarantees. Common examples of such schedulers are WFQ, WF$^2$Q, SCFQ, and SFQ. These schedulers implement the so-called fluid fair queueing model (FFQ), in which packet flows are treated as a hypothetical fluid that can simultaneously serve multiple sessions and traffic is infinitely divisible. Obviously, FFQ is not directly applicable to real traffic flows in which only one session can receive service at a time and the entire packet has to be scheduled before picking the next packet. However, FFQ has been implemented in GPS, which is presented in Section 2.7.1.

### 2.7.1 GENERALIZED PROCESSOR SHARING

The GPS scheme is an ideal (unimplementable), work conserving scheduling discipline. It assumes that each connection has a separate logical queue. It serves each nonempty queue on a turn-by-turn basis while serving an infinitesimal amount of data from each queue in a finite amount of time; thus it exactly achieves max-min fairness.

To describe GPS, let a set of positive real numbers, i.e., $\phi_1, \phi_2, ..., \phi_N$, be associated with $N$ connections that are being served at a fixed rate $r$. A session $1 \leq i \leq N$ is said to be backlogged at time $t$ if a positive amount of traffic has been enqueued. Let $S_i(\tau, t)$ and $S_j(\tau, t)$ denote the amount of data that has been served in the interval $(\tau, t]$ for connection $i$ and $j$, respectively. The GPS server is defined as: $\frac{S_i(\tau,t)}{S_j(\tau,t)} \geq \frac{\phi_i}{\phi_j}$, [6,7]; where $S_i(\tau, t) \sum_j \phi_j \geq (t - \tau)r\phi_i$ and session $i$ is guaranteed a rate of $g_i = \frac{\phi_i}{\sum_j \phi_j} r$. The fairness characteristics of GPS can be expressed as: $\left| \frac{S_i(\tau,t)}{\phi_i} - \frac{S_j(\tau,t)}{\phi_j} \right| \geq 0$, which provides an ideal bound for comparing the performance of other scheduling algorithms.

In forthcoming sections we shall observe that GPS serves as a benchmark for making scheduling decisions in various practical schedulers, both in wireline and wireless networks, some of which can also be used for WiMAX. In such schedulers, a GPS server is simulated, for error-free environments, in which the arrival rate of the traffic is set equal to the real value of the packet-traffic. It will provide a time sequence, called *virtual time*, for servicing enqueued packets. Thus, GPS constitutes a framework for designing schedulers that are based on FFQ.

Further, we can categorize schedulers as designed specifically for wireline or wireless networks. It can be seen that some of the scheduling principles, invented originally for wireline networks, have also been employed in wireless networks, either with or without modifications.

---

* A scheduling algorithm for which the number of operations needed to select the next packet is constant with respect to the number of traffic flows.

## 2.7.2  WIRELINE SCHEDULERS

### 2.7.2.1  Weighted Round Robin

Weighted Round Robin (WRR) is the simplest implementation of GPS, which serves at least a packet of data instead of serving an infinitesimal amount of data from the backlogged queue. It is a very popular scheduler, due to its simplicity and ease of implementation, $\mathcal{O}(1)$. It is used for providing differentiated bandwidth guarantees to heterogeneous flows of IP traffic. It works quite well when all connections have equal weights and all packets have the same size.

To ensure that WRR can emulate GPS correctly for variable sized packets, the source's average packet size must be known in advance; whereas, in many applications this is hard to predict. Also WRR does not provide good fairness properties over timescales shorter than a round trip time [21,22]. At shorter timescales, if some connection has a smaller weight or the number of connections is very large, it will lead to long periods of unfairness.

In WiMAX, the WRR-based algorithms cannot be used for higher priority traffic classes at the SSs, due to their poor fairness properties. However, they can be used for nrtPS and BE traffic classes at the SS.

### 2.7.2.2  Deficit Round Robin

To overcome the problems of WRR, the Deficit Round Robin (DRR) scheduler has been proposed in Ref. [23]. It is a simple, $\mathcal{O}(1)$, and easily implementable scheme that can provide good fairness characteristics. It can handle variable size packets without knowing the average packet size.

DRR attaches a counter, which is initialized to zero, with each of the enqueued connections. The scheduler visits each connection in a round robin fashion and tries to serve a *quantum* , $\psi_i$, worth of data bits from each queue. The packet at the head of the queue is served if its size is less than $\psi_i$, otherwise a value of $\psi_i$ is added to its deficit counter. In the next round, if the packet size is less than or equal to the previously computed sum and $\psi_i$ (deficit counter = $2*\psi_i$), then it is served and the deficit counter is decreased by the packet size [21]. The size of a quantum should be at least equal to the largest packet size, i.e., $\forall i, \psi_i \geq L_{max_i}$, otherwise the complexity of DRR can be as large as $\mathcal{O}(N)$ [36]. Also, the latency of DRR depends on the number of active flows and the sum of the maximum packet lengths for all queues [36]. Hence, DRR cannot be employed for those traffic classes (at SSs) which need to provide strict delay and throughput guarantees.

Furthermore, DRR is also unfair for timescales shorter than the round trip times [21]. However, due to its extreme simplicity, DRR (or its variants) is typically implemented in high speed routers, such as in Cisco's 12000 GSR and Juniper's M-series, for wireline networks. Similarly, despite their shortcomings, WRR and DRR can still be employed for scheduling at the BS of a WiMAX network [29].

### 2.7.2.3  Weighted Fair Queueing

Weighted Fair Queueing (WFQ), [8], and packet-by-packet GPS (PGPS), [6], are the practical implementations of the ideal GPS scheduler, cf. Section 2.7.1. Though discovered independently, both PGPS and WFQ are essentially the same concept. They are also referred to as practical implementations of an FFQ model developed in Ref. [6].

One of the main drawbacks of the WFQ scheduling discipline is that packets can leave much earlier than in a GPS system, i.e., WFQ can be far ahead of GPS in terms of the number of bits served for a given traffic session [25]. It leads to sending bursts of packets across the network, which can cause undesired oscillations in feedback-based congestion control algorithms; such as in TCP/IP traffic.

It has been employed by Cisco as one of the premier techniques in providing QoS for wireline networks. In WiMAX-based networks, WFQ can be used at SSs for scheduling the traffic belonging to the nrtPS class [11]. It has been also employed for all traffic classes in Ref. [62].

#### 2.7.2.4  Worst-Case Fair Weighted Fair Queueing

Worst-case fair weighted fair queueing WF$^2$Q [25] has been proposed to overcome the weaknesses of WFQ over GPS. The fairness measure adopted in Ref. [25] is a refined notion and is called worst-case fairness, which can be defined as follows: a service discipline is worst-case fair for a session $i$, if at any time $\tau$ the delay of an arriving packet, $D_i$, is bounded above by

$$D_i < a_i + \frac{Q_i}{r_i} + C_i, \tag{2.4}$$

where
   $r_i$ is the throughput guarantee
   $Q_i$ is the queue size at time $a_i$ for session $i$
   $C_i$ is a constant independent of the queues of other sessions sharing the same multiplexer

A service discipline is called worst-case fair if it is worst-case fair for all sessions.

In WFQ, the next packet for transmission is chosen among all of the backlogged packets, the first packet that would complete service in the corresponding GPS system. In WF$^2$Q, the next packet for transmission is chosen among the set of packets that have started and possibly finished receiving service in GPS. However, the computation of virtual time in both WFQ and WF$^2$Q scheduling disciplines adds a considerable amount of complexity, thus making their implementation a difficult task. WF$^2$Q can be employed at SSs for scheduling the nrtPS traffic class. Also, in Ref. [63] it has been adopted for all traffic classes.

#### 2.7.2.5  Self-Clocked Fair Queueing

The Self-Clocked Fair Queueing (SCFQ) discipline modifies the notion of virtual time in WFQ and ties it up with the actual queueing system rather than a hypothetical system based on the FFQ model of GPS [26]. The use of GPS as a reference system for fairness is the main reason of complexity in WFQ and WF$^2$Q. Also, to add further complexity, several packets will be receiving service at the same time in a GPS system.

SCFQ can be explained as: on arrival, each packet is marked with a service tag before placing it in the queue. The packets in the queue are serviced in an increasing order of associated tags. For each session $k$, the tags for the arriving packets are computed iteratively by $F_i^k = L_i^k/r_k + \max\{F_{i-1}^k, v(a_i^k)\}$ with $F_0^k = 0$, for $i = 1, 2, ..., k$; whereas $v(t)$ is regarded as virtual time at $t$ and is defined as being equal to the service tag of the packet being served at that instant of real time.

Furthermore, it has been conjectured in Ref. [26] that due to the logarithmic computational complexity and bounded end-to-end delay property, the SCFQ algorithm is well suited for broadband applications. However, it has been contended in Ref. [25], that both delay and fairness properties of SCFQ are worse than the WF$^2$Q algorithm. A comparison between WFQ, WF$^2$Q, and SCFQ has been presented in Table 2.3. In addition, a hierarchical packet fair queueing algorithm has been proposed in Ref. [41], which has lower value for the worst fair index, Equation 2.4, than SCFQ. A variation of SCFQ is start time fair queueing (SFQ), where, for each arriving packet, two tags are computed: a start and a finish tag [13]. Unlike SCFQ, packets are serviced in increasing order of their start tags and ties are broken arbitrarily. The virtual time, $v(t)$, is defined as the start tag of a packet in service at time $t$. Because, the computation of $v(t)$ is inexpensive, the computational complexity of this algorithm is the same as for SCFQ, i.e., $\mathcal{O}(\log N)$. Due to better characteristics than both WFQ and WF$^2$Q, SCFQ and SFQ can also be experimentally used for various traffic classes as in Refs. [11,62,63]. Several other variations of fair queueing have been omitted from this discussion.

**TABLE 2.3**

**A Comparison between WFQ, WF²Q, and SCFQ**

| Scheduler Type | E2E Delay Bound | E2E Delay-Jitter Bound | Buffer Size | Complexity |
|---|---|---|---|---|
| WFQ | $\frac{\sigma_j + nL_{\max}}{r_j} + \sum_{i=1}^n \frac{L_{\max}}{C_i}$ | $\frac{\sigma_j + nL_{\max}}{r_j}$ | $\sigma_j + hL_{\max}$ | $\mathcal{O}(N)$ |
| WF²Q | $\frac{\sigma_j + nL_{\max}}{r_j} + \sum_{i=1}^n \frac{L_{\max}}{C_i}$ | $\frac{\sigma_j + nL_{\max}}{r_j}$ | $\sigma_j + hL_{\max}$ | $\mathcal{O}(N)$ |
| SCFQ | $\frac{\sigma_j + nL_{\max}}{r_j} + \sum_{i=1}^n \frac{K_i L_{\max}}{C_i}$ | $\frac{\sigma_j + nL_{\max}}{r_j} + \sum_{i=1}^n \frac{(K_i - 1)L_{\max}}{C_i}$ | $\sigma_j + hL_{\max}$ | $\mathcal{O}(\log N)$ |

*Source:* Zhang, H., *Proc. IEEE*, 83, 1374, October 1995. With permission.

*Note:* $C_i$ is the link speed, $h$ is a switch number, $K_i$ is the number of connections sharing the link, $r_j$ is the guaranteed rate, $\sigma_j$ is the maximum burst size of traffic, and $L_{\max}$ is a maximum packet size.

#### 2.7.2.6 Earliest Deadline First Scheduling

The earliest deadline first (EDF) is a scheduling policy in which the enqueued packets are transmitted on the link on the basis of smallest deadline first [28]. It is a dynamic scheme, in which the priority of scheduling a packet increases with the amount of time it spends in the queue. Define a nonnegative vector $D = [D_1, D_2, ..., D_i, ..., D_N]$ that denotes a list of required upper bounds on delay, i.e., for session $i$, no packet is delayed by more than $D_i$ units of time in the scheduler. The deadline for a packet $k$ in session $i$, arriving at time $t_k$ is defined as $d_k^i = t_k + D_i$; also for a finish time of $f_k^i$, the lateness is defined as $l_k^i = f_k^i - d_k^i$. EDF can be categorized into preemptive and nonpreemptive types, i.e., PEDF and NPEDF, respectively. The PEDF is a policy which, at any instant, schedules packets with the smallest deadline first. Ties are resolved by picking one of the packets in an arbitrary fashion. It is a delay optimal scheduling policy [28]. On the other hand, NPEDF is a scheduling policy that behaves like PEDF but takes decisions only at the time of completion of packet transmission or arrival of a new packet in an empty system.

For $M$ packets waiting in a queue, with appropriate deadlines, the complexity of EDF is $\mathcal{O}(\log M)$. An optimization of EDF can be obtained by maintaining a priority list consisting of a deadline for only a single packet from each flow, thereby reducing the complexity to $\mathcal{O}(N)$. The EDF-based scheduler has been suggested for the rtPS traffic class in Ref. [11].

### 2.7.3 WIRELESS SCHEDULERS

#### 2.7.3.1 Idealized Wireless Fair Queueing

The Idealized Wireless Fair Queueing (IWFQ), [4], is an application of the FFQ concept in wireless networks. It is based on a model that incorporates the time-varying characteristics of channels.

In IWFQ the granularity of each flow is a bit. A service tag is associated with each bit, which is a virtual time for the error-free FFQ service. The service tag of a backlogged flow is the same as that of the first bit, whereas the service tag of a non backlogged flow is set to $\infty$. IWFQ allows the lagging flows to make their lag by causing leading flows to give their lead, within the specified bounds [4].

There are several issues in the implementation of IWFQ, which include unpredictability of the channel state, at the BS the lack of complete knowledge about the number of backlogged packets on various links, and bursty errors will give precedence to lagging flows as those channels will be polled more often. Several of these issues can be addressed by cooperation of the MAC layer. Thus, practical implementation of wireless fair scheduling, i.e., IWFQ, requires a close coupling with the MAC layer. IWFQ scheduling principle can be experimentally adopted for use with rtPS and ertPS traffic classes at SSs of WiMAX network [42].

### 2.7.3.2   Channel-Independent Fair Queueing

Channel-Independent Fair Queueing (CIFQ) algorithm is another well-known wireless scheduling principle, which aims to provide throughput and delay guarantees for error-free sessions, long-term fairness guarantee among the sessions in error state of channel, short-term guarantee among error-free sessions, and graceful degradation in QoS for sessions having already received excessive service due to the presence in the error-free channel state [27].

The CIFQ algorithm employs the SFQ algorithm for reference purposes and classifies each session as leading, lagging, or satisfied. It does not employ GPS as in the case of IWFQ. Each session $i$ is associated with a virtual time and a parameter $lag_i$, where for $\mathcal{A}$ active sessions we have $\sum_{i \in \mathcal{A}} lag_i = 0$ as the system is work conserving. A session is lagging if $lag_i > 0$, leading if $lag_i < 0$, and satisfied if $lag_i = 0$. Also, it has been derived that for $n$ number of active sessions, CIFQ has a complexity of $\mathcal{O}(n \log n)$ for the packet leaving operation, which is larger than WFQ. However, due to its optimum design for dynamically changing wireless channels, CIFQ can be experimentally deployed for either rtPS or ertPS traffic classes in WiMAX.

It should be noted that due to a lack of space, several other wireless schedulers, e.g., channel state dependent packet scheduling (CSDPS) [14], CSDPS with class based queueing (CSDPS+CBQ) [32,33], and server based fair approach (SBFA) [9,34], have been omitted from this section. They are not directly applicable to WiMAX, but can be studied for better understanding of various issues involved in wireless packet scheduling.

We now discuss some existing scheduling frameworks for WiMAX networks. These experimental frameworks attempt to provide a complete scheduling infrastructure for each of the five classes of WiMAX, Table 2.1. It has been identified that some of the work available in the literature has been done for the (now) obsolete GPC mode. However, it has been summarily presented here for completion and comparison purposes, though GPC is absent in the current WiMAX standards.

## 2.8   PROPOSED SCHEDULING FRAMEWORKS FOR WiMAX

Now we overview some major existing scheduling frameworks for WiMAX networks. This is not a complete survey, but we report some major works in this area. In a similar manner to Ref. [48], we have categorized these proposed frameworks as Homogenous, Heterogenous, and Miscellaneous types; where the first two types are based on legacy scheduling principles, described at length in Section 2.7, and the last type covers several novel disciplines. This taxonomy is presented in Figure 2.6.

### 2.8.1   HOMOGENOUS FRAMEWORKS

These frameworks employ a single kind of scheduler for all types of WiMAX traffic.

In Ref. [24], simulations have been used to evaluate the QoS performance of WiMAX [19]. DRR has been selected as a DL scheduler and WRR has been chosen as an UL scheduler. Similarly, both WRR and DRR have been selected, independently, to be employed at SSs. This chapter does not consider the application of a wide variety of other schedulers for the various traffic classes, as done in Refs. [11,42].

In Ref. [62], two sets of simulation experiments are performed: first EDF for all traffic classes and second WFQ for all traffic classes. It has been reported in these studies that, if the total traffic load is under 100%, both schedulers can satisfy traffic class performance requirements. However, when one traffic class starts consuming more than its fair share of bandwidth, EDF favors streams with more crucial deadlines, whereas WFQ strives for fairness and punishes the flow that is unfair. It has been concluded that using either EDF or WFQ for all traffic classes is not optimal.

Based on the fair queueing model presented in Refs. [41,63] proposes a hierarchical scheduling model for WiMAX traffic classes. It specifies three types of scheduling servers, viz., hard-QoS,

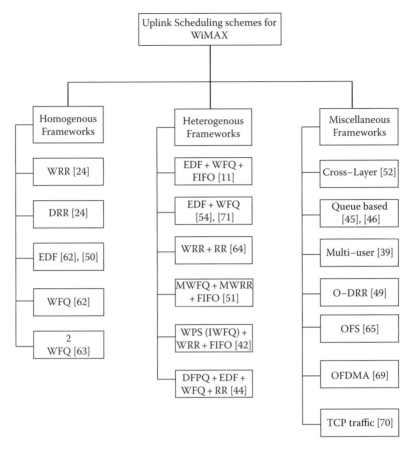

**FIGURE 2.6** A summary of proposed scheduling schemes at SSs for uplink transmission. (Adapted from Dhrona, P., A performance study of uplink scheduling algorithms in point to multipoint WiMAX networks, Master's thesis, Queen's University, Ontario, Canada, January 2008, http://hdl.handle.net/1974/973.) FIFO, First In First Out; RR, Round Robin; WRR, Weighted Round Robin; DRR, Deficit Round Robin; WFQ, Weighted Fair Queueing; $W_2FQ$, Worst Case Fair Weighted Fair Queueing; EDF, Earliest Deadline First; MUFSS, Multi-class Uplink Fair Scheduling Structure; MWRR, Modified Weight Round Robin; WPS, Wireless Packet Scheduling; IWFQ, Idealized Wireless Fair Queueing; DFPQ, Deficit Fair Priority Queue; O-DRR, Opportunistic DRR; OFS, Opportunistic Fair Scheduling; OFDMA, Orthogonal Frequency Division Multiple Access.

soft-QoS, and best effort servers. UGS traffic is mapped into the hard-QoS server and rtPS traffic into the soft-QoS server; whereas nrtPS can be mapped into either. All servers implement the $WF^2Q$ algorithm [25].

A token bucket based call admission control and an UL packet scheduler employing EDF, as in Ref. [11], has been proposed in Ref. [50]. The focus of their algorithm is to meet the bandwidth and delay requirements of rtPS flows. It employs the arrival service curve and a two-dimensional rtPS database as also proposed in Ref. [11]. However, this algorithm does not consider the ertPS traffic class. A token rate estimation model, for Poisson traffic fed to both an infinite and a finite queue, has been developed by employing well-known Markov chain models. The integration of admission control and scheduling has been left as future work. Thus, this approach needs to be investigated further before it can be adopted for use in WiMAX. It is also known that a potential problem with token bucket-based schemes involves allowing large bursts of traffic, for which a leaky bucket can be arranged after the token bucket [58].

### 2.8.2   HETEROGENOUS FRAMEWORKS

These frameworks employ more than one type of scheduler for the various traffic classes of WiMAX.

In Ref. [64], WRR has been adopted for rtPS and nrtPS traffic classes, whereas simple RR has been employed for BE. It does not consider the ertPS traffic class and nor does it specify clearly whether or not GPSS is used.

Based on priority scheduling and dynamic bandwidth allocation, a QoS architecture for WiMAX has been proposed in Ref. [42]. The contention slot allocator has been proposed to be used by the BS to dynamically adjust the ratio of the bandwidth allocated to the contention and reservation slots. WRR has been used as an upstream scheduler at the BS. At the SS level, there are two levels of scheduler, viz., (1) strict priority scheduling for distribution of allocated bandwidth among various classes and (2) schedulers for each of four classes of Ref. [19]. At the first level of scheduling at SS, the Multi-class Priority Scheduler has been suggested [43]; whereas, at the second level of scheduling at the SS IWFQ has been proposed for higher priority, WRR for middle priority, and first-in-first-out (FIFO) for lower priority traffic classes of WiMAX. No experimental or simulation results have been presented.

In Ref. [11], a scheduling framework has been proposed for Ref. [19], which is based on a centralized mechanism similar to the GPC mode of operation. The proposed UL packet scheduler requires three modules at the BS: an information module, a scheduling database module, and a service assignment module. The information module performs the following functions: retrieving the queue size information for each connection from bandwidth request messages, determining the number of packets that have arrived from an rtPS connection in the previous time frame by using arrival service concept, determining rtPS packets' arrival time/deadlines and updating the scheduling database, queueing information from nrtPS, and BE request messages are directly passed to the scheduling database. The proposed scheduling database module will provide scheduling information to all of the connections; whereas the service assignment module determines UL-MAP by using information in the scheduling database [11]. A strict priority scheduler is used for allocation of bandwidth from UGS, rtPS, nrtPS, and BE. Further, it has been proposed that there is no scheduler for UGS, EDF for rtPS, WFQ for nrtPS, and FIFO for BE. This chapter also presents some brief experimental results to support their scheduling architecture.

In Ref. [44], deficit fair priority queueing (DFPQ) has been proposed for the first layer of scheduling at SSs. It has been derived from the DRR discipline. At the second layer of scheduling, EDF has been proposed for rtPS, WFQ for nrtPS, and WRR for BE. Some simulation results have also been presented. It is not clear whether GPSS or GPC has been employed for this work. However, the idea of DFPQ can be easily adopted for other advanced scheduling frameworks involving WiMAX.

A multiclass uplink fair scheduling structure (MUFSS), to support delay and bandwidth require-ments of IEEE 802.16 QoS, has been proposed in Ref. [51]. The BS adopts modified WRR, but these details are not provided. At the SS, a modified WFQ has been suggested for UGS and rtPS classes; whereas, modified WRR and FIFO have been suggested for nrtPS and BE classes. It has been concluded in Ref. [51] that although MUFSS can provide guarantees for delay sensitive real-time traffic, it has less throughput overall and higher processing time.

### 2.8.3   MISCELLANEOUS FRAMEWORKS

#### 2.8.3.1   Cross-Layer Approach

Recently, the design of cross-layer optimized wireless multimedia communication networks has gained significant popularity in the research community. In such systems, the biggest challenges are limited and highly varying channel bandwidth, stringent QoS requirements, limited battery power, and a wide variety of protocols and standards. Cross-layer methodologies depend on the interaction of various protocols and hold great promise for reliable end-to-end wireless communication. Thus, it is useful to consider and investigate the application of cross-layer techniques in WiMAX networks.

A cross-layer scheduler has been proposed in Ref. [52], where each connection employs adaptive modulation and a coding scheme at the physical layer. A dynamically updated priority, based on channel quality and service requirements, is assigned to each connection and the highest priority is scheduled first. Priority functions have been defined for each of rtPS, nrtPS, and BE traffic classes; whereas nothing has been mentioned concerning the ertPS traffic class. The proposed scheduler allocates all time slots to only one connection each time, which is not very efficient. Scheduling of multiple connections have been left as future work. Hence, it is an experimental scheme and needs further analysis and evaluation, before it can be adopted for WiMAX.

### 2.8.3.2  Queue Length Aware Approach

A queue length aware scheme for bandwidth allocation and rate control mechanisms has been proposed in Ref. [45]. The polling service traffic (rtPS and nrtPS) in WiMAX has been modeled as a Markov Modulated Poisson Process and an analytical model is developed to investigate the steady state and transient states. Also, an approximate model for BE has been developed. The proposed rate control mechanism is similar to the well-known random early detection (RED) algorithm [47]. The results reported are similar to those obtained in wired network scenarios, i.e., RED helps to achieve higher throughput with smaller buffers as well as providing lower queueing delays [5,47]. The issue of dropping PDUs using the drop function has not been discussed, which makes this scheme less attractive for use in WiMAX. The loss of PDUs at SSs' queue may cause degradation of services offered by various polling services. Also, no packet level simulation results have been presented and therefore, further experimentation is needed to evaluate this approach for bandwidth allocation.

Recently, another queue length-based scheduler for allocating the granted bandwidth by a BS to the polling traffic classes in SSs has been proposed in Ref. [46]. This scheme considers only UGS, rtPS, and nrtPS, while omitting ertPS and BE traffic classes. The operation of this algorithm depends on the ratio of the delay of a new MAC PDU and the maximum latency of rtPS flows. This ratio, defined as $\alpha$, attempts to control the BW divided among rtPS and nrtPS flows. The available frames after allocation to UGS, $F_{poll} = F_{tot} - F_{ugs}$, are divided among $N^{rt}$ and $N^{nr}$ numbers of real-time and nonreal-time flows by the following two simple relations:

$$F^{rt} = F_{poll} \cdot \left( \frac{N^{rt} \cdot \alpha}{N^{rt} \cdot \alpha + N^{nr}} \right) \tag{2.5}$$

$$F^{nr} = F_{poll} \cdot \left( \frac{N^{nr}}{N^{rt} \cdot \alpha + N^{nr}} \right) \tag{2.6}$$

However, despite its apparent simplicity, this scheme is silent about allocations to ertPS and BE traffic. Therefore, to accommodate all of the five classes of the WiMAX, 1.1, the basic idea of the algorithm needs to be modified appropriately.

### 2.8.3.3  Multiuser Diversity Approach

In wireless networks, the channel characteristics are always time varying and thus so is the available bandwidth to mobile users. The propagation of electromagnetic waves is affected by three major phenomena: path-loss, slow log-normal shadowing, and fast multipath fading [38]. In traditional communication systems, these effects are usually considered as harmful as they degrade the quality of the transmission/receiving of signals. However, recently, in 3G cellular networks an effort has been made to exploit these time variations in channel characteristics, rather than to mitigate them. The time-varying channels in a multiuser environment provide diversity, referred to as multiuser diversity, that can be exploited by opportunistic schedulers. Thus, opportunistic scheduling refers to a modern concept in wireless communication networks.

A simple example of opportunistic scheduling has been given in Figure 2.7 [39]. Three SSs are shown in three different time slots with different channel conditions. In $T_a$ and $T_c$, SS1 is experiencing a deep fade in the channel, while SS2 and SS3 are in deep fade in $T_b$ and $T_c$. A scheduler that does not consider channel conditions, such as RR, will have zero throughput, whereas a channel aware packet scheduler could give full throughput [39].

| Time slots | Backlogged mobile users | | | User scheduling order | |
|---|---|---|---|---|---|
| | SS1 | SS2 | SS3 | OS | RR |
| $T_a$ | Fade | Clear | Clear | SS3 | SS1 |
| $T_b$ | Clear | Fade | Clear | SS1 | SS2 |
| $T_c$ | Fade | Clear | Fade | SS2 | SS3 |

**FIGURE 2.7** A comparison of data transfer between BS and SS, in a fading channel employing OS and RR Scheduling. (Adapted from Gyasi-Agyei, A., *IEEE Comm. Lett.*, 9, 670, 2005.)

Multiuser diversity scheduling can be adapted to WiMAX networks, with some additional complexity and a signaling scheme between the PHY and the MAC layers. Fundamentally, opportunistic scheduling is not applicable to flows that require QoS guarantees, as exactly the available bandwidth and delays are unpredictable due to channel dynamics. However, one way to enhance the existing QoS paradigm in WiMAX is to use opportunistic scheduling for nrtPS and BE traffic classes. It would free up some bandwidth for higher priority traffic classes. The deployment of multiuser diversity-based opportunistic scheduling will require novel scheduling mechanisms. For brevity further discussion is omitted; readers interested in more details are referred to Refs. [38,39].

### 2.8.3.4 Opportunistic DRR

A polling-based opportunistic DRR (O-DRR) scheduler has been proposed in Ref. [49]. It uses the DRR idea of maintaining a quantum and a deficit counter for each SS and considers the fairness measures defined in Ref. [23]. Based on the traffic requirements, a list of schedulable SSs is maintained. Then, an expression for the polling interval in which subscribers for each traffic class must be polled has been analytically derived. One of the limitations of this approach is that even if an SS becomes eligible for scheduling, due to an improvement in the channel conditions, it is not scheduled until the next polling interval [48]. Hence, this scheme needs further refinement and performance evaluation before it can be adopted in WiMAX networks.

### 2.8.3.5 Opportunistic Fair Scheduling

Inspired from work done on proportional fair scheduling in 3G wireless networks, an opportunistic fair scheduling (OFS) mechanism for a WiMAX DL has been proposed in Ref. [65]. The OFS deployed at BS will consider the history of transmission to each SS, before making allocation decisions. It is performed by maintaining an exponential weighted moving average of transmission rate of each SS. In OFS, better is the channel quality for a SS, the higher will be its priority for transmission in the next frame. Thus, transmission to SSs in bad channel state will be deferred until the channel quality improves.

### 2.8.3.6 Adaptive Request Approach

In Ref. [53], an adaptive request mechanism for rtPS traffic has been proposed; wherein an SS predicts the arrival rate of real-time traffic and requests the required bandwidth in advance. Through simple simulations, it has been observed that the proposed scheme works better than weighted schemes, in which packets wait at the SS for allocation. The idea of adaptive scheduling for rtPS traffic, [53], has been slightly extended in Ref. [57] by defining an estimation function that considers current and past bandwidth requests. These schemes, [53,57], do not provide a complete scheduling framework for all traffic classes of WiMAX; for instance, nothing has been proposed for ertPS. Thus, these novel techniques are not readily available for deployment into WiMAX. However, similar adaptive methods can be experimented with for development of intelligent schedulers for WiMAX.

### 2.8.3.7  Scheduling for VoIP

An UL scheduling algorithm for VoIP on rtPS traffic class has been proposed in Ref. [66]. However, it does not consider the ertPS traffic class, which has been specifically introduced in Ref. [20] for VoIP services. This algorithm can be employed for IEEE 802.16-2004 type fixed WiMAX profiles. For codecs with a silence detector, an SS can know whether its state is in the ON or OFF condition. Such higher layer information can be known in the MAC layer by using primitives of the convergence sublayer. One of the reserved bits in the generic MAC header has been used to inform the BS when the voice state of the SS is ON. Simulation results have confirmed that performance of the proposed algorithm is better than both rtPS and UGS. It should be noted that this approach relies on the voice activity detection capability of the codec and timely reporting of the state transition to the BS. It may increase the complexity and thus the cost of the SS.

In Ref. [69], four scheduling algorithms, for throughput optimization of fixed WiMAX, have been presented. The concept of proportional fairness has been applied in deriving these scheduling schemes. Their application at the BS and SSs has also been discussed. However, no simulation or test-bed results have been presented. Thus, the efficacy of these disciplines for various traffic classes in WiMAX needs to be investigated further.

### 2.8.3.8  Scheduling for BE Traffic

The BE traffic in WiMAX can be treated in a manner similar to the ordinary best effort wireline Internet, which mostly uses FIFO queues. The FIFO or Droptail queue is the simplest scheduling discipline where packets are served as they arrive and start getting dropped once the buffers become full. It cannot support QoS and there is no protection against misbehaving flows.

To avoid packet losses due to buffer overflow, an obvious choice would be to increase the buffer size of a queue. However, an increase in buffer size will improve throughput, by reducing packet loss, but will also cause an undesirable increase in the end-to-end packet delay. Thus, with a FIFO-based scheduling discipline there exists an inherent tradeoff between throughput and delay, influenced by the buffer size.

In wireline networks, the RED algorithm is usually employed to overcome the drawbacks of the FIFO [47]. RED defines a minimum and a maximum threshold for the average queue size. The average queue size is computed by an exponential weighted moving average process. If the queue size is in between two thresholds, the packets are dropped or marked probabilistically by using a piece-wise linear probability function. Packets are dropped or marked if the queue size exceeds the maximum threshold. For the design of a rate controller for polling service queues, the concept of RED has been employed in Ref. [45]. Similarly, RED can also be experimented for BE traffic.

In Ref. [70] a scheduling scheme for providing bandwidth to BE flows has also been proposed. It strives to achieve max-min fairness while maintaining high link utilization. It measures the current sending rate of each flow and allocates the available bandwidth accordingly. Their simulation setup only considers the BE traffic class. The operation of such a TCP-aware scheduling mechanism for BE, while working along with the other four traffic classes in a complete WiMAX setup, needs to be further investigated.

## 2.9  A MODEL SCHEDULING ARCHITECTURE FOR WiMAX

For a WiMAX network, packet scheduling will be required both at SSs and the BS. The UL request/grant scheduling is carried out by the BS, with the aim of providing each subordinate SS with a bandwidth for UL transmission or opportunities to request bandwidth [19]. Thus by specifying scheduling service and its QoS parameters, the BS scheduler can anticipate throughput and latency requirements and provide polls or grants at suitable intervals. An UL scheduling algorithm is difficult to design, as it does not have complete information (e.g., queue size, traffic type etc.) about all SSs. Also, it has to coordinate itself with all SSs, in making correct decisions. On the other hand, the DL is only concerned with communicating the decisions made at the BS to various SSs.

### 2.9.1 SCHEDULING AT BASE STATION

To meet the stringent QoS guarantees for various traffic classes of WiMAX, an efficient scheduler is required at the BS. It should work equally well for both UL and DL transmission of data. The requirements of each traffic class, Table 2.2, will be translated into an appropriate of number of time slots at the BS, which will be used by the bandwidth requesting SSs. The scheduler at the BS will intimate its scheduling decisions to SSs by issuing the UL-MAP and DL-MAP messages at the start of each frame. By means of these MAP messages, each SS will know exactly that which slots have been allocated for the operation of the UL and DL channels. Thus, a proper choice of scheduler at the BS is very crucial for optimum performance of a WiMAX network.

In Ref. [71], a three-level hierarchical scheduler has been proposed for the BS. It employs EDF at the first level, WF$^2$Q at second, and WFQ at third levels of the architecture. However, due to the nature of QoS provisioning in WiMAX, it is difficult to maintain a hierarchy of schedulers for fulfilling the necessary requirements of the various traffic classes, Table 2.2. Also, the bandwidth request of an SS is dynamically changing. Thus, it is not sufficient to configure the BS scheduler only once, i.e., when an SS joins or leaves the WiMAX network. Therefore, a single level round-robin based scheduler has been suggested for BS in Ref. [29].

The QoS requirements of each connection will be translated into the number of time slots needed, that can be served by employing the simple WRR discipline [21]. It can be seen that all allocated time slots in UL-MAP and DL-MAP are of the same size, thus the same packet size at the queue of BS scheduler, there is no need for complex schedulers like WFQ, WF$^2$Q, and their variants. The fair queueing-based algorithm has been designed for variable sized packets encountered in the ordinary Internet. Another reason that can be given to avoid WFQ at the BS is that the number of slots allocated by a BS is an integer value whereas the weights in WFQ are floating point values, which will require conversion from floating point to integer. Similarly, hindrances can be encountered for the DRR scheduler.

However, WRR is a work conserving scheduling discipline which is in contrast to the nonwork conserving nature of WiMAX. The WRR will cyclically start serving the next nonempty queue after finishing the current one, thus serving each nonempty queue in a periodic manner. In WiMAX, after a set of DL flows has been served in their particular subframe, no additional DL flows can be served until the end of the subsequent UL subframe. Thus, the output link can be idle even if there are packets waiting in certain queues [31]. Another issue is that, for WRR we can specify the number of packets to be served, for any connection, by changing the weight, but the order of packets cannot be changed, which may be necessary in certain high priority traffic, as WiMAX allows for the explicit assigning of time slots. This allows for controlling delay and jitter for certain connections. After considering the above issues, the following three-stage scheduling procedure has been suggested in Refs. [29,30]:

- Allocation of a minimum number of time slots to each connection in the various traffic classes. It will allot a minimum number of slots for fulfilling the necessary QoS requirement as guaranteed by WiMAX.
- Allocation and utilization of the unused time slots, if possible. This process will make the operation of the BS scheduler as work conserving. It will ensure higher levels of efficiency by avoiding the wastage of unused slots and thus bandwidth.
- Reordering of time slots may be needed for providing QoS guarantees.

It can be seen that the first requirement is mandatory whereas the other two are optional in nature. The work conserving discipline can be useful when performance metrics are average delay and average throughput; but it distorts the traffic pattern due to fluctuations in load. However, the nonwork conserving discipline can be useful in QoS systems, such as WiMAX, where bounds on end-to-end delay are more important than the mean values. The problem of distortion of the traffic pattern can be handled in a better way by employing nonwork conserving schedulers at each switch [3]. Further, the choice of being work conserving at BS schedulers has to be made by the WiMAX provider.

### 2.9.1.1 Time Slots Allocation at BS Scheduler

For a connection $i$ with a bandwidth requirement of $B_i$, slot size $\kappa_i$, i.e., the number of bytes that can be sent in a time slot, and $\psi$ number of frames/s, the number of time slots $\eta_i$ allocated by BS scheduler are $\eta_i = \lceil B_i/\kappa_i\psi \rceil$ [29]. It can be interpreted as the total number of slots required per frame for a given bandwidth requirement of SS. This idea of slot allocation has been further generalized in Refs. [29,30], for each of the five classes of WiMAX. In summary, it can be explained as follows:

Let $B_i^{\min}$ and $B_i^{\max}$ be the minimum and maximum bandwidth requirements of a connection $i$ in a set of connections $C_i$ for request size of $\theta_i$, the number of slots for each class is determined by the following:

- UGS: It does not involve bandwidth requests and thus, does not participate in contention mode. It provides periodic and fixed sized grants which eliminate overhead and latency due to SS requests.
- rtPS: The number of allocated slots is based on the bandwidth requirements and size of the request, $\theta_i$. The SS is prevented from using contention requests for connection under this traffic class.
- ertPS: The slot allocation for ertPS is similar to UGS and rtPS as it combines the efficiency of both traffic classes. It has been proposed for VoIP with silence suppression. In this application, data is sent at a constant rate when speech is active, that is similar to UGS. It can request bandwidth during a silence period.
- nrtPS: For this traffic class, the slot allocation is similar to rtPS, but with a difference that it can participate in the bandwidth contention process.
- BE: For this traffic class, no slot will be allocated and the maximum number of allocated slots should not exceed the bandwidth request

The maximum and minimum number of slots, $\eta_{\max}$ and $\eta_{\min}$, according to the above scheme has been tabulated in Table 2.4.

### TABLE 2.4
### Slot Allocations at the BS for Various Traffic Classes in WiMAX

| | WiMAX Traffic Classes | | | | |
|---|---|---|---|---|---|
| | UGS | rtPS | ertPS | nrtPS | BE |
| $\eta_{\min}$ | $\left\lceil \frac{B_i}{\tau_i \psi} \right\rceil$ | $\begin{cases} 1 & \text{for } \theta_i = 0, \\ \left\lceil \frac{B_i^{\min}}{\tau_i \psi} \right\rceil & \text{for } \theta_i > 0 \end{cases}$ | $\begin{cases} 1 & \text{for } \theta_i = 0, \\ \left\lceil \frac{B_i}{\tau_i \psi} \right\rceil & \text{for } \theta_i > 0 \end{cases}$ | $\begin{cases} 0 & \text{for } \theta_i = 0, \\ \min\left( \left\lceil \frac{B_i^{\min}}{\tau_i \psi} \right\rceil, \left\lceil \frac{\theta_i}{\tau_i} \right\rceil \right) & \text{for } \theta_i > 0 \end{cases}$ | $0$ |
| $\eta_{\max}$ | $\left\lceil \frac{B_i}{\tau_i \psi} \right\rceil$ | $\min\left\{ \left\lceil \frac{B_i^{\max}}{\tau_i \psi} \right\rceil, \left\lceil \frac{\theta_i}{\tau_i} \right\rceil \right\}$ | $\begin{cases} 1 & \text{for } \theta_i = 0, \\ \left\lceil \frac{B_i}{\tau_i \psi} \right\rceil & \text{for } \theta_i > 0 \end{cases}$ | $\min\left\{ \left\lceil \frac{B_i^{\max}}{\tau_i \psi} \right\rceil, \left\lceil \frac{\theta_i}{\tau_i} \right\rceil \right\}$ | $\left\lceil \frac{\theta_i}{\tau_i} \right\rceil$ |

*Source:* Sayenko, A., Alanen, O., Karhula, J., and Hamalainen, T., *ACM MSWiM'06*, Terromolinos, Spain, 108, 2006. With permission.

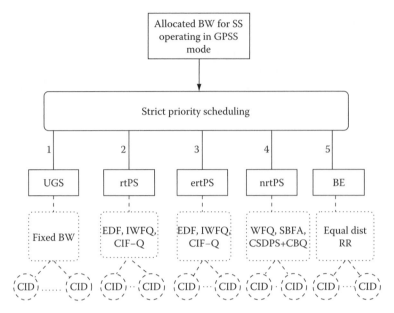

**FIGURE 2.8**   Bandwidth allocation at SS. (From Gyasi-Agyei, A., *IEEE Comm. Letters*, 9, 670, 2005. With permission.)

### 2.9.2 SCHEDULING AT SSS

After getting a bandwidth grant from the BS, the SSs have to divide it among all traffic classes. Each SS is free to choose a scheduler at its output interface, thus controlling the allocation of bandwidth in a desired manner to the individual connections, while operating in GPSS mode. As an example, consider an SS that can be a router in an office that connects a LAN with the Internet through WiMAX, in that case scheduling at the SS will control services offered to various departments. Similarly, BS can choose a scheduler at the interface used for connecting to the wired Internet. However, it can be envisaged that the wired Internet will have a higher capacity than WiMAX, hence a simple FIFO policy will be fine [29].

The allocated grant per flow at each SS should follow a strict priority from highest to lowest, i.e., UGS, rtPS, ertPS, nrtPS, and BE [11]. An obvious disadvantage of strict priority scheduling is that it can starve the lower priority traffic classes from their share of bandwidth. Thus, a traffic policing module will be needed at each SS that will force bandwidth allocation within their limits. Such an architecture for UL scheduling has been proposed in Ref. [11], which consists of a strict priority service discipline, EDF for rtPS and WFQ for nrtPS. For BE, the bandwidth is equally distributed among all connections. One such hierarchical structure is depicted in Figure 2.8. It shows that two-layer scheduling will be needed at SSs in GPSS mode. The first level of scheduling will distribute the total allocated bandwidth (grant) among each of four/five traffic classes in a strict priority manner, whereas the second level scheduler will distribute the class bandwidth among all connections in a fair manner. Such mechanisms have already been surveyed in Section 2.8.

### 2.10   PERFORMANCE MEASUREMENTS FOR WiMAX NETWORKS

It has been noted that, at the time of writing, there is not much published work available regarding real-life performance evaluation of WiMAX networks. However, some of the currently available literature will be briefly described in this section.

In Refs. [31,67], WiMAX test-bed based results for measurements in the field have been reported. The Alvarion test bed, BreezeMax http://www.alvarion.com/, operating in the 3.5 GHz licence band and fully IEEE 802.16-2004 compliant, has been employed for measurements. Experimental data has been collected for four nodes operating in PMP mode in a rural residential environment. A sectorial antenna with a gain of 14 dbi, covering all three SS (FDD half duplex), has been deployed. All nodes run using a Linux distribution and are attached to WiMAX equipment through the Ethernet. Data flow and CBR VoIP are generated by the freely available tool known as D-ITG, http://www.grid.unina.it/software/ITG/; for more details about equipment and experimental setup please see Ref. [67].

It has been observed that the performance of a G.711 codec is far too low to be acceptable and SS cannot support more than two high quality calls. The G.723.1 codec outperforms G.729.2. It has been pointed out that UL measurements contradict simulation results reported in Ref. [24] and its earlier version, where larger delays in UL are ascribed to bandwidth request mechanisms and PHY overhead. However, Ref. [67] ascribes this to be due to activation of piggybacking for bandwidth reservation. It has also been pointed out that the R-factor, E-model, needs to be considered for scheduler design, though, nothing has been mentioned about the type of scheduler that was being used at SSs and BS during the test-bed measurements.

Recently, a performance study of UL Scheduling algorithms for PMP WiMAX has been carried out in Ref. [48]. It has studied major scheduling algorithms using the NS-2 simulator [59]. Also, it provides pseudocode for various schedulers in a simple and accessible manner. The existing scheduling mechanisms have been divided into three categories: homogenous, heterogenous, and opportunistic types. It has been reported that EDF and (EDF+WFQ+FIFO) result in the lowest average delay for rtPS and ertPS; WRR, WFQ, and (EDF+WFQ) provide a fair distribution of bandwidth; the (EDF+WFQ) hybrid setup is fairer than (EDF+WFQ+FIFO). In addition, it has been concluded that most of the legacy schedulers are not very suitable for WiMAX applications. However, WFQ, cross-layer, and queueing theoretic-based schedulers are seen to be promising candidates for applications in a WiMAX network.

## 2.11  SOME OPEN RESEARCH ISSUES

In the foregoing discussion we have alluded to various incomplete studies and opportunities for further research. In this section, we shall list some open research issues related to QoS scheduling in WiMAX, viz.,

- One of the most attractive features of WRR and DRR is their $\mathcal{O}(1)$ complexity and ease of implementation. An open issue is to how to provide tight delay and throughput guarantees in rtPS and ertPS traffic classes while using round robin schedulers, i.e., WRR and DRR. Also, the applications of other variations of round robin, such as smoothed round robin [15] and stratified round robin [18], need to be further investigated.
- Integration of time-stamped schedulers, such as $WF^2Q$, SCFQ, and SFQ, with IWFQ type of algorithms for wireless networks, thus designing schedulers that are specific for high priority traffic classes of WiMAX.
- The use of IWFQ within rtPS and ertPS traffic classes needs to be investigated further, as it requires significant support from the MAC layer [4].
- The CIF-Q algorithm has higher complexity, and authors have argued that the wireless part has a lower bandwidth, which is obviously not true in the case of WiMAX. Thus, an interesting area for further work is to investigate the behavior of CIF-Q with rtPS and ertPS traffic classes.
- Design of dynamic and efficient priority-based mechanisms for distribution of the allocated grants among the various classes at SS. The issues of inter- and intraclass fairness need to be investigated further.
- Development of intelligent bandwidth requests, admission control, and bandwidth allocation schemes at the BS. Also, an investigation of efficient and adaptive techniques for traffic policing and shaping at BS/SSs is required.

- Transforming the schedulers designed for PMP mode into the mesh mode of operation for WiMAX.
- It is important to investigate the use of RED-based buffer management for BE at SSs. Also, the use of modern scheduling techniques, such as multiuser diversity, that exploit variations in wireless channels must be investigated. Similarly, cross-layer approach, adaptive bandwidth requests, and queue length based scheduling are needed to be investigated further. One of the basic issues with the cross-layer approach is how to cater for multiple SSs in the same time frame.
- Performance measurements of WiMAX based services is also very important. For this purpose a test bed has been developed in Ref. [31], which employs the well-known Packet-E-Model for estimation of voice quality. It is also important to develop models that can estimate performance of various services (voice/video/data) offered in WiMAX. Hence, this area also needs to be investigated further.
- One of the important issues is cooperation between various current technologies, for the benefit of customers. In this regard, the interoperation WiMAX with 3GPP's High Speed Packet Access based technologies would be very worthwhile to investigate.

## 2.12   DISCUSSION AND CONCLUDING REMARKS

This chapter has dealt with QoS-based scheduling for WiMAX networks. It started by introducing the essential aspects of QoS provisioning in various traffic classes and identified a related set of parameters that a scheduler should satisfy in delivering appropriate service guarantees. Traffic policing and shaping are required at the BS or SSs. These functions will ensure that each connection conforms to a negotiated set of QoS parameters related to its particular class of traffic and that service level agreements cannot be violated by misbehaving users.

We started our discussion with GPS, which is an ideal but unimplementable scheduler. It sets the benchmark for performance measurement for practical schedulers. We summarized two important types of round robin schedulers, WRR and DRR, which have $\mathcal{O}(1)$ complexity. They can be employed both at the BS and at SSs. However, the lack of tight delay bounds and throughput guarantees makes both WRR and DRR unsuitable for serving UGS, rtPS, and ertPS traffic classes at these SSs. We then studied WFQ, WF$^2$Q, SCFQ, and SFQ scheduling algorithms in the context of adopting existing wireline schedulers for wireless operation. Their application to WiMAX has been discussed. The WFQ and WF$^2$Q are based on an FFQ model of GPS, whereas the SCFQ and SFQ derive their virtual time from the packet arrival process. These sets of scheduling algorithms represent two orthogonal approaches to QoS scheduling.

Another important scheduler was EDF, which operates on the principle of scheduling packets on the basis of earliest deadline first. It can be adopted for rtPS and ertPS traffic classes. We also discussed the basic concepts of IWFQ and CIFQ as wireless schedulers. They can serve as a good starting point for designing newer scheduling policies for WiMAX networks.

In general, designing of fair wireless schedulers is a difficult task. Similarly, scheduling in WiMAX is a very challenging problem due to the very stringent QoS requirements of various traffic classes. However, as a first step, homogenous frameworks have been proposed. They largely consist of a single type of legacy wireline schedulers. The problems associated with such algorithms are already well known. Thus, only a few homogenous proposals exist. They can be considered only of interest for experimental purposes. The next step is to use a combination of legacy schedulers, referred to as heterogenous frameworks in this chapter and in Ref. [48]. These hybrid architectures attempt to employ the good characteristics of various scheduling disciplines for different traffic classes. The hybrid architectures are relatively more efficient, but at the same time they are more complex.

Legacy scheduling algorithms, both of the homogenous and heterogenous framework type, attempt to fulfill the QoS requirements of each traffic class in an implicit and approximate manner, which is not very efficient and desirable. For instance, rtPS and ertPS depend upon the max-min sustained traffic rate and maximum latency, (Table 2.2), but WRR has been suggested in Ref. [24]. An important consideration for some situations is that we have a limited amount of battery power in certain mobile devices. Recently, several newer proposals have been put forward as scheduling architectures; cf. Section 2.8.3. These include cross-layer, queue length aware, and multiuser diversity approaches. They hold the promise of fulfilling some of the strict QoS requirements of both static and mobile profiles for WiMAX standards. It can be safely conjectured that the future scheduling architectures of WiMAX will consist of a hybrid approach between the legacy and modern concepts.

The WiMAX standard allows a connection to change its QoS parameters during its life time; cf. Section 2.6, a two- or three-way handshake process initiated by a BS or an SS with DSA, DSC, and DSD messages. Therefore, it would be necessary for scheduling algorithms to adapt to changes in QoS requirements by reassigning slots to various traffic classes. For instance, in the case of WFQ, cf. Section 2.8.1, dynamic weight assignment can be one solution, although it will tend to make the system too complicated. Also, the WiMAX standard requires that the frame size should be of fixed size, which will result in some slots not being utilized in the current work conserving scheduling frameworks, cf. Section 2.9.1. Hence, resulting in performance similar to nonwork conserving scheduling disciplines [3]. Therefore, most of the scheduling frameworks available in the current literature, cf. Section 2.8, have made a significant assumption about scheduler performance.

Furthermore, it can be seen that an efficient scheduling architecture depends heavily upon admission control and bandwidth allocation mechanisms. Thus, it is very crucial to determine the performance of scheduling algorithms in a complete setup consisting of efficient bandwidth requests, admission control, and bandwidth allocation mechanisms. In addition, we emphasize the importance of performing real-life and test-bed measurements for various scheduling architectures proposed for WiMAX networks. Finally, we conclude that current scheduling frameworks need to improve their overall performance and also an adaptive behavior is required to fulfill customer demands in a scenario of widely varying traffic characteristics.

## REFERENCES

[1] T. Nandagopal, S. Lu, and V. Bharghavan, A unified architecture for the design and evaluation of wireless fair queueing algorithms, *ACM Mobicom*, Seattle, WA, pp. 132–142, August 1999.

[2] H. Fattah and C. Leung, An overview of scheduling algorithms in wireless multimedia networks, *IEEE Wireless Communications*, 9(5): 76–83, October 2002.

[3] H. Zhang, Service disciplines for guaranteed performance service in packet-switching networks, *Proceedings of IEEE*, 83(10): 1374–1396, October 1995.

[4] S. Lu, V. Bharghavan, and R. Srikant, Fair scheduling in wireless packet networks, *IEEE/ACM Transactions on Networking*, 7(4): 473–489, August 1999.

[5] R. Guerin and V. Peris, Quality-of-service in packet networks: Basic mechanisms and directions, *Computer Networks*, 31: 169–189, 1999.

[6] A. K. Parekh and R. G. Gallager, A generalized processor sharing approach to flow control in integrated services networks: The single-node case, *IEEE/ACM Transactions on Networking*, 1(3): 344–357, June 1993.

[7] A. K. Parekh and R. G. Gallager, A generalized processor sharing approach to flow control in integrated services networks: The multiple node case, *IEEE/ACM Transactions on Networking*, 2(2): 137–150, April 1994.

[8] A. Demers, S. Keshav, and S. Shenker, Analysis and simulation of a fair queueing algorithm, *ACM SIGCOMM Computer Communication Review*, 19(4): 1–12, September 1989.

[9] Y. Cao and V. O. K. Li, Scheduling algorithms in broad-band wireless networks, *Proceedings of IEEE*, 89(1): 76–87, January 2001.

[10] T. S. Eugene Ng, I. Stoica, and H. Zhang, Packet fair queueing algorithms for wireless networks with location-dependent errors, *Proceedings of IEEE INFOCOM*, 3: 1103–1111, March–April 1998.

[11] K. Wongthavarawat and A. Ganz, Packet scheduling for QoS support in IEEE 802.16 broadband wireless access systems, *International Journal of Communication System*, 16: 81–96, 2003.

[12] M. Hawa and D. W. Petr, Quality of service scheduling in cable and broadband wireless access systems, *10th IEEE International Workshop on QoS*, pp. 247–255, May 2002.

[13] P. Goyal, H. M. Vin, and H. Cheng, Start-time fair queueing: A scheduling algorithm for integrated services packet switching networks, *IEEE/ACM Transactions on Networking*, 5(5): 690–704, October 1997.

[14] P. Bhagwat, P. Bhattacharya, A. Krishna, and S. K. Tripathi, Enhancing throughput over wireless LANs using channel state dependent packet scheduling, *IEEE INFOCOM'96*, 3: 1133–1140, March 1996.

[15] G. Chuanxiong, SRR: An O(1) time complexity packet scheduler for flows in multi-service packet networks, *ACM SIGCOMM*, 3: 211–222, August 2001.

[16] Westech Communications Inc., Can WiMAX address your applications ? http://www.wimaxforum.org/technology/downloads/Can_WiMAX_Address_Your_Applications_final.pdf, October 2005.

[17] Senza Fili Consulting, Fixed, nomadic, portable and mobile applications for 802.16-2004 and 802.16e WiMAX networks, http://www.wimaxforum.org/technology/downloads/Applications_for_802.16-2004_and_802.16e_WiMAX_networks_final.pdf, November 2005.

[18] S. Ramabhadran and J. Pasquale, Stratified round robin: A low complexity packet scheduler with bandwidth fairness and bounded delay, *ACM SIGCOMM'03*, Karlsruhe, Germany, pp. 239–249, August 2003.

[19] IEEE-802.16-2004, IEEE standard for local and metropolitan area networks—Part 16: Air interface for fixed and mobile broadband wireless access systems, October 2004.

[20] IEEE 802.16e-2005, IEEE standard for local and metropolitan area networks—Part 16: Air interface for fixed and mobile broadband wireless access systems, February 2006.

[21] S. Keshav, *An Engineering Approach to Computer Networks*, Addison-Wesley Longman Publishing Co., Boston, MA, September 1997.

[22] A. Francini, F. M. Chiussi, R. T. Clancy, K. D. Drucker, and N. E. Idirene, Enhanced weighted round robin schedulers for accurate bandwidth distribution in packet networks, *Elsevier Computer Networks*, 37: 561–578, 2001.

[23] M. Shreedhar and G. Varghese, Efficient fair queuing using deficit round robin, *Proceedings of ACM SIGCOMM'95*, Boston, MA, pp. 231–242, 1995.

[24] C. Cicconetti, A. Erta, L. Lenzini, and E. Mingozzi, Performance evaluation of the IEEE 802.16 MAC for QoS support, *IEEE Transactions on Mobile Computing*, 6(1): 26–38, January 2007.

[25] J. C. R. Bennett and H. Zhang, WF$^2$Q: Worst-case Fair Weighted Fair Queueing, *IEEE INFOCOM'96*, 1: 120–128, March 1996.

[26] S. J. Golestani, A self-clocked fair queueing scheme for broadband applications, *IEEE INFOCOM'94*, 2: 636–646, 1994.

[27] T. S. Eugene, I. Stoica, and H. Zhang, Packet fair queueing algorithms for wireless networks with location-dependent errors, *IEEE INFOCOM'98*, 3: 1103–1111, 1998.

[28] L. Georgiadis, R. Guerin, and A. Parekh, Optimal multiplexing on a single link: Delay and buffer requirements, *IEEE Transactions on Information Theory*, 43(5): 1518–1535, 1997.

[29] A. Sayenko, O. Alanen, J. Karhula, and T. Hamalainen, Ensuring the QoS requirements in 802.16 scheduling, *ACM MSWiM'06*, Terromolinos, Spain, pp. 108–117, 2006.

[30] A. Sayenko, O. Alanen, and T. Hamalainen, Scheduling solution for the IEEE 802.16 base station, *Computer Networks*, 52: 96–115, 2008.

[31] F. D. Pellegrini, D. Miorandi, E. Salvadori, and N. Scalabrino, QoS support in WiMAX Networks: Issues and experiemntal measurements, Create-Net Technical Report, N. 200600009, June 2006. http://www.create-net.org/ždmiorandi/CN_techrep_200600009.pdf

[32] C. Fragouli, V. Sivaraman, and M. B. Srivastava, Controlled multimedia wireless link sharing via enhanced class-based queueing with channel-state-dependent packet scheduling, *Proceedings of IEEE INFOCOM'98*, 2: 572–580, 1998.

[33] S. Floyd and V. Jacobson, Link-sharing and resource management models for packet networks, *IEEE Transactions on Networking*, 3(4): 365–386, 1995.

[34] P. Ramanathan and P. Agrawal, Adapting packet fair queueing algorithms to wireless networks, *ACM Mobicom'98*, Dallas, TX, pp. 1–9, 1998.

[35] A. Haider and R. Harris, A novel proportional fair scheduling algorithm for HSDPA in UMTS networks, *Proceedings of 2nd IEEE AusWireless*, 43–50, 2007.

[36] L. Lenzini, E. Mingozzi, and G. Stea, Tradeoffs between low complexity, low latency and fairness with deficit round-robin schedulers, *IEEE/ACM Transactions on Networking*, 12(4): 681–693, 2004.

[37] Deploying License-Exempt WiMAX Solutions, Intel White paper http://www.intel.com/netcomms/technologies/wimax/306013.pdf, 2005.

[38] X. Liu, E. K. P. Chong, and N. B. Shroff, Opportunistic transmission scheduling with resource-sharing constraints in wireless networks, *IEEE Journal on Special Areas in Communications*, 19(10): 2053–2064, 2001.

[39] A. Gyasi-Agyei, Multiuser diversity based opportunistic scheduling for wireless data networks, *IEEE Communication Letters*, 9(7): 670–672, 2005.

[40] M. S. Kuran, B. Yilmaz, F. Alagoz, and T. Tugcu, Quality of service in mesh mode IEEE 802.16 networks, *International Conference on Software in Telecommunications and Computer Networks (SoftCOM)*, Splilt-Dubrovnik, Croatia, pp. 107–111, 2006.

[41] J. C. R. Bennett and H. Zhang, Hierarchical packet fair queueing algorithms, *IEEE/ACM Transactions on Networking*, 5(5): 675–689, Oct. 1997.

[42] G. Chu, D. Wang, and S. Mei, A QoS Architecture for the MAC protocol of IEEE 802.16 BWA system, *IEEE International Conference on Communications, Circuits and Systems and West Sino Expositions*, 1: 435–439, 2002.

[43] J. R. Moorman and J. W. Lockwood, Multiclass priority fair queuing for hybrid wired/wireless quality of service support, *Proceedings of 2nd ACM International Workshop on Wireless Mobile Multimedia*, Seattle, WA, pp. 43–50, 1999.

[44] J. Chen, W. Jiao, and H. Wang, A service flow management strategy for IEEE 802.16 broadband wireless access systems in TDD mode, *Proceedings of IEEE ICC*, 5: 3422–3426, 2005.

[45] D. Niyato and E. Hossain, Queue-aware uplink bandwidth allocation and rate control for polling service in IEEE 802.16 broadband wireless networks, *IEEE Transactions on Mobile Computing*, 5(6): 668–679, 2006.

[46] K. R. Raghu, Sanjay K. Bose, and M. Ma, Queue based scheduling for IEEE 802.16 wireless broadband, *Proceedings of ICICS*, 5(6): 1–5, December 2007.

[47] S. Floyd and V. Jacobson, Random early detection, *IEEE/ACM Transactions on Networking*, 1(4): 397–413, 1993.

[48] P. Dhrona, A performance study of uplink scheduling algorithms in point to multipoint WiMAX networks, Master's thesis, Queen's University, Ontario, Canada, January 2008, http://hdl.handle.net/1974/973.

[49] H. K. Rath, A. Bhorkar, and V. Sharma, An opportunistic uplink scheduling scheme to achieve bandwidth fairness and delay for multiclass traffic in WiMAX (IEEE 802.16) broadband wireless networks, *Proceedings of IEEE Globecom'06*, 1–5, November 2006.

[50] T.-C. Tsai, C.-H. Jiang, and C.-Y. Wang, CAC and packet scheduling using token bucket for IEEE 802.16 networks, *Journal of Communications*, 1(2): 30–37, May 2006, http://www.academypublisher.com/jcm/vol01/no02/jcm01023037.pdf

[51] J. Lin and H. Sirisena, Quality of service scheduling in IEEE 802.16 broadband wireless networks, *Proceedings of ICIIS*, pp. 396–401, 2006.

[52] Q. Liu, X. Wang, and G. B. Giannakis, Cross-layer scheduler design with QoS support for wireless access networks, *Proceedings of 2nd International Conference on Quality of Service in Heterogeneous Wired/Wireless Networks (QShine'05)*, 8, 2005, DOI: 10.1109/QSHINE.2005.16.

[53] R. Mukul, P. Singh, D. Jayaram, D. Das, N. Sreenivasulu, K. Vinay, and A. Ramamoorthy, An adaptive bandwidth request mechanism for QoS enhancement in WiMAX real time communication, *Proceedings of IFIP International Conference on Wireless and Optical Communications Networks*, 5, 2006, DOI: 10.1109/WOCN.2006.1666583.

[54] K. Vinay, N. Sreenivasulu, D. Jayaram, and D. Das, Performance evaluation of end-to-end delay by hybrid scheduling algorithm for QoS in IEEE 802.16 network, *Proceedings of IFIP International Conference on Wireless and Optical Communications Networks*, 5, 2006.

[55] J. He, K. Guild, K. Yang, and H.-H. Chen, Modeling contention based bandwidth request scheme for IEEE 802.16 networks, *IEEE Communications Letters*, 11(8): 698–700, August 2007.

[56] O. Alanen, Multicast polling and efficient VoIP connections in IEEE 802.16 networks, *ACM MSWiM'07*, pp. 289–295, October 2007.

[57] Z. Peng, Z. Guangxi, L. Hongzhi, and S. Haibin, Adaptive scheduling strategy for WiMAX real-time communication, *Proceedings of ISPACS*, pp. 718–721, Nov.–Dec. 2007.

[58] A. S. Tanenbaum, *Computer Networks*, Fourth Edition, Prentice Hall, Upper Saddle River, NJ, 2003.

[59] J. Chen, C. Wang, F. Tsai, C. Chang, S. Liu, J. Guo, W. Lien, J. Sum, and C. Hung, The design and implementation of WiMAX module for ns-2 simulator, *Proceedings of ACM Workshop on ns-2: The IP Network Simulator*, 5 pp., 2006.

[60] M. Venkatachalam, Optimizing the bandwidth request latency, IEEE 802.16 BroadbandWireless Access-Working Group, November 2005, http://wirelessman.org/netman/contrib/C80216g-05_055.pdf

[61] C. Nie, M. Venkatachalam, and X. Yang, Adaptive polling service for next-generation IEEE 802.16 WiMAX networks, *Proceedings of IEEE Globecom'07*, pp. 4754–4758, November 2007.

[62] N. Ruangchaijatupon, L. Wang, and Y. Ji, A study on the performance of scheduling schemes for broadband wireless access networks, *International Symposium on Communications and Information Technologies*, pp. 1008–1012, 2006.

[63] Y. Shang and S. Cheng, An enhanced packet scheduling algorithm for QoS support in IEEE 802.16 Wireless Network, Springer *LNCS*, 3619: 652–661, 2005.

[64] M. Settembre et al., Performance analysis of an efficient packet-based IEEE 802.16 MAC supporting adaptive modulation and coding, *Proceedings of IEEE ISCN'06*, pp. 11–16, 2006.

[65] M. Mehrjoo, M. Dianati, X. Shen, and K. Naik, Opportunistic fair scheduling for the downlink of IEEE 802.16 wireless metropolitan area networks, *Proceedings of QShine'06*, Article No. 52, 2006.

[66] H. Lee, T. Kwon, and D. Cho, An enhanced uplink scheduling algorithm based on voice activity for VoIP services in IEEE 802.16d/e system, *IEEE Communications Letters*, 9(8): 691–693, August 2005.

[67] N. Scalabrino, F. D. Pellegrini, R. Riggio, A. Maestrini, C. Costa, and I. Chlamtac, Measuring the quality of VoIP traffic on a WiMAX testbed, *Proceedings of TridentCom*, pp. 1–10, May 2007.

[68] S. Cheng, P. Lin, D. Huang, and S. Yang, A study on distributed/centralized scheduling for wireless mesh networks, *Proceedings of ACM IWCMC'06*, pp. 599–604, July 2006.

[69] V. Singh and V. Sharma, Efficient and fair scheduling of uplink and downlink in IEEE 802.16 OFDMA networks, *Proceedings of IEEE WCNC*, 2: 984–990, 2006.

[70] S. Kim and I. Yeom, TCP-aware uplink scheduling for IEEE 802.16, *IEEE Communication Letters*, 11(2): 146–148, February 2007.

[71] N. Liu, X. Li, C. Pei, and B. Yang, Delay character of a novel architecture for IEEE 802.16 systems, *IEEE PDCAT'05*, pp. 293–296, 2005.

[72] D. Poulin, The challenges of WiMAX design, *IET Communications Engineer*, 27–28, 2006.

[73] M. Marques, J. Ambrsio, C. Reis, D. Gouveia, D. Robalo, F. Velez, R. Costa, and J. Riscado, Design and planning of IEEE 802.16 networks, *IEEE PIMRC'07*, pp. 1–5, September 2007.

[74] H. Shi and H. Sethu, An evaluation of timestamp-based packet schedulers using a novel measure of instantaneous fairness, *Proceedings of IEEE IPCCC*, pp. 443–450, April 2003.

[75] N. Bayer, B. Xu, V. Rakocevic, and J. Habermann, Improving the performance of the distributed scheduler in IEEE 802.16 mesh networks, *Proceedings of IEEE VTC*, pp. 1193–1197, April 2007.

# 3 QoS and Fairness in WiMAX

*Antonio Iera, Antonella Molinaro, and Sara Pizzi*

## CONTENTS

Differentiated traffic treatment is a key aspect in providing heterogeneous media flows with quality of service (QoS) at a network level. Heterogeneous media may have very different QoS requirements and may ask the network for very different levels of throughput, delay, jitter, and packet losses. This gives rise to the significant challenge of designing a QoS-aware broadband wireless network, and worldwide interoperability for microwave access (WiMAX) is one of the best candidates to play this role. It follows that the success of WiMAX strongly relies on how the underlying network can meet individual QoS requirements of a wide variety of multimedia applications. Integration of a QoS framework into the Medium Access Control (MAC) layer of the IEEE 802.16 protocol [1] is the WiMAX choice for such an endeavor. In the WiMAX QoS framework, a very critical role is played by the packet scheduler within the base station (BS). This component is expected to coordinate all QoS-related functions to provide differentiated QoS guarantees and fair resource allocation to multiple users and multiple media flows. The most challenging aspect in this context originates from the wireless nature of the WiMAX channel, which may impair the scheduler QoS-support capability and may invalidate its theoretical fairness in assigning the available bandwidth. This suggests a cross-layer approach that exploits the time-varying nature of the wireless channel to take decisions on the data to deliver to each user. According to this approach, based on channel status awareness, the scheduler performs an opportunistic packet selection and allocates resources to users experiencing good channel conditions. On the one hand, by allocating resources to users with better channel quality, the scheduler can maximize the overall system throughput. On the other hand, it may degrade other QoS metrics such as delay, because the scheduler postpones transmissions toward users sensing low-quality channels until the relevant channel conditions recover.

The design of channel-aware schedulers in WiMAX networks and the impact of channel-awareness on WiMAX QoS support and fairness provisioning capabilities are the topics investigated in this chapter.

## 3.1  INTRODUCTION

The intrinsic ability in delivering services despite of the user location makes wireless communication a widespread paradigm for exchanging information. The nature of such an information is going to become more and more interactive and bandwidth-demanding. Real-time video streaming and videoconferencing, video on demand, voice over IP, and online gaming are just a few examples of the wide portfolio of multimedia applications which are bringing about novel interests into the user market. This is where the request for a network able to provide broadband access through a wireless medium comes from. Broadband wireless access (BWA) represents a cost-effective solution to provide last-mile access, and in some cases it may represent the only viable solution to the digital divide.

As a consequence of the wide interest in wireless broadband access, in 1999 the IEEE (Institute of Electrical and Electronics Engineers, Inc.) Standards Board activated the 802.16 Working Group on BWA standards with the aim of developing standards and recommended practices to support the development and deployment of broadband wireless metropolitan area networks. In 2004, this standardization activity resulted in the specification of an air interface for fixed BWA systems [1] operating at frequency bands ranging between 2 and 66 GHz. Since its birth, the IEEE 802.16 standard has been promoted by the WiMAX forum [2] as the leading technology for wireless broadband service provisioning. The WiMAX forum is an industry-led, not-for-profit organization established to certify and promote the compatibility and interoperability of broadband wireless products based on the harmonized IEEE 802.16/ETSI HiperMAN standard.

The success of WiMAX, in its competition with alternative wireless technologies (such as cellular networks) to become the technology for high-speed multimedia service delivery, strongly relies on its capability of meeting quality of service (QoS) needs of the promising variety of applications that WiMAX is called to support. Each traffic flow is expected to ask the network for a specific treatment in terms of allocated bandwidth, maximum delay, jitter, and packet loss. Traffic differentiation and proportional fairness in resource assignment are, thus, the crucial features on which to concentrate the research efforts to allow WiMAX winning the competition.

Currently, the IEEE 802.16 standard leaves QoS related features (e.g., traffic policing and shaping, connection admission control, and packet scheduling) open to the vendors' algorithm design and implementation. In designing a QoS framework capable of providing differentiated traffic treatment and fair resource allocation, a special attention needs to be put on the centralized packet scheduler at the BS, which is expected to coordinate all other QoS-related functions for WiMAX.

Scheduling of users' downlink (DL) data transmissions at a BS has attracted a substantial amount of research attention in the last few years; however, no scheduling scheme for WiMAX has been standardized or commercialized yet. What makes the scheduling process very critical is the promising variety of multimedia applications, the traffic volume to be managed by a WiMAX BS, and the fluctuating wireless channel status. Especially, the main obstacle to an effective scheduler design for WiMAX comes from the variability, both in time and in space, of the radio links between BS and the distributed subscriber stations (SSs). Such variability might adversely affect the scheduler performance, so invalidating the theoretical fairness of its resource assignment policy and its QoS-support capability. This is the reason why scheduling schemes should be channel-aware and take opportunistic decisions according to the channel conditions. This means that a scheduler allocates resources to subscribers with good channel conditions, and it postpones transmissions to subscribers with poor channel conditions until their channels recover. This opportunistic approach is expected to maximize the overall system throughput, but it may degrade individual QoS metrics (especially delay) of some "unlucky" flows waiting for relevant channel conditions to improve. This may result in an unfair allocation of radio resources unless the channel-aware scheduling is coupled with some compensation techniques that account for missed transmission opportunities and look for fulfilling individual QoS requirements and for providing proportional fairness in sharing the bandwidth.

The aim of this chapter is to analyze the extent to which the wireless channel nature of WiMAX can impair the scheduler's fair and QoS-capable assignment of the available bandwidth, and to demonstrate how channel awareness, coupled with compensation for missed transmission opportunities, can help in recovering the ideal scheduler performances. Also, some indications are given on how to design a packet scheduler for users' DL traffic delivery, which (1) adheres to the WiMAX philosophy of "class-based" QoS differentiation and "flow-based" resource assignment, (2) is aware of the channel conditions and accounts for wireless impairments when assigning resources, and (3) applies a compensation technique to safeguard from the drawbacks of pure opportunistic scheduling.

## 3.2  QoS ARCHITECTURE IN WiMAX NETWORKS

The IEEE 802.16 standard gives specifications for the MAC and physical (PHY) layers [1] of a fixed BWA system.

With the aim of accounting for different environmental conditions, multiple PHY layers are specified, based on single carrier (SC) or multi-carrier modulations, like orthogonal frequency division multiplexing (OFDM) and orthogonal frequency division multiple access (OFDMA). The WirelessMAN-OFDM air interface is the one targeted by the WiMAX consortium and considered in this chapter. The 802.16 PHY specifications allow for adaptive burst profiling, i.e., transmission parameters, including modulation and coding schemes, may be adjusted individually to each SS on a frame-by-frame basis. WiMAX networks can operate either in time division duplexing (TDD) or frequency division duplexing (FDD) mode. Both alternatives support adaptive burst profiles. In the remainder of the chapter we will refer to the TDD case. The TDD frame, which is shown in Figure 3.1, has a fixed duration, but uplink (UL) and DL separation can vary under the control of the BS. Time division multiple access (TDMA) is used in the UL subframe, while time division multiplexing (TDM) is used in the DL subframe. A DL-map specifies when PHY layer transitions, i.e., change of the modulation-coding parameters (burst profile), have to occur. An UL-map specifies the UL bandwidth assigned to the SSs.

The IEEE 802.16 standard supports both point-to-point (PP) and point-to-multipoint (PMP) topologies, and an optional mesh configuration. In a fixed PMP WiMAX network, a BS communicates with multiple stationary SSs, as shown in Figure 3.2. The MAC of a PMP WiMAX network is centrally

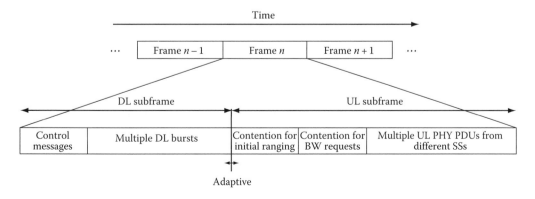

**FIGURE 3.1**  OFDM frame structure with TDD.

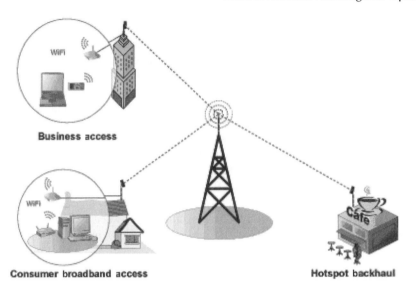

Business access

Consumer broadband access                                    Hotspot backhaul

**FIGURE 3.2**    PMP WiMAX network.

managed by the BS, which offers connection-oriented services to individual traffic flows. Each traffic flow is uniquely identified by a connection identifier (CID), and belongs to one of the listed QoS classes, which are defined at the WiMAX MAC layer [1] to provide differentiated traffic treatment:

- Unsolicited Grant Service (UGS) supports real-time constant-bit-rate (CBR) applications that generate fixed size data packets on a periodic basis. The BS assigns them unsolicited fixed UL bandwidth grants at periodic intervals, based on their maximum sustained traffic rate.
- Real-time Polling Service (rtPS) is appropriate to real-time applications that produce variable size data packets on a periodic basis. BS provides SSs the possibility of issuing bandwidth requests (BRs) on a periodic basis by means of a polling mechanism.
- Extended real-time Polling Service (ertPS) is similar in nature to rtPS service, but the BS can assure to ertPS a default bandwidth (corresponding to the maximum sustained traffic rate, like for UGS) and dynamically provide additional resources.
- Nonreal-time Polling Services (nrtPS) is appropriate to delay-tolerant applications that generate variable size data packets and require a minimum data rate. nrtPS flows can use contention slots to ask for bandwidth grants, but typically the BS polls them for issuing BRs on a regular basis (if possible). The polling interval is not guaranteed but may depend on the network traffic load.
- Best Effort (BE) is appropriate to traffic with weak QoS requests. The flows of this class can only use contention slots to deliver their BRs.

Bandwidth can be asked to/assigned by the BS through request/assignment messages, polling mechanisms, and contention slots. Bandwidth can be claimed by an SS, through a request message, on a "CID basis" (i.e., by referring to the requirements of a single connection), although it is granted by the BS on an "SS basis", i.e., the BS aggregates the reply to all the CIDs requests from the same SS. SS is then in charge of distributing the assigned aggregated resources to its currently active connections. Polling is the process through which the BS allocates to SSs some control resources finalized to convey their BRs. Polling is done on an SS basis. Flows belonging to nrtPS and BE classes may ask for bandwidth grants over contention slots at the beginning of each UL subframe.

The IEEE 802.16 standard foresees the presence of a few components with QoS supporting functionalities at the MAC layer of BS and SSs:

- Admission Control is necessary at the BS to avoid network overload or starvation of some flows. This block decides the admission of a new flow on the basis of both its traffic class/requirements and the network resource availability.
- Traffic Shaping and Policing is required at the SS to ensure the conformance between actually generated traffic and negotiated traffic at connection setup.
- Traffic Scheduling and Buffer Management provide the desired QoS to the traffic crossing the air link, while optimally (and fairly) utilizing link resources. Packet transmission scheduling is foreseen at both the BS and SSs to select packets to be transmitted on the DL and UL, respectively. Buffer management is used for queue length control and packet dropping (if necessary).

Downlink scheduling runs at the BS for the transmission of users' information (DL data packets) and the management of bandwidth claims (UL requests) from SSs. These two tasks need to be executed by the BS at the beginning of each frame for formatting the control information messages (DL-map and UL-map) needed for the management of both the DL and UL subframes. An SS must schedule packets from the queues of its (multiple) active connections into the transmission opportunities allocated to it within a frame. UL scheduling is necessary to leave the SS the flexibility to redistribute the bandwidth assigned by the BS among its active connections, according to self-defined parameters (e.g., the status of its queues, the expiry time of buffered packets, and so on).

## 3.3  PROBLEM OF SCHEDULING IN WIRELESS NETWORKS

The topic of scheduling has been a long debated in the literature. From the very beginning, wired networks were the reference environment where applying the scheduling proposals, and fairness and complexity were the two main concerns in the design of a high-performing scheduler. In wireline networks, fair queuing algorithms have long been a popular paradigm for providing bounded delay channel access and fairness among packet flows over a shared unidirectional link. This paradigm guarantees that in any time window $T$, two generic backlogged flows, $f_i$ and $f_j$, are, respectively, serviced in proportion to their weights $w_i$ and $w_j$. Among fair queuing algorithms, the generalized processor sharing (GPS) scheduler [3] is a useful benchmark against which other service disciplines can be measured. GPS is unrealizable because it assumes fluid traffic model with infinitesimal packet sizes, but there are several fair queuing schemes that track the GPS performance quite closely; one of this is weighted fair queuing (WFQ) [4]. Expensive implementation cost is the critical aspect that has been pointed out against fair queuing schemes, which achieved nearly perfect fairness but were processing intensive, especially at high speed. This is the reason why additional efforts in the scheduling design followed the direction to trade off between fairness and complexity. Among packet fair queuing algorithms proposed for wired networks, two schemes can be chosen as representatives of these two trends. On the one hand, worst-case fair weighted fair queuing (WF$^2$Q+) [5], derived from WF$^2$Q [6], is the one achieving near perfect fairness and best approximation of the idealized GPS behavior [3]. It simultaneously supports guaranteed real-time, adaptive BE, and controlled link-sharing services with a relatively low complexity $O(\log(n))$, where $n$ is the number of active flows, compared to $O(n)$ of WF$^2$Q. This algorithm's complexity is due to timestamping the incoming packets, sorting packets, selecting the one with minimum timestamp, and updating the virtual time function. On the other hand, deficit round robin (DRR) [7] has been proposed as a cheaper approximation of fair queuing, which achieves nearly perfect fairness in terms of throughput at a very low processing work per packet, $O(1)$. Differently from previous round-robin scheme, DRR keeps fairness even if different flows use different packet sizes; this is because it uses round-robin service with a quantum of service assigned to each queue. If a queue is not able to be served in a given round because of its large packet size, the remainder of the quantum (deficit) is added to the quantum for the next round.

Most of the scheduling algorithms proposed for a generic wireless environment derive from analogous proposals in wireline networks. Nevertheless, conventional scheduling algorithms designed for wired networks are supposed to operate in an error-free channel; for this reason they cannot be directly extended to a wireless environment, where the high variability of radio links (subject to time- and location-dependent signal attenuation, fading, interference, and noise) can result in bursty errors and time-varying bandwidth availability, which can neutralize the effectiveness of scheduling decisions. This means that, even if the scheduler fairly allocates resources among traffic flows, some flows could not actually exploit them due poor channel qualities. Therefore, in wireless networks, only a subset of backlogged flows (i.e., flows with packets available for transmission) can be scheduled for transmission at any time instant, and this subset dynamically changes.

All of this call for a wireless scheduler that takes decisions relying on the availability of channel state information [8–11] and uses some forms of compensation for lost transmission opportunities of those traffic flows sensing a poor channel. Compensation for channel errors is a mandatory feature to guarantee a service comparable to the "error-free" service, which is provided by an equivalent scheduler in the wired network. Basically, the idea of compensation is to swap channel access between a backlogged flow $f_i$, which currently perceives a bad channel during a time window $T_1$, and a backlogged flow $f_j$, which, instead, perceives a good channel during the same window $T_1$. The intention is of reclaiming the channel access for flow $f_i$ when it perceives a clean channel during a later time window $T_2$. This additional channel access is granted to $f_i$ at the expense of flows $f_j$, which were granted additional channel access during $T_1$ although $f_i$ was unable to transmit any data. At any given time, a flow is said to be leading if it has received channel allocation in excess of its error-free service; it is lagging if its received channel allocation is less than its error-free service. A flow, which has received a channel allocation that is exactly the same as its error-free service, is said to be in sync.

Regardless of the characteristics of the error-free scheduler, it is worth saying that compensation requires monitoring of the channels conditions and keeping track of the leading/lagging status of each traffic flow. This causes an increase in the complexity of any channel-aware wireless scheduler; such a complexity must be summed up to the complexity of the equivalent error-free scheduler. Even the complexity of a very simple algorithm, like DRR, can increase from $O(1)$ to $O(n)$ due to compensation, as demonstrated in Ref. [12]. In Ref. [13], compensation is also applied to WF$^2$Q+; the BS of a generic wireless system detects leading/lagging status of each flow, and executes additional features (like graceful degradation and readjusting) to cope with wireless scheduling. All these features raise the complexity from $O(\log(n))$ to $O(n)$.

Another class of schedulers, which rely on availability of channel status information to decide which packet to transmit over the radio interface, is the class of opportunistic schedulers. The basic idea behind opportunistic scheduling is to gain advantage from the multiuser diversity effect caused by the wireless channel nature (subject to time and location dependent signal attenuation, fading, interference, and noise), which makes different users experiencing different channel conditions at any given time. A pure opportunistic scheduler selects the packet with the best perceived channel to be sent over the air interface, and postpones transmissions toward users sensing low-quality channels until their channel conditions recover. By selecting the user with the best channel conditions to transmit at any time, opportunistic scheduling is effective in maximizing the overall system throughput. Nevertheless, a pure opportunistic scheduling decision does not consider service differentiation and individual QoS guarantee; this means that it may cause degradation of some QoS metrics, such as delay of the packets waiting for their low-quality channels to recover. To enhance opportunistic scheduling algorithms' performance, the challenge is also to address fairness and users' service requirements while exploiting the multiuser diversity gain. References [14–16] are some interesting examples testifying of this research effort.

## 3.4 OVERVIEW OF SCHEDULING IN WiMAX NETWORKS

Although the topic of scheduling in WiMAX networks has recently attracted a substantial amount of research attention, the design of WiMAX schedulers is currently an open issue, and no scheme has been standardized or commercialized yet. Without any pretension of being exhaustive, some relevant references of scheduling in WiMAX networks are cited in this section.

Typical approaches to the WiMAX scheduling of DL data traffic prefer simplicity to QoS performance; e.g., classical algorithms like weighted round robin (WRR) [17] or DRR [7] have been recommended. On the contrary, hybrid scheduling is suggested for UL data traffic, with different disciplines tailored to the traffic requirements of each service class; e.g., either WFQ [4] or earliest deadline first (EDF) [18] for more demanding flows (such as UGS and rtPS), either WFQ or WRR for nrtPS class, and either FIFO or round robin for BE traffic. This is the case in Refs. [19–21]. Also in Ref. [22] hybrid scheduling is proposed, but the scheduling disciplines are modified to support adaptive modulation and coding (AMC) and packets fragmentation. Many proposals consider UGS connections as granted by a constant periodic amount of bandwidth negotiated at connection setup and not subject to scheduling. Other proposals, such as Refs. [23,24], aim at guaranteeing a minimum rate to all classes, including BE connections, utilizing a one-level scheduler; this is one of the reasons why they implement a DRR policy into the DL scheduler and WRR into the UL scheduler. The proposal in Ref. [25] considers a scheduling discipline for the WiMAX BS that is conceptually similar to the WRR scheduler. The number of slots allocated to each connection depends on its traffic class; but if there are unused slots, they are allocated to other active connections in a priority order, which gives advantages to rtPS, then to nrtPS, and finally to BE connections. An interesting proposal, which elaborates on the concept of proportional fairness, can be found in Ref. [26]. The main innovation is that the delay constraint is embedded into the original proportional fairness formulation by introducing a new parameter (the time window for calculating the short-term throughput of each queue), which can be manipulated for differentiating the queue's delay behavior.

Aspects related to impairments on the WiMAX radio channel have been often ignored in most of the studies in literature, and information on the channel status has rarely been exploited in the view of a more effective scheduling design. Only recently, some works started considering PHY layer issues and awareness of the channel status when designing the WiMAX scheduling policy. Some of these works rely on AMC to combat bad effects of the wireless fading channel, by adjusting transmission modes to channel variations while maintaining an agreed packet error rate. Among AMC-based proposals is Ref. [27]. AMC has also been proposed in Ref. [28] within an analytical framework for service differentiation in IEEE 802.16 networks with SC air interface, which unifies fair scheduling, based on packetized GPS, and connection admission control functions. Other recent proposals exploit [29] channel state awareness and inter-cell interference estimation to design jointly scheduling and resource allocation in a multicellular scenario. In Ref. [30], the authors analyze the adverse effects of taking scheduling decisions, which are unaware of the channel status, at the BS of a WiMAX network. They demonstrate how channel awareness and compensation of missed transmission opportunities are successful in satisfying individual flow's QoS requirements and in providing fairness among traffic classes sharing the DL bandwidth.

Another very recent interesting trend in the literature is to design opportunistic scheduling for WiMAX networks. Opportunistic decisions must be combined with additional features to consider QoS and fainess requirements. In Ref. [31], the authors devise an opportunistic deficit round-robin (O-DRR) scheme for UL traffic scheduling, which accounts for both channel status and flows' delay constraints, and they demonstrate that their scheduler, combined with a polling-based scheme at the BS, is able to provide a choice of balancing fairness with delay. In Ref. [32] an opportunistic stable queue scheduling scheme for non real-time applications in the DL of a WiMAX PMP network has been proposed. Scheduling decisions are taken accounting for the channel status and the BS queue length to improve the overall throughput, and at the same time, to avoid queues build-up for those SSs suffering from long-term bad channel conditions. In Ref. [33] the impact of opportunistic scheduling on QoS provisioning is investigated in the course of a performance analysis of multiuser OFDM-TDMA and OFDMA, conducted in a cross-layer QoS framework similar to that in IEEE 802.16. It is also observed that the opportunistic assignment can be employed more effectively in OFDMA.

## 3.5  CHANNEL- AND QoS-AWARE SCHEDULER DESIGN FOR WiMAX NETWORKS

In this section some design guidelines are given for a "channel-aware" and "QoS-aware" scheduler to be implemented at the BS of a WiMAX PMP network for the delivery of DL traffic to a set of distributed SSs with active connections of different traffic nature. The scheduling algorithm, which is running at the WiMAX base station, needs to fulfill the following requirements:

- Efficient link utilization: The scheduler shall take opportunistic decision and not assign a transmission opportunity to a flow with a currently low-quality link, because the transmission will be wasted.
- Delay bound: The algorithm shall be able to provide delay bound guarantees for individual flows, to support delay-sensitive applications; besides, it shall prevent too late packet transmissions from wasting bandwidth.
- Fairness: The algorithm shall redistribute available resources fairly among flows; thus providing short-term fairness to error-free flows and long-term fairness to error-prone flows.
- Throughput: The algorithm shall provide guaranteed short-term throughput to error-free flows and guaranteed long-term throughput to all flows.
- Implementation complexity: A low-complexity algorithm is necessary to take quick scheduling decisions.
- Scalability: The algorithm shall operate efficiently when the number of flows sharing the channel increases.

To match all the above-mentioned needs, (1) a class-based wireless scheduling is recommended to fulfill the WiMAX service class differentiation needs, and also to simplify interworking with the Internet (supporting class-based differentiated services); (2) "per class" service differentiation must be achieved, and simple "per flow" mechanisms (e.g., lead/lag counters per each flow) must be provided to guarantee fair service to traffics within a class; (3) channel awareness must be exploited to make efficient use of wireless resources, and a compensation technique for missed transmission opportunities must be provided to guarantee proportional fairness in sharing the bandwidth; (4) useless compensation must be avoided through simple measures, e.g., periodic buffer cleaning of over-delayed packets could be used; and (5) simple expedients must be provided to avoid monopolization of the scheduler by a lagging flow after the recovery of its channel, and to guarantee graceful throughput degradation of leading flows; this could come at a very low cost by using combination of lead/lag counters and queue parameters.

In Figure 3.3, the reference channel- and QoS-aware scheduling architecture is illustrated. It is considered to be implemented at the MAC layer of a WiMAX BS, which manages local traffic queues for packet delivery over the DL channel. The illustrated framework must be able to provide per-class differentiated QoS and fair service to flows in the same class through the operation of the following QoS-support modules:

- An error-free service scheduler that decides how to provide service to traffic flows based on an error-free channel assumption.
- A lead/lag counter for each traffic flow that indicates whether the flow is leading, in sync with, or lagging its error-free service model and in which extent.

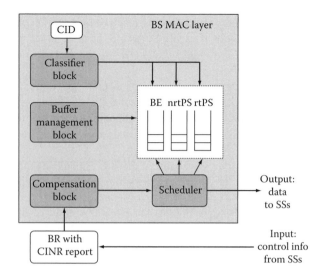

**FIGURE 3.3**   Channel-aware QoS scheduling architecture in the BS.

- A compensation technique that is used to improve fairness among flows. A lagging flow is compensated at the expense of a leading flow when its link becomes error free again. The system maintains credit/debit counters for each lagging/leading flow.
- Separate per-class packet queues used to support rtPS, nrtPS, and BE traffic flows.
- A means for monitoring and predicting the channel state of each backlogged flow.

### 3.5.1   PACKET CLASSIFIER

Packets received by the BS and destined to the DL are sorted by the packet classifier and buffered into one of the per-class queues in the BS, according to the class of service they belong to. In the queues, packets wait for transmission over the air toward a given SS. Queues are managed by the scheduler on a frame basis.

The main task of the classifier is to match the CID of the incoming packets with the traffic class they belong to, and to queue such packets in their relevant buffers. Only the service received by rtPS, nrtPS, and BE queues will be analyzed, because flows belonging to the UGS and ertPS service classes are assumed to receive a constant reserved amount of bandwidth frame-by-frame and not need to be scheduled. This choice allows high demanding (UGS and ertPS) flows to save up the time required for the BR process.

Another important task of the classifier is to timestamp each incoming packet according to its arrival time. This information is exploited by the buffer manager to recognize when a queued packet's timeout expires, i.e., when its elapsed waiting time exceeds the maximum tolerated latency.

### 3.5.2   BANDWIDTH REQUESTS AND CINR REPORTS

To make the channel-aware scheduler effectively operate, information about the quality of the wireless links toward each SS must be gathered by the BS. To this aim, the wireless channel state is continuously monitored by each SS, which gets CINR (carrier to interference and noise ratio) measurements, updates estimates of the CINR's mean and standard deviation values, and reports them back to the BS through specific signalling messages provided by the IEEE 802.16 standard [1], e.g., through BR messages (normally used by SSs for requesting bandwidth), which can also include the CINR report.

The compensation block is in charge of collecting received information on the channel qualities and making the BS aware of the status of each link toward the SSs. Two assumptions hold in our reference scenario:

- Wireless channel quality of each connection changes on a per-frame basis, but it remains constant for a frame duration; i.e, the BS assumes that the CINR report received by the SS at the frame beginning is valid for the entire subsequent frame duration. This is equivalent to modelling a block-fading channel, which is a feasible assumption for the slowly varying radio links of the fixed WiMAX PMP scenario.
- Perfect channel state information is available at the BS. This means that the CINR reports must be correctly received by the BS over an error-free feedback channel.

The monitored channel quality at the SS is compared against the allowed values of signal-to-noise ratio and receiver sensitivity for each transport mode (burst profile) at a given bit error rate (BER). As soon as the received power level becomes lower than the receiver sensitivity threshold for a given DL transport mode and target BER, then a more robust profile must be chosen. This change in the transport mode can proceed until the most robust modulation/coding scheme (BPSK modulation with $1/2$ coding rate) is chosen. A channel is considered "good" as long as a more robust transport mode can be selected, which is capable to cope with the channel adverse conditions. Otherwise, the channel is finally considered as "bad."

### 3.5.3 Channel Error Compensator

At the beginning of each frame, the compensation block classifies a flow (i.e., a CID) as sensing a good or a bad channel on the basis of the CINR reports received by the relevant target SS. Accordingly, CIDs destined to an SS sensing a lossy channel are marked as "banned" from being transmitted. When the error-free service scheduler selects a head-of-line (HOL) packet to be extracted from queue $i$, it interacts with the compensator to check whether the CID of the selected packet is banned or not. If it is banned (the packet would risk to be corrupted if transmitted over the radio link), then the scheduler looks for a packet belonging to an unmarked flow of the same class (i.e., in the same queue) to transmit instead of the HOL packet. To help the scheduler in selecting the substitute flow, the compensator manages a debit/credit counter for each active (leading or lagging) flow. Each time a flow substitutes another one for transmission, it is considered as a leading flow, and its credit counter increases by one; at the same time, the replaced flow is considered as a lagging flow, and its debit counter increases by one. Each time a HOL packet cannot be transmitted due to poor channel conditions, the substitute flow is the one with the highest debit counter among unmarked flows of the same class. If no unmarked packets can be found in the queue, then the entire queue is banned and the turn passes to the successive queue.

The pseudocode listed below briefly shows how the compensator works and interacts with the scheduler, regardless of the specific algorithm ($WF^2Q+$ or DRR) this latter uses.

```
1 ban flow if CINR > CINRthr;
2 until available resources {
3       until a pkt is dequed || all queues examined {
4             scheduler selects queue x : x !banned i.e. scheduler selects flow i = head(x);
5             if i is marked {
6                   select less serviced flow j : j !banned;
7                   if !found {
8                         mark x;
9                         goto 3; }
10                  else {
```

```
11              deque pkt;
12              modify scheduler variables;
13              modify credit/debit counter of i and j;
14              exit; } }
15      else {
16          modify scheduler variables;
17          deque pkt; } } }
```

At the beginning of each frame, the compensator bans the flows directed to SSs, which currently sense a lossy channel (line 1), according to the received channel state information reports. For the duration of the DL subframe, packets within the queues are examined by the scheduler until at least one queue contains a packet belonging to an unbanned CID (lines 2–3). The scheduler selects the queue to be served among all backlogged (and unbanned) queues; this corresponds to select the first packet in the queue (line 4). If this packet belongs to an unbanned flow, then it gains service and the variables for the specific scheduling algorithm (WF$^2$Q+ or DRR) are updated accordingly (lines 15–17). These variables can be the quantum and deficit counter values for DRR, and $(S, F, V)$ values for WF$^2$Q+. Otherwise, if the HOL packet is banned, the scheduler starts searching for the unbanned flow (in the same queue) that has received the worst service in the past (lines 5–6). If no unbanned flow exists, the Scheduler considers the entire queue as no longer servable and selects another queue among the backlogged ones, if any (lines 7–9). If at least an unbanned flow exists, the first enqueued packet belonging to this flow is selected and the scheduling variables are updated accordingly (lines 10–14). Also credit/debit counters for the originally selected flow and the substitute flow are consequently updated.

### 3.5.4  BUFFER MANAGER

Per-class queues at the BS are assumed infinite in size. Nevertheless, persisting poor channel conditions over some radio links could delay relevant packets in the BS queues for a long time. Then, to avoid bandwidth waste with the transmission of over-delayed packets, it is necessary to keep the queues clean and periodically drop over-delayed packets. Thus at each frame beginning, the buffer manager purges queues of all those packets buffered for a time longer than their maximum tolerated delay. The purpose of purging is avoiding a meaningless compensation over already expired packets and a bandwidth waste to transmit timely invalid packets. This periodic buffer cleaning achieves the same objective of the readjusting technique in Ref. [13], but without the need of tuning the maximum leading/lagging amount per flow.

### 3.5.5  PACKET SCHEDULER

This module is responsible for the selection of packets to be transmitted over the airlink and has a central role to provide proportional fairness and QoS differentiation in the network. The core of this module implements an error-free service scheduler, which shall interact with the compensator block to take opportunistic decisions. With the aim of reflecting the per-class traffic differentiation envisioned by the IEEE 802.16 standard [1], the scheduler operates on per-class queues. These queues are managed according to two basic scheduling algorithms, WF$^2$Q+ [5] and DRR [7], which are enhanced with a per-flow channel error compensation technique. WF$^2$Q+ and DRR have been chosen as the representatives of two approaches in the design of fair scheduling algorithms; the former aiming at achieving strict fairness but with higher complexity, the latter very simple but less strict in guaranteeing fairness. Details about WF$^2$Q+ and DRR scheduling algorithms will be given later in this chapter.

### 3.5.5.1  WF²Q+-Based Algorithm

WF²Q+ [5] belongs to the class of fair queuing scheduling algorithms and it is one of the best approximation of the ideal GPS [3] scheduling performance. As already mentioned, GPS cannot be implemented in a real world, but it serves as a benchmark for comparison of real packet scheduling algorithms, in terms of end-to-end delay bounds and fair bandwidth allocation. WF²Q+ has the main advantage of providing almost perfect fairness in the worst case scenario and tight delay bounds. It maintains all the properties of its predecessor WF²Q, but it reduces the implementation complexity. In the WiMAX reference scenario, the WF²Q+ algorithm has to be adapted to manage per-class queues instead of per-flow queues.

The scheduler maintains a virtual time function $V_{WF^2Q+}(t)$, which can be interpreted as the marginal rate at which backlogged queues receive service in GPS. Each queue $q_i (i = 1, ..., 3)$ is assigned a virtual start service time $S_i$ and a virtual finish service time $F_i$, and it is said to be eligible at time $t$ if its $S_i$ is not greater than the system virtual time $V_{WF^2Q+}(t)$. WF²Q+ always selects the eligible queue with the smallest virtual finish time $F_i$ and schedules its HOL packet for transmission.

When a packet $p_i^k$, of length $L_i^k$ (bit), reaches the head of queue $q_i$ at time $a_i^k$, then $S_i$ and $F_i$ are updated according to the following equations:

$$F_i = S_i + \frac{L_i^k}{r_i} \tag{3.1}$$

$$S_i = \begin{cases} \max\left(F_i, V_{WF^2Q+}\left(a_i^k\right)\right) & \text{if } q_i \text{ is empty} \\ F_i & \text{if } q_i \text{ is not empty} \end{cases} \tag{3.2}$$

where $r_i$ (bit/s) is the rate guaranteed to $q_i$.

Each time a packet is extracted from queue $i$ to be transmitted, even if it is not the HOL packet, then the $S_i$ and $F_i$ values of that queue are updated with information related to the transmitted packet. If no substitute flow is available in the current queue, then the queue keeps its $S_i$ and $F_i$ values unchanged and is bypassed by the scheduler. This latter selects another queue to serve in the current frame, according to its error-free service policy. No debit/credit counter is used for the bypassed (lagging) queue, because the intrinsic compensation capability of WF²Q+ is exploited. In fact, when no packets can be transmitted from queue $i$, due to bad channel conditions, the virtual start time of queue $i$ will become relatively small compared to the other serviced (leading) queues; this assuring that, in the next frame, packets from queue $i$ will be serviced ahead of packets in other queues (definitely, only after recovery of the channel conditions for some flows in queue $i$).

### 3.5.5.2  DRR-Based Algorithm

DRR [7] is a scheduling algorithm that approximates less strictly the fairness performance of GPS, but with a low computational complexity, which is $O(1)$ processing work per packet. It uses round-robin service with a quantum of service assigned to each queue. Differently from traditional round-robin scheme, if a queue was not able to send a packet in a given round because of its large packet size, the remainder of the quantum (deficit) is added to the quantum for the next round. The use of the deficit variable makes DRR suitable to the management of queues with different packet sizes.

DRR maintains for each queue $q_i$ ($i = 1, ..., 3$) a deficit counter $DC_i$ and a quantum $Q_i$ measured in bytes. At the beginning, all $DC_i$ variables are set to zero. The algorithm works serving queues in turn; each time a queue is selected, its $DC_i$ is incremented by the $Q_i$ value; then packets are sent out if their size (in bytes) is less than $DC_i$. Each time a packet is extracted from queue, $DC_i$ is decremented by the packet size. To avoid examining empty queues, the scheduler keeps an auxiliary list, active list, which includes the indices of queues that contain at least one packet. The round-robin selector points to the head of the active list. Whenever a packet arrives to a previously empty queue, the index of that queue is inserted into the active list.

DRR has to be adapted to provide differentiated treatment to per-class queues (rtPS, nrtPS, and BE) by providing a different quantum per queue. As previously said, traditional DRR scheduler in turn serves the queues belonging to the active list which includes queues that have at least one packet to send. In a wired environment a nonempty queue can be served with success in all cases. Instead, in a wireless scenario, a nonempty queue could not be served if all the queued packets are destined to SSs, which currently do not sense a clear channel. With the aim of making DRR scheduling suitable to the WiMAX environment, we redefine the concept of active list by including in it only the queues that have at least one packet to be sent over a good channel. This implies removing from the list those backlogged queues, which are not able to serve any of the queued packets because of the bad channel conditions.

## 3.6  PERFORMANCE ANALYSIS

With the aim of validating the channel and QoS-aware scheduling framework described so far, we have used the NS2 network simulator [34] tool to carry out a simulation campaign over a reference WiMAX cell operating in the 3.5 GHz band. The BS transmits heterogeneous traffic flows to four SSs, which are placed in a $350 \times 350$ m grid and are located at different distances from the BS; this causes SSs sensing different channel conditions. Each simulation has been repeated several times to get a 95 percent confidence interval in all the achieved results.

The simulated scenario is depicted in Figure 3.4, which shows the SSs distances from BS, and the flows to be delivered to such stations, respectively. No UGS and ertPS flow is simulated, because these two traffic classes are supposed to get a constant reserved amount of bandwidth frame-by-frame. All the SSs are expected to receive the same three-flow configuration set (i.e., one rtPS, one nrtPS, and one BE flow), but they differ because of the channel conditions and distances from the BS. These simulation settings allow us appreciating how dissimilar channel conditions differently impact on the same traffic configurations, and how compensation is efficient in all cases.

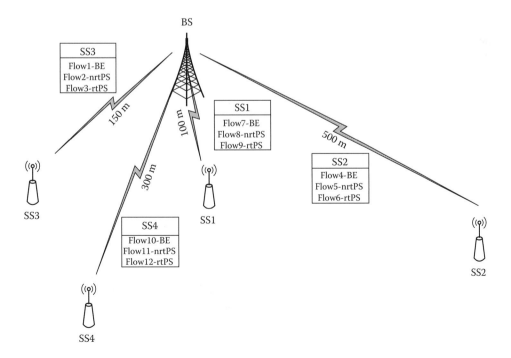

**FIGURE 3.4**  Investigated scenario.

According to the ITU-T recommendations Y.1541, rtPS, nrtPS, and BE traffic are, respectively, mapped over ITU QoS Classes 1, 4, and 5 with different maximum latency values (400 ms, 1 s, and no limit, respectively). These values have been considered as per-class delay thresholds. To appreciate clearly the behavior of the proposed compensation technique, CBR flows with the same characteristics in terms of maximum sustained rate and the packet size is simulated for each traffic class. Weights used by the $WF^2Q+$ scheduler to manage the queues are arbitrarily chosen as 4, 2, and 1, for rtPS, nrtPS, and BE class, respectively, and they represent different shares of the available channel bandwidth. Quantum values for the DRR scheduler are chosen as 600, 300, and 150 (bytes) for rtPS, nrtPS, and BE class, respectively, to match different priorities in accessing the radio channel. With this choice of weights and quantum values the relative proportion between classes is the same both in $WF^2Q+$ and DRR. This allows performing a fair comparison between the two algorithms.

The DL subframe is assigned a maximum number of 120 OFDM symbols out of the 729 symbols/frame available for each TDD frame [1], to account for the presence of UL traffic. We assume a frame duration of 10 ms, a packet size of 100 bytes for all traffic types, and the most robust burst profile adoption for all SSs. Under these simulated circumstances, we calculate the theoretical maximum achievable throughput in the DL direction as the ratio between the amount of bits carrying data information and the frame duration. In the computation we take into account the overhead introduced by both the control information (i.e., long preamble, DL-map, and UL-map) and the packet headers. These settings correspond to a theoretical maximum throughput of 1.7 Mbps.

Channel characteristics are simulated by using the most widely used path-loss models for signal strength prediction in fixed broadband wireless networks. These empirical time-dispersive models are the Stanford University Interim (SUI) channel models [35] developed under the IEEE 802.16 advices. A set of 6 SUI channels have been selected for three typical terrain types (A, B, and C). Type A is associated with maximum path loss and is for hilly terrain with moderate-to-heavy tree densities; Type C is associated with minimum path loss and applies to flat terrain with light tree densities; and Type B is for intermediate situations of flat and hilly terrains. Simulations are carried out by assuming the worst environment, which is modelled through SUI-A channels between BS and SSs. Furthermore, we account for macroscopic channel effects, such as path loss and shadowing, according to the following equation (valid for $d > d_0$):

$$ \text{PL} = 20 \, \log_{10} \left( \frac{4\pi d_0}{\lambda} \right) + 10\gamma \, \log_{10} \left( \frac{d}{d_0} \right) + X_f + X_h + s \tag{3.3} $$

where $d$ (m) is the distance between the SS and BS antennas, $d_0 = 100$ m, and $s$ is a log-normally distributed factor that accounts for the shadow fading and has a standard deviation value between 8.2 and 10.6 dB. We use 9 dB for simulations. The parameter $\gamma$ is the path loss exponent; it depends on the given terrain type and the BS antenna height $h_b$ above ground (between 10 and 80 m); $\gamma = (a - bh_b + c/h_b)$, where the constants used for $a$, $b$, and $c$ depend on the terrain type for the simulated case. Corrections factors $X_f$ and $X_h$ have been used that account of the operating frequency outside the 2.5 GHz and the given terrain and SS antenna height above ground ($h_r$). Path loss values computed through Equation 3.3 are assumed to be constant for a frame duration, and are newly generated frame-by-frame for each link between BS and SSs. These values are supposed to be measured by SSs and used to estimate channel quality and to format periodic CINR reports to the BS. Without loosing in generality, all connections are assumed to use the most robust DL operational profile (BPSK modulation and 1/2 coding rate). Therefore, when referring to a bad channel, we mean a channel, which not even the most robust transmission mode could satisfactorily cope with. We used results in the IEEE 802.16 standard [1] to fix at $-83.22$ dB the receiver sensitivity needed to have a BER lower than $10^{-6}$ when using BPSK (1/2 coding rate).

### 3.6.1  Monitored Performance Parameters

The simulation campaign has the aim of assessing the scheduling framework behavior in terms of both (1) class-based QoS differentiation capability and fair share of the channel bandwidth and (2) fairness in the treatment of traffic flows in the same class. The monitored performance parameters are the following:

- Throughput: represents the achieved bit rate for a given traffic flow, accounting for the totality of packets delivered at the target SS. The totality of delivered packets include both useful and unuseful packets. Unuseful packets are either errored or expired packets.
- Goodput: represents the achieved bit rate for a given traffic flow, accounting for the successfully delivered packets at the target SS. Successfully delivered packets are those packets received without errors and within the maximum tolerated latency. The goodput index intrinsically accounts for both packet losses and delays.
- Average latency: is the delay accumulated by packets of each flow from their arrival time at the BS to the delivery time at the target SS.
- Total packet loss percentage: accounts for packet losses occurred both at the BS, due to deadline expiration of queued packets waiting for an error-free channel, and at the receiving SS due to channel errors or to the reception of over-delayed packets.
- Goodput fairness: is evaluated through the Jain fairness index [36], defined as $\left(\sum_{i=1}^{N} x_i\right)^2 \big/ \left(N \cdot \sum_{i=1}^{N} x_i^2\right)$, where $x_i$ is the goodput achieved by flow $i$ and $N$ is the number of competing flows in the same class. This index is equal to 1 in the case of perfect fairness.

### 3.6.2  Numerical Results

The simulation campaign has been organized with a first initial aim to analyze the extent to which the wireless channel nature of WiMAX can impair the scheduler's fair and QoS-capable assignment of the available bandwidth, and then to demonstrate how channel awareness, coupled with compensation for missed transmission opportunities, can help in recovering the ideal scheduler performances, regardless of the specific implemented error-free scheduling algorithm (WF$^2$Q+ or DRR).

Results in Figure 3.5 refer, respectively, to the system behavior in the presence of (a) an error-free channel (i.e., no packet loss occurs) and (b) a "real" channel. They show the average throughput of each of the 12 active flows delivered to SSs when the "per-flow" maximum sustained rate increases

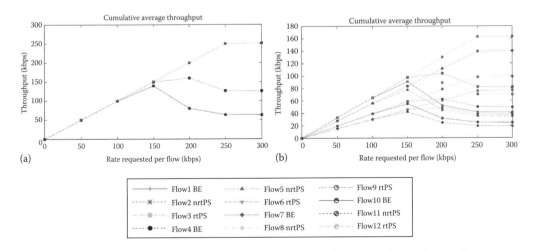

**FIGURE 3.5**  Average throughput in presence of (a) an error-free channel and (b) a real channel.

up to 300 Kbps. In Figure 3.5a, under ideal channel conditions, all flows belonging to the same traffic class are shown to achieve the same performance (curves perfectly overlap). And this hold for both $WF^2Q+$ and DRR algorithms. Furthermore, the available bandwidth is proportionally fairly shared among the three traffic classes, according to the weights of $WF^2Q+$ or the quantum values of DRR. Below the network saturation level (i.e., for an overall offered traffic load lower than 1.7 Mbps), each flow in each class achieves the requested bit rate. When the traffic load exceeds the maximum available network capacity, the situation changes: initially (for a flow data rate higher than 100 kbps) only BE traffic requests are not fully satisfied; then (for a flow data rate higher than 150 kbps) also nrtPS flows' requests are not matched any longer, because of congestion in the network. Only rtPS traffic flows can achieve the requested bit rate, and this holds up to a flow data rate equal to 250 Kbps.

Looking at Figure 3.5b, the impact of channel errors on the service received by each flow is manifest. In this case, the scheduler is supposed to be unaware of the channel status; this means that packets are transmitted regardless of the channel quality, and resources can be wasted due to useless packet transmissions over poor channels. Thereby, the fair bandwidth sharing among traffic classes cannot be longer guaranteed by the scheduler. Furthermore, also fairness in the treatment of flows in the same class cannot be guaranteed. Traffic flows affected more from channel errors are those suffering from more severely degraded throughput. As an example, the rtPS flow 6 delivered to SS2 suffers from degraded service compared to other rtPS flows. This happens because SS2, which is the farthest SS from the BS, is the one that has the highest probability to suffer from adverse channel propagation conditions. A similar reasoning holds for nrtPS flow 5 against flows 2, 8, and 11.

In the remainder of this section, we will show some results that highlight how opportunistic decision and channel error compensation mechanisms effectively mitigate the adverse effects of a real channel. The ultimate aim is to demonstrate that channel awareness and compensation are able to provide, for any scheduling algorithm ($WF^2Q+$ or DRR) and in any channel condition, the target QoS (in terms of throughput and bounded delay) to each WiMAX service class, while achieving fairness in the treatment of traffic flows belonging to the same class. For each analyzed performance parameter, we will show the obtained results both for (a) $WF^2Q+$-based scheduler and (b) DRR-based scheduler.

All the curves presented in the following show the comparison of the following three cases:

- Ideal: The radio channel between BS and SSs is ideal; no packet loss occurs.
- No Comp: The radio channel between BS and SSs is not ideal, but the scheduler is channel-unaware, thereby waste of resources may occur for useless packet transmissions over poor channels.
- Comp-Drop: The radio channel between BS and SSs is not ideal, but the scheduler is channel-aware; it makes opportunistic decisions and applies compensation techniques.

The maximum sustained rate for each traffic flow is fixed at 250 kbps (corresponding to a simulated traffic load of 3 Mbps over a DL with a maximum capacity of 1.7 Mbps). This means that the simulated network is overloaded, and this makes the job of the channel-aware scheduler even harder in trying to get back to the ideal performance. This is also the reason why the achieved performance figures are very low, even in the ideal case; but our aim was to appreciate the capability of the designed framework to react also under stressing conditions.

Figure 3.6 shows the goodput values achieved by each traffic flow. In the Ideal case, only rtPS flows can obtain the requested bit rate (250 Kbps); nrtPS and BE flows suffer from a significant goodput degradation; this is because of the very high offered traffic load compared to the DL channel capacity. Nevertheless, interclass traffic differentiation is perfectly achieved; rtPS flows get higher bandwidth than nrtPS flows, which on their turn get higher bandwidth than BE flows. Moreover, fairness in the goodput achieved by flows of the same class is guaranteed as well. In the No Comp case, the goodput values achieved by any traffic flow significantly degrade compared to the Ideal

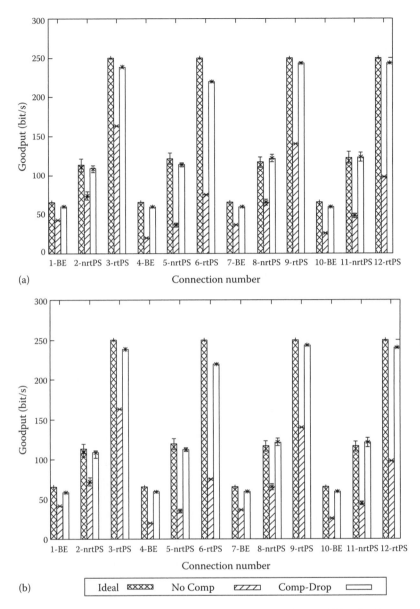

**FIGURE   3.6**  Goodput comparison—percentage of packets received within the latency thresholds. (a) $WF^2Q+$-based scheduler and (b) DRR-based scheduler.

case, because of packet losses due to channel errors. The degree of goodput degradation depends on the channel state conditions; higher degradation is suffered over lower-quality channels. Neither interclass differentiation nor intra-class fairness is achieved any longer; therefore, the theoretical performance of the error-free schedulers (both $WF^2Q+$ and DRR) is neutralized by the channel errors. The bars labelled as Comp-Drop show the efficiency of the compensation mechanism in restoring a situation similar to the Ideal case, both in terms of interclass traffic differentiation and intraclass fairness. This is confirmed by the values in Table 3.1, which summarizes the Jain's fairness index for goodput. This holds for both $WF^2Q+$ and DRR schedulers.

The introduction of the compensation mechanism can increase the delay accumulated by queued packets waiting for a clean channel. This is due to the compensator that refrains transmissions of

**TABLE 3.1**

**Jain Fairness Index**

| | WF²Q+ | | | DRR | | |
|---|---|---|---|---|---|---|
| | Ideal | No Comp | Comp-Drop | Ideal | No Comp | Comp-Drop |
| rtPS | 1 | 0.92 | 0.99 | 1 | 0.92 | 0.99 |
| nrtPS | 0.99 | 0.93 | 0.99 | 0.99 | 0.93 | 0.99 |
| BE | 0.99 | 0.92 | 0.99 | 0.99 | 0.92 | 0.99 |

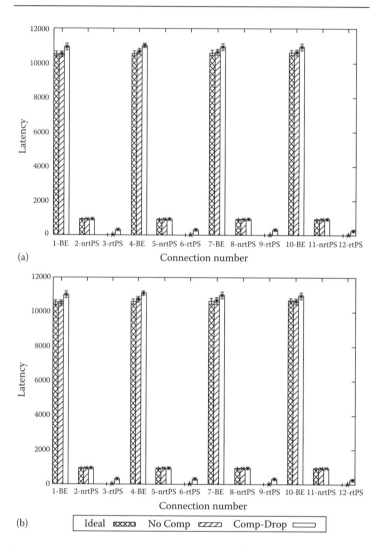

**FIGURE 3.7** Average packet latency (ms) comparison. (a) WF²Q+-based scheduler and (b) DRR-based scheduler.

packets over error-prone radio links. Queue purging (i.e., dropping of expired packets) is especially crucial when compensation introduces too long delays. Figure 3.7 shows the increase in the latency values incurred in the Comp-Drop case compared to the Ideal case. The long suffered delays of BE traffic flows are due to the simulated unconstrained delay requirements of this service class (and to

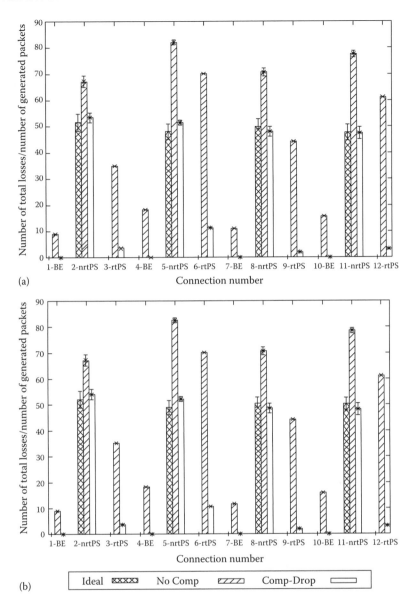

**FIGURE 3.8**  Total packet losses percentage. (a) WF$^2$Q+-based scheduler and (b) DRR-based scheduler.

the overloaded network situation). Delays suffered by each flow of rtPS and nrtPS classes slightly increase compared to the Ideal case; however, they are kept below their maximum tolerated values, thanks to the action of the buffer management module. This is again true for both WF$^2$Q+ and DRR algorithms.

Packet loss percentages are shown in Figure 3.8 for each traffic flow. Packet losses in the Ideal case are caused uniquely by the action of the buffer management module, which is activated because of the overloaded network condition. The worse effect of queue purging is on the most delay-sensitive service class. The highest values of packet losses are found in the No Comp case, because they count also for losses due to channel errors. In the Comp-Drop case, packet losses are due to the queue purging; these losses are slightly higher than the Ideal case, because of the additional delay introduced by the compensation mechanism.

## 3.7   CONCLUSIONS AND FUTURE WORK

WiMAX networks promise to offer an easy deployable and relatively low cost solution for the wireless broadband access. In usual operating conditions, WiMAX will likely support traffic belonging to a wide range of broadband applications, and it is claimed to provide differentiation among heterogeneous demanding flows. Traffic scheduling is the key component to provide QoS capability and proportional fairness in the bandwidth sharing over a changing radio environment. For these reasons, channel- and QoS-aware traffic scheduling is a feature of utmost importance for the success of WiMAX architectures.

In this chapter, we have analyzed the problem of designing a channel-aware scheduling framework capable of achieving QoS differentiation and proportional fairness in a WiMAX network. We have analyzed the extent to which the WiMAX channel nature can impair the scheduler's fair and QoS-capable bandwidth assignment, and we have given some design guidelines of an effective packet scheduler to run at the WiMAX BS for users' DL traffic delivery. We have worked on a technique for channel error compensation, which can safeguard from the drawbacks of a pure opportunistic scheduling, and, as far as possible, can be generalized to be applied to different schedulers to the purpose of preserving QoS differentiation and fairness in the treatment of traffic flows when delivering DL data, even in nonideal channel conditions. To demonstrate the general validity of our approach, we have chosen to test it with two scheduling algorithms, $WF^2Q+$ and DRR. They have been chosen as the main representatives of two different approaches in the design of scheduling algorithms; the former aims at achieving strict fairness but at the cost of a higher complexity, the latter is very simple but is less strict in guaranteeing fairness.

We have validated, through simulations, the opportunity of introducing a channel- and QoS-aware scheduler, enhanced with a compensation mechanism, which dynamically reassigns resources virtually allocated by the error-free scheduler to each traffic flow, according to the knowledge of the channel status, and can help in recovering the ideal scheduler performances. Followed approach proved to be effective and a set of encouraging simulation results witnessed to the positive effect that the transmission of heterogeneous WiMAX traffic could derive from its adoption under nonideal channel conditions.

We are currently investigating on the performance of the proposed channel-aware QoS scheduling architecture over more realistic traffic models for rtPS and nrtPS sources (like MPEG video traffic, FTP, or web traffic with Pareto distribution), and various channel conditions, propagation models, and heterogeneous traffic combination at the SSs. As a future work, we will investigate on scheduling algorithms which decouple bandwidth and delay requirements, and we will combine the DL scheduling framework with a scheduling technique at the SS, which dynamically decides how to distribute the received bandwidth grant among its active connections. Also the impact of proposed framework over the optimal use and assignment of control channels in both DL and UL subframes will be considered, together with analysis of optimal adaptation of DL/UL slot allocation.

## REFERENCES

[1] IEEE 802.16 - Standard for local and metropolitan area networks, Part 16: Air interface for fixed broadband wireless access systems, December 2005. http://www.wirelessman.org/.

[2] http://www.wimaxforum.org

[3] A. Parekh and R. Gallager, A generalized processor sharing approach to flow control in integrated service networks: The single node case, *IEEE/ACM Transaction on Networking*, 1(3), 344–357, June 1993.

[4] A. Demers, S. Keshav, and S. Shenker, Analysis and simulation of a fair queueing algorithm, *SIGCOMM 89*, New York, 1989.

[5] J. C. R. Bennett and H. Zhang, Hierarchical Packet Fair Queuing algorithms, *IEEE/ACM Transactions on Networking*, 5(5), 675–689, October 1997.

[6] J. C. R. Bennett and H. Zhang, $WF^2Q$: Worst-case fair weighted fair queuing, *IEEE INFOCOM'96*, San Francisco, CA, pp. 120–128, March 1996.

[7]   M. Shreedhar and G. Varghese, Efficient fair queuing using deficit round-robin, *IEEE/ACM Transactions on Networking (TON)*, 4(3), 375–385, June 1996.

[8]   H. Fattah and C. Leung, An overview of scheduling algorithms in wireless multimedia networks, *IEEE Wireless Communications*, 9(5), 76–83, October 2002.

[9]   S. Lu, V. Bharghavan, and R. Srikant, Fair scheduling in wireless networks, *IEEE/ACM Transactions on Networking*, 7(4), August 1999.

[10]  V. Bharghavan, S. Lu, and T. Nandagopal, Fair queuing in wireless networks: Issues and approaches, *IEEE Personal Communication*, 6(1), February 1999.

[11]  S. Lu, V. Bharghavan, and T. Nandagopal, Design and analysis of an algorithm for fair service in error-prone wireless channels, *ACM/Baltzer Journal of Wireless Network*, 6, 323–343, August 2000.

[12]  H. Fattah and C. Leung, An efficient scheduling algorithm for packet cellular networks, Vehicular Technical Conference (VTC-Fall), 2002.

[13]  Y. Yi, Y. Seok, T. Kwon, Y. Choi, and J. Park, $W^2F^2Q$: Packet fair queuing in wireless packet networks, Workshop on Wireless Mobile Multimedia (WowMom), August 2000.

[14]  P. Bender, P. Black, M. Grob, R. Padovani, N. Sindhushyana, A. Viterbi, CDMA/HDR: A bandwidth efficient high speed wireless data service for nomadic users, *IEEE Communications Magazine*, 38(7), 70–77, July 2000.

[15]  X. Liu, E. K. P. Chong, and N. B. Shroff, A framework for opportunistic scheduling in wireless networks, *The International Journal of Computer and Telecommunications Networking*, 41, 451–474, March 2003.

[16]  S. S. Kulkarni and C. Rosenberg, Opportunistic scheduling policies for wireless systems with short term fairness constraints, Globecom 2003.

[17]  M. Katevenis, S. Sidiropoulos, and C. Courcoubetis, Weighted round-robin cell multiplexing in a general-purpose ATM switch chip, *IEEE Journal on Selected Areas in Communications*, 9(8), 1265–1279, October 1991.

[18]  L. Georgiadis, R. Guerin, and A. Parekh, Optimal multiplexing on a single link: Delay and buffer requirements, *Proceedings of IEEE INFOCOM 94*, 2, 524–532, 1994.

[19]  K. Wongthavarawat and A. Ganz, Packet scheduling for QoS support in IEEE 802.16 broadband wireless access systems, *International Journal of Communication Systems*, 16, 81–96, 2003.

[20]  J. Chen, W. Jiao, and H. Wang, A service flow management strategy for IEEE 802.16 broadband wireless access systems in TDD Mode, *IEEE International Conference on Communications*, 5, 2005.

[21]  K. Vinay, N. Sreenivasulu, D. Jayaram, and D. Das, Performance evaluation of end-to-end delay by hybrid scheduling algorithm for QoS in IEEE 802.16 network, *IFIP International Conference on Wireless and Optical Communications Networks*, April 2006.

[22]  M. Settembre, M. Puleri, P. Testa, R. Albanese, M. Mancini, and V. Lo Curto, Performance analysis of an efficient packet-based IEEE 802.16 MAC supporting adaptive modulation and coding, *IEEE International Symposium on Computer Networks*, 2006.

[23]  C. Cicconetti, L. Lenzini, E. Mingozzi, and C. Eklund, Quality of service support in IEEE 802.16 Networks, *IEEE Network*, 20(2), 50–55, 2006.

[24]  C. Cicconetti, A. Erta, L. Lenzini, and E. Mingozzi, Performance evaluation of the IEEE 802.16 MAC for QoS support, *IEEE Transactions on Mobile Computing*, 6(1), 26–38, January 2007.

[25]  A. Sayenko, O. Alanen, and T. Hamalainen, Scheduling solution for the IEEE 802.16 base station, *The International Journal of Computer and Telecommunications Networking*, 52(1), 96–115, January 2008.

[26]  F. Hou, P. Ho, X. Shen, and A. Chen, A novel QoS scheduling scheme in IEEE 802.16 networks, *WCNC* 2007.

[27]  L. Qingwen, W. Xin, and G. B. Giannakis, A cross-layer scheduling algorithm with QoS support in wireless networks, *IEEE Transactions on Vehicular Technology*, 55(3), 839–847, May 2006.

[28]  D. Niyato and E. Hossain, Service differentiation in broadband wireless access networks with scheduling and connection admission control: A unified analysis, *IEEE Transactions on Wireless Communications*, 6(1), January 2007.

[29]  L. Badia, A. Baiocchi, A. Todini, S. Merlin, S. Pupolin, A. Zanella, and M. Zorzi, On the impact of physical layer awareness on scheduling and resource allocation in broadband multicellular IEEE 802.16 systems, *IEEE Wireless Communications*, 14(1), 36–43, February 2007.

[30]  A. Iera, A. Molinaro, and S. Pizzi, Channel-aware scheduling for QoS and fairness provisioning in IEEE 802.16/WiMAX broadband wireless access systems, *IEEE Network*, 21(5), 34–41, September 2007.

[31] H. K. Rath, A. Bhorkar, and V. Sharma, An opportunistic uplink scheduling scheme to achieve bandwidth fairness and delay for multiclass traffic in Wi-Max (IEEE 802.16) broadband wireless networks, Globecom 2006.

[32] M. Mehrjoo, Xuemin (Sherman) Shen, and Kshirasagar Naik, A joint channel and queue-aware scheduling for IEEE 802.16 wireless metropolitan area networks, *WCNC*, 2007.

[33] Y.-J. Chang, F.-T. Chien, and C. C.-J. Kuo, Cross-layer QoS analysis of opportunistic OFDM-TDMA and OFDMA Networks, *IEEE Journal on Selected Areas in Communications*, 25(4), 657–666, May 2007.

[34] http://www.isi.edu/nsnam/ns

[35] Channel models for fixed wireless applications, IEEE 802.16 Broadband Wireless Access Working Group, June 2003.

[36] R. Jain, D. Chiu, and W. Hawe, A quantitative measure of fairness and discrimination for resource allocation in shared computer systems, DEC res. rep.TR-301, 1984.

# 4 Random Access and Contention Strategies in WiMAX

*Yousry Abdel-Hamid, Fayez Gebali, and T. Aaron Gulliver*

## CONTENTS

The key issue in designing the Medium Access Control (MAC) layer of a communication system is multiple access. This is gaining importance with the dramatic growth in the number of users and the significant increase in data rates due to multimedia and real-time applications. The WiMAX standard defines different scheduling services to deliver resources efficiently. This includes initial synchronization, and ranging and bandwidth request mechanisms in the uplink (UL). In addition to network entry procedures that rely on contention, nonreal-time, and best-effort services for bandwidth requests, contention-based mechanisms are employed for subscriber station resource requests for UL transmission. Therefore, random access plays a major role in the WiMAX MAC layer, especially considering that unsolicited grants and unicast polling services are uneconomical when the amount of data is small relative to the required resources. In this chapter, we focus on random access in

WiMAX and present an analysis of the contention process and contention resolution mechanisms. The results are presented to illustrate the analysis.

## 4.1 INTRODUCTION

Multiple access is required in communication systems because multiple users need simultaneous access to the same channel. When a Carrier Sense Multiple Access (CSMA)-based network, e.g., a Wireless LAN such as WiFi [1] has less than ten users, collisions occur occasionally, so backoff and retransmissions only add marginal overhead which can be tolerated. If the number of users, and hence access points, rises to dozens or hundreds, many more users will collide, backoff, and retransmit data. As a result, network efficiency is degraded, capacity is reduced, and users experience noticeable delays. WiMAX is designed to serve a metropolitan area with thousands of users, where it manages delays and minimizes collisions by utilizing a request–grant mechanism. Bandwidth (BW) is granted by the base station (BS) to a subscriber station (SS) on a per connection basis in response to SS requests. This is why WiMAX is termed as demand-assigned multiple access (DAMA) system. The request–grant mechanism is designed to be scalable, efficient, and self-correcting [2,3]. The WiMAX access system must deal with multiple connections having multiple quality of service (QoS) levels and a large number of multiplexed users. WiMAX allows a wide variety of request mechanisms. In the downlink, resources are allocated based on the scheduling algorithm. In the UL, resources may be requested on a contention basis [4].

In this chapter, we provide an overview of random access in WiMAX. We discuss contention and resolution in the WiMAX standard. An analytical study of the contention process and contention resolution algorithms is given. We focus on several contention parameters and evaluate backoff performance by modeling the random access with discrete-time Markov chains. The performance is evaluated based on throughput and average packet delay.

### 4.1.1 ORTHOGONAL FREQUENCY DIVISION MULTIPLEXING AND ORTHOGONAL FREQUENCY DIVISION MULTIPLE ACCESS

WiMAX has three distinct PHY layers [2,5] in the 2–11-GHz air interface:

1. WirelessMAN-SCa using a single carrier modulation format.
2. WirelessMAN-orthogonal frequency division multiplexing (OFDM) with 256 subcarriers. Multiple access is achieved via TDMA.
3. WirelessMAN-orthogonal frequency division multiple access (OFDMA) with 512, 1024, or 2048 subcarriers. Multiple access is by OFDMA resource (subchannel) allocation, where a subchannel is a group of subcarriers. OFDMA is also called multiuser OFDM [6].

OFDMA is a combination of modulation and multiple access. It is an extension of OFDM, merging time division multiple access (TDMA) with the inherent frequency division multiple access (FDMA) of OFDM [2,6]. With a much larger number of subcarriers than OFDM, OFDMA provides for more diverse multiple access and scheduling techniques. OFDM transmits the same amount of energy on each subcarrier, although OFDMA may transmit different amounts of energy in each subchannel [7]. OFDMA provides greater flexibility by offering more allocated channels, including high-BW channels, allowing a large variety of services. Because of this, OFDMA is the multiple access technique of choice for future WiMAX systems including Mobile WiMAX IEEE802.16e.

### 4.1.2 SUBCHANNELIZATION

Subchannelization divides the available subcarriers into subchannels. Each subchannel consists of a subset of the available subcarriers. In WiMAX OFDM-PHY, subchannelization is primarily employed for UL transmission. If subchannelization is active in OFDM-PHY, the 192-data-only subcarriers

**TABLE 4.1**
**OFDM-PHY and OFDMA-PHY Parameters**

| Parameters | OFDM-PHY | OFDMA-PHY |
|---|---|---|
| Number of subcarriers | 256 | 512 |
| Number of data subcarriers | 192 | 360 |
| Sampling frequency $f_s$ | 15.6 | 10.94 |
| OFDM symbol duration | 72 | 102.9 |
| Number of OFDM symbols in a 5 ms frame | 69 | 48 |

(excluding guard and pilot subcarriers) for UL transmission are divided into 16 subchannels each containing 12 subcarriers. These subchannels are divided into four groups of three adjacent subcarriers [2,3]. An SS can be assigned 1, 2, 4, 8, or 16 subchannels by the BS in the UL. A 5-bit index is given to each subchannel and each group of assignable subchannels. Subchannelized transmission in OFDM-PHY (fixed WiMAX) allows the SS to transmit on only a part of the BW available from the BS (as low as 1/16).

In the IEEE 802.16e (Mobile) WiMAX standard, subchannelization is also used for downlink transmission using OFDMA. In OFDMA, different subcarriers are allocated to different users providing very flexible multiple access. This is the reason the IEEE 802.16e WiMAX PHY layer is named OFDMA-PHY. OFDMA subcarrier allocation is an open problem that constitutes several allocation techniques and scheduling algorithms, e.g., full usage of subcarriers (FUSC), partial-usage of subcarriers (PUSC), and adaptive modulation and coding (AMC). Table 4.1 provides the OFDM and OFDMA parameters used in WiMAX from [4].

The allocation of subcarriers within a subchannel is strictly a BS management process. Subcarriers are allocated in a subchannel according to the QoS and the chosen scheduling algorithm. One physical slot (PS) equals four OFDM or OFDMA modulation symbols [2]. A contiguous group of PSs in OFDMA WiMAX allocated to an individual user (SS) by the BS is termed a data region.

## 4.2 RANDOM ACCESS IN WiMAX

Random access in IEEE 802.16 involves the request portion of the request–grant process for network users. A portion of each UL frame is allocated to the contention-based initial access. This contention interval (channel) is divided into ranging (RNG) and BW request regions. These regions are used for initial network entry, ranging, power adjustments, and BW requests for UL transmission. In addition, best-effort data may be sent on the contention channel, but this is only suitable for the transmission of small amounts of data. This data may also include additional requests for resources [2,3].

The major tasks that employ the contention channel are initial ranging (IR) and BW requests. IR comprises channel synchronization and ranging procedures during network entry, such as closed loop time–frequency and power adjustments. The contention channel is often termed the Ranging Channel in the IEEE802.16e-2005 standard because BW requests are not necessarily performed on a contention basis, e.g., unsolicited granted service (UGS) and real-time polling services (rtPS) can be employed, as will be shown in Section 4.2.2. In this case, the contention interval is mainly used for IR. A user enters the network by sending a request to the BS to allocate UL resources for data transmission.

The BS evaluates an SS request in the context of the SS service level agreement, and grants resources accordingly. The granted resources are in the form of variable-sized time by frequency bursts in the UL subframe. A burst is a group of subchannels over some PSs. These bursts constitute the major part of the WiMAX frame structure. Burst allocation information is included in the UL-MAP MAC message of the broadcast downlink subframe, specifically in the information element

(IE) field. The IE is a data structure that contains complete information about a burst, such as the dimensions in time and frequency, start and end of each burst, and the corresponding physical channel information.

In Section 4.2.1, we explain the contention mechanisms and scheduling services for the UL contention channel as defined in the IEEE 802.16d-2004 and IEEE 802.16e-2005 standards.

## 4.2.1 WiMAX QoS and Scheduling Services

The scheduling service employed plays a key role in defining the QoS in WiMAX. It determines the data-handling mechanisms supported by the MAC scheduler for data transport on a connection. Each connection is characterized by a connection identifier (CID) and a set of QoS parameters. The scheduling service determines the number and quality of the UL and DL transmission opportunities, in addition to BW allocation mechanisms, as will be explained in Section 4.2.2.

The five QoS categories in WiMAX are:

- UGS: This class is designed to support delay-intolerant and real-time services with fixed-size data packets such as VOIP. UGS allocates fixed grants to the specified SSs on a periodic real-time basis. Thus an SS using the UGS class does not need to request resources because the size and amount of resources granted are defined at connection setup.
- rtPS: The rtPS class is designed to support variable-size data packets such as Motion Picture Expert Group (MPEG) video traffic. This service offers real-time unicast request opportunities on a periodic basis as with UGS, but with more request overhead than UGS. rtPS supports variable grant sizes [2]. The BS provides periodic unicast request opportunities during the connection setup phase. This allows the SS to also use unicast request opportunities to obtain UL transmission opportunities. The unicast polling opportunities are typically frequent enough to meet the latency and real-time services requirements [4].
- Nonreal-time polling service (nrtPS): nrtPS is similar to rtPS except that the SS uses contention-based polling in the UL to request BW. The BS provides timely unicast opportunities, but the interarrival time of adjacent opportunities is large compared to rtPS. The SSs in a polled group contend for resource request opportunities. The contention environment can result in collisions which require a resolution strategy as will be explained in Section 4.5.
- Best-effort service (BE): This service class is provided for applications not requiring QoS. Transmission is contention-based so users compete for transmission opportunities and only send data when resources are available.
- Extended real-time polling service (ertPS): ertPS combines UGS and rtPS so the SS can use UL allocations for both data transmission and resource requests. This allows the SS to accommodate time-varying BW requirements. The SS is only allowed to use this service on non-UGS-related connections (ertPS was introduced only recently in the IEEE802.16e standard for mobile WiMAX).

From the above categories, we see that only the nrtPS and BE scheduling services involve random access and contention-based mechanisms for BW requests. Performance evaluation on the different QoS categories described above can be found in Refs. [8,9].

## 4.2.2 Uplink Resource Requests and Grant Mechanisms

During initial network entry, the BS assigns up to three dedicated CIDs to the SS for transmitting and receiving MAC control messages. Connection begins using the basic CID. As mentioned previously, WiMAX DAMA services are given resources on a demand assignment basis (as the need arises).

The downlink and UL request–grant mechanisms are distinct. In the downlink, allocation of resources to an SS is done on a CID basis. The BS scheduler allocates BW according to predetermined QoS levels. As MAC PDUs arrive for a CID, the BS determines the resources based on the corresponding QoS and the scheduling algorithm. The BS indicates these allocations in the DL-MAP control message of the downlink subframe. In the UL, the SS is controlled by the overall demand for resources, taking into account the scheduling as defined in Section 4.2.1. Requests can be either standalone (occupying a dedicated MAC PDU for request purposes), or piggybacked on a generic MAC PDU. The UL/BW requests are either incremental or aggregate. When the BS receives an incremental request, it increases the BW granted according to the BW requested. The BW request type field of the MAC PDU indicates whether the request is incremental or aggregate. Because piggybacked requests are on a generic PDU with no type field, they are always incremental. Due to the possibility of collisions, BW requests sent in broadcast or multicast modes must be aggregate. The standard defines three BW request–grant mechanisms. These mechanisms are defined below. Evaluation and performance for various BW request–grant techniques based on QoS mechanisms can be found in Ref. [10].

- Unsolicited BW grants: In the UGS mechanism, BW requests are primarily allocated in dedicated slots in the UL subframe. The requests can also be piggybacked on a generic PDU.
- Unicast polling: Unicast polling or simply polling is the process where dedicated resources are provided to an individual SS in the UL to make BW requests. The BS indicates to the SS the request-allocated slots in the UL-MAP MAC message in the DL subframe to send standalone request PDUs. The BS assigns the polling allocation for requests to the polled SS using the primary CID (one of the three assigned during network entry and initialization). An SS being polled should not remain silent if no BW is needed during polling. Instead, the SS can send a dummy request PDU using padding (all zeros), to fill the allocation field in the current CID. A data grant information element (IE) is associated with the basic CID. Note that implicit UGS users will not be polled unless the Poll-Me bit is set in the header of the packet in the UGS connection [2].
- Multicast/broadcast polling—If the SS does not acquire sufficient BW when polled individually, multicast or broadcast polling can be used to poll a group of SSs. As with individual polling, the SS can join a group of SSs in multicast polling to obtain additional BW, but uses the BW allocated in the UL-MAP using a multicast/broadcast CID. All SSs in a polling group contend for request opportunities during the multicast/broadcast polling interval. Intuitively, multicast polling saves BW compared with unicast polling by grouping SSs. To reduce the likelihood of collisions, only SSs with a BW request will reply. However, if several SSs are requesting BW resources, collisions may occur and therefore contention resolution is necessary.

Contention-based requests are allocated in the random access channel. The size of the contention slot is assigned by the BS as a request IE. SSs belonging to a multicast polling group contend for the BW request slots. A collision occurs if two or more SSs simultaneously make requests in the same slot. Therefore, the requesting SSs must employ a contention resolution algorithm to acquire slots to send BW requests. The WiMAX standard does not define an explicit algorithm or strategy by which the SS knows the status of a request sent. In fact, if the SS does not receive the resources sought in the very next DL subframe, the request may be deemed lost or collided. Several strategies have been proposed to acknowledge the status of a request and also for resolution if a collision occurs. To better understand the contention phase and contention resolution in WiMAX, in the next section we define the WiMAX frame allocations and MAC management messages involved with contention and contention resolution mechanisms.

## 4.3  RELEVANT FIELD PARAMETERS

### 4.3.1  UPLINK CONTENTION REGION

The UL subframe in both time division duplexing (TDD) and frequency division duplexing (FDD) has the same format. TDD is usually the duplexing technique of choice in OFDM/OFDMA because it automatically supports asymmetrical data rates between the BS and SS [3]. The SS uses the ranging request (RNG-REQ) MAC management message for IR during the IR interval using the basic CID of the SS. There is no explicit MAC management message for BW requests. BW requests are issued by every SS during the dedicated BW request interval in the contention region of the UL subframe. The BS assigns the interval based on the BW allocation scheme assigned to the SS associated with the QoS class as described in Section 4.2.2. An IE is a mandatory subsection of the MAC management message. It contains control information explicitly dedicated to a corresponding burst. The request IE is the BS specified interval at the beginning of every UL subframe within which SSs seek BW grants. The request IE is CID-dependent to indicate a unicast, multicast, or broadcast transmission. In the unicast case, it is an invitation for a single SS to request resources (BW) for transmission. In the multicast or broadcast case, the request IE is an opportunity for contention among a group of SSs to request BW. Similarly, IR IE indicates the interval in which new SSs contend to join the network and perform ranging. The standard specifies an interval equivalent to the maximum round trip delay plus the transmission time of the RNG-REQ to allow new stations to contend for resources and perform IR. Upon the termination of a successful (noncollided) IR or BW request, the BS allocates IR or BW opportunities to the SS accordingly. The exact value of the allocated interval is published in the uplink channel descriptor (UCD) MAC message to the corresponding SS in the downlink subframe [2].

In Section 4.3.2, we show the relevant MAC messages and provide details of the structures related to the contention process and resolution in WiMAX. Fields and control information beyond the scope of this chapter are shown for the sake of completeness but are not explained.

### 4.3.2  UPLINK CHANNEL DESCRIPTOR MAC MANAGEMENT MESSAGE

The UCD is a MAC management message transmitted by the BS periodically every 10 s. The UCD contains the essential contention information needed by the SS to define the contention resolution parameters in case of a collision. In addition, the UCD provides the UL physical channel properties.

Table 4.2 shows the UCD MAC management message layout. The initial and final backoff window sizes for IR and BW request contention resolution algorithms are expressed as a power of 2 from the least significant bit position (higher order unused bits are set to 0). Contention resolution is explained in detail in Section 4.5.

**TABLE 4.2**
**UCD MAC Management Message Layout**

| Field | Size | Description |
|---|---|---|
| MAC message type (type = 0) | 8 bits | Message type identifier (UCD = 0) |
| Configuration change count (CCC ) | 8 bits | SS prediction of subsequent field change |
| Ranging Backoff Start | 8 bits | IR backoff window starting size |
| Ranging Backoff End | 8 bits | IR backoff window final size |
| Request backoff start | 8 bits | Initial BW request window starting size |
| Request backoff end | 8 bits | Last backoff BW request window final size |
| TLV encoding | Variable | PHY specific descriptors |
| UL burst profile | Variable | UL-MAP IE specific |

**TABLE 4.3**
**RNG-REQ MAC Management Message Layout**

| Field | Size | Description |
|---|---|---|
| MAC message type (Type = 4) | 8 bits | Message type identifier |
| Downlink channel ID | 8 bits | Downlink channel identifier |
| TLV | Variable | PHY specific |

**TABLE 4.4**
**OFDM UL-MAP MAC Management Message Layout**

| Field | Size | Description |
|---|---|---|
| MAC Message type (Type = 3) | 8 bits | MAC message type (UL-MAP = 3) |
| Uplink channel ID | 8 bits | Channel identifier this UL-MAP belongs to |
| UCD count | 8 bits | Matching the value of CCC in UCD |
| Allocation start time | 32 bits | States the start time of UL frame allocation (bursts) in PS In OFDM-PHY, one PS = 4 OFDM symbols |
| UL-MAP_IE | Variable | The $i$th IE corresponding to the $i$th burst PHY dependent |
| Padding | 4 bits | Padding bits |

### 4.3.3 RANGING REQUEST MAC MANAGEMENT MESSAGE

Table 4.3 shows the RNG-REQ MAC message. This message is transmitted by the SS on a contention basis at initial network entry to perform IR, and periodically to determine network delay and power adjustments. The downlink channel ID should not be confused with the CID. The former identifies the downlink channel on which this SS received the UCD. The UCD contains the UL channel properties which the SS uses to send the request.

### 4.3.4 OFDM UPLINK MAP MAC MANAGEMENT MESSAGE AND INFORMATION ELEMENTS

Table 4.4 shows the uplink MAP (UL-MAP) message layout. As explained earlier, the UL-MAP message contains all the necessary information for UL access to the system. The effective start time of the UL allocation bursts in the UL subframe is included in every UL-MAP. Every UL-MAP must contain the UL-MAP_IE that describes the corresponding UL burst, and at least one UL-MAP_IE marking the start and duration of the last UL burst.

The MAC CID determines the type of BW allocation described in Section 4.2.2 (unicast, multicast, or broadcast). In the unicast case, the CID is the basic CID of the particular SS to be granted the allocation.

## 4.4 CONTENTION-BASED BANDWIDTH REQUESTS

The WiMAX standard defines the following contention-based mechanism for the OFDM and OFDMA physical layers.

### 4.4.1 CONTENTION-BASED BANDWIDTH REQUESTS FOR OFDM-PHY

In OFDM-PHY, WiMAX supports two contention-based BW request mechanisms, full and focused. The full contention-based mechanism is either with or without subchannelization. The default is

**TABLE 4.5**
**OFDM UIUC Allocation Values**

| UIUC value | Function |
|---|---|
| 0 | Reserved |
| 1 | Initial ranging |
| 2 | REQ region full |
| 3 | REQ region focused |
| 4 | Focused contention IE |
| 5–12 | Burst profile |
| 13 | Subchannelized network entry |
| 14 | End of this UL-MAP |
| 15 | Extended UIUC |

**TABLE 4.6**
**OFDM-Focused-Contention_IE Layout**

| Field | Size | Description |
|---|---|---|
| Frame number index | 4 bits | Specifies the frame index |
| Transmit opportunity index | 3 bits | Transmit opportunity index |
| Contention channel index | 6 bits | Index of the contention channel |
| Contention code index | 3 bits | Index of the contention code |

full with no subchannelization. However, if the SS is capable of subchannelization, it can choose between full and subchannelized contention. The default is then full with no subchannelization. In this case, the SS is allowed to send a full contention BW request using a full request IE in the UL subframe referred by the BS as REQ region full with no subchannelization. This is indicated by setting UIUC = 2 (see Table 4.5). Each transmission opportunity (TO) is one preamble field and one OFDM symbol having the most robust modulation and coding in the burst profile (BPSK, code rate = 1/2). If subchannelization is active, the TO in the UL REQ region consists of a two dimensional time–frequency unit. Frequency (width) in allocated subchannels and time (length) in OFDM symbols. The dimension of every TO is defined in the UCD PHY specific field (type = 150) REQ Region-Full parameters (see Ref. [2], Section 8.3.7).

A REQ region-focused mechanism is denoted by UIUC = 3 in the UL broadcast by the BS. The SS contends for the TO represented by a contention code [2]. The contention code [2] is sent on four subcarriers of two OFDM symbols. termed the contention channel. The TOs are sequentially coded, the first TO is coded 0. The SS randomly selects a contention code from the eight possible codes, and randomly selects a contention channel from those available in the REQ region. The SS sends this contention code to the BS on the selected channel during the REQ region-focused phase in the UL TO.

Upon detection of the selected code, the BS sends an allocation to the SS in the DL subframe to transmit its MAC PDU and optionally data (only in BE class). Instead of the basic CID, the BS sends the broadcast CID in combination with an OFDM-Focused_IE (UIUC = 4). The OFDM-Focused_IE specifies the frame index, TO index, contention channel index, and contention code index (see Table 4.6). The SS then matches these parameters with the parameters it sent and can easily determine the status of its request and whether it has been granted an allocation as a result. This focused contention transmission is a better strategy than full contention because there are fewer collisions. During the TO, the power of each subcarrier can be boosted above the normal level, i.e.,

**TABLE 4.7**
**OFDM-Subchannelized-Network-Entry_IE Layout**

| Field | Size | Description |
|---|---|---|
| Frame number index | 4 bits | Specifies the frame index |
| Transmit opportunity index | 4 bits | Transmit opportunity index pointed to by the frame index |
| Contention subchannel | 4 bits | Index of the subchannel used (ranges from 0 to 0xF) |

the power of a noncontention OFDM symbol. The power boost in dB is equal to the value of the focused contention power boost parameter in the UCD in effect.

A specific type of subchannelization is to be specified for network access in WiMAX OFDM-PHY. If the SS is capable of subchannelized contention, the UL-MAP_IE subchannelized network entry signal is identified by UIUC = 13 (see Table 4.5). The number of subchannelized codes is denoted by $C_{SE}$. The contention code is selected at random from the subset of codes indicated by $C_{SE}$. The OFDM Subchannelized-Network-Entry_IE is shown in Table 4.7. If the BS supports subchannelization, only the last $C_{SE}$ codes among the eight available will be used by a subchannelization-capable SS. The value of $C_{SE}$ is defined in the UCD TLV encoding field. The default value is 0.

### 4.4.2 CONTENTION-BASED CDMA BANDWIDTH REQUESTS FOR OFDMA-PHY

CDMA-based requests in OFDMA-PHY are very similar to those for subchannelized OFDM-PHY. The OFDMA-PHY defines a Ranging Subchannel and a subset of Ranging Codes. An SS requiring BW randomly selects a Ranging Code from the subset allocated for BW requests in the Ranging Channel. In OFDMA-PHY, the Ranging Code is transmitted on the Ranging Subchannel during the corresponding UL allocation. A similar procedure is employed for IR, in which case a subset of Ranging Codes are defined for this use. The subset is allocated in the UCD channel encoding for IR, periodic Ranging, and BW Requests. The BS determines the purpose of the code by the subset to which the code belongs.

An SS wishing to make a BW request or IR employs the following procedure. The SS randomly chooses a Ranging Subchannel and a Ranging Code from the corresponding subset and transmits to the BS using this code. The BS cannot identify which SS sent the CDMA Request Code; therefore, the BS broadcasts a Ranging Response message RNG-RSP that advertises the received code and the Ranging Subchannel on which the code is sent. The corresponding SS can determine that the response belongs to it by identifying the request parameters. The SS sends the Ranging Code to the BS in the appropriate UL allocation. Upon detection, the BS provides an UL allocation for the SS. Instead of a basic CID, the BS will broadcast a CID in combination with CDMA-Allocation_IE. The IE specifies the information and the Ranging Code that were used by the SS. The SS can then determine whether or not its request was granted. The layout of CDMA-Allocation_IE is shown in Table 4.8

## 4.5 CONTENTION RESOLUTION STRATEGIES IN WiMAX

Collisions occur during contention-based events, i.e., during IR and BW requests, when two or more stations choose the same resource simultaneously. The conventional truncated Binary Exponential Backoff (BEB) algorithm is the default method of contention resolution in WiMAX. The initial and final backoff window values are indicated in the UCD MAC message (see Table 4.2). The backoff window values are represented by a power-of-two minus 1 ($2^n - 1$). For example Ref. [2], a value of 3 indicates a window between 0 and 7. The range of window values is often termed the size of

**TABLE 4.8**

**CDMA-Allocation_IE Layout**

| Field | Size | Description |
|---|---|---|
| Duration | 6 bits | Duration in OFDMA symbols |
| Repetition code | 2 bits | Burst specified |
| Ranging code | 8 bits | The CDMA code sent by the SS |
| Ranging symbol | 8 bits | The OFDMA symbol used by the SS |
| Ranging subchannel | 7 bits | Indicates the ranging subchannel used by the SS to send the CDMA code |
| Bandwidth request inquiry | 1 bit | 1 = true, 0 = false indicates whether the SS requested BW or not |

the backoff window. The maximum possible value of $n$ is 16, resulting in a window size between 0 and 65535. When an SS enters the contention process, it sets its initial BW request or IR backoff window equal to the backoff start value in the UCD MAC message referenced by the UCD count in the UL-MAP MAC message currently in effect. After the window size is determined, the SS selects a random number within the range (size) of the backoff window. This number indicates the number of transmission opportunities the SS will defer before transmission. The SS only considers the opportunities for which it would have been eligible to transmit. A TO is defined as a request IE for BW request or an Initial-Ranging_IE for IR. A request IE or Initial-Ranging_IE may contain multiple contention transmission opportunities.

Deferring the correct number of transmissions is mandatory. For example Refs. [2,3], consider an SS that has chosen a window size of 11 from an initial window size of 15. The SS must defer a total of 11 transmissions. If the first request IE contains six opportunities, the SS ignores these and defers for five more transmissions. If the next request IE has two opportunities, they are ignored and the SS has three more transmissions to defer. If the third has seven opportunities, the SS transmits on the fourth.

The SS must wait for a Data-Grant-Type_IE or a RNG-RSP MAC message in the UL MAP of the subsequent frame. Upon its arrival, the contention process is complete and successful. The SS considers the request lost or collided if no opportunity (for a BW request) or RNG-DSP response (for IR) is granted in the next frame, or upon the expiration of the T66 or T3 timers as defined in Ref. [2]. If a collision occurs, the SS will start retrial after increasing its window size by a factor of 2, as long it is less than the maximum, in which case the maximum is used. The SS select a random number from this new backoff window and repeats the process until the maximum number of retries has been reached. Note that the maximum number of retries is independent of the initial and maximum window sizes given by the BS in the UCD message. If during unicast, the SS receives a unicast Data-Request–Grant_IE, the SS ends the contention process and uses the explicit TO.

### 4.5.1 CONTENTION PERFORMANCE

The performance and optimization of the contention system have been previously studied. In Ref. [11], the relationship between the maximum and minimum backoff window values was examined. The transmission attempts were uniformly distributed over the available random access slots. The authors derived the conditional probability of success as a two-dimensional stochastic process for the transmission and backoff time functions. In Ref. [12], the optimal length of the contention channel was introduced as a cost function optimization problem using a delay-throughput trade-off. In Refs. [13,14], the transmission collision probability was studied to resolve the unfairness problem using a CSMA protocol with propagation delay.

The contention performance depends on the behavior of the SSs when collisions occur. In this section, we investigate performance metrics that have a significant effect on system behavior during the contention process. In addition, we propose new parameters such as the retransmission probability to improve the contention resolution procedure. In Ref. [15], we used a simple adaptation technique based on an access opportunity window. In Ref. [16], we analyzed the contention system using two backoff strategies for contention. We also presented an alternative method to compensate for the increase in retransmission stages due to the backoff procedure. A discrete-time Markov chain was used to develop analytical models for the contention and resolution processes. These models were then employed to evaluate the performance of the system during the contention process based on the performance metrics throughput and delay according to the analysis in Ref. [17].

### 4.5.2 SINGLE-STAGE BACKOFF MODEL

Because the state of an SS request only depends on the previous state, its behavior can be modeled as a discrete-time Markov chain with three possible stages: transmit: only one station transmits a request in the window, collision: two or more stations transmit a request in the same window, and idle: no stations transmit a request. In Figure 4.1, a Markov state diagram with single stage backoff is shown. This tristate model represents a simple single-stage backoff strategy where it is assumed that a collided SS retries to send its request until it is successfully transmitted. Although an infinite number of retrials is unrealistic, the model provides an accurate measure of the effect of user retransmission probability on the system performance, regardless of the number of retrials.

Consider a finite population of $N$ users and a limited number of available slots, $K$. Each user has traffic to send (a RNG-REQ or a BW request), during the contention interval with probability $a$. A collided user resends its request during the contention interval with probability $\gamma$. Having a request to send, an idle user at state $S_i$ enters the transmission state $S_t$ if its request is successfully transmitted (with probability $x$), and returns to the idle state after transmission. In case of collision, a collided user stays in the collision stage until the request is successful. Therefore, the probability that a user remains in the collided state is the sum of the two conditional transitions

$$(1 - \gamma) + (1 - x)\gamma = 1 - x\gamma. \tag{4.1}$$

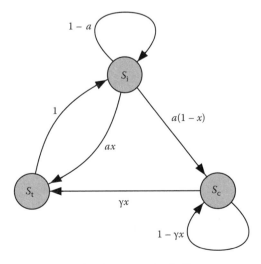

**FIGURE 4.1** Markov chain state diagram for single stage backoff.

A successful user in the transmission state $S_t$ returns to the idle state $S_i$ after every transmission, i.e., with transition probability 1. The system state probabilities in steady state are [16,17]

$$S_i = \frac{\gamma x}{D_s} \tag{4.2}$$

$$S_t = \frac{a\gamma x}{D_s} \tag{4.3}$$

$$S_c = \frac{a(1-x)}{D_s} \tag{4.4}$$

where

$$D_s = a(1-x) + \gamma x(1+a) \tag{4.5}$$

The system performance is measured using the following metrics [17]:

1. Throughput: The throughput Th is defined as the number of successful requests per frame, which is given by
$$\text{Th} = \min[(N \, S_t), K] \tag{4.6}$$

2. Delay: A request is delayed due to retransmissions caused by collisions. The average number of retransmissions is given by [17]

$$\text{Delay} = \frac{1 - p_a}{p_a} \tag{4.7}$$

where $p_a$ is the probability of acceptance given by $p_a = \text{Th}/Na$.

The following results are obtained based on traffic intensity, i.e., the number of requests sent per contention interval, for various retransmission probabilities $\gamma$. Note that some users will not retransmit during a given contention interval. Three different values of $\gamma$ are considered, 0.1, 0.5, and 1.0. Based on the results in Ref. [16], these values provide sufficient information on the system behavior due to $\gamma$.

Figure 4.2 presents the throughput vs. input traffic for different retrial probabilities $\gamma$. It shows that at intermediate traffic levels, the performance starts to degrade at $\gamma$ values over 0.5. From

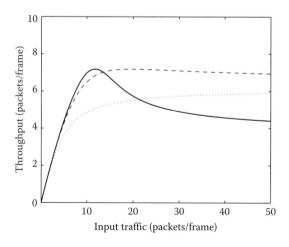

**FIGURE 4.2**   Throughput vs. input traffic for a single collision stage with $\gamma = 0.1$ (dotted), 0.5 (dashed), and 1.0 (solid), $N = 50$ users and $K = 20$ slots.

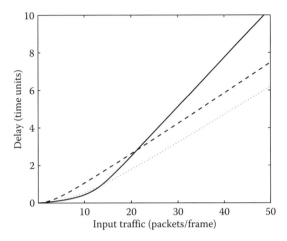

**FIGURE 4.3** Delay vs. input traffic for a single collision stage with $\gamma = 0.1$ (dotted), 0.5 (dashed), and 1.0 (solid), $N = 50$ users and $K = 20$ slots.

Equation 4.6, the throughput is proportional to the transmission probability $S_t$. The figure indicates a maximum throughput at $\gamma$ of approximately 0.5. The contention interval decreases when $\gamma$ exceeds 0.5. A similar effect on delay is shown in Figure 4.3. A value of $\gamma = 0.5$ provides a reasonable delay response. At higher values of $\gamma$, the delay increases sharply indicated by the crossover of the $\gamma = 1$ curve at traffic values around 20 packets/frame.

### 4.5.3 MULTIPLE-STAGE BACKOFF MODEL

We now introduce a model of the contention process which is more practical than the single-stage backoff model. In this model, the factors introduced are the SS probability of retrial $\gamma_j$, where $0 \leq j \leq m$, and the backoff radix factor $r = \gamma_j / \gamma_{j+1}$. If $r = 2$, the model represents the truncated binary exponential backoff (BEB) contention resolution presented in Ref. [2]. We develop an analytical model using Markov chains to study the effect of these parameters on system performance. The same performance metrics as in Section 4.5.2 are considered.

A multiple-stage backoff Markov state diagram is shown in Figure 4.4. When the SS has a request to send, it enters the contention process to acquire an empty slot in the frame contention period assigned by the BS. Having a request to send, an idle user at stage $S_i$ enters the transmission stage $S_t$ if its request is successfully transmitted (with probability $x$), and returns to the idle stage after transmission. In case of collision, the collided SS retries sending the request by entering a finite multiple-retrial stage. Each retrial stage is represented by a stage in the chain. The reader may refer to Ref. [17] for a more detailed explanation and an analysis of the algorithm.

The Markov steady state probabilities for multiple collision state backoff are given by Ref. [16]

$$S_i = \frac{1}{\Delta} \tag{4.8}$$

$$S_c = \frac{a\sigma_1}{\Delta} \tag{4.9}$$

$$S_t = \frac{ax(1 + \sigma_2)}{\Delta} \tag{4.10}$$

where

$$\Delta = a\sigma_1 + ax(1 + \sigma_2) + 1 \tag{4.11}$$

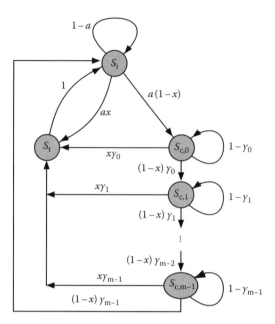

**FIGURE 4.4** Markov chain state diagram for multiple collision stage backoff.

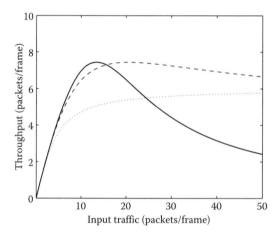

**FIGURE 4.5** Throughput vs. input traffic for a multiple-stage model with $\gamma_0 = 0.1$ (dotted), 0.5 (dashed), and 1.0 (solid), $N = 50$ users, $K = 20$ slots and $m = 8$ stages.

$$\sigma_1 = \frac{1-x}{\gamma_0}\left[\frac{r - r[(1-x)/r]^m}{r - x - 1}\right] \tag{4.12}$$

$$\sigma_2 = \frac{1-x}{x}[1 - (1-x)^m)] \tag{4.13}$$

and $m$ is the maximum number of backoff stages. We first consider a finite number of stages with the conventional backoff radix $r = 2$ and different retrial probabilities $\gamma_0$.

In Figure 4.5, we use a maximum number of backoff stages $m = 8$ with $r = 2$. Performance degrades when the retrial probability increases beyond $\gamma_0 = 0.5$. This is due to the finite number of retrials which results in packets being lost. Figure 4.6 shows the higher delay resulting from $\gamma_0 > 0.5$ at traffic levels of 10–20 packets/frame.

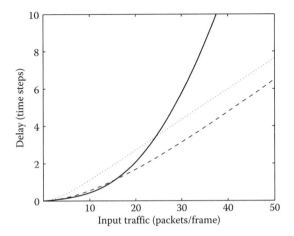

**FIGURE 4.6** Delay vs. input traffic for a multiple-stage model with $\gamma_0 = 0.1$ (dotted), 0.5 (dashed), and 1.0 (solid), $N = 50$ users, $K = 20$ slots and $m = 8$ stages.

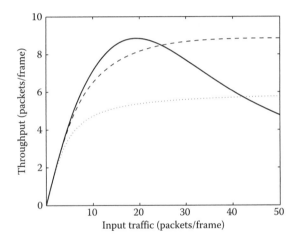

**FIGURE 4.7** Throughput vs. input traffic for a multiple collision stage model with $\gamma_0 = 0.1$ (dotted), 0.5 (dashed), and 1.0 (solid), $N = 50$ users, $K = 20$ slots, and $m = 16$ stages.

Figure 4.7 shows the effect of increasing the number of collision retrial stages to $m = 16$ while using the same values of $\gamma_0$ as before with $r = 2$. The result shows only a marginal improvement in performance compared to those with $m = 8$ in Figure 4.5.

Figure 4.8 shows that increasing the radix greatly improves the performance even at the highest value of $\gamma_0 = 1$. This indicates that a larger radix can be employed instead of adding more backoff states. In addition, at $\gamma_0 = 1.0$, the response is only slightly degraded at higher traffic levels. The shift in the maximum throughput to a higher traffic value provides an improvement in system response at intermediate traffic levels of about 20 request packets/frame.

We now increase the number of backoff stages to $m = 16$ and the radix to $r = 4$. Figure 4.9 shows that increasing both the maximum number of retrials and the radix dramatically improves the performance. The throughput is now monotonic for all values of $\gamma$, even at higher traffic levels. This indicates that there is no loss in performance as the traffic level increases to the maximum.

Next, we compare the system throughput for different radix values under worst-case conditions. Transmission retrial is guaranteed in case of collision, i.e., $\gamma_0 = 1.0$, and only $K = 10$ slots are

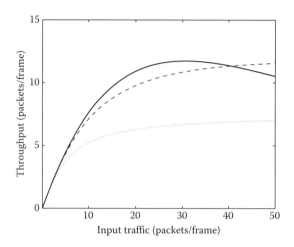

**FIGURE 4.8** Throughput vs. input traffic for a multiple collision stage at values of $\gamma_0 = 0.1$ (dotted), 0.5 (dashed), and 1.0 (solid). $N = 50$ users, $K = 20$ slots, and $r = 4$.

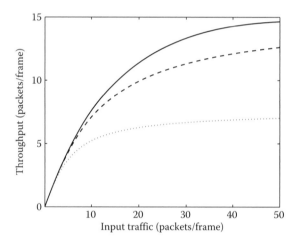

**FIGURE 4.9** Throughput vs. input traffic for a multiple collision stage model with $\gamma_0 = 0.1$ (dotted), 0.5 (dashed), and 1.0 (solid), $N = 50$ users, $K = 20$ slots, $m = 16$, and $r = 4$.

available. Figure 4.10 shows that increasing the radix from 2 to 4 has the greatest impact on system performance. Any further increase in the radix ($r > 4$), only marginally affects performance, which indicates that increasing the radix to values higher than $r = 4$ is unnecessary.

### 4.5.4 Reservation Window with Adaptive Backoff

In Ref. [15], we proposed a simple backoff algorithm based on an adaptive backoff window strategy. This provides an organized controllable contention technique for the IEEE 802.16 standard by assigning a reservation window size within the DL-MAP bursts. Thus additional hardware is not required. A discrete-time Markov chain analysis was applied to this protocol in Ref. [15]. Each user maintains an estimate of the average traffic and adapts its backoff window size accordingly.

We consider the same performance measures as before. The results obtained show that this technique is very effective in improving contention access in the WiMAX UL contention channel. The system assumes $N$ equal priority SSs randomly accessing $K$ available slots. The BS provides $L$ reservation windows. The probability that a user has a packet to transmit in the UL is $a$. We assume

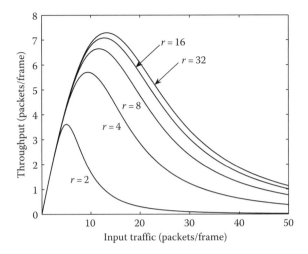

**FIGURE 4.10**   Throughput vs. traffic intensity for different radix values $r$.

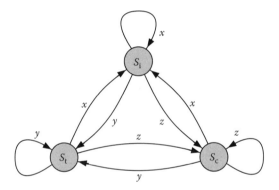

**FIGURE 4.11**   Markov chain state diagram for the reservation window.

that each request sent fits in one resource (slot). A user with a packet to send picks one of the $L$ available windows at random with probability $l = 1/L$.

The adaptive backoff strategy implies that the probability a collided user transmits a request during the contention interval equals the packet arrival probability $a$. A collided user employs a fixed adaptive backoff strategy where each user keeps an estimate of its traffic intensity and adapts its backoff strategy accordingly. Therefore, both successful and collided users are treated equally from the system perspective as state transitions are equally probable. Figure 4.11 shows the system Markov state diagram of the reservation window in effect. $S_i$, $S_t$, and $S_c$ represent the idle, success, and collided states, respectively.

The reader may refer to Ref. [15] for details of the Markov chain analysis. Using the same metrics as before, the performance of the system is examined for $N = 50$ and $K = 10$. The reservation window size is varied by integer multiples of the available resources, i.e., $L = K$, $2K$, $3K$, and $5K$. With transition probabilities $x$, $y$, and $z$, the Markov steady state probabilities are given by

$$S_i = x = (1 - \alpha)^N \tag{4.14}$$

and the probability that exactly one user issues a successful request in a window is

$$S_t = y = N\alpha(1 - \alpha)^{N-1} \tag{4.15}$$

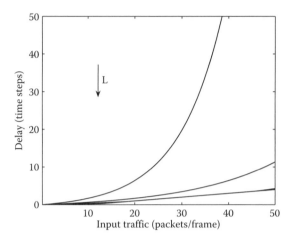

**FIGURE 4.12**   Delay vs. traffic intensity with $L = K, L = 2K, L = 3K$, and $L = 5K$.

If $z$ is the probability that a collision has taken place, it is simply given by

$$S_c = z = 1 - S_t - S_i \qquad (4.16)$$

where $\alpha = a/L$ is the probability of a user issuing a request in a reservation window.

Figure 4.12 shows the system delay when $N = 50$ and $L = K, 2K, 3K$, and $5K$. This shows that increasing the window size $L$ significantly reduces the delay due to fewer retransmissions, even at higher traffic rates.

## 4.6   CONCLUSIONS

Random access is the most reliable and widely used technique to establish a wireless connection. This is because UGS and rtPS (contention-less) services require knowledge of the wireless link to grant resources with minimal loss. Moreover, most real-time and delay-intolerant applications work efficiently using contention-based mechanisms. In Section 4.5.1, we presented a model of two backoff strategies for IEEE802.16 random access. We employed discrete-time Markov chains to analyze the system performance. The performance metrics considered were throughput and delay.

A simple single collision state model was developed where a collided user enters an infinite retrial phase until successful transmission takes place. This model can provide an accurate measure of the effect of user retransmission probability on the system performance, regardless of the number of retrials. We then considered the performance of a multiple-stage collision model using the same retransmission probabilities. In this model, a collided user retransmits until a maximum number of retrials, after which the packet is deemed lost. By knowing the optimal retransmission probabilities for the single stage backoff model, we were able to determine the impact of these probabilities on the performance of a practical multiple stage backoff system as adopted in the WiMAX standard. These values can be used by the BS to determine network load conditions. We also presented a method to mitigate the increase in the number of finite backoff stages by introducing the radix parameter into the contention algorithm. We found that increasing the radix by a factor of 2 dramatically improves system performance and compensates for the increase in retransmissions. An increased radix can be employed instead of a higher number of backoff stages. This is especially important at the maximum retrial probability ($\gamma = 1.0$). Thus allowing a radix value in the range 2 to 4 will save adding extra hardware to provide more backoff stages.

Finally, we considered modeling partitioning of the random access channel into variable-sized reservation windows. Using the same performance metrics as before, it was found that the reservation window protocol can increase system performance by allowing a suitable window of opportunity to access the available resources in the contention channel. This protocol provides an organized and controllable technique for the IEEE 802.16 standard by assigning a reservation window size within the DL-MAP bursts. Thus no additional hardware is required.

## REFERENCES

[1] Wireless LAN Medium Access Control (MAC) and Physical Layer (PHY) Specification, IEEE 802.11, IEEE, Piscataway, NJ, 1997.

[2] IEEE standard for local and metropolitan area networks part 16: Air interface for fixed broadband wireless access systems, IEEE, Piscataway, NJ, 2004–2005.

[3] L. Nuaymi, *WiMAX Technology For Broadband Wireless Access*, Wiley, Chichester, UK, 2007.

[4] J. Andrews, A. Gosh, and R. Mohammed, *Fundamentals of WiMAX, Understanding Broadband Wireless Access Networking*, Pearson-Prentice Hall, 2007.

[5] C. Eklund, R. B. Marks, K. L. Stanwood, and S. Wang, IEEE standard 802.16: A technical overview of the WirelessMAN air interface for broadband wireless access, *IEEE Communications Mag.*, 40(6), 98–107, June 2002.

[6] I. Koffman and V. Roman, Broadband wireless access solutions based on OFDM access in IEEE 802.16, *IEEE Communications Mag.*, 40(4), 96–103, April 2002.

[7] OFDMA—Theory, principles, design considerations and applications, White Paper, Runcom Technologies Ltd., 2004.

[8] C. Cicconetti, L. Lenzini, E. Mingozzi, and C. Eklund, Quality of service support in IEEE 802.16 networks, *IEEE Network*, 20, 50–55, March 2006.

[9] D. Staehle and R. Pries, Comparative study of the IEEE 802.16 random access mechanisms, *Proceedings International Conference Next Generation Mobile Applications, Services and Technologies*, pp. 334–339, September 2007.

[10] C. Cicconetti, A. Erta, L. Lenzini, and E. Mingozzi, Performance evaluation of the IEEE 802.16 MAC for QoS support, *IEEE Trans. Mobile Computing*, 6(1), 26–38, January 2007.

[11] A. Vinel, Y. Zhang, M. Lott, and A. Turlikov, Performance analysis of the random access in IEEE 802.16, *Proceedings IEEE International Symposium on Personal, Indoor and Mobile Radio Communications*, pp. 1596–1600, Sept. 2005.

[12] S.-M. Oh and J.-H. Kim, The analysis of the optimal contention period for broadband wireless access network, *Proceedings IEEE International Conference on Pervasive Computing and Communications*, pp. 215–219, March 2005.

[13] V. Kabliakov, A. Turlikov, and A. Vinel, Distributed queue random multiple access algorithm for centralized data networks, *Proceedings IEEE International Symposium on Consumer Electronics*, pp. 1–6, June 2006.

[14] B. A. Guo and J. Sydor, Performance analysis of the P2MP metropolitan and broadband wireless access network, *Proceedings IEEE International Symposium on Personal, Indoor and Mobile Radio Communication*, pp. 260–264, September 2003.

[15] Y. Abdel-Hamid, F. Gebali, and T. A. Gulliver, A reservation-based multiple access protocol for OFDMA networks with adaptive backoff strategy, *Proceedings IEEE Pacific Rim Conference on Communications, Computers and Signal Processing*, pp. 161–165, August 2007.

[16] Y. Abdel-Hamid, F. Gebali, and T. A. Gulliver, Performance analysis of two backoff strategies for the IEEE 802.16 random access channel, (submitted).

[17] F. Gebali, *Computer Communication Networks: Analysis and Design*, Northstar Digital Design, Victoria, B.C. Canada, 2004.

# 5 Enhanced Hybrid ARQ for WiMAX

*Melody Moh, Teng-Sheng Moh, and Yucheng Shih*

## CONTENTS

One major challenge in broadband wireless access networks is to provide fast, reliable services for real-time communication. This chapter describes the hybrid automatic repeat request (HARQ) scheme for the worldwide interoperability for microwave access (WiMAX). It also proposes an enhanced HARQ that would greatly reduce the latency while providing reliable transmissions. The new scheme follows the multiple-channel stop-and-wait chase combining HARQ adopted by WiMAX, yet it enables the base station (BS) to proactively react to poor channel conditions. The BS would send multiple copies of the same data burst to the subscriber station (SS) over multiple HARQ channels available to the SS, and thereby effectively reduces the total time needed for a successful transmission. Simulation results show that on comparing with the original WiMAX HARQ the new scheme significantly reduces latency while maintaining a comparable level of throughput. This work is most significant for time-critical applications that need fast, reliable services, such as VoIP, interactive communications, and media streaming.

## 5.1   INTRODUCTION

Worldwide interoperability for microwave access (WiMAX) is a wireless digital communication system that can provide an entire metropolitan area with broadband wireless services [1,2]. Furthermore, the mobile WiMAX supports not only fixed subscribers but also subscribers that are moving at vehicular speed [2], which further demonstrates the great potential of WiMAX as the next generation broadband wireless technology.

Data transmitted through the air is likely to be corrupted because of channel interferences, background noises, etc., and automatic repeat request (ARQ) and hybrid ARQ (HARQ) both have been used to ensure that packets are successfully received in sequence. Implementing orthogonal frequency division multiple access (OFDMA) in the physical layer, WiMAX uses multiple-channel stop-and-wait HARQ [1] for error control. During a noisy channel condition, however, the original WiMAX HARQ may cause a long latency for data transmission. This is undesirable especially for time-sensitive network services.

Time-sensitive network services are mostly real-time, multimedia applications that have stringent delay requirements. They include voice over Internet Protocol (VoIP) and Internet multiplayer gaming that require fast, interactive contents and live news and concerts that require real-time services. The principal motivation of the work is to design a fast, reliable HARQ scheme for WIMAX to support these time-sensitive services.

This chapter first describes the main scheme of the HARQ adopted by WiMAX [2]. Next, it presents an enhanced HARQ scheme that uses a multiple-copy approach. The proposed scheme dynamically reacts to poor channel conditions. Under noisy environments, a BS would proactively send multiple copies of the same data burst over the subsequent channels available to the SS. As a result, the new scheme greatly reduces latency of successfully transmitted data bursts.

The chapter is organized as follows. Section 5.2 describes background and related studies, including the WiMAX HARQ. Section 5.3 presents the proposed multiple-copy HARQ, including a simple example, the data structures, and detailed algorithms with flowcharts of the operations both in the BS and in the SS. The performance evaluation is illustrated in Section 5.4. Finally Section 5.5 concludes the chapter.

## 5.2   BACKGROUND AND RELATED STUDIES

This section first presents a brief overview of ARQ and HARQ technologies, including three major types of HARQ. Next, WiMAX HARQ is introduced, together with an example illustrating its BS operation; this is followed by a description of some related studies.

### 5.2.1   ARQ and HARQ

ARQ has long been used in wireless communications to ensure a reliable, in-sequence data transmission by retransmitting corrupted data [3]. There are three types of basic ARQ schemes: stop-and-wait, go-back-N, and selective repeat. The stop-and-wait scheme is the simplest to implement, yet it has the lowest throughput especially when the propagation delay is long. To take its advantage of simple implementation while avoiding the long, wasteful delays, the multiple-channel (or parallel) stop-and-wait ARQ has been proposed. One famous example is the Advanced Research Projects Agency Network (ARPAnet) that supports multiplexing of eight logical channels over a single link and runs stop-and-wait ARQ on each logical channel [4].

The performance of ARQ schemes suffers quickly because more retransmissions are required. To reduce the frequency of retransmissions, a system can adopt a forward error correction (FEC) functionality to correct errors that occur during transmissions. Combining an ARQ scheme with the FEC functionality is called HARQ.

In general, there are three types of HARQ, described below. In type I HARQ, the receiver simply discards erroneously received packets after it fails to correct errors, and sends a negative acknowledgement (NAK) back to the transmitter to request a retransmission. There is thus no need to have a buffer storing erroneous received packets. A fixed code rate is used for error correction, type I HARQ therefore cannot effectively adapt to changing channel conditions. (Note that code rate is defined as the ratio of the total number of information bits over the total bits transmitted. Thus, the higher the code rate, the lower the redundancy.) Using a code rate too high may cause too many retransmissions in high packet error rate (PER) conditions; on the other hand, using a code rate too low may cause too much redundancy in low PER conditions. The throughput may therefore be degraded by either a high frequency of retransmissions or too many redundant data in transmissions. Accordingly, choosing a suitable code rate is crucial for type I HARQ. Type I HARQ is therefore best suited for a channel that has a consistent level of signal-to-noise ratio (SNR).

In type II HARQ, in addition to packets being coded with ARQ and FEC as in type I HARQ, each receiver also keeps the erroneously received packets in the buffer in order to combine them with the retransmitted packets. There are two major FEC categories of coding for type II HARQ: chase combining (CC) [5] and incremental redundancy (IR) [6], introduced below.

In CC, the receiver combines received copies of the same packet to get diversity gain. All the redundant bits for each retransmitted packet are the same as the first transmission; therefore, it is relatively simple but less adaptive to the channel condition because the decoder may just need a smaller number of redundant bits to correct errors. The buffer needed is the number of coded symbols of one coded packet.

In IR, it adapts to changing channel conditions by retransmitting redundant bits gradually to a receiver. This is done while waiting for a positive ACK until all redundancies are sent. At the beginning, a transmitter using IR sends coded packets with a small number of FEC redundant bits or even without any. If a retransmission is needed, different redundant bits, derived from different puncturing patterns, are retransmitted depending on the base coding rate. The information data will not be retransmitted unless a receiver still cannot successfully decode the packet after a transmitter has sent all the redundant bits. This approach increases the receiver's coding gain one retransmission at a time. The IR scheme therefore allows the system to adjust the channel encoder rates to the channel quality. Comparing with CC, a bigger buffer is required to store all retransmitted data, including the first transmitted data.

Type II HARQ needs a larger buffer size than type I HARQ, but it has higher performance in terms of throughput. The drawback of type II IR HARQ is that a receiver has to receive the first transmitted data to combine it with subsequently received redundant bits. To overcome this drawback, a type III HARQ has been proposed [7]. It may be viewed as a special case of type II HARQ such that each retransmitted packet is self-decodable. A receiver can correctly decode information data by either combining the first transmitted packet with a retransmitted packet or use only one of the retransmitted packets.

## 5.2.2  WiMAX HARQ

HARQ is supported in WiMAX that uses OFDMA physical layer. As mentioned before, WiMAX uses a basic stop-and-wait HARQ protocol. Using multiple HARQ channels can compensate the propagation delay of the stop-and-wait scheme, that is, one channel transmits data while others are waiting for feedbacks. Therefore, using a small number of HARQ channels (e.g., using 6 channels), multichannel stop-and-wait HARQ is a simple, efficient protocol that simultaneously achieves high throughput and low memory requirement [2].

In this scheme, each HARQ channel is independent of each other; that is, a data burst can only be retransmitted by the HARQ channel that initially sent the data burst. Each HARQ channel is distinguished by a HARQ channel identifier (ACID – ARQ channel identifier). Data reordering in an SS is done by referring to Protocol Data Unit (PDU) sequence numbers that are enabled when the HARQ operation is used. The HARQ Downlink Map Information Element (DL MAP IE) contains the information about the DL HARQ Chase sub-burst IE. It specifies the location of HARQ sub bursts, the ACID, the HARQ Identifier Sequence Number (AI_SN), etc. By referring to MAP IE, an SS can correctly retrieve a given data burst. HARQ feedbacks are sent by the SS after a fixed delay (this is called synchronous feedbacks). To specify the start of a new transmission on each HARQ channel, one-bit HARQ AI_SN is toggled on each successful transmission [1].

When CC is used in WiMAX HARQ, the total buffer capability needed in a specific SS is determined by the maximum data size and may be transmitted by a single channel at a time, and the total number of HARQ channels provided for the SS (i.e., the product of multiplying these two values). In the IEEE 802.16 Standard, this buffer capacity is used by the BS as a rate control mechanism when sending data to an SS.

### 5.2.2.1 A Simple Example for WiMAX HARQ

In this section, the multiple-channel stop-and-wait HARQ used by WiMAX is illustrated by a simple example. As mentioned above, these HARQ channels are independent of each other; retransmissions of a data burst can only be done by its initial sending HARQ channel. Thus, a negative acknowledged (NAKed) data burst can only be resent via the initial sending channel until it is successfully received.

Figure 5.1 shows the channel activities versus time frames of a HARQ scheme, from a BS to a particular SS, assuming that four channels are available for the SS. The first row shows the consecutive time frames. The next four rows show the data bursts sent and the ACK or NAK received from/by each channel at the corresponding time frame. The last row shows the data bursts storing in the SS buffer at the corresponding time frame (waiting to be forwarded to the upper layer). It is assumed that data bursts have either been received correctly or erroneously. Erroneously received data bursts are placed in the receiver's buffer waiting for the correct copies. On the other hand, correctly received data bursts are placed in the receiver's buffer waiting to be forwarded further (either to the next hop or to the upper layer) in the right sequence, that is, the receiver is still waiting for a correct copy of some earlier data bursts.

Figure 5.1 may be explained below. Data bursts 1 through 4 are sent via channels 1 through 4, at time frames 1 through 4, respectively. A corresponding ACK or NAK will be received by the BS by the end of the 4th time frame after the BS sent out the data burst. So, at the end of 4th time frame, a NAK for data burst 1 is received, and thus the second copy of data burst 1 is sent at time frame 5, via channel 1 (the same channel it was initially sent). Similarly the second copy of data burst 2 is sent at time frame 6 via channel 2. Since an ACK for data burst 3 is received at time frame 6, a new data burst (5) is sent via channel 3 at time frame 7.

| Channel \ Time frame | 1 | 2 | 3 | 4 | 5 | 6 | 7 | 8 | 9 | 10 | 11 | 12 |
|---|---|---|---|---|---|---|---|---|---|---|---|---|
| Channel 1 | 1 | | | NAK 1 | 1 | | | ACK 1 | 7 | | | NAK 7 |
| Channel 2 | | 2 | | | NAK 2 | 2 | | | ACK 2 | 8 | | |
| Channel 3 | | | 3 | | | ACK 3 | 5 | | | NAK 5 | 5 | |
| Channel 4 | | | | 4 | | | ACK 4 | 6 | | | NAK 6 | 6 |
| Data bursts in receiver's buffer | | 1 | 1,2 | 1,2,3 | 1,2,3,4 | 2,3,4 | | 5 | 5,6 | 5,6,7 | 5,6,7,8 | 6,7,8 |

**FIGURE 5.1** A simple example of the WiMAX HARQ scheme.

The data bursts stored at the receiver's buffer may be explained as follows. At time frame 4, data bursts 1, 2, and 3 are in the receiver's buffer. Data bursts 1 and 2 have been erroneously received (and are waiting for correct copies) while data burst 3, even though has been correctly received, is also waiting so that it may be forwarded further in the correct order. Similarly, at time frame 5, data bursts 1 through 4 are in the buffer. At time frame 6, only data bursts 2 through 4 are left in the buffer since data burst 1 has been correctly forwarded. At time 7, after data burst 2 is correctly forwarded, data bursts 3 and 4 are also forwarded, so nothing is left in the buffer. The rest of the time frames may be explained similarly.

### 5.2.3  RELATED STUDIES

Using multiple-channel stop-and-wait HARQ with CC has been widely used in wireless networks, such as the High-Speed Downlink Packet Access (HSDPA), that have been reported for the 3rd Generation Partnership Project [8], in which a detailed protocol description and simulation results were included. The HSDPA uses HARQ channels to transmit data as in WiMAX, that is, each channel is independent of each other. On the other hand, the proposed multiple-copy HARQ scheme makes data transmissions more flexible and adaptive to channel conditions.

Timing of an $N$-channel stop-and-wait HARQ operation can be in fully asynchronous, partially asynchronous, and synchronous modes [9]. The transmitter can retransmit data at any time in the fully asynchronous mode. In the partially asynchronous mode, retransmissions can only be done at $i + n^*N$ frame intervals (where $i$ is the frame in which the first transmission takes place, $n$ is a positive integer, and $N$ is the feedback delay in frames). In the synchronous mode, retransmissions can take place only at every fixed time interval. The proposed multiple-copy scheme is designed to work in the synchronous mode: Each HARQ channel receives feedback and transmits data at every fixed time interval.

Multiple-channel stop-and-wait HARQ with CC can also be used with multiple input multiple output (MIMO) technology. A new such scheme has been proposed [10]. By attaching each substream a separate cyclic redundancy check (CRC) code, and employing one HARQ entity with three processes in each transmission antenna, the throughput is greatly improved. In a similar way, MIMO technology may also be applied to the proposed multiple-copy HARQ scheme. This would not only improve throughput but also reduce waiting time of erroneously received data bursts in a receiver's buffer.

## 5.3  MULTIPLE-COPY HARQ

In this section, we first give a simple example to illustrate the high-level mechanism of the proposed scheme. Next, the major data structures are presented. This is followed by a detailed description of the proposed multiple-copy HARQ.

### 5.3.1  A SIMPLE EXAMPLE FOR MULTIPLE-COPY HARQ

In the original WiMAX HARQ scheme, channels are independent of each other. Retransmissions can only be done via the channel through which the initial data burst was sent (Figure 5.1). In the proposed multiple-copy HARQ scheme, the BS is allowed to transmit multiple copies of the same data burst to an SS in the subsequent HARQ channels available to the SS. The BS decides to transmit multiple copies when it learns that the channel condition is poor via receiving NAKs. This may be illustrated by the example shown in Figure 5.2, which is similar to the example given in Figure 5.1 (Section 5.2.2.1), and may be described as follows.

Activities in time frames 1 through 6 are similar to those in Figure 5.1. By then, the BS has received two NAKs and learnt that channels are noisy. So, in time frames 7 and 8, it sends two copies of data burst 5, via channels 3 and 4. Similarly, in time frames 9 and 10 it sends two copies of data burst 6, via channels 1 and 2.

| Time frame / Channel | 1 | 2 | 3 | 4 | 5 | 6 | 7 | 8 | 9 | 10 | 11 | 12 |
|---|---|---|---|---|---|---|---|---|---|---|---|---|
| Channel 1 | 1 | | | NAK 1 | 1 | | | ACK 1 | 6 | | | NAK 6 |
| Channel 2 | | 2 | | | NAK 2 | 2 | | | ACK 2 | 6 | | |
| Channel 3 | | | 3 | | | ACK 3 | 5 | | | NAK 5 | 7 | |
| Channel 4 | | | | 4 | | | ACK 4 | 5 | | | ACK 5 | 7 |
| Data bursts in receiver's buffer | 1 | 1,2 | 1,2,3 | 1,2,3,4 | 2,3,4 | | | 5 | | 6 | | 7 |

**FIGURE 5.2** A simple example of the multiple-copy HARQ scheme.

Data bursts stored in the receiver buffer are shown on the last row, and may be explained similarly as in Example 1. Comparing Figure 5.2 versus Figure 5.1, one can make some simple observations: (1) there are fewer data bursts in the receiver buffer, which implies that (2) data bursts are being correctly forwarded more quickly (smaller latency). The latter is the main motivation of, and the major performance improvement achieved by the new scheme.

### 5.3.2 Data Structures

First, unlike the original WiMAX HARQ scheme, multiple copies of the same data burst will now be sent through different channels [11]. It is therefore crucial for an SS to correctly identify these copies. Two parameters are therefore added onto each transmitted data burst (more specifically, onto the DL HARQ Chase sub-burst IE [1]), as shown in Part (A) of Table 5.1.

Next, because an unacknowledged (unACKed) data burst may be retransmitted over different HARQ channels, the BS stores each unACKed data burst, D_x, and its associated four parameters,

---

**TABLE 5.1**

**Summary of Parameters**

**(A) Parameters associated with each data burst transmission**

| | |
|---|---|
| MC | Multiple copy ("False" only for first transmission) |
| ICN | Initial channel number (Initial sending channel number) |

**(B) Parameters kept in the BS associated with each unacknowledged data burst D_x**

| | |
|---|---|
| C_x | Initial sending channel's channel number |
| NS_x | Number of copies that have been transmitted (initially 0) |
| NR_x | Number of NAKs that have been received (initially 0) |
| S_x | Data burst status |
| | (**T** = **T**ransmitting multiple copies |
| | **F** = **F**eedback waiting—multiple copies have been sent |
| | **R** = **R**etransmission waiting—all feedbacks are NAKs) |

**(C) Parameters used in the BS for channel condition feedback**

| | |
|---|---|
| M_i | Channel condition for channel i, use ACK and NAK received to estimate the number of multiple copies needed to be sent over channel i for a successful transmission. (Initially 1) |
| M_avg | Global channel condition, an average over all M_i used by the proposed scheme to determine the number of multiple copies to be sent initially for each data burst. (Initially 1) |

---

as explained in Part (B) of Table 5.1. Note that the status S_x is set to T when the transmission of multiple copies is still needed, F when not all the feedbacks have been received (NS_x > NR_x), and R when all the feedbacks have been received, but they are all NAK (NS_x = NR_x), that is, D_x still has not been successfully transmitted.

Finally, two parameters, M_i (for channel i) and M_avg (for all channels used for the SS) are used in the BS to indicate channel conditions, which determine the number of multiple copies needed to be sent.

### 5.3.3 PROPOSED HARQ SCHEME

The two major objectives of the new scheme are (1) to accurately estimate the number of multiple copies needed and (2) to prioritize transmissions and retransmissions of data bursts so that the average time needed for a successful transmission is reduced; they are described in Sections 5.3.3.1 and 5.3.3.2. Then, the HARQ operations at the BS are presented by both an algorithm and a flowchart in Section 5.3.3.3; they are followed by a similar description for the SS part in Section 5.3.3.4.

#### 5.3.3.1 Estimating the Number of Multiple Copies via Channel Condition Feedback

The core of the proposed scheme is to adjust M_i and M_avg, which represent channel conditions, and are used to determine the number of multiple copies to be sent. Their values are adjusted as described in the following:

Whenever an ACK of the first copy of a data burst is received over channel i, M_i is set to be 1. When a NAK is received, M_i is adjusted to NR_x + 1 (number of NAKs received + 1). M_avg is the average of all the M_i values. More formally

Estimation of the Number of Multiple Copies:

1. Initially, M_i is set to be 1.
2. When a BS receives an ACK of the first copy of a data burst, D_x, sent by channel i, M_i is adjusted to 1.
3. Whenever a NAK is received for a data bust D_x with C_x = i, then NR_x is incremented, and M_i is adjusted to the new NR_x + 1.
4. M_avg is updated whenever any M_i is updated. It is calculated as the average of all M_i, adjusted to the nearest integer, as shown in Equation 5.1:

$$M\_avg = Floor\ [Avg\ (M\_1 + M\_2 + \cdots + M\_N) + 0.5] \tag{5.1}$$

M_avg determines the number of multiple copies of a data burst to be transmitted initially by the BS. Before the transmission of each new data burst D_x, the BS checks M_avg. When the number of transmission of D_x is smaller than M_avg (NS_x < M_avg), then the status remains T (i.e., S_x = T), until NS_x ≥ M_avg in which the status becomes F (i.e., S_x = F).

#### 5.3.3.2 Prioritized Transmissions and Retransmissions

When a channel is available for transmitting a data burst, to reduce the time needed for a successful transmission, the proposed scheme decides which data burst to transmit according to the following prioritized order:

Transmission and retransmission order:

1. A data burst with status T (Transmission; not all multiple copies have been transmitted).
2. A data burst with status R (Retransmission, all feedbacks have been received but all are NAK).
3. A data burst with status F (More Feedbacks are still waited upon while all multiple copies have been sent).
4. A new data burst if the SS buffer capacity has not been exceeded.

### 5.3.3.3 Scheme Description: BS Operations

The proposed multiple-copy HARQ scheme operating in the BS for a particular SS is given in Figure 5.3, we briefly explain the procedure below. For easier understanding, the corresponding flow is given in Figure 5.4.

(I). Initially for channel i := 1 to N{//N: no. of channels
   (1). Send a new data burst D_x to the SS along with
   MC = false and wait for the feedback (synchronous). // first copy of the data burst
   (2). Store D_x entry in the BS.
}//end_for of (I)
(II). For channel i := 1 to N{
   (1). If channel i receives a feedback of a data burst D_x:
     (a). If data burst D_x was initially sent by channel **i**:
       i. If receiving an ACK:
         1. If D_x is the first copy and M_i is not equal to 1:
           Set M_i = 1 //set M_i back to 1. Update M_avg according to Equation 5.1
         2. Delete D_x entry from the BS.
       ii. Else // receiving a NAK
         1. NR_x++, M_i = NR_x + 1.
         2. Update M_avg according to Equation 5.1
         3. If all feedbacks for D_x are NAK: set S_x as R //NS_x = NR_x
     (b). Else // D_x was not initially sent by channel i
       i. If receiving an ACK: Delete D_x entry from the BS
       ii. Else // receiving a NAK
         1. NR_x++, M_j = NR_x + 1. (j ≠ i)
         2. Update M_avg according to Equation 5.1
         3. If all feedbacks for D_x are NAK: set S_x as R //NS_x = NR_x
   (2). If there is a data burst D_y with status T:
     (a). Transmit one more copy of D_y to the SS along with ICN = C_y, MC = true
     (b). Update parameters associate with D_y:
       i. NS_y++
       ii. If M_avg > NS_y: set S_y = T; Else set S_y = F
     Else // no data burst with status T
   (3). If there is any data burst with status R:
     Retransmit first data burst D_y with status R
     along with ICN = C_y, MC = true; set S_y = F, NS_y++
     Else // no data burst in status T or R
   (4). If there is any data burst of status F with initial sending channel = i:
     // i.e., any D_x such that C_x = i, S_x = F
       Retransmit the first data burst with status F, say D_y,
       along with ICN = C_y, MC = true; set NS_y++
     Else // no data burst of status F with initial sending channel = i
   (5). If maximum buffer size has not reached:
     (a). Send a new data burst D_x to the SS, along with MC = false.
     (b). Add D_x's entry in the BS
     (c). Update parameters associate with D_x:
       i. If M_avg > NS_x: Set S_x as T
       ii. Else set S_x as F
}//end_for of (II)

**FIGURE 5.3**  Description of multiple-copy HARQ operation in the BS.

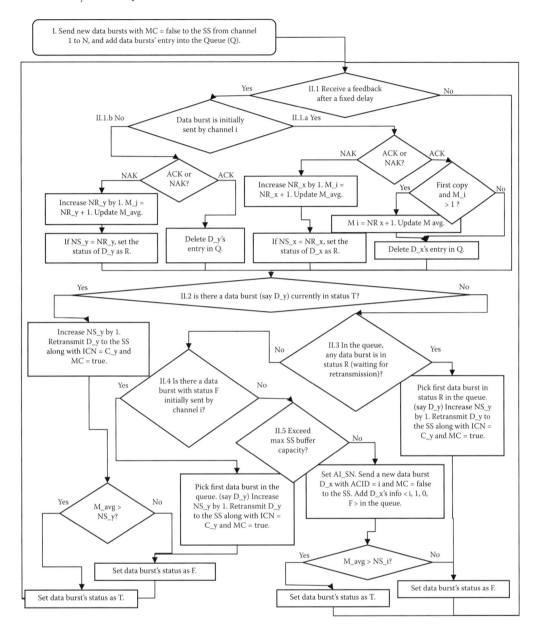

**FIGURE 5.4** Flowchart of multiple-copy HARQ operation in the BS.

Initially, in Step I, a new data burst is sent over each and every channel available to the SS, and a data burst entry (with the data burst and its associated parameters shown in Table 5.1(B)) is stored in the BS.

Step II is the main body of the scheme that contains five major steps. First, when a feedback is received (Step II.1), depending on whether it is an ACK or a NAK, relevant data structures such as $M_i$, $M\_avg$, $NR_x$, and $S_x$ are updated accordingly (refer to Table 5.1, Sections 5.3.2 and 5.3.3.1).

Next (Step II.2 to Step II.5), following the order given in Section 5.3.3.2, a data burst is transmitted. Note that each step starts by testing for the right condition; then, a data burst is sent accordingly, this is followed by updating the relevant data structures (refer to Table 5.1 and Section 5.3.3.2).

(1). While the SS receives a data burst D_x from channel i

   (1) If MC = false for the receiving data burst D_x // D_x is the first copy

      (a). If D_x has the same AI_SN as that of channel i (i.e., AI_SN_i)// old data burst

         then ignore D_x

      (b). Else // new data burst

         i. Set AI_SN for channel i the same as that of D_x

         ii. Decode D_x. If success then

            1. Send an ACK back to the BS at a pre-scheduled time (synch. feedback)

            2. If D_x is in-sequence

               then forward D_x and other in-sequence data bursts to the upper layer

            3. otherwise store D_x in the buffer // waiting for in-sequence forwarding

         iii. Else// decoding not successful

            1. Send a NAK back to the BS

            2. Store D_x for later combined decoding

   (2). Else// MC = true; second or later copies

      (a). Check the ICN (initial channel number) of D_x to match with stored copies

      (b). If D_x has been successfully decoded

         then ignore D_x

      (c). Else (D_x not successfully decoded yet)

         i. Combine D_x with the previously stored copies of D_x

         ii. If success then

            1. Send an ACK back to the BS

            2. If D_x is in-sequence

               then forward D_x and other in-sequence data bursts to the upper layer

            3. otherwise store D_x in the buffer // waiting for in-sequence forwarding

         iii. Else// decoding not successful

            1. Send a NAK back to the BS

            2. Store D_x for later combined decoding

**FIGURE 5.5**  Description of multiple-copy HARQ operation in the SS.

### 5.3.3.4  Scheme Description: SS Operations

The proposed multiple-copy HARQ scheme operating in an SS is given in Figure 5.5, we also briefly explain the procedure below. For easier understanding, the corresponding flowchart is given in Figure 5.6.

When receiving a data burst D_x, the SS first checks its parameter MC (Step I.1). If MC = false, that is, the data burst is the first copy, then (in Step I.1.a) the SS checks its AI_SN to see if it is a new transmission (i.e., if the AI_SN is the same as that of channel i). If they are the same, then D_x is an old data burst, so it is ignored (Step I.1.a). Otherwise, D_x is a new data burst (Step I.1.b), then the AI_SN of channel i is set to be that of D_x, and SS starts to decode D_x. The SS sends an ACK back to the BS if the decoding is successful (Step I.1.b.ii), otherwise it sends back a NAK and stores the erroneously received data burst in the buffer to combine with subsequent retransmissions (Step I.1.b.iii). Successfully decoded data bursts are sent to the upper layer in the right sequence.

On the other hand, if MC = true in the receiving data burst D_x (Step I.2), that is, it is a retransmission, then the initial channel number (ICN) of D_x is first checked. This is used to identify the data burst in the buffer whose AI_SN matches with this ICN (note that a match implies that they are the duplicate copies of the same data burst). If D_x has been successfully decoded, the received copy is ignored (Step I.2.a). Else, the received copy is combined with the stored copy (copies), and the usual steps as described in the previous paragraph (Steps I.2.b.ii and I.2.b.iii) apply.

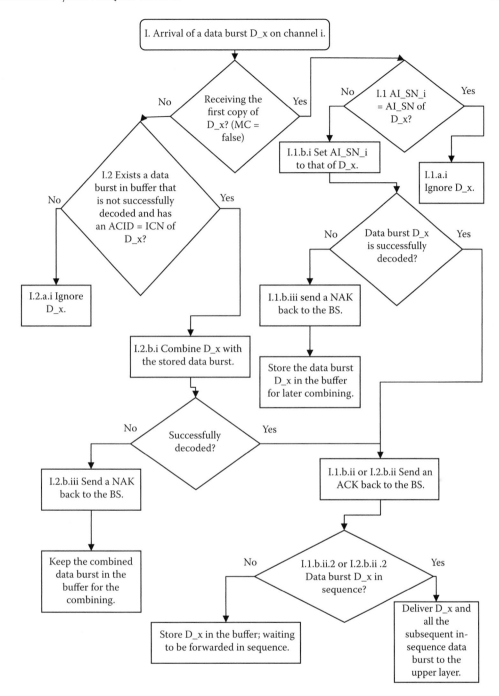

**FIGURE 5.6**   Flowchart of HARQ operation in the SS.

## 5.4   PERFORMANCE EVALUATION

This section first describes simulation settings including the modeling of noise bursts and transmissions needed in CC, and the evaluation criteria. Next, detailed simulation results are presented including the effects of buffer size, number of available channels, and rate control.

### 5.4.1 Simulation Settings

The simulation demonstrates the HARQ operation running between the BS and the SS. The uplink feedback channel is assumed to be error free, and the BS is assumed to have a full traffic load for the SS. Both the original WiMAX scheme and proposed multiple-copy HARQ scheme are simulated. All simulation parameters are summarized in Table 5.2. Note that the data burst size is determined by assuming a 64 Mbps DL data rate, a frame size of 5 msecs [1], and a total of 100 SS served by the BS.

#### 5.4.1.1 Modeling Noise Bursts and Chase Combining

Noise bursts make a channel condition unpredictable. They may prevent an adaptive modulation and coding scheme chosen by a BS from achieving a low-targeted block error rate (BLER) [12]. To simulate noise burst situations, three factors need to be considered: (1) the occurrence rate of noise bursts (N_rate), that is, the rate at which a noise burst happens, (2) the duration of a noise burst (N_dur), and (3) the BLER during noise bursts. The overall error rate can be expressed as Equation 5.2, where BLER_n and BLER_b are the BLER in normal and noisy-burst conditions, respectively.

$$\text{Overall BLER} = (1 - \text{N\_rate} * \text{N\_dur}) * \text{BLER\_n} + (\text{N\_rate} * \text{N\_dur}) * \text{BLER\_b} \quad (5.2)$$

The above three factors should be properly adjusted, so that their impacts on the performances of HARQ schemes may be observed. Four combinations of these three factors are used in the simulation, and listed in Table 5.3.

Simulation of CC HARQ is explained below. The erroneously received data bursts are stored in the SS for later combining with retransmissions. The size of the data busts are the same for both first transmission and retransmissions. In general, only one retransmission is needed in a normal channel condition [13], which is also assumed in the simulation. Furthermore, the number of required retransmissions is lower in CC than in an ordinary ARQ (and not a HARQ) [13]. Accordingly, the

**TABLE 5.2**
**Simulation Parameters**

| | | | |
|---|---|---|---|
| Transmission and feedback timing | Synchronous | Duplexing Mode | TDD (Time Division Duplex) |
| HARQ type | Chase combining | Number of channels | 4, 6, 8, 12, 16 |
| DL (downlink) data rate | 64 Mbps | Buffer size (various) | 4 or more data bursts |
| Frame size | 5 msecs | Duration of each simulation run | 100,000 frames |
| Data burst size | 1536 bytes | Each data point | Average of 1000 runs |

**TABLE 5.3**
**Noise Burst Modeling**

| Channel Condition | Noise Burst Occurrence Rate | Noise Burst Duration | BLER During Noise Bursts (percent) | Overall BLER |
|---|---|---|---|---|
| A | 1/1000 | 100 | 60 | 15/100 |
| B | 1/1000 | 100 | 90 | 18/100 |
| C | 5/1000 | 100 | 60 | 35/100 |
| D | 5/1000 | 100 | 90 | 50/100 |

**TABLE 5.4**

**Chase Combining Modeling**

| BLER | Number of Retransmissions Needed |
| --- | --- |
| 10 percent BLER | 1 time |
| 60 percent BLER | 80 percent 1 time, 20 percent 2 times |
| 90 percent BLER | 75 percent 4 times, 25 percent 3 times |

number of retransmissions required for a successful transmission in CC is assumed to be relatively small, as summarized in Table 5.4.

### 5.4.1.2 Evaluation Criteria

Three performance metrics are measured: (1) *Waiting time*, defined as the duration from the time the first copy of a data burst is received by the SS until the correct data burst is sent, in sequence, to the upper layer, (2) *throughput*, defined to be the total number of bits (of successfully received data bursts) per second, and (3) *maximum buffer occupancy*, which is defined as the maximum number of erroneous and out-of-sequence data bursts stored in the SS buffer (for all the channels) during one simulation run.

### 5.4.2 SIMULATION RESULTS

In the following, we present simulation results of the two HARQ schemes: the ORiginal WiMAX (OR) and the proposed multiple copy (MC). Four channel conditions, as shown in Table 5.3, are used.

### 5.4.2.1 Effects of Buffer Size

The waiting time of both HARQ schemes, OR and MC, using 4 channels and with a buffer size ranging from 4 to 7 (in terms of number of data bursts) is shown in Figure 5.7. Note that when the overall BLER is low (conditions A and B), both OR and MC have the similar waiting time. But, under severe noisy channel conditions (C and D), MC achieves a significantly lower waiting time (only about 50 percent of that of OR). This is so because MC proactively sends out multiple copies of the same data bursts, the SS is therefore able to quickly reconstruct the correct data. In OR,

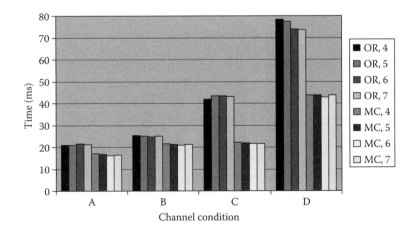

**FIGURE 5.7**   Average waiting time with various buffer sizes (OR = original WiMAX HARQ, MC = multiple-copy HARQ).

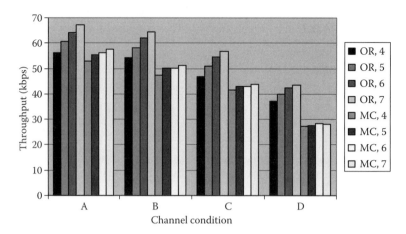

**FIGURE 5.8**   Throughput with various buffer sizes (OR = original WiMAX HARQ, MC = multiple-copy HARQ).

retransmissions may only be made on the same channel. The SS therefore needs to wait for a longer time for all the retransmissions until a correct data burst may be reconstructed. Changing buffer size, however, does not seem to have much effect for both schemes in this case when the channel number is small.

Throughput results are shown in Figure 5.8. In general, comparing with OR, MC has slightly lower throughput, and both schemes suffer significantly when the channel condition has a high overall BLER, as in condition D. Note that when the buffer size increases, OR throughput has a noticeable improvement while MC does not. As explained above, OR needs a longer waiting time (and therefore a larger buffer space). MC, on the other hand, proactively sends multiple copies and needs a smaller buffer space, as shown in Figure 5.13.

### 5.4.2.2   Effects of Available Channels (Limited Buffer Size)

When a BS has more bandwidth resource, it can assign more HARQ channels to an SS. The IEEE 802.16e standard uses four bits for HARQ channel ID [1]; thus, up to 16 channels may be supported. Figure. 5.9 and 5.10 show the comparative performance (MC/OR, or the result of MC relatively to

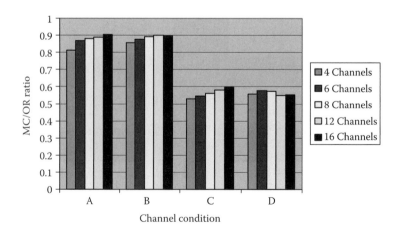

**FIGURE 5.9**   Average waiting time (MC/OR ratio) with different number of channels (buffer space = number of channels).

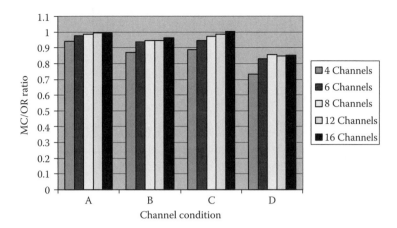

**FIGURE 5.10**  Throughput (MC/OR ratio) with different number of channels (buffer space = number of channels).

OR) when the number of available HARQ channels increases from 4 to 16. Buffer space is limited; the number of available buffer space in terms of data bursts is equal to the number of channels.

Figure 5.9 shows the comparative waiting time. It is clear that, under bad channel conditions (C and D), MC achieves a significantly lower waiting time (only about 50–60 percent of that of OR). This is consistent with those shown in Figure 5.7. There is no noticeable difference when the number of available channels increases.

When more channels are available, it allows MC to utilize them in a more flexible way; the OR scheme however cannot take advantage of big number of channels. This can be seen from Figure 5.10; increasing available channels greatly improves the relative throughput of MC; the throughput of MC/OR is almost 100 percent when using 12 or 16 channels (Conditions A, B, and C). In Condition D, using more than 4 channels also improves the throughput of MC.

### 5.4.2.3  Effects of Available Channels (Unlimited Buffer Space)

So far we have observed that using more channels improves the throughput of MC, while using more buffer space increases that of OR. In this section, we compare their waiting time and throughput when the buffer space is not limited, as shown in Figures 5.11 and 5.12. Because their waiting time

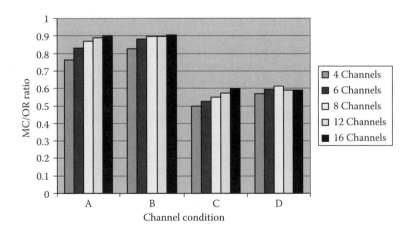

**FIGURE 5.11**  Average waiting time (MC/OR ratio) with different number of channels (buffer is unlimited).

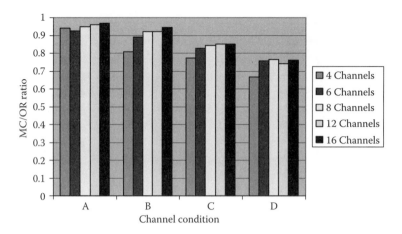

**FIGURE 5.12**   Throughput (MC/OR ratio) with different number of channels (buffer is unlimited).

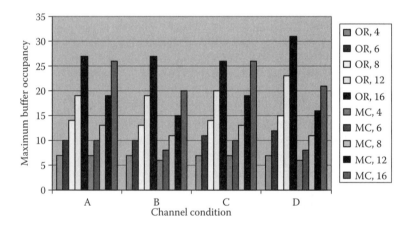

**FIGURE 5.13**   Maximum buffer occupancy comparison of both HARQ schemes.

is not significantly affected by buffer space, as can be seen from Figure 5.7, we focus on throughput here. We also measure the maximum buffer occupancy, as shown in Figure 5.13.

Comparing Figure 5.12 (unlimited buffer space) with Figure 5.10 (limited buffer space), the MC/OR relative throughput is now slightly lower, as an unlimited buffer increases the throughput of OR. Note that when the number of available channels increases, the relative MC/OR throughput improves as before, again because MC is able to utilize the increased number of channels.

### 5.4.2.3.1   Maximum Buffer Occupancy

In Figure 5.13, when the number of retransmissions needed is small (as in channel conditions A and C), both schemes have the similar maximum buffer occupancy. On the other hand, in conditions B and D, the number of retransmissions needed is high (up to 4); MC sends multiple copies continuously on subsequent channels. This allows the SS to quickly recover a data burst preventing the buffer from building up by erroneous and out-of-sequence data bursts. Its buffer occupancy is therefore significantly lower than that of OR.

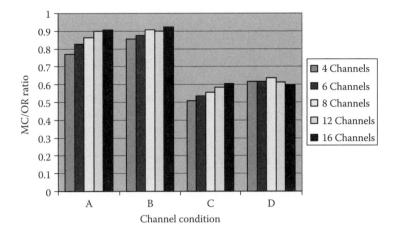

**FIGURE 5.14**   Average waiting time (MC/OR ratio) with different number of channels (buffer space = number of channels, non-IEEE approach).

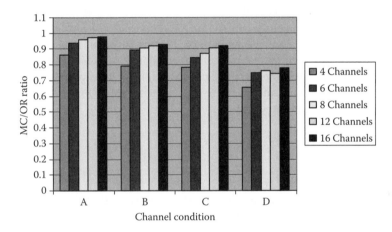

**FIGURE 5.15**   Throughput (MC/OR ratio) with different number of channels (buffer space = number of channels, non-IEEE approach).

#### 5.4.2.4   An Approach without IEEE 802.16e Rate Control

All the above experiments assumed the sending rate in the BS is controlled as recommended in the IEEE 802.16 Standard, such that the BS stops sending any data to an SS when buffer space of the SS is exhausted [1]. Here we investigate the performance in a situation where the BS keeps sending new data bursts. When the SS encounters a buffer overflow, it simply sends back a NAK. We repeat the experiments described in Section 5.4.2.2 (limited buffer, varying channel number), results are shown in Figures 5.14 and 5.15. The results in waiting time and throughput are both very similar to those shown in Section 5.4.2.2 (Figures 5.9 and 5.10). That is, comparing with OR, MC effectively reduces the waiting time, while achieves a very comparable level of throughput.

### 5.5   CONCLUSION

This chapter has presented an extensive study of the WiMAX HARQ. First, the original HARQ scheme adopted for the IEEE 802.16 WiMAX standard, the multiple-channel stop-and-wait scheme has been described. Then, an enhanced HARQ is proposed. It aims to provide fast, reliable transmissions over noisy channels. The original scheme requires all the retransmissions of a data burst to be sent over the same channel used for the original transmission, which results in a long waiting time for successful transmissions. The new scheme utilizes the multiple channels available for an SS, such that it allows the BS to continuously send multiple copies of a data burst onto the subsequent channels of the SS. This significantly reduces the total time needed for a successful transmission. The improvement is especially large when the channels are noisy, when the buffer space is limited, or when the number of available channels is larger than the number of retransmissions needed. Note that a WiMAX BS may choose to implement both the original and the enhanced HARQ schemes so that, under different circumstances, it may switch between the two schemes. This may achieve the best balance between delay and throughput, and between performance and implementation overhead. Future works may include generalizing the new scheme for the fully and partially synchronized timing modes, and extending it for the MIMO technology.

## ACKNOWLEDGMENTS

The authors appreciate many helpful discussions with Dr. Wei-Peng Chen of Fujitsu Laboratories America. Some preliminary results of this work was presented at 2nd IEEE Broadband Wireless Access Workshop, held in conjunction with *The 5th IEEE Consumer Communications and Networking Conference (CCNC)*, held in Las Vegas, Nevada, January 2008, and appeared in the conference proceedings. The authors would like to thank the anonymous reviewers of the workshop for their helpful comments. Melody Moh (email: moh@cs.sjsu.edu) is the contact author and was supported in part by Fujitsu Laboratories of America. Teng-Sheng Moh was supported in part by 2007/08 SJSU Junior Faculty Career Development Grant.

## REFERENCES

[1] IEEE, 802.16 IEEE Standard for Local and Metropolitan Area Networks, Part 16: Air Interface for Fixed and Mobile Broadband Wireless Access Systems, 2005.

[2] WiMAX Forum, *WiMAX System Evaluation Methodology*, version 1.0, 1/20/2007.

[3] P. Kosut and J. Polec, Investigation into optimum go-back-N ARQ strategy of Bruneel and Moeneclaey, *IEEE Electronics Letters*, 36(4): 381–382, 17, February 2000.

[4] L. Kleinrock and H. Opderbeck, Throughput in the ARPANET—Protocols and measurement, *IEEE Transactions on Communications*, 25(1): 95–104, January 1977.

[5] K. D. Chase, Code combining: A maximum-likelihood decoding approach for combining an arbitrary number of noisy packets, *IEEE Transactions on Communications*, 33: 593–607, May 1985.

[6] M. W. El Bahri, H. Boujernaa, and M. Siala, Performance comparison of type I, II, and III hybrid ARQ schemes over AWGN channels, *Proceedings of IEEE International Conference on Industrial Technology*, 3: 1417–1421, December 8–10, 2004.

[7] S. Kallel, Complementary punctured convolutional (CPC) codes and their applications, *IEEE Transactions on Communication*, 43(6): 2005–2009, June 1995.

[8] 3rd Generation Partnership Project, *Technical Specification Group Radio Access Network; Physical Layer Aspects of UTRA High Speed Downlink Packet Access*, V. 4.0.0, March 2001. Retrieved on August 1, 2007, from http://www.3gpp.org/ftp/Specs/html-info/25848.htm.

[9] 3rd Generation Partnership Project, TSG-RAN Working Group 1, *Text Proposal for the TR 25.848*, January 15–19, 2001. Retrieved on July 10, 2007, from http://www.3gpp.org/ftp/tsg_ran/WG1_RL1/TSGR1_18/Docs/PDFS/R1-010124.pdf.

[10] H. Zheng, A. Lozano, and M. Haleem, Multiple ARQ processes for MIMO systems, *Proceedings of the 13th IEEE International System Symposium on Personal, Indoor and Mobile Radio Communication*, 3: 1023–1026, September 15–18, 2002.

[11] Y. Shih, *The Design and Evaluation of an Enhanced WiMAX HARQ: A Multiple-Copy Approach*, Master Writing Project, Department of Computer Science, San José State University, San Jose, CA, September 2007.

[12] F. Frederiksen and T. E. Kolding, Performance and modeling of WCDMA/HSDPA transmission/ H-ARQ schemes, *Proceedings of 2002 IEEE 56th Vehicular Technology Conference*, 1: 472–476, September 24–28, 2002.

[13] WINNER, *Test Scenarios and Calibration Cases Issue 2*, IST-4-027756 WINNER II, D6.13.7 v1.00, December 31, 2006, Retrieved on July 15, 2007, from https://www.ist-winner.org/WINNER2-Deliverables/D6.13.7.pdf.

# 6 Resource Allocation in OFDM-Based WiMAX

*Mehri Mehrjoo, Mohamad Khattar Awad, and Xuemin (Sherman) Shen*

## CONTENTS

Worldwide interoperability for microwave access (WiMAX) extends the transmission rate and range of wireless communication beyond the limits of existing technologies while allowing for heterogeneous traffic transmission. To achieve all these goals, qualified protocols for WiMAX should effectively utilize the spectrum and overcome the deficits of wireless channel while maintaining a satisfactory level of heterogeneous services for users. WiMAX supports air interfaces based on orthogonal frequency division multiplexing (OFDM), which is a robust and flexible technique for transmission and resource allocation, over a wireless channel. The basic characteristics of OFDM that mitigate the wireless channel impairments are stated in this chapter. Moreover, several resource allocation schemes for subcarrier and power allocation problem in OFDM-based networks are surveyed that provides a deep insight into the problem. Besides, a resource allocation scheme for WiMAX is presented that considers the heterogeneous requirements of WiMAX users and the utilization of scarce resources simultaneously.

## 6.1 INTRODUCTION

WiMAX is expected to provide high data-rate services over a service area as large as a metropolitan area network. Broad spectrum and large coverage area usually cause severe interference and multipath transmissions, unless an appropriate design takes effect. WiMAX deploys multicarrier transmission based on OFDM and multiple access based on the orthogonal frequency division multiple access (OFDMA). OFDM mitigates noise, multipath, and interference effects, which are the main challenges of wireless communication. OFDMA is very flexible in allocating resources which are very critical for wireless networks.

The main scarce resources in wireless networks are spectrum and power. In spite of the large frequency band of the broadband networks, network designers allocate it very efficiently to serve as many satisfied users as possible while maintaining a reasonable level of revenue for service providers. Mobile and portable devices are required to consume minimal power to extend their battery lifetime. Furthermore, fixed equipment and base stations (BSs) are expected to consume as low power as possible due to health and global energy concerns. The resource constraints of the wireless medium become more critical when resource demanding applications are dealt with. According to the IEEE 802.16 standard [1], the coexistence of real-time and nonreal-time traffic in WiMAX is promising. Therefore, existing resource allocation schemes that support one type of the traffic fails and the development of new schemes that simultaneously satisfy diverse quality-of-service (QoS) demands of the heterogeneous traffic and fairly manage resources becomes necessary.

This chapter focuses on the resource allocation in OFDM-based WiMAX networks. We investigate how the flexibility and granularity of OFDM can be incorporated in a resource allocation scheme to improve network performance and resources utilization. We review the OFDM spectrum and power allocation schemes for centralized and decentralized networks, respectively. In a centralized infrastructure, a central station known as BS allocates the resources to users based on either perfect or imperfect channel states information (CSI). In a decentralized infrastructure, users compete or coordinate to capture the resources; BS does not exist or has a minimal role in the network management. Since centralized schemes have a slightly classical form, we present a general model for the resource allocation problem and some proposed solutions. The limitations of each problem are discussed and a problem formulation that conforms the diverse QoS requirements of WiMAX is described. Unlike the centralized schemes, decentralized schemes are quiet different in implementation, so we briefly review some of the proposed schemes and discuss their specifications. Some of the reviewed schemes in this chapter have not been essentially designed for WiMAX. However, reviewing them will give a deep insight of the challenges of the OFDM resource allocation in WiMAX. Besides, these can be considered in some specific applications of WiMAX, e.g., delay tolerant networks.

The remainder of this chapter is organized as follows. We present the details of the wireless channel and the OFDM transceiver in Section 6.2. First, the wireless radio channel impairments and the basics of OFDM and OFDMA are explained. Then, the medium access (MAC) sublayer of the IEEE 802.16 standard [1] is briefly described. Sections 6.3 and 6.4 are devoted to reviewing the resource allocation schemes for centralized and decentralized networks, respectively. Open research issues are stated in Section 6.5. The chapter is concluded in Section 6.6.

## 6.2 OFDM-BASED WiMAX

The channel impairments significantly degrade the performance of a broadband wireless communication network. Multicarrier transmission is selected as a promising technique for future communication due to its robustness against the frequency selectivity of broadband communications [2].

In this section, we explain the fading channel's characteristics and how OFDM can improve communication over fading channels. An overview of the OFDM and OFDMA transceiver structures along with a detailed discussion of their operations is presented. In addition, the required knowledge of PHY and MAC relevant to the resource allocation problem formulation is described such as the relation among transmission rate, power, channel gain, and bit error probability. In this chapter, MAC specification of WiMAX is based on the IEEE 802.16 standard.

### 6.2.1 RADIO CHANNEL

The wireless propagation channel constrains the information communication capacity between a transmitter and a receiver. The design of a wireless communication system's coding, modulation, signal-processing algorithms, and multiple access scheme is predicated on the channel model. The wireless channel is also generally time varying, space varying, frequency varying, and polarization varying, dependent on the particular environment and the transmitter and receiver's location. Each type of variation presents randomness and unpredictability. Nevertheless, the communication channel modeling has been well established to characterize the channel by various time and frequency metrics [3].

A channel's impulse response to $\delta(\tau)$, Dirac impulse function transmitted at the moment $\tau$, looks like a series of impulses, because of the multipath reflections, represented by a time-variant function [4]

$$h(\tau, t) = \sum_{p=0}^{Np-1} a_p(t) e^{j(2\pi f_{D,p} t + \varphi_p)} \delta(\tau - \tau_p(t)). \tag{6.1}$$

$a_p, f_{D,p}, \varphi_p$, and $\tau_p$, respectively refer to the $p$th multipath's complex-valued arrival amplitude, Doppler frequency, phase, and arrival excess delay, i.e., the delay measured with respect to the arrival of the first multipath component. $N_p$ symbolizes the number of multipaths whose amplitudes exceed the detection threshold. In practice, the number of multipath components that can be distinguished is very large. Therefore, only those multipaths that are temporally resolvable, i.e., their difference in arrival time to the receiver is greater than the inverse of the input signal bandwidth, are considered in detection.

The multipath propagation mechanisms (reflection, diffraction, and scattering) result in delay dispersion, which corresponds to frequency selectivity in the spectral domain. Each multipath power and delay is given by the power delay profile (PDP) denoted by $|h(\tau)|^2$, where $h(\tau)$ denotes the temporally stationary discrete-time channel's impulse response:

$$h(\tau) = \sum_{p=0}^{Np-1} a_p \delta \left( \tau - \tau_p \right). \tag{6.2}$$

The channel's PDP is a mathematical function whose complete characterization necessitates a long signal. However, the scalar metrics mean delay

$$\bar{\tau} = \frac{\int_0^\infty |h(t)|^2 \, t \mathrm{d}t}{\int_0^\infty |h(t)|^2 \, \mathrm{d}t} = \frac{\sum_{p=0}^{Np-1} |a_p|^2 \tau_p}{\sum_{p=0}^{Np-1} |a_p|^2} \tag{6.3}$$

and root mean square (RMS) delay spread

$$\tau_{\mathrm{RMS}} = \frac{\int_0^\infty |h(t) \, (t - \bar{\tau})|^2 \, \mathrm{d}t}{\int_0^\infty |h(t)|^2 \, \mathrm{d}t} = \frac{\sum_{p=0}^{Np-1} |a_p \left( \tau_p - \bar{\tau} \right)|^2}{\sum_{p=0}^{Np-1} |a_p|^2} \tag{6.4}$$

characterize the PDP temporal dispersion [5]. Figure 6.1a shows the graphical representations of the mean delay and RMS delay spread. When the channel delay dispersion is greater than the signal reciprocal bandwidth, i.e., the symbol duration $T_s \ll \tau_{\mathrm{RMS}}$, the transmitted train of symbols overlaps at the receiver. This phenomenon is known as inter-symbol interference (ISI) [6] which is illustrated in Figure 6.1a.

Whereas the preceding metrics are in the relative-delay domain, channel fading may also be characterized in the spectral domain. Specifically, the coherence bandwidth, $B_c$, offers an alternative metric to the RMS delay spread, $\tau_{\mathrm{RMS}}$, to measure the channel's delay dispersion. The channel impulse response's Fourier transform gives the time-variant channel transfer function [4]:

$$H(f, t) = \sum_{p=0}^{Np-1} a_p(t) e^{j(2\pi(f_{D,p} t - f \tau_p(t)) + \varphi_p)}. \tag{6.5}$$

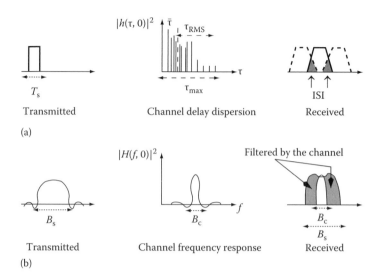

**FIGURE 6.1** Wireless channel effect: (a) delay dispersion (b) frequency selectivity.

We define the channel frequency response's autocorrelation function as [7]

$$R(\Delta f) = E\{H(f,0)H^*(f - \Delta f, 0)\} \tag{6.6}$$

where $(\cdot)^*$ denotes the complex conjugate. The coherence bandwidth $B_c$ measures the spectral width of $|R(\Delta f)|$ over which the channel is considered frequency flat. Note that the frequency selectivity is relative to the transmitted signal bandwidth. In particular, if the channel's $B_c$ is less than the transmitted signal bandwidth, the channel distorts the received signal at selected frequencies as shown in Figure 6.1b. On the other hand, the channel does not affect the received signal, if its $B_c$ is greater than the transmitted signal bandwidth.

Independently from delay dispersion and frequency selectivity, user mobility causes frequency dispersion which in turn results in channel time selectivity. The time correlation function [7,8]

$$R(\Delta t) = E\{H(0,t)H^*(0, t - \Delta t)\} \tag{6.7}$$

quantifies the time-varying nature of the channel. From $R(\Delta t)$, the channel coherence time $T_c$ can be obtained, and it is defined as the time duration over which the channel is essentially flat [4]. $R(\Delta t)$ Fourier transform is the channel Doppler power spectrum that its correlation width is the Doppler spread $B_D$. If the channel impulse response changes rapidly within the symbol duration, i.e., $T_c < T_s$, the transmitted signal undergoes fast fading which leads to signal distortion [6], or the Doppler spread $B_D$ is greater than the transmitted signal bandwidth. This effect induces frequency offset and possibly inter-carrier interference (ICI) in dense spectrums. In summary,

- Delay dispersion results in frequency selective fading that alters the received signal waveform and hence causes performance degradation. The channel effect can be avoided by transforming the broadband signal into parallel narrowband signals with bandwidth smaller than the channel's $B_c$.
- Frequency dispersion smears the signal spectrum in the frequency domain. It causes time selectivity that varies the signal at a rate higher than the rate at which the channel can be accurately estimated.

## 6.2.2 AN OVERVIEW OF OFDM

Frequency division multiplexing (FDM) appeared in 1950s [2]; however, its implementation required multiple analog radio frequency (RF) modules in each transceiver that made FDM impractical [8]. Recently, the implementation of inverse fast Fourier transform/fast Fourier transform (IFFT/FFT) and FDM ability in mitigating the channels ISI brought FDM back under light. While FDM major advantage is eliminating the ISI effect, it does not eliminate the ICI that rises due to closely packed multicarriers. Alternatively, data symbols can be modulated on orthogonal multiple carriers to reduce ICI, which is termed OFDM [9].

The high data-rate stream is partitioned into $B$ data blocks of $N_{sc}$ length at the transmitter. The symbols are serial-to-parallel converted which increases the source symbol duration $T_s$ to

$$T_s' = N_{sc}T_s. \tag{6.8}$$

As the symbol duration increases, the ISI effect significantly decreases (Section 6.2.1). Thus, the need for an equalizer at the receiver is eliminated, which reduces the complexity of the receiver. After serial-to-parallel conversion, block $b$ represents the OFDM symbol which consists of $N_{sc}$ complex data symbols denoted by $\{s_n, b\}$, $n = 0, \ldots, N_{sc} - 1$, $b = 1, \ldots, B$. For simplicity, the $b$th index is dropped and we only refer to the $s_n$, $n = 0, \ldots, N_{sc} - 1$ sequence in each block, i.e., the OFDM symbol, unless it is necessary. Figure 6.2 depicts a typical multiuser OFDM/OFDMA transmitter and receiver block diagram. OFDM is implemented by applying inverse discrete Fourier transform (IDFT) to the data sequence $s_n$ which gives the following samples $x_v$, $v = 0, \ldots, N_{sc} - 1$, of each OFDM symbol [4]:

$$x_v = \frac{1}{N_{sc}} \sum_{n=0}^{N_{sc}-1} s_n e^{j\frac{2\pi nv}{N_{sc}}} \quad v = 0, \ldots, N_{sc} - 1. \tag{6.9}$$

Whereas serial-to-parallel conversion only reduces the ISI effect, cyclic extension of the symbol by inserting a guard interval $T_g$ that is longer than the maximum channel dispersion time $\tau_{max}$ eliminates the residual ISI effect [4]. The guard interval is a copy of $T_s'$ last $L_g = \left\lceil \frac{\tau_{max}N_{sc}}{T_s'} \right\rceil$ samples (practically, the symbol source is continuous and guard insertion is archived by adjusting the starting phase and making the symbol period longer [10]). After cyclic extension of the OFDM symbol, the time domain sampled sequence becomes

$$x_v' = \frac{1}{N_{sc}} \sum_{n=0}^{N_{sc}-1} S_n e^{j\frac{2\pi nv}{N_{sc}}} \quad v = -L_g, \ldots, N_{sc} - 1. \tag{6.10}$$

Then, the sequence $x'$ is passed through digital-to-analog converter, and its output is transmitted through the wireless channel. Figure 6.3 depicts the time and frequency representation of an OFDM frame.

By implementing the OFDM multicarrier modulation, the continuous channel transfer function (Equation 6.5) is sampled in time at the OFDM symbol rate $1/T_s''$ and in frequency at spacing $F_s$. The discrete channel transfer function adapted to multicarrier signals is given by [4]

$$H_{n,i} = H(nF_s, iT_s'')$$

$$= \sum_{p=0}^{N_p-1} a_p(t) e^{j(2\pi(f_{D,p}iT_s'' - nF_s\tau_p(t)) + \varphi_p)}$$

$$= a_{n,i} e^{j\varphi_{n,i}}. \tag{6.11}$$

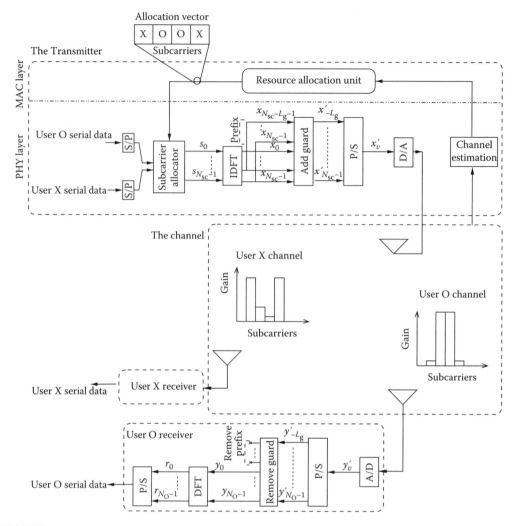

**FIGURE 6.2**   OFDMA transmitter and receiver's PHY and MAC structure.

A transmitted symbol on subcarrier $j$ of the OFDM symbol $b$ is multiplied by the resulting fading amplitude $a_{n,b}$ and rotated by random phase $\varphi_{n,i}$. The subcarrier gains can be represented by the following $N_{sc} \times N_{sc}$ channel matrix for the OFDM symbol $i$

$$\mathbf{H} = \left[ H_{0,0} \; H_{1,1} \; \ldots \; H_{N_{sc}-1,N_{sc}-1} \right] \times \mathbf{I}, \tag{6.12}$$

where $\mathbf{I}$ is the identity matrix. The OFDM symbol index $i$ has been dropped for simplicity. Let $h_v$ be the sampled $L$ sequence of the channel impulse response $h(\tau, t)$ given in Equation 6.1 at a particular time instant $t$, i.e., $h_v = h(lT_s'', bT_s'')$, $l = 0, \ldots, L = (\tau_{\max}/T_s'')$, and represented by the vector $\mathbf{h}$. Then, the matrix $\mathbf{H}$ diagonal elements are the discrete Fourier transform (DFT) of channel discrete impulse response.

After analog-to-digital conversion at the receiver, the received sampled sequence $y_v'$, $v = -L_g, \ldots, N_{sc} - 1$ contains ISI in the first $L_g$ samples that are discarded. The remaining sequence $0, \ldots, N_{sc} - 1$ is demodulated by the DFT. The DFT demodulated multicarrier sequence $r_n$, $n = 0, \ldots, N_{sc} - 1$, consists of $N_{sc}$ complex valued symbols [4]:

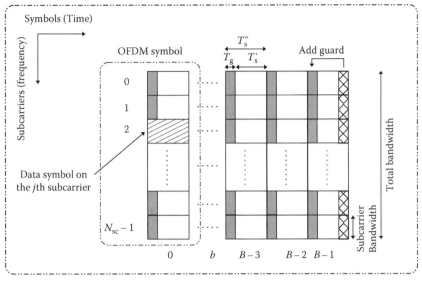

OFDM frame

**FIGURE 6.3**  Time and frequency representation of an OFDM frame.

$$r_n = \sum_{n=0}^{N_{sc}-1} y_v e^{-j\frac{2\pi n v}{N_{sc}}}, \quad n = -0, \ldots, N_{sc} - 1. \quad (6.13)$$

Since $x_v$ and $y_v$ are the sampled sequences of the transmitted and received signals, respectively, the vector representation of the received data symbols is given by

$$\mathbf{r} = \begin{bmatrix} r_0 \ r_1 \ldots r_{N_{sc}-1} \end{bmatrix}^{\mathrm{T}}$$
$$= \mathbf{W}^{\mathrm{H}}\mathbf{y}. \quad (6.14)$$

The operator $(\cdot)^{\mathrm{H}}$ denotes the matrix Hermitian, $\mathbf{y}$ is defined as $\begin{bmatrix} y_0 \ y_1 \ldots y_{N_{sc}-1} \end{bmatrix}^{\mathrm{T}}$, and $\mathbf{W}$ is the $N_{sc} \times N_{sc}$ IDFT matrix. The received signal in the frequency domain cannot be represented by the multiplication of the transmitted and the channel frequency domain representation because $y_v$ is related to $x_v$ and $h_v$ by linear convolution rather than circular convolution [8]. However, the cyclic convolution is created for OFDM by appending the $L_g$ samples $(x_{N_{sc}-L_g-1}, x_{N_{sc}-L_g}, \ldots, x_{N_{sc}-1})$ to the sequence $x_v$. This circular convolution of the two periodic sequences is transformed into the product of their DFTs [8]:

$$r_n = H_n s_n \quad n = 0, \ldots, N_{sc} - 1 \quad (6.15)$$

which can be alternatively written as

$$\mathbf{r} = \mathbf{H}\mathbf{s}. \quad (6.16)$$

$\mathbf{r}$ and $\mathbf{s}$, respectively, are the received symbols and transmitted symbols matrix representation. Ignoring the additive noise effect and substituting Equation 6.14 in the left-hand-side of Equation 6.16 result in

$$\mathbf{W}^{\mathrm{H}}\mathbf{y}^{\mathrm{T}} = \mathbf{H}\mathbf{s}$$
$$\mathbf{H}^{-1}\mathbf{W}^{\mathrm{H}}\mathbf{y}^{\mathrm{T}} = \mathbf{s}, \quad (6.17)$$

where $\mathbf{H}^{-1}$ is the matrix inverse (the inverse of a diagonal matrix is a matrix with diagonal elements $\frac{1}{H_{n,n}}$ $n = 0, \ldots, N_{sc} - 1$). Therefore, based on the availability of the channel estimation matrix $\mathbf{H}$ and by the implementation of DFT, the transmitted symbols can simply be recovered [11].

### 6.2.3 TRANSMISSION RATE

Multicarrier modulation transforms a wide band channel experiencing selective fading onto multiple bands that experience flat fading. The flat fading channel is assumed to be static over the OFDM symbol duration. In addition, a perfect CSI is assumed to be available at the transmitter. Under these assumptions, the normalized transmission rate (bits/seconds/hertz) on the $j$th subcarrier is given by [12]

$$r_j = \log_2\left(1 + p_j \frac{|H_j|^2}{N_0}\right),$$  (6.18)

where $p_j$, $|H_j|^2$, and $N_0$, respectively are, the allocated power, the channel gain, and the additive white Gaussian noise (AWGN) variance. The Shannon capacity in Equation 6.18 is an upper bound that asymptotically approaches the transmission rate over wireless channels. Nevertheless, this upper bound is hard to achieve in practice especially in the network under consideration where adaptive modulation and coding (AMC) is adopted. Particularly, WiMAX [1] supports the adaptive constellation size change of M-OAM and M-PSK following the channel gains changes. From a resource allocation perspective, given the required $P_b$ and channel gains, the allocated power $p_j$ and transmission rate $r_j = \log_2 M$, where $M$ denotes the modulation level, can be adapted. This adaptation is performed by inverting the modulation schemes' $P_b$ approximation functions. The exact approximations of the M-QAM and M-PSK $P_b$, respectively are, given by [13]

$$P_b \approx \frac{4}{\log_2 M} Q\left(\sqrt{\frac{3p_j \frac{|H_j|^2}{N_0} \log_2 M}{M - 1}}\right)$$  (6.19)

$$P_b \approx \frac{2}{\log_2 M} Q\left(\sqrt{2p_j \frac{|H_j|^2}{N_0} \log_2 M} \sin\left(\frac{\pi}{M}\right)\right).$$  (6.20)

In Refs. [14–16] Equations 6.19 and 6.20 are inverted to obtain the constellation size and power adaptation for a specific $P_b$. However, the $Q(\cdot)$ function cannot be easily inverted in practice, because numerical inversions are necessary [13]. Alternatively, the exact approximation can be written in a form that is easy to invert [17–20]. Because both modulation schemes are special cases of the M-ary modulation techniques [21], Equations 6.19 and 6.20 can be written as

$$P_b \approx c_1 \exp\left[\frac{-c_2 p_j \frac{|H_j|^2}{N_0}}{2^{c_3 r_j} - c_4}\right],$$  (6.21)

where $c_1 = 0.2$, $c_2 = 1.5$, $c_3 = 1$, and $c_4 = 1$ for M-QAM and $c_1 = 0.05$, $c_2 = 6$, $c_3 = 1.9$, and $c_4 = 1$ for M-QPSK [13]. The constants for different bounds can be found in [22]. By assuming "=" instead of "≈" in Equation 6.21 and solving for $M$, we obtain

$$M = \sqrt[c_3]{\left(\frac{c_2}{-\ln\left(\frac{P_b}{c_1}\right)} P_j \frac{|H_j|^2}{N_0} + c_4\right)}$$  (6.22)

The adaptive modulation transmission rate as a function of $P_b$ can be obtained by substituting Equation 6.22 in $r_j = \log_2 M$:

$$r_j = \frac{1}{c_3} \log_2 \left( c_4 + \frac{c_2}{-\ln\left(\frac{P_b}{c_1}\right)} P_j \frac{|H_j|^2}{N_0} \right).$$                (6.23)

Note that the transmission rates Equations 6.23 and 6.18 are similar. Thus, a resource allocation scheme that maximizes one of them maximizes the other [23]. This result broadens the applicability of resource allocation schemes to networks that adopt different modulation schemes.

### 6.2.4 MAC SUBLAYER

Resource allocation is one of the major tasks of MAC because the medium access mechanisms of MAC directly affect the spectrum and power utilization. Accordingly, the MAC sublayer specifications in the IEEE 802.16 standard, relevant to our discussion in the next section, are introduced in the following.

Despite the advantages of OFDM in mitigating the channels's impairments as mentioned before, underutilization of transmitter power and network subcarriers is its disadvantage. When an OFDM transmitter accesses the channel in a time division manner, e.g., time division multiple access (TDMA), the transmitter is forced to transmit on all available subcarriers $N_{sc}$, although it may require a less number of subcarriers to satisfy its transmission rate requirement. Consequently, the transmitter power consumption increases as the number of subcarriers increases. This disadvantage motivates the development of a PHY technology where transmitters are multiplexed in time and frequency, i.e., OFDMA. In such a technology, the users are exclusively assigned a subset of the network available subcarriers in each time slot [8,24]. The number of both time slots and subcarriers can be dynamically assigned to each user; this is referred to as dynamic subcarrier assignment which introduces multiuser diversity. The multiuser diversity gain arises from the fact that the utilization of given resources varies from one user to another. A subcarrier may be in deep fading for one user (e.g., the second subcarrier for user X in Figure 6.2) while it is not for another user (e.g., the same subcarrier for user O). Allocating this particular subcarrier to the user with higher channel gain permits higher transmission rate. To achieve multiuser diversity gain, a scheduler at the MAC sublayer is required to schedule users in appropriate frequency and time slots.

Point-to-multipoint (PMP) as well as mesh topologies are supported in the IEEE 802.16 standard. PMP mode operations are centrally controlled by the BS, but mesh mode can be either centralized or decentralized, i.e., distributed. In a centralized mesh, a mesh BS (a node that is directly connected to the backbone) coordinates communications among the nodes. Decentralized mesh is similar to multihop adhoc networks in the sense that the nodes should coordinate among themselves to avoid collision or reduce the transmission interference.

In PMP mode, the uplink (UL) channel, transmissions from users to BS, is shared by all users, i.e., UL is a multiple access channel. On the other hand, downlink (DL) channel, transmissions from BS to subscriber stations (SSs), is a broadcast channel. The duplexing methods of UL and DL include time division duplexing (TDD), frequency division duplexing (FDD), and half-duplex FDD (HFDD). Unlike PMP mode, there is no clear UL and DL channel defined for mesh mode.

An outstanding feature of WiMAX is heterogeneous traffic support over wireless channels. IEEE 802.16 provides service for four traffic types known as service flows. The mechanism of bandwidth assignment to each SS depends on the QoS requirements of its service flows. The service flows and their corresponding bandwidth request mechanisms are as follows:

- Unsolicited grant service (UGS): This service supports constant bit rate traffic. Bandwidth is granted to this service periodically or in case of traffic presence by the BS.
- Real-time polling service (rtPS): This service has been provided for real-time service flows with variable-size data packets issued periodically. rtPS flows can send their bandwidth request to the BS after being polled.
- Nonreal-time polling service (nrtPS): This service is for nonreal-time traffic with variable-size data packets. nrtPS can gain access to the channel using monocast or multicast polling mechanisms. Upon receiving a multicast polling, the nrtPS service can take part in a contention in the bandwidth contention range.
- Best effort (BE) service: This service provides the minimum required QoS for nonreal-time traffic. The channel access mechanism of this service is based on contention.

## 6.3  CENTRALIZED SUBCARRIER AND POWER ALLOCATION SCHEMES

In a network with a centralized resource allocation scheme, the BS allocates OFDM subcarriers and power to the users in both UL and DL based on the perfect knowledge of CSI. The users estimate the CSI and report it to the BS at the beginning of each MAC frame interval. It is assumed that the estimation error is negligible and the CSI remains constant during the next frame duration [25]. The BS assigns the resources, subcarrier, and power based on the CSI and broadcast the allocation vector on a signaling channel at the beginning of the next MAC frame transmission. The main difference between UL and DL resource allocation is the power limitation. In DL, the maximum allocated power is limited by the BS power, $P_{BS}$. However, in UL allocated power to the subcarriers of each user is limited by the user's device transmission power which is assumed to be totally devoted to transmission unless it causes an unacceptable interference among neighbors. In this section, first, we present a general resource allocation model for DL. Then, we discuss how the model changes for a different objective of the resource allocation problem. Finally, we present the subcarrier and power allocation model for UL.

### 6.3.1  PROBLEM FORMULATION

DL resource allocation is usually modeled as an optimization problem whose objective function and constraints are determined based on the users' requirements and network specifications. Depending on the definition of the objective functions, different utilization performance are expected. Resource allocation algorithms are available in the literature focus on two general objectives; either data rate maximization or power minimization subject to constraints based on the network model. Using a general objective function of rate, $F(r)$, we present a model for the subcarrier and power allocation optimization problem constrained by the BS maximum power. The problem formulation of power minimization in DL are not discussed here. Interested users are referred to Refs. [16,26].

Mixed integer nonlinear programming (MINLP) model is appropriate where a discrete network structure and continuous parameters are simultaneously formulated [27]. In DL, the subcarrier and power allocation problem is usually addressed as an MINLP optimization problem. The feasible region of the MINLP model contains integer variables representing subcarriers allocated to the users and continuous variables representing the power allocated to the subcarriers. The network parameters used in the optimization model are given in Table 6.1.

To show the subcarrier assignment to user $i$, a $K \times 1$-vector $c_i$ of binary variables, called the subcarrier allocation vector of user $i$th, is defined with elements as follows:

$$c_{ij} = \begin{cases} 1 & \text{if } j\text{th subcarrier is assigned to } i\text{th user} \\ 0 & \text{otherwise.} \end{cases} \tag{6.24}$$

Each user can be allocated several subcarriers, but each subcarrier is exclusively allocated to one user. A subcarrier may not be assigned to any user due to its severe channel gain. This constraint is mathematically shown by

**TABLE 6.1**

**Notions Description**

| Notion | Description |
|---|---|
| $U$ | Total number of users in the network |
| $K$ | Total number of the OFDM subcarriers in the network |
| $\mathcal{K} := \{1, 2, ..., K\}$ | The set of subcarriers |
| $\mathcal{U} := \{1, 2, ..., U\}$ | The set of users |
| $i$ | $i$th user |
| $j$ | $j$th subcarrier |
| $|H_{ij}|^2$ | The channel gain of the $i$th user on the $j$th subcarrier |
| $N_0$ | AWGN noise variance |
| $p_{ij}$ | Allocated power to the $i$th user on the $j$th subcarrier |
| $r_{ij}$ | Allocated rate to the $i$th user on the $j$th subcarrier |
| $P_{BS}$ | BS total power budget |
| $R^i_{min}$ | Minimum service rate requirement of the $i$th user |

$$\sum_{i=1}^{U} c_{ij} \leq 1 \quad \forall j \in \mathcal{K}. \tag{6.25}$$

If the $j$th subcarrier is not assigned to the $i$th user, the allocated power to the $i$th user on the $j$th subcarrier must be zero. Therefore, for every user $i = 1, \ldots, U$ and every subcarrier $j = 1, \ldots, K$, we must have

$$\text{If } c_{ij} = 0 \text{ then } p_{ij} = 0. \tag{6.26}$$

We include this restriction in the model through the following equation:

$$p_{ij} \leq P_{BS} c_{ij} \quad \forall i \in \mathcal{U}, \ \forall j \in \mathcal{K}. \tag{6.27}$$

Note that, if $c_{ij} = 0$, the Equation 6.27 implies $p_{ij} \leq 0$ that along with the non-negativity constraint $p_{ij} \geq 0$ yields $p_{ij} = 0$ and satisfies the assumption Equation 6.26. When $c_{ij} = 1$, Equation 6.27 is reduced to the redundant constraint $p_{ij} \leq P_{BS}$, because we have the following equation to assure that the total allocated power to all subcarriers in each time slot is limited by $P_{BS}$:

$$\sum_{i=1}^{U} \sum_{j=1}^{K} c_{ij} p_{ij} \leq P_{BS}. \tag{6.28}$$

In the presence of the set of Equation 6.27, that guarantees the restriction Equation 6.26, the variables $c_{ij}$ can be removed from the Equation 6.28 as follows:

$$\sum_{i=1}^{U} \sum_{j=1}^{K} p_{ij} \leq P_{BS}.$$

The transmission rate to each user depends on the number and index of allocated subcarriers to the user and allocated power to each subcarrier. If continuous rate adaptation is assumed, the approximate rate of the $i$th user, $r_i$, can be obtained by either Equation 6.18 or Equation 6.23 as follows:

$$r_i = \sum_{j=1}^{K} r_j \quad \text{bits/s/Hz}. \tag{6.29}$$

The heterogeneous traffic in the network inquires different QoS. The minimum service rate requirement of the $i$th user, $R^i_{\min}$, is guaranteed through the following equation:

$$r_i \geq R^i_{\min} \quad \forall i \in \mathcal{U}.$$

Respecting the equation listed above and the chosen objective function, $F_i(r_i)$, for user $i$, the resource allocation optimization problem can be modeled as follows:

$$(P_1) : \max_{c_{ij}, p_{ij}} \sum_{i=1}^{U} F_i(r_i) \tag{6.30}$$

$$\text{s.t.} \quad r_i = \sum_{j=1}^{K} r_{ij} \qquad \forall i \in \mathcal{U}, \tag{6.31}$$

$$r_i \geq R^i_{\min} \qquad \forall i \in \mathcal{U}, \tag{6.32}$$

$$\sum_{i=1}^{U} \sum_{j=1}^{K} p_{ij} \leq P_{\text{BS}}, \tag{6.33}$$

$$\sum_{i=1}^{U} c_{ij} \leq 1 \qquad \forall j \in \mathcal{K}, \tag{6.34}$$

$$0 \leq p_{ij} \leq P_{\text{BS}} c_{ij} \qquad \forall i \in \mathcal{U}, \forall j \in \mathcal{K}, \tag{6.35}$$

$$c_{ij} \in \{0, 1\} \qquad \forall i \in \mathcal{U}, \forall j \in \mathcal{K}. \tag{6.36}$$

Problems $(P_1)$ is an MINLP problem. Solving MINLP problems can be very challenging, due to the combination of mixed integer and nonlinear program difficulties in MINLP problems [27]. It is proved in Ref. [28] that the integer variables in problem $(P_1)$ are redundant and can be eliminated. Accordingly, a nonlinear programming (NLP) model is proposed that unifies the subcarrier and power allocation in a rate allocation problem. Equation 6.34, defined below, is replaced by Equation 6.37 for this purpose. For every $\hat{i} \in \mathcal{U}$ and for every $j \in \mathcal{K}$:

$$r_{ij} \cdot r_{\hat{i}j} = 0 \qquad \forall j \in \mathcal{K}, \forall i \in \mathcal{U}, \forall \hat{i} \in \mathcal{U}, i \neq \hat{i} \tag{6.37}$$

It is proved that Equation 6.37, the same as Equation 6.34, guarantees exclusive rate allocation to the $i$th user on the $j$th subcarrier and the optimal value of problem $(P_1)$ equals the optimal value of problem $(P_2)$ stated as follows [28].

$$(P_2) : \quad \max_{r_{ij}} \sum_{i=1}^{U} F_i(r_i)$$

$$\text{s.t.} \quad r_i = \sum_{j=1}^{K} r_{ij} \qquad \forall i \in \mathcal{U},$$

$$r_i \geq R^i_{\min} \qquad \forall i \in \mathcal{U}, \tag{6.38}$$

$$\sum_{i=1}^{U} \sum_{j=1}^{K} \frac{1}{r_{ij}} (2^{r_{ij}} - 1) \leq P_{\text{BS}},$$

$$r_{ij} r_{\hat{i}j} = 0 \qquad \forall i \in \mathcal{U} \setminus \{\hat{i}\} \, \forall j \in \mathcal{K},$$

$$0 \leq r_{ij} \qquad \forall i \in \mathcal{U}, \forall j \in \mathcal{K}.$$

The new model will be easier to deal with, both by optimal/suboptimal and heuristic approaches. As the objective function is continuous over the range of allocated rates and the feasible region is closed and bounded, the extreme value theorem (Weierstrass theorem) implies that problem $(P_2)$ has global optimal solution(s). Extreme value theorem: Let $f$ be a continuous real-valued function whose domain, $D_f$, is bounded and closed. Then, there exist $x_1$ and $x_2$ in $D_f$ such that

$$f(x_1) \leq f(x) \leq f(x_2) \quad \forall x \in D_f.$$

In Section 6.3.2, we review several optimal/suboptimal and heuristic solutions proposed in the literature for simple forms of problem $(P_1)$ that relax some constraints and consider linear/concave objective functions. However, none of the literature available schemes includes the total requirement of heterogeneous traffic of WiMAX simultaneously. We justify a utility-based resource allocation, based on $(P_2)$, for WiMAX, discuss the challenges, and apply genetic algorithm to solve the problem and to show the effectiveness of the model.

## 6.3.2 RELATED WORKS

In the following, similar models to $(P_1)$, each with its own definition of objective function and a variation of constraints, are reviewed. We discuss each model in terms of the proposed solution, and practical advantages and disadvantages.

Bit rate maximization problems, maximizing total users' data rate for a given power budget, is the most common objective function deployed in Refs. [20,29–37]. References [29,30] present a framework for the joint subcarrier and power allocation with power constraint modeled as an MINLP problem. Reference [31] formulates the problem by allowing a subcarrier to be shared by multiple users. The optimization problem is decoupled into two subproblems, subcarrier allocation to users and power allocation to subcarriers. A set of users whose transmission on a specific subcarrier maximizes achievable data rate of that subcarrier are determined in subcarrier allocation step. Then, the subcarriers allocated power is determined to maximize the overall data rate. A suboptimal solution in Ref. [32] allocates uniform power to subcarriers. Given the channel gain and the fixed power allocation $(P_{BS}/K)$, the subcarriers rates $(r_{ij})$ are known. The problem is converted into a linear integer programming (LIP) problem with integer variables $c_{ij}$. Then, a reduced computational complexity algorithm is deployed to solve LIP by, first, allocating the subcarriers to maximize the total users' data rates, irrespective of the users' minimum required data rate constraint, and, second, adjusting subcarrier allocation to satisfy the constraint of the users' minimum required rate. A more general form of the bit rate maximization problems are weighted rate maximization problems with the objective function $\max \sum^U \sum^K \omega_i r_{ij}$. A geometric programming (GP), a special form of convex optimization, has been proposed in Ref. [38] for weighted rate maximization or weighted power minimization. There exist several algorithms to solve GP efficiently and optimally. However, the challenge is to convert or approximate the objective and constraints to be recognized as problems compatible with GP [39].

Unfair rate allocation among users is the major drawback of bit rate maximization problems. The resource allocation is in favor of the users with good channel status. To resolve this issue, Ref. [33] formulates the problem to balance between capacity and fairness. A set of nonlinear constraints are added into the optimization problem to assure proportional users' data rates. The primal solution of the constrained fairness problem is computationally complex to obtain, so a low-complexity suboptimal algorithm that separates subchannel and power allocation is proposed. The decoupled allocation algorithm, first allocates subcarriers assuming uniform power allocation. Then, an optimal power allocation algorithm maximizes the sum capacity while maintaining proportional fairness. Max-min fairness solution is addressed in Ref. [34] by maximizing the minimum users' data rates, i.e., $\max \min r_i$. A convex feasible region is obtained for the problem by relaxing the constraint of exclusively allocating one subcarrier to only one user. Assuming equal amount of power is allocated to each subcarrier, Ref. [34] proposes an algorithm to assign subcarriers to the users. Ref. [35] addresses the unfairness issue of OFDMA transmissions and combines the resource allocation problem with a fair scheduling scheme (generalized processor sharing) to compensate for unfairness of bit rate maximization scheme. Also, the principle of generalized processor sharing is deployed as a constraint of the optimization problem in Ref. [20] to allocate the subcarriers fairly among the users.

The definition of the objective function $\sum_{j=1}^{K} c_{ij} r_{ij}$ (bit rate maximization) does not reflect the heterogeneous traffic specification and application requirements. An appropriate form of the objective function in networks with heterogeneous traffic is to maximize aggregate utility functions of all users in the network. A utility function characterizes a user's satisfaction of an application level QoS requirement. A utility function of rate defined as $U(r) = r$ represents that the user's satisfaction increases linearly by allocating more rate to the user, or a step utility function of rate represents that the user expects a threshold rate, allocating less rate is not useful at all, and allocating more rate is wasteful.

Assuming concave or linear utility functions, Ref. [40,41] investigate the utility-based resource allocations in OFDMA networks for both discrete and continuous adaptive rate. The optimization problem is decomposed into two problems: dynamic subcarrier assignment (DSA) and adaptive power allocation (APA). The DSA problem is modeled as a uniform power allocation problem, and the APA problem is modeled as a fixed subcarrier assignment. Different approaches for solving DSA, APA, and joint DSA/APA problems are proposed. DSA is relaxed to a nonlinear integer (binary) problem. A sorting search algorithm is proposed for subcarrier assignment. When all utility functions are linear or subcarriers bandwidth is small enough to be considered infinitesimal (rate region is concave), sorting search algorithm gives optimal solutions. Otherwise, the solution is suboptimal, but it reduces the computation complexity. A sequential-linear-approximation water-filling algorithm is proposed to solve the APA continuous rate adaptation. The relaxed nonlinear concave problem is approached by a series of linear optimization problems derived by a sequential-linear-approximation algorithm named Frank–Wolfe method. For APA with discrete rate adaptation, a greedy algorithm is deployed to allocate bits and the corresponding power. In each bit loading iteration, the greedy algorithm allocates power to some subcarriers that maximize the utility argument per power. Assuming concave utility functions, the greedy algorithm results in optimal bit loading and power allocation. Finally, a joint DSA and APA solution is proposed for the original problem. For continuous rate adaptation, a combination of iterative subcarrier assignment, power allocation, and the update of marginal utility is deployed. A new subcarrier assignment is derived based on the property of the subgradient property of concave utility function; the corresponding power allocation is determined by linear approximation of the objective function; the algorithm stops when the marginal utility function is negligible. For discrete rate adaptation, a combination of sorting-search DSA and the greedy APA algorithm is deployed.

The suboptimal or heuristic algorithms from previous papers have considered specific scenarios, such as homogeneous traffic or concave objective functions. These assumptions are not valid for modeling heterogeneous services with nonconcave utility functions as in WiMAX. Commonly obtained application level utility functions for real-time and nonreal-time traffic are sigmoid and logarithm functions as shown in Figure 6.4. While a logarithm function is concave, sigmoid functions and most of application level utility functions of real-time traffic are not concave. Such utility functions yield maximization problems whose objective functions are not concave. When the set of feasible solutions that satisfy all constraints, also known as feasible region, is a convex set, and the objective function is concave, any local optimum will be a global optimum solution. Moreover, for many concave utility functions the optimization problem can be solved efficiently. In the problem of utility maximization for heterogeneous traffic, some of the utility functions are not concave (and consequently nonlinear), so finding a global optimum is computationally difficult, and the available softwares can only offer a local solution. Search algorithms that span the entire feasible region can obtain the near optimal solutions, but the solution time is not polynomial. We have applied genetic

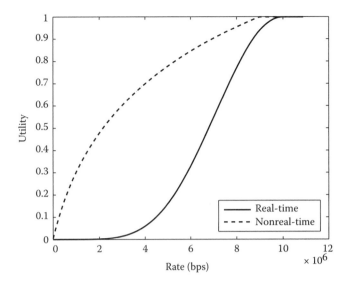

**FIGURE 6.4**  Typical utility functions for real-time and nonreal-time traffic.

algorithm to solve the problem [28]. Simulation results show the genetic algorithm effectiveness in utilizing resources and the convergence of obtained solution.

### 6.3.3 Uplink Subcarrier and Power Allocation

Although resources are user specific in UL, such as user devices power, users rely on the BS for signaling CSI, and resource allocation in PMP and centralized mesh. In PMP mode, users inform the BS of their CSI in each MAC frame, or average CSI of each user over several MAC frame. In centralized mesh, the nodes send their available resources (in terms of the quality or capacity of their links to their neighbors) and their bandwidth requests to the BS. The BS determines the allocation vector and distributes it to the nodes in PMP and centralized mesh. Similar to DL, an objective function of rate may be maximized to allocate spectrum and power in UL, except the power is constrained by users' specific power limitation. Therefore, only the total power constraint of the BS, $\sum_{i=1}^{U} \sum_{j=1}^{K} p_{ij} \leq P_{BS}$, is replaced by the power constraint of individual users, $\sum_{j=1}^{K} p_{ij} \leq P_i \ \forall i \in \mathcal{U}$, where $P_i$ is the $i$th user power constraint.

Due to the limited power of the users' devices, power minimization is a well-considered objective in UL resource allocation problems. The problem of power minimization in UL has the same objective function as the ones in DL, except a weighted power objective function may be used, i.e., $\min \sum^{U} \sum^{K} \lambda_i p_j$ [38]. In other words, minimizing the total power is an appropriate objective in DL where BS is the only source of power. Users in UL may have diverse power constraint, so a weighted power minimization is more appropriate.

## 6.4  DECENTRALIZED SUBCARRIER AND POWER ALLOCATION SCHEMES

The distributed infrastructure of WiMAX mesh and relay networks, and the need for reducing communication overhead between the BS and network nodes are the motivations behind proposing decentralized resource allocation schemes. Potentially, either no central controller exists or it does not influence the allocation decision in a decentralized resource allocation. These facts make the decentralized schemes more scalable.

Resource allocation in decentralized networks is essentially different from centralized one. In centralized schemes, the BS collects CSI from all users, allocates the subcarriers or power to the users, and informs the users of allocated resources. On the other hand, users may not need to know the CSI of the other parties in decentralized schemes. Besides, the parties may not be aware of the decision of each other, so a collision is probable. Accordingly, in each proposed distributed resource allocation scheme for OFDM-based networks, the following questions should be answered:

- How does a user achieve the required CSI? The hidden terminal and exposed node problems are common problems in self-organized and decentralized networks, because it is assumed that no central controller exists to assist in signaling. Besides, the signaling overhead should be reasonable for a practical implementation.
- How should the node coordinate or compete with the other nodes to attain the resources? A node does not know the requirements of the other nodes, at each instant, so competence or coordination is a "must" for a node which wants to start capturing the resources.

An interference aware subchannel allocation scheme that overcomes the drawbacks of decentralized schemes, i.e., hidden and exposed node problems is proposed in Ref. [42]. As the scheme uses updated CSI at the beginning of each MAC frame, the channel does not need to be assumed time-invariant over multiple MAC frames. The scheme is appropriate for OFDM-TDD networks. The MAC frame is divided into mini-slots; at the first mini-slot of each MAC frame the ongoing receiver nodes broadcast a busy signal to inform the respected transmitters of the quality of the allocated subchannel. The inactive nodes does not send any busy signal. The ongoing transmitter nodes listen to the busy signal and adapt their subcarrier allocation to their specific receiver nodes according to the information on the busy signal. Also, a node that wants to start transmission listens to the busy signal and chooses the subcarriers that are not interfered by ongoing transmissions and their interference. In other words, it selects those subcarriers with a received busy signal power less than a threshold. The advantages of the scheme are as follows:

- Signaling overhead is low compared to other radio resource allocation schemes
- The co-channel interference is reduced significantly since the scheme is interference aware
- Full frequency reuse is possible

A decentralized power allocation problem for a cooperative transmission is formulated in Ref. [43]. The objective is to minimize the total transmission power of all users subject to providing minimum rate requirement of each user. The network model uses a time division multiple access with the OFDM multiplexing (TDMA-OFDM), so only one user accesses the total OFDM subcarriers in each time slot. The user may allocate some of the subcarriers to transmit its traffic and the rest of subcarriers to relay the other users' traffic to minimize total transmission power. In other words, the resource allocation problem determines if a user should cooperate with other users, and in case of cooperation it determines the subcarriers and power allocation. A cooperative user may be allocated more power than a noncooperative user, because its location and channel gain allows it to cooperate with other users. However, the total network power is minimized by cooperation among users. A decoupled subcarrier and power allocation scheme is proposed to solve the problem.

Assuming an access point in the network, Ref. [44] proposes a distributed decision-making scheme for the resource allocation. Each user measures its CSI upon receiving a beacon signal from the access point. The subcarriers are divided into several groups (equal to the number of users), and the approximate channel gain of each group for each user is estimated. Then, the users contend with each other to achieve the group with the best channel quality of their own. A backoff mechanism is proposed to avoid collision. Each user start contending for the best group of subcarriers after a backoff time which is proportional to the best group gain for that user. After contention, the access point informs the users of the winners which can transmit in the next transmission interval. A frame is divided into three subframes, contention, acknowledge, and transmission subframes. This scheme is actually a distributed resource allocation scheme for a centralized network that aims to reduce the signaling overhead and processing task of the access point.

A collaborative subcarrier allocation using the swarm intelligent is proposed in Ref. [45]. The scheme relies on the central controller to achieve the updated information of available subcarriers and the highest and lowest demands for each subcarrier. In other words, the nodes negotiate with the central controller iteratively in a negotiation phase. In each iteration, the nodes inform the central controller of their demand for a specific subcarrier. Then, the central controller broadcast a message indicating the highest and lowest demands for each subcarrier. Based on the feedback messages that occur several times in each negotiation phase, the nodes intelligently decide upon subcarriers.

A comparison between the capacity performance of OFDM-TDMA and OFDM-FDMA in a two-hop distributed network is performed in Ref. [46]. The time frame is divided by two in OFDM-TDMA. In each half of the time frame, all subcarriers are devoted for transmission over one hop. A power allocation algorithm is proposed to maximize the end-to-end capacity subject to the overall transmit power constraint of the two hops. For OFDM-FDMA, the subcarriers are assigned to the two hops without overlapping and a joint subcarrier and power allocation algorithm is proposed. The simulation performance results of the algorithms show that OFDM-FDMA achieves a higher end-to-end capacity than OFDM-TDMA.

## 6.5  OPEN RESEARCH CHALLENGES

Although the resource allocation for OFDM-based networks has been well studied in the literature, few schemes have been specifically designed for WiMAX. These schemes should be modified or new schemes should be defined for OFDM-based WiMAX to effectively utilize the network resources and improve the network performance for integrated voice, video, and data services over fixed, nomadic, portable, and fully mobile users. An appropriate resource allocation scheme for OFDM-based WiMAX should consider diverse QoS requirement of heterogeneous traffic and mobility issues simultaneously, because a scheme that guarantees QoS for one type of traffic in a fixed network may not perform well for a different type of traffic in a fully mobile network. Moreover, the scheme should balance between users requirements and service providers revenue.

The proposed scheme should carefully consider channel characteristics; ignoring some channel characteristic to make resource allocation easier may result in an impractical scheme, e.g., interference may be a strict constraint in a distributed scheme. A signaling mechanism for reporting CSI and allocation vectors among correspondent nodes, such as the BS and users, should be identified. The CSI measurement may cause an unacceptable overhead when a large number of clients exist. It is usually assumed that perfect CSI is achievable for all users at the same time and at regular intervals, while the wireless channel impairment can make this assumption totally impractical [47]. Overall, a resource allocation scheme should be robust to imperfect CSI and balance between performance gain achieved by network adaptivity and performance loss due to network complexity.

## 6.6  CONCLUSION

The main objective of this chapter is to investigate the resource allocation schemes for WiMAX networks that adopt OFDM and OFDMA as their multicarrier transmission technologies. We investigate how a resource allocation scheme can take advantage of the OFDM technique flexibility in terms of subcarrier and power allocation to assign resources to achieve a specific objective. We categorize the schemes based on centralized, where a central controller allocates resources, and decentralized, where the users coordinate or compete to achieve resources. Variants of a classical model for centralized resource allocation schemes are reviewed, and an appropriate scheme for OFDM-based WiMAX networks that conforms the IEEE 802.16 standard and its diverse service flows is proposed.

The reviewed decentralized resource allocation schemes outline the basic objectives and challenges of the OFDM resource allocation schemes when the central node is not available or users are coordinate or compete to share the available resources.

## REFERENCES

[1] IEEE Std 802.16-2004, IEEE standard for local and metropolitan area networks part 16: Air interface for fixed broadband wireless access systems, October 2004.

[2] S. Hara and R. Prasad, *Multicarrier Techniques for 4G Mobile Communications*, Artech House, Norwood, MA, 2003.

[3] M. K. Awad, K. T. Wong, and Z. Li, An integrative overview of the open literature's empirical data on the indoor radiowave channel's temporal properties, *IEEE Transactions on Antenna and Propagation*, 56(5), 1451–1468, 2008.

[4] K. Fazel and S. Kaiser, *Multi-Carrier and Spread Spectrum Systems*, Wiley, New York, 2003.

[5] P. Vandenameele, L. Perre, and M. Engels, *Space Division Multiple Access for Wireless Local Area Networks*, Kluwer Academic Publishers, 2002.

[6] J. W. Mark and W. Zhuang, *Wireless Communications and Networking*, Upper Saddle River, NJ: Pearson Education International, 2003.

[7] S. Haykin and M. Moher, *Modern Wireless Communications*. Pearson/Prentice Hall, 2005.

[8] H. Liu and G. Li, *OFDM-Based Broadband Wireless Networks: Design and Optimization*, Hobken, NJ: Wiley-Interscience, 2005.

[9] J. Cai, J. W. Mark, and X. Shen, ICI cancellation in OFDM wireless communication systems, in *Proceeding of IEEE Global Telecommunications Conference, GLOBECOM '02*, Taipei, Taiwan, November 2002.

[10] C. Langton, Orthogonal frequency division multiplex (OFDM) tutorial, October 2007 [Online]. Available at: http://www.complextoreal.com/chapters/ofdm2.pdf

[11] J. Cai, X. Shen, and J.W. Mark, Robust channel estimation for OFDM wireless communication systems— An H-infinity approach, *IEEE Transactions on Wireless Communications*, 3(6), 2060–2071, 2004.

[12] K. Kumaran and H. Viswanathan, Joint power and bandwidth allocation in downlink transmission, *IEEE Transactions on Wireless Communications*, 4(3), 1008–1016, 2005.

[13] A. Goldsmith, *Wireless Communications*. Cambridge University Press, U.K., 2005.

[14] I. Kim, H. Lee, B. Kim, and Y. Lee, On the use of linear programming for dynamic subchannel and bit allocation in multiuser OFDM, in *Proceeding of the IEEE Global Telecommunications Conference, GLOBECOM '01*, vol. 6, 3648–3652, San Antonio, TX, November 2001.

[15] S. Pietrzyk, G. J. M. Janssen, A. N. Unit, P. T. C. Sp, and P. Warsaw, Radio resource allocation for cellular networks based on OFDMA with QoS guarantees, in *Proceeding of the IEEE Global Telecommunications Conference, GLOBECOM '04*, vol. 4, 2694–2699, Dallas, TX, 2004.

[16] C. Wong, R. Cheng, K. Lataief, and R. Murch, Multiuser OFDM with adaptive subcarrier, bit, and power allocation, *IEEE Journal on Selected Areas in Communications*, 17(10), 1747–1758, 1999.

[17] Y. Chang, F. Chien, and C. Kuo, Performance comparison of OFDM-TDMA and OFDMA with cross-layer consideration, in *Proceeding of the IEEE 64th Vehicular Technology Conference, VTC-2006*, Montréal, Canada, Fall 2006.

[18] G. Song and Y. Li, Adaptive resource allocation based on utility optimization in OFDM, in *Proceeding of the IEEE Global Telecommunications Conference, GLOBECOM '03*, vol. 2, Singapore, 2003.

[19] X. Zhang, E. Zhou, R. Zhu, S. Liu, and W. Wang, Adaptive multiuser radio resource allocation for OFDMA systems, in *Proceeding of the IEEE Global Telecommunications Conference, GLOBECOM '05*, vol. 6, St. Louis, MI, 2005.

[20] D. Niyato and E. Hossain, Adaptive fair subcarrier/rate allocation in multirate ofdma networks: Radio link level queuing performance analysis, *IEEE Transactions on Vehicular Technology*, 55(6), 1897–1907, 2006.

[21] S. Haykin, *Digital Communications*. John Wiley & Sons, Inc., New York, 1988.

[22] S. Chung and A. Goldsmith, Degrees of freedom in adaptive modulation: A unified view, *IEEE Transactions on Communications*, 49(9), 1561–1571, 2001.

[23] X. Qiu and K. Chawla, On the performance of adaptive modulation in cellular systems, *IEEE Transactions on Communications*, 47(6), 884–895, 1999.

[24] Z. Abichar, Y. Peng, and J. Chang, Wimax: The emergence of wireless broadband, *IEEE IT Professional*, 8(4), 44–48, 2005.

[25] J. Cai, X. Shen, and J. W. Mark, EM channel estimation algorithm for OFDM wireless communication systems, in *Proceeding of the IEEE 14th International Symposium on Personal, Indoor and Mobile Radio Communication*, Beijing, China, 2003.

[26] L. Chen, B. Krongold, and J. Evans, An adaptive resource allocation algorithm for multiuser OFDM, in *Proceedings of the 7th Australian Communications Theory Workshop*, Perth, Western Australia, February 2006.

[27] M. Bussieck and A. Pruessner, Mixed-integer nonlinear programming, *Society for Industrial and Applied Mathematics/Optimization Activity Group Newsletter: Views and News*, 14(1), February 2003.

[28] M. Mehrjoo, S. Moazeni, and X. Shen, Opportunistic fair scheduling for the downlink of IEEE 802.16 wireless metropolitan area networks, in *Proceedings of IEEE ICC '08*, Beijing, China, May 2008.

[29] K. Letaief and Y. Zhang, Dynamic multiuser resource allocation and adaptation for wireless systems, *IEEE Wireless Communications Magazine*, 13(4), 38–47, August 2006.

[30] M. Ergen, S. Coleri, and P. Varaiya, QoS aware adaptive resource allocation techniques for fair scheduling in OFDMA based broadband wireless access systems, *IEEE Transactions on Broadcasting*, 49(4), 362–370, December 2003.

[31] J. Jang and K. Lee, Transmit power adaptation for multiuser OFDM systems, *IEEE Journal on Selected Areas in Communications*, 21(2), 171–178, February 2003.

[32] Y. Zhang and K. Letaief, Multiuser adaptive subcarrier-and-bit allocation with adaptive cell selection for OFDM systems, *IEEE Transaction on Wireless Communications*, 3(5), 1566–1575, September 2004.

[33] Z. Shen, Multiuser resource allocation in multichannel wireless communication systems, PhD thesis, The University of Texas, Austin, TX, May 2006.

[34] W. Rhee and J. Cioffi, Increase in capacity of multiuser OFDM system using dynamic subchannel allocation, *IEEE VTC2000*, Tokyo, Japan, May 2000.

[35] J. Cai, X. Shen, and J. Mark, Downlink resource management for packet transmission in OFDM wireless communication systems, *IEEE Transactions on Wireless Communications*, 4(4), 1688–1703, 2005.

[36] R. Iyengar, K. Kar, and B. Sikdar, Scheduling algorithms for point-to-multipoint operation in IEEE 802.16 networks, in *Proceeding of 4th International Symposium on Modeling and Optimization in Mobile, Ad Hoc and Wireless Networks*, Boston, MA, April 2006.

[37] M. K. Awad and X. Shen, OFDMA based two-hop cooperative relay network resources allocation, in *Proceeding of IEEE ICC '08*, Beijing, China, May 2008.

[38] K. Seong, D. Yu, Y. Kim, and J. Cioffi, Optimal resource allocation via geometric programming for OFDM broadcast and multiple access channels, in *Proceeding of IEEE Global Telecommunications Conference, GLOBECOM '06*, San Francisco, CA, November 2006.

[39] S. Boyd, S. Kim , L. Vandenbergh, and A. Hassibi, Tutorial on geometric programming, Revised for optimization and engineering, July 2005 [Online]. Available at: www.stanford.edu/ boyd/gp-tutorial.html

[40] G. Song and Y. Li, Cross-layer optimization for OFDM wireless networks—Part I: Theoretical framework, *IEEE Transactions on Wireless Communications*, 4(2), 614–624, 2005.

[41] G. Song and Y. Li, Cross-layer optimization for OFDM wireless networks—Part II: Algorithm development, *IEEE Transactions on Wireless Communications*, 4(2), 625–634, 2005.

[42] H. Haas, V.D. Nguyen, P. Omiyi, N. Nedev, and G. Auer, Interference aware medium access in cellular OFDMA/TDD networks, in *Proceeding of IEEE ICC '06*, Istanbul, Turkey, June 2006.

[43] Z. Han, T. Himsoon, W. Siriwongpairat, and K. J. R. Liu, Energy-efficient cooperative transmission over multiuser OFDM networks: Who helps whom and how to cooperate, in *Proceeding of IEEE Wireless Communications and Networking Conference*, New Orleans, LA, March 2005.

[44] D. Wang, H. Minn, and N. Al-Dhahir, A distributed opportunistic access scheme for OFDMA systems, In *Proceeding of IEEE Global Telecommunications Conference, GLOBECOM '06*, San Francisco, CA, November 2006.

[45] J. C. Dunat, Collaborative allocation of ortogonal frequency division multiplex sub-carrier using the swarm intelligent, *Journal of Communications*, 1(1), 68–76, April 2006.

[46] J. Shi, G. Yu, Z. Zhangt, and P. Qiu, Resource allocation in OFDM-based multihop wireless networks, in *Proceding of Vehicular Technology Conference, VTC2006*, Melbourne, Australia, May 2006.

[47] M. K. Awad and X. Shen, Uplink ergodic mutual information of OFDMA two-hop cooperative relay networks based on imperfect CSI, Submitted to the *IEEE Global Telecommunications Conference, GLOBECOM '08*, November 2008.

# 7 Handoff Management in WiMAX

*Neila Krichene and Noureddine Boudriga*

## CONTENTS

The mobile WiMAX version represents the future of the wireless metropolitan networking as it combines mobile and fixed broadband networks while addressing mobility, guaranteeing high data rates, supporting advanced applications, and enhancing the provided quality of service (QoS). This chapter intends to address the handoff management in WiMAX networks; it aims at demonstrating how will the WiMAX handoff process offer services continuity while providing enhanced QoS guarantees and preserving security to meet the subscriber demands. First, we will survey the mobility management concepts. Second, we will study the mobile WiMAX physical and MAC layers specifications while presenting the end-to-end WiMAX architecture. Third, we will detail the WiMAX handoff management by presenting the supported architecture, then detail the functional decomposition adopted by this architecture and address the supported handoff schemes. Next, we will give an overview of the handoff control while specifying the adopted handoff monitoring functions and the different parameters that should be taken into account to optimize the handoff operation. After that, we will address the heterogeneous handoff by presenting the IEEE 802.21 standard, then show how mobile WiMAX can be interoperable with other mobile networking technologies. Finally, we will survey the different mechanisms implemented to guarantee security during handoff.

## 7.1  MOBILITY MANAGEMENT IN WIRELESS NETWORKS

### 7.1.1  HANDOFF VERSUS ROAMING

#### 7.1.1.1  Handoff: Overview

The recent years have been marked by a growing need for advanced applications and Internet-related services at high throughput and low costs while guaranteeing continuous and open access to such services. Besides, mobile access to Internet services requires access to packet-switched wireless networks, which are generally divided into an access part and a core part. The access part contains the link-layer devices while the core part contains network-layer devices mainly acting at the network layer [1]. A mobile node (MN) wishing to access Internet services needs to attach to the first link-layer device in the access network, also called, link-layer point of attachment (PoA). Nevertheless, the communicating MN may change the link-layer PoA during an active session due to particular reasons. This process is called handover or handoff. Handoff may be mobility driven when link conditions change due to mobility. In such case, the handoff occurs when the MN leaves the radio coverage of the actual PoA and becomes threatened by losing the network connectivity. On the other hand, handoff may be policy driven when the change in the PoA is argued by higher data rates, better services, lower costs, etc. After selecting the new PoA, a particular mobility mechanism implementing the handover process is executed. Such a mobility mechanism updates the mobile device and the network states to reflect the new PoA and redirects the ongoing data packets that were addressed to the old PoA during the handoff. If the MN has switched to a new link-layer PoA belonging to the same access network, a link-layer mobility mechanism will be executed. However, if the mobile device changes between link-layer PoAs belonging to different access networks, a network-layer mobility mechanism will be executed. When both PoAs (the old and the new) implement the same physical layer technology, the handoff is referred to as intratechnology handoff or horizontal handoff. Besides, when the new PoA implements a different physical layer technology compared to the old PoA, an intertechnology handoff, vertical handoff, or roaming takes place. In the latter case, the MN should be equipped with more than one network interface card (NIC). On the other hand, the handoff, which involves connecting to a new link and disconnecting from an old one, may be hard, soft, smooth, or seamless depending on the implemented mobility solution. Hard handoff takes place when the MN receives data packets from only one PoA during the handoff as the current link is disconnected first and then the connection to a new link is established. Controversially, soft handoff enables the MN to receive data packets by more than a PoA simultaneously as first a connection is established to a new link and then the old link is disconnected. Finally, a smooth handoff is a handoff where packet loss is minimized while a seamless handoff is a handoff transparent to the application. Handoff is always triggered by the occurrence of a certain event at the mobile device or in the network. If the handoff trigger occurs in the mobile device, the handoff is mobile initiated; otherwise, it is network initiated. As there is more than one candidate PoA to switch to, the decision to select the new PoA may be taken by the MN or the network; thus the handoff may be mobile controlled or network controlled, respectively. The handoff is mobile controlled and network assisted when the handoff decision is taken by the mobile based on information received from the network. Controversially, if the handoff decision is taken by the network upon the reception of information from the mobile such as the signal quality of neighboring PoAs, the handoff is network controlled and mobile assisted.

### 7.1.1.2   Roaming: Overview

Roaming appeared with the Global System for Mobile (GSM) technology. It may be defined as the set of mechanisms that allows extending the connectivity service to a location that is not covered by the home network (HN) where the service is registered but is covered by a visited network (VN). The VN always asks the HN for the authentication and authorization data associated with the visiting MN to allow or deny service access. Roaming also refers to changing the access network while on move. For instance, an MN which is GSM enabled and equipped with a wireless local area network (WLAN) NIC may roam between the two networks depending on its location without interrupting an active session. Roaming support is provided through the implementation of mobility management, authentication, authorization, and billing procedures. Different service providers managing different networks need to negotiate a roaming agreement to define the legal aspects related to service availability and billing. Often, the roaming process consists of three steps. First, the HN detects the MN as an unknown device which is not registered; therefore, the VN tries to identify the MN's HN. If there is no roaming agreement between the HN and the VN, the MN will be not able to access VN services. However, if a roaming agreement exists, the VN requests MN's service information to deduce whether the MN is allowed to roam. If it is the case, the VN maintains a subscriber record for the device while the HN updates the MN-related information so that any information destined to the MN will be correctly routed. The subscribers activity is registered in a file maintained by the VN. This file includes the details of the initiated calls, the subscribers visited locations, the calling parties, the volume of the exchanged data in case of data calls, the time of the calls and their duration, etc. Based on such information, the HN will be able to perform billing. Interstandards roaming allows MNs to seamlessly move between mobile networks with different access technologies. Nevertheless, this roaming type is particularly challenging as communication technologies have evolved independently across continents and were implemented by different industry bodies.

### 7.1.2   THE OSI MODEL AND HANDOFF

A layer 2 handoff consists in a change of the layer 2 information related to the current attachment between the MN and the network. This information may consist in a layer 2 address or a layer 2 PoA. A trivial example of layer 2 handoff is performed when a laptop is associated with a WLAN AP and then switches to a second WLAN AP in the same subnet [1]. Layer 2 handoff induces messages exchange between the affected MN and the previous and target PoAs. The time period of the first and the last layer 2 signaling message is referred to as layer 2 handoff delay, after which higher-layer protocols can proceed with their signaling procedure. This delay is a built-in value for a particular access technology.

Layer 3 handover consists in a change of layer 3 information related to the current attachment between the MN and the network. This information may consist in a layer 3 address such as an IP address, a routing table entry, or a point of service (PoS) such as an access router in WLANs. A trivial example is when a laptop is associated with an AP in one subnet and then switches to a second AP in a different subnet. When an MN changes its PoS during an active data session, all packets will continue to be routed to the old address unless particular mobility solutions are implemented. To address this issue, the IP packets may be reassociated with the same transport layer session; thus inducing important delays. IP tunneling may also be adopted. More specifically, a second address will be used as the tunnel end point while the old IP address will be maintained. IP tunneling, however, induces important overheads along with important delays. Inter-subnet mobility solutions addressing layer 3 handoff may be classified with respect to the protocols they use. One can distinguish IPv4-based, IPv6-based, or MPLS-based mobility solutions. Besides, it is worth noticing that some mobility solutions also called reactive solutions trigger the layer 3 handoff after the layer 2 handoff. Contrarily, predictive mobility solutions trigger the layer 3 handoff before the layer 2 handoff. Last but not least, some hierarchical mobility solutions define one global (interdomain) location update point and one local (intradomain) location update point to keep the intradomain handoff transparent to the global location update point and reduce the signaling overhead.

The Session Initiation Protocol (SIP) [2] was designed to support mobility at the application layer; it is also used as the signaling protocol for real-time multimedia calls including voice over IP ones. As it acts at a higher layer, SIP can be combined with Mobile IP protocols, with the possibility of end-to-end adaptation for vertical handoff. A mobile host (MH) registers with a SIP server in its home domain. When a correspondent host (CH) sends an INVITE message to the MH to initiate a call, the redirect servers (RS), which have the current location of the MH, redirect the INVITE to the new location. When the MH moves during a session, it should send a new INVITE to the CH using the same call identifier in the original call setup, but it indicates the new IP address in the contact field of the SIP signaling message. The MH should also perform a new registration at the SIP server with its unique URI (uniform resource identifier) for all the new incoming calls. When both MH and CH are mobile, the reinvite messages are sent through the SIP server because it keeps track of the current CHs location.

## 7.2  MOBILE WiMAX OVERVIEW

The fixed WiMAX version based on the IEEE 802.16d standard has proved its efficiency as a backhauling solution and a fixed wireless alternative to cable and DSL [3]. 802.16d WiMAX adopts the orthogonal frequency division multiplexing (OFDM) modulation technique and operates in the line of sight (LoS) and nonline of sight (NLoS) environments; it may also operate at cost-effective license-exempt bands and offer higher throughput to cope perfectly to enterprise users' requirements [4]. The mobile WiMAX version based on the IEEE 802.16e standard is an amendment of the 802.16d WiMAX solution; it is also referred to as 802.16e WiMAX. 802.16e WiMAX implements mobility management services and supports mobile users moving at vehicular speed ranging from 73 to 93 mi/h; it adopts the scalable-OFDMA (SOFDMA) modulation technique to enhance the multi-access capabilities and enriches the MAC layer by addressing handoff and power saving. Mobile WiMAX is designed with respect to an end-to-end network architecture based on an all-IP platform.

### 7.2.1  Physical Layer Overview

802.16e WiMAX system profiles will cover 5, 7, 8.75, and 10 MHz channel bandwidths for licensed spectrum allocations in the 2.3, 2.5, 3.3, and 3.5 GHz frequency bands [3]. Targeting a worldwide coverage, the WiMAX forum intends to promote the allocation of frequency bands inferior to 6 GHz for civilian applications while considering the available spectrum all over the world. For instance, the 3.5 GHz frequency band is already assigned to fixed services in many countries, the 2.3 GHz band was reserved for the deployment of the WiBro solution in South Korea, while the 2.5 and

2.7 GHz bands have been assigned by the United States for fixed and mobile WiMAX deployment. Considering the modulation technique, mobile WiMAX is based on the orthogonal frequency division multiple access (OFDMA) technique and particularly on the SOFDMA variant [3]. The frequency division multiplexing (FDM) principle consists in using multiple frequencies to transmit different signals in parallel. This is achieved by assigning a frequency range or subcarrier to each signal then modulating it by data. The multiple subcarriers used for transmitting different signals should be separated by guard bands to prevent interferences [5]. Orthogonal FDM (OFDM) eliminates the guard bands by using overlapping subcarriers that are spaced apart at precise frequencies; thus achieving orthogonality. With OFDM, the center of the modulated carrier coincides with the edge of the adjacent carrier so that the independent demodulators performing a discrete Fourier transform see only their own frequencies. Redundancy may be guaranteed by scattering some bits over some sets of distant subcarriers. The OFDMA is a multiple access and multiplexing scheme that multiplexes data streams generated by multiple users onto the downlink (DL) subchannels and fulfils uplink (UL) multiple access by means of UL subchannels [3]. The OFDMA symbol structure is made up of data subcarriers for data transmission, pilot subcarriers for estimation and synchronisation purposes, and null subcarriers used for guard bands and DC carriers. Data and pilot subcarriers are organized in a subset of subcarriers called subchannels. The minimum frequency–time resource unit of subchannelization is one slot equal to 48 data tones. It is worth noticing that there exist two types of subcarrier permutations for subchannelization which are the diversity permutation and the contiguous permutation. The diversity permutation organizes the subcarriers in a pseudorandom fashion to form a subchannel while the contiguous permutation organizes a block of contiguous subcarriers forming a bin. The bin is formed by nine contiguous subcarriers in a symbol with eight assigned for data, and one assigned for a pilot. Generally speaking, diversity subcarriers permutation achieves a high performance with the mobile applications while the contiguous subcarrier permutation fits well to fixed, portable, or low mobility environments. The SOFDMA technology optimizes the mobile access by assigning a set of subcarriers to particular users. For instance, subcarriers 1, 3, and 7 may be assigned to user 1 while subchannels 2, 5, and 9 to user 2 and so on [5]. Users close to the base station (BS) will benefit from a larger throughput by getting an important number of subchannels with a high modulation scheme. SOFDMA offers multiple bandwidths to allow multiple spectrum allocation and fulfil different usage-model requirements. In fact, the scalability is guaranteed by adjusting the fast Fourier transform (FFT) size with respect to the available bandwidth while fixing the subcarrier spacing at 10.94 kHz. As the resource unit-subcarrier bandwidth and the symbol duration are fixed, scaling bandwidth will not affect higher layers [3]. The first release of mobile WiMAX will adopt a system channel bandwidth of, respectively, 5 and 10 MHz, with a sampling frequency of, respectively, 5.6 and 11.2 MHz. The FFT size will be either 512 or 1024 while the number of subchannels will be either 8 or 16 and the useful symbol time will be fixed to 91.4 ms. The OFDMA symbol duration is 102.9 ms while the number of OFDMA symbol within a 5 ms frame is 48 [3]. The modulation scheme will gradually vary from 16 quadrature amplitude modulation (QAM) to quaternary phase shift keying (QPSK) (four channels) and even binary phase shift keying (BPSK) (two channels) at longer ranges while the power allotted to each channel will be increased. The adaptation of multiple-input multiple-output (MIMO) antenna technique along with advanced coding and modulation achieve peak DL data rates up to 63 Mbps per sector and peak UL data rates up to 28 Mbps per sector in a 10 MHz channel.

IEEE 802.16e supports both time division duplex (TDD) and frequency division duplex (FDD) duplexing modes. TDD is adopted when the license-exempt spectrum is used because a unique channel is shared between the UL and the DL traffics which occupy different time slots. Contrarily to TDD, FDD operates with two channels, one is dedicated to the UL traffic and the second is reserved to the DL traffic [4]. Mobile WiMAX in its first release supports only the TDD duplexing mode although the WiMAX forum intends to address particular market opportunities by supporting FDD in future releases. With TDD, it becomes feasible to adjust the DL/UL ratio with respect to the nature of the ongoing traffic so that the DL/UL asymmetric traffic is efficiently supported. Besides, TDD guarantees channel reciprocity for better support of link adaptation, MIMO, and other advanced antenna technologies. Meanwhile, TDD offers a greater flexibility for adaptation to varied spectrum allocations as it uses the same channel for both UL and DL traffics. Last but not least, transceivers designed for TDD implementations are less complex and less expensive [3]. An OFDM frame structure for a TDD implementation is divided into DL and UL subframes separated by transmit/receive and receive/transmit transition gaps (TTG and RTG) to eliminate DL and UL transmission collisions. The frame control information is carried by the preamble, the frame control header (FCH), the DL-MAP, and UL-MAP, the UL ranging, and the UL fast channel feedback (UL CQICH) fields [3]. More specifically, the preamble field appears as the first OFDM symbol in the frame and it is used for synchronization. The FCH field identifies the frame configuration information including the MAP message length and coding scheme and usable subchannels. DL-MAP and UL-MAP provide subchannel allocation for the DL and the UL subframes while the UL ranging subchannel, which is allocated for the mobile stations (MSs), is used for closed-loop time, frequency, and power adjustments as well as bandwidth requests. Finally, the UL CQICH enables the MSs feedbacking the channel-state information to the managing entities.

The WiMAX forum included advanced physical layer features to enhance the mobile WiMAX network coverage and capacity. For instance, mobile WiMAX supports the mandatory QPSK, 16 quadrature amplitude modulation (QAM), 64 QAM schemes in the DL, and the optional 64 QAM in the DL. Meanwhile, both convolutional code (CC) and convolutional turbo code (CTC) coding schemes are supported along with two other optional coding schemes, which are the block turbo code (BTC) and low density parity check code (LDPC). The combination of various modulation and code rates results in a fine resolution of data rates so that the BS scheduler may determine the best suited data rate for each burst allocation, based on the buffer size and the channel propagation conditions at the receiver. Moreover, a channel quality indicator (CQI) channel is used for providing channel-state information from the MSs to the base station scheduler while other channel-state information can be provided to the BS by the CQICH, which includes the physical carrier to interference plus noise ratio (CINR), the effective CINR, the MIMO mode selection, and the frequency selective subchannel selection [3]. Hybrid auto repeat request (HARQ) is supported to provide fast response to packet errors and improve the cell edge coverage through using $N$ channel Stop and Wait protocol. The previously stated adaptive modulation and coding, CQICH and HARQ, guarantee robust link adaptation in mobile environments at vehicular speeds exceeding 120 km/h.

### 7.2.2 MAC LAYER OVERVIEW

The WiMAX forum has designed the physical and MAC layers of the mobile WiMAX based on the amendments of the IEEE 802.16e to offer broadband services including voice, data, and video at the metropolitan scale for mobile users. The mobile WiMAX network comprises BSs and subscriber stations (SSs). Each SS is assigned a 48-bit MAC universal address that it used to uniquely identify it toward a BS. Mobile WiMAX uses UL and DL maps to prevent collisions. More specifically, SSs implement the time division multiple access (TDMA) to share the UL while BSs use the time division multiplexing [6]. UL and DL schedules are exchanged between the BS and the managed SSs in every frame using the UL-MAP and the DL-MAP messages. The MAC layer is connection-oriented; besides, all data communication is associated with a connection. Each connection with its QoS parameters forms a service flow and is identified by a 16-bit connection identifier (CID). MAC layer connections can be compared to TCP connections. In fact, thanks to TCP, a computer may have simultaneously different active connections for different applications using different ports. With MAC connections, an SS may have many connections to a BS for different services such as network management or user data transport; every connection is characterized by its own bandwidth, security, and priority parameters [6]. When a new SS joins the network, the managing BS assigns to it three CIDs with different QoS requirements used by different management levels which are the basic, the primary, and the secondary management connections. The basic connection enables the transfer of short, time-critical MAC and radio link control messages; the primary management connection is used to transfer larger but more delay-tolerant messages while the secondary management connection is used to transfer standards-based management messages such as DHCP ones. Note that a CID may carry traffic for many different higher-layer sessions. As stated in Section 7.2.1, the mobile WiMAX version adapts dynamic modulation and forward error codes to correctly serve the SSs located far from the BS in the rural areas and resists the weather conditions [6]. Both the BS and the SS can adapt the transmission of the burst profiles by lowering the bandwidth for higher robustness. For instance, the BS always begins by adopting the most robust modulation and forwards error code scheme so that all SSs in the coverage area can correctly receive the DL-MAP and the UL-MAP messages. Meanwhile, an SS may ask its BS to have a longer UL window when needed. The BS and the managed SSs exchange MAC protocol data units (PDUs) carrying MAC management messages or convergence sublayer MAC service data units (MAC SDUs). The MAC PDU has a fixed MAC header, a variable length payload and a cyclic redundancy check (CRC) field. The MAC header may be a generic MAC header (GMH) or a bandwidth request header. The GMH is used to transfer the standard MAC management messages while the bandwidth request header is a header sent without payload to request additional bandwidth. Both the payload and the CRC fields are optional. It is worth noticing that the MAC header and the MAC management messages are never encrypted to facilitate registration, ranging, and normal operation of the MAC sublayer; however, this decision has opened the door to eavesdropping and other serious attacks. An SS which enters the network may be programmed to register with a certain BS; but generally speaking, it begins by scanning its frequency to detect an operating channel. Scanning consists in listening to each possible frequency until the frame preamble is heard. After detecting that channel, the SS tries to synchronize to the DL transmission by waiting for the DL-MAP stating the map of the timeslot locations in use for the frame. After that, the MS waits for the downlink channel descriptor (DCD) and the uplink channel descriptor (UCD) messages that are periodically broadcasted to specify the modulation and the FEC schemes used on the carrier. After gathering the required information describing the parameters needed for initial ranging transmission, the SS scans the UL-MAP to find an opportunity to perform the ranging. Initial ranging is used to determine the transmit power requirements of the MS to reach the BS. It is worth noticing that each SS should be informed about when to send the ranging request as many SSs may try to join simultaneously the network and highly affect the networks efficiency. Therefore, each new SS will send a ranging request (RNG-REQ) message and wait for the corresponding ranging response (RNG-RSP) message, indicating the timing advance, the power adjustment, and the basic and the primary management CIDs. After correctly determining the timing advance of SS transmissions, the SS and the BS will continue exchanging RNG-REQ and RNG-RSP messages until an acceptable radio link is established. Then, the SS should perform the authentication and the registration processes to enter the network. First, the BS asks the SS for strong authentication. Upon successful authentication, the SS will be able to register to the network. The SS will then establish a secondary management CID to receive secondary messages for different services. For instance, the SS will get an IP address through DHCP and a Trivial File Transfer Protocol (TFTP) address to request configuration files when needed [6].

QoS is guaranteed thanks to five QoS classes implementing different scheduling mechanisms. Those classes are the unsolicited grant service (UGS), the real-time polling service (rtPS), the extended rtPS (ErtPS), the nonreal-time polling service (nrtPS), and the best effort (BE). UGS fulfils the requirements of real-time communications occurring at periodic intervals such as voice over IP (VoIP); it is given a grant by BS that accommodates the maximum sustained traffic rate of such applications. rtPS uses unicast polling to support periodic real-time variable-size transmissions such as MPEG video streams, but generates more overhead than UGS. ErtPS is a combination of UGS and rtPS; in fact, the bandwidth is allocated without solicitation, but the allocation is done in a dynamic fashion. ErtPS fits well in the case of voice with activity detection like VoIP with silence suppression applications. nrtPS uses unicast polling regularly and allows contention requests to guarantee a minimum data rate for delay-tolerant applications, which generate variable-sized traffic such as FTP. Finally, the BE class guarantees a minimum QoS level for the associated traffic. Therefore, the SSs are never polled individually but are rather permitted to use contention requests and unicast requests [3].

### 7.2.3 MOBILE WiMAX END-TO-END ARCHITECTURE

IEEE 802.16 standards family designed the physical and the MAC layers to provide broadband services access at the metropolitan scale for fixed, nomadic, portable, and mobile subscribers. The WiAMX forum, whose mission is to promote the interoperability of the broadband wireless access equipments implementing the IEEE 802.16 and ETSI HIPERMAN standards, has created a Network Working Group and a Service Provider Working Group to address higher-layer specifications such as intervendor inter-network interoperability for roaming, multivendor access networks, and intercompany billing [3]. The result of these standardization efforts is the mobile WiMAX end-to-end architecture. The architecture is based on an all-IP platform implementing packet switching; it defines an access network and a common core network while decoupling the access from connectivity IP service. Moreover, the end-to-end architecture is designed to support loosely coupled interworking with existing wireless networks such as UMTS and existing wired networks such as DSL. Besides, a global roaming across WiAMX operator networks is achieved through the support for credential reuse, common billing and settlement, and consistent use of authentication, authorization, and accounting (AAA) services. The WiMAX forum designed the WiMAX network reference model (NRM), which aims at achieving interoperability through identifying the functional entities and reference points and the corresponding communication protocols and data plane treatment within a logical representation of the network architecture. As depicted in Figure 7.1, the MS, the access service network (ASN), and the connectivity service network (CSN) represent a set of functional entities that may be implemented by a single physical device or by different physical devices. The ASN is formed by at least one BS and one ASN gateway (ASN GW); it implements the access services and represents a boundary for functional interoperability with WiMAX clients and WiAMX connectivity service functions. For instance, the BS manages the MSs in its coverage while the ASN

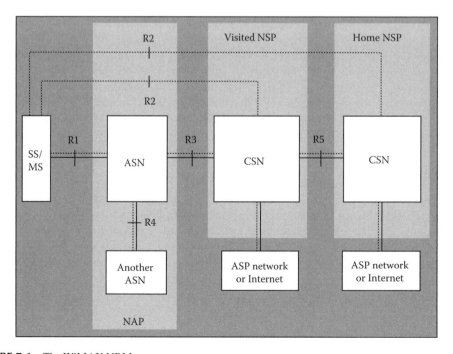

**FIGURE 7.1**  The WiMAX NRM.

GW relays data to the CSN. On the other hand, the CSN may be defined as a network of Internet gateways, user databases, routers, servers, and proxies providing IP connectivity services to WiMAX subscribers.

Mobile WiMAX end-to-end architecture is based on a security framework which provides basic security services and particularly authentication and confidentiality. In fact, each MS authenticates itself to the WiMAX network while the WiMAX network authenticates itself to the MS by implementing consistent and extensible authentication mechanisms. Besides, data confidentiality and integrity, replay protection and nonrepudiation services are guaranteed using applicable key length. Last but not least, MSs have the possibility to initiate and terminate specific security mechanisms such as virtual private networks (VPNs). The end-to-end architecture supports advanced IPv4- or IPv6-based mobility management mechanisms. For instance, vertical handovers can occur with wireless LANs or third generation wireless networks while roaming between different network service providers (NSPs) is supported. Seamless handover is also supported at up to vehicular speed while optimizing the overall network resources through the implementation of dynamic and static home address configurations, dynamic assignment of the home agent in the service provider network, and in the home IP network based on policies, etc. WiMAX mobility management will be further detailed.

## 7.3 WiMAX HANDOFF

IEEE 802.16e standard defined the required procedures and functions that should be implemented at the physical and MAC layers to perform handoff. The mobile WiMAX version inherits from the IEEE 802.16e standard, but it also defines the protocols that should be implemented at the higher layers to support intra-/inter-ASN handoff, roaming, seamless handoff, and micro-/macro-mobility [7].

### 7.3.1 SUPPORTED ARCHITECTURE

As described earlier, an ASN includes at least one ASN GW responsible for communicating with the CSN and a BS managing the connections to the MSs in its coverage. An ASN GW may be associated with one or more BSs while a BS can be managed by one or more ASN GWs so that multiple vendors can simultaneously interoperate within the same ASN. The BS may be a serving BS or a target BS depending on its task during the handoff process. In fact, the serving BS is the BS related to the MS before handoff while the target BS is the BS associated with the MS after handoff. On the other hand, we distinguish the serving ASN GW, the target ASN GW, and the anchor ASN GW. The serving ASN GW is the ASN GW corresponding to the serving BS; the target ASN GW is the ASN GW connected to the target BS while the anchor ASN GW is the ASN GW receiving the CSN data addressed to the MS and relaying them to the serving ASN GW. Thanks to the anchoring ASN GW, the MSs mobility is transparent to the CSN that does not need to know which ASN GW is managing the BS that is serving the MS. Therefore, the anchoring function prevents the CSN from frequently changing IP addresses. If the serving ASN GW is directly receiving data from the CSN, it is also considered as the anchor ASN GW. Nevertheless, the anchor ASN GW does not need to be a serving ASN GW or a target ASN GW. The intra-ASN handoff is processed between BSs within the same ASN; it does not induce important delay and minimizes data loss. Besides, intra-ASN handoff does not result in a change of the MSs IP address because the mobility is transparent to the outside of the ASN. Contrarily, an inter-ASN handoff is processed between BSs belonging to different ASNs and involves ASN GWs associated with separate ASNs. These ASN GWs need to coordinate their actions by adopting either anchoring or reanchoring to make the handoff smooth to the MS.

### 7.3.2 FUNCTIONAL DECOMPOSITION

The ASN-anchored mobility management is defined as mobility of an MS not involving a change in the CoA; it applies to mobility in networks not based on MIP. The specifications identify three functions responsible for the handoff, the MS context, and the data delivery control [7]. More specifically,

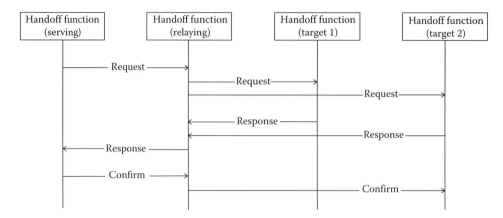

**FIGURE 7.2** Handoff function network transaction.

the Handoff function, which is implemented on the serving, the relaying, and the target peers, manages the signaling messages exchange and takes decisions associated with the handover. Figure 7.2 illustrates a possible handoff scenario. First, the serving-handoff function sends a handoff request (HO_Req) and waits for the corresponding reply. That HO_Req should include at least the MS_ID identifying the MS that requests the handoff, the list of the candidate target BS identifiers (IDs), possibly the MS/Session information content, and the first requested bicast SDU sequence number [8]. The relaying handoff function relays the HO_Req to multiple target handoff functions, which are in charge of analyzing the request, formulating, and sending the correspondent handoff responses (HO_Rsp). The HO_Rsp primitive includes at least the MS_ID and the list of the recommended target BS IDs; it may also carry other optional information. The received responses are forwarded by the handoff-relaying function to the serving-handoff function. The latter should send back a handoff confirmation (HO_Cnf) to the chosen target stating the final handoff action that may either be an initiation, a cancellation, or a handoff rejection. The HO_Cnf should at least indicate the MS_ID, the DL ARQ synchronization information per service flow describing the context necessary to restore communication from the point it has been interrupted and the UL ARQ synchronization information per service flow describing the context necessary to restore communication from the point it has been interrupted.

On the other hand, the context function manages the MS context and related information while handling their exchange in the backbone to set up any state or retrieve any state in network elements. For instance, the MSs context in the context function associated with the serving/anchor handoff function needs to be updated. More specifically, the MSs context in the context function associated with the serving handoff-function will be transferred to the context function associated with the target handoff function. Context information transfer may be triggered to populate a new MSs security context at a target BS, inform the network of an MSs initial network entry, or inform the network of the MSs idle mode behavior. The specifications identify relaying context functions, context functions acting as context servers, and context functions acting as context clients [8]. The relaying context function mediates information delivery between context-client and context-server functions. The context-server function stores the most updated session context information for the MS while the context-client function, which is associated with the functional entity having the 802.16 physical link, retrieves session context information stored at the context-server level during handoff processes.

Finally, the data path (DP) function, also referred to as the bearer function, establishes the routes and manages the current data packets transmission between two functional entities. More specifically, the DP function controls the setup of the bearer plane between two BSs, two gateways, or a gateway and a BS; it may implement the setup of tunnels and support multicast and broadcast [8]. The specifications distinguish four DP functions with respect to their roles in the handoff process. First, the anchor DP function anchors the data path associated with the MS across handoffs by forwarding the received data packets toward the serving DP function; it may buffer some of the packets and maintain some state information regarding bearer for the MS during handoffs. Second, the serving DP function is implemented at the end of the DP and associated with the serving PHY (physical)/MAC function (e.g., the serving BS) to handle the transmission of all data packets destined to the MS. Third, the target DP function is associated with a target BS that has been selected as the target of the handoff; it communicates with the anchor DP function to establish the DP that will replace the current path after the termination of the handoff. If the handoff succeeds, the target DP function becomes the serving DP function. Fourth, the relaying DP function mediates message exchange between serving, target, and anchor DP functions.

### 7.3.3  HANDOFF SCHEMES

Mobile WiMAX specifications support five types of accesses that reflect five mobility-related usage scenarios [9]. First, fixed broadband access assumes a subscriber being in the same geographic location during the whole duration of access to the network services. Second, nomadic access supports the movement between different cells without managing handoff. Third, portable access provides nomadic access to a portable device with the expectation of a BE handoff. Fourth, simple mobility access supports subscribers moving at speeds up to 60km/hour; it provides service continuity despite mobility and fulfils brief interruptions during handoff. Both portable and simple mobility accesses implement the hard handoff concept. Finally, full mobility access guarantees service continuity at high speeds up to 160km/hour and achieves seamless handoff with less than 50 ms latency and less than 1 percent packets loss ratio. To support full mobility access, IEEE 802.16e specifications define three handoff methods, which are the hard handover (HHO), the fast base station switching (FBSS) handover, and the macrodiversity handover (MDHO). HHO is mandatory while both FBSS and MDHO are optional. Initially, HHO is the only type required to be implemented by certified mobile WiMAX equipments.

### 7.3.3.1  Hard Handoff

Hard handoff results in a sudden connection transfer from one managing BS to a second one as the MS can communicate with only one BS each time. Therefore, all connections with the serving BSs are broken before a new connection with the target BS is established. In Figure 7.3, the red thick line at the border of the cells indicates the place where HHO is executed. The threshold level hysteresis is used in practice to avoid the repeated switching of neighbors BS during a movement lengthwise

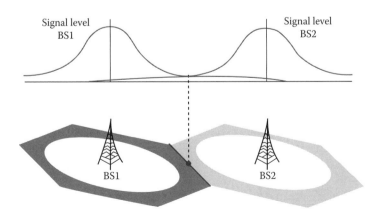

**FIGURE 7.3**  The HHO process.

to the cell boundaries. HHO is a less complex handoff type but it induces high latency; therefore, it is used for data as it is not suitable for real-time latency-sensitive applications such as VoIP.

The handoff decision may be taken by either the BS, the MS, or a third entity in light of the periodic measurements done by the MS [9]. In fact, each MS periodically processes a radio frequency scan during scanning intervals allocated by the serving BS and measures the signal quality of the neighboring BSs. Scanning consists in monitoring each possible frequency until a DL signal is received. The number of scanned frequencies depends on the regulatory-provisioned bandwidth, the physical specification, and the chosen bandwidth per channel, which depends on the physical specification. The MS is allowed to perform the initial ranging process to associate with one or more neighboring BSs. Once the handoff is decided, the MS starts the synchronization with the DL transmission of the target BS; it then performs ranging if this was not done during scanning, and it finally ends the connection with the previous BS. The undelivered MAC protocol data unit (MPDU) is stored at the BS until the timer expires [9].

### 7.3.3.2 Macrodiversity Handoff

MDHO is a form of soft handoff as the MS is allowed to maintain a valid connection simultaneously with more than one BS. When the MDHO is supported by both the MS and the BS, a diversity set, also referred to as active set, is maintained. The active set is a list of BSs that may be involved in the handoff procedure. Such BSs list is updated through the exchange of MAC management messages. These messages are sent based on the long-term CINR of BSs, which depends on two threshold values broadcasted in the DCD: the Add Threshold H_Add_Threshold and the Delete Threshold H_Delete_Threshold. A serving BS is dropped from the diversity set when the long-term CINR is less then H_Delete_Threshold. A neighbor BS is added to the diversity set when its long-term CINR is higher than H_Add_Threshold. The MS continuously monitors the BSs in the diversity set and defines an anchor BS among them. The MS synchronizes and registers to the anchor BS and performs ranging while monitoring the DL channel for control information. The MS communicates with anchor BS and active BSs in the diversity set as depicted in Figure 7.4. Two or more BSs transmit data on the DL so that multiple copies are received by the MS which needs to combine them using any of the well-known diversity-combining techniques.

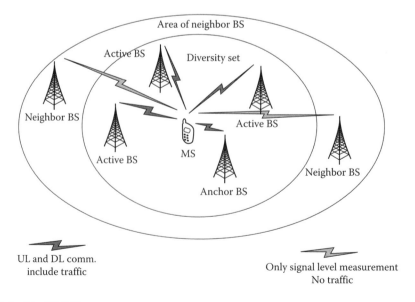

**FIGURE 7.4** The MDHO process.

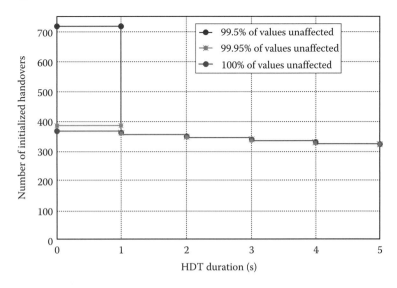

**FIGURE 7.5**   Number of initialized handovers in function of HDT duration.

To evaluate the performances, authors in Ref. [9] suggest the implementing of a handover delay timer that adds a short delay between the time when the handover conditions are met and the handover initialization is started. They assumed a MDHO handover and a periodic reporting of 1 s and then evaluated the number of handovers between 4 BSs and 42 MSs. The principle was to select the BS with the best signal strength for each MS and then create the diversity set for each MS based on defined thresholds and signal strengths. Three different cases were simulated. In the first case (variety I), the probability of affected CINR value is 0.5 percent (it means that the 99.5 percent of values are unaffected). In the second case (variety II), the probability of affected value is 0.05 percent (99.95 percent of values are unaffected). In the third case (variety III), no values were affected (100 percent of values are unaffected). There has been simulated 30 min. time interval and the values were evaluated with 1 s step for all varieties. Results are depicted in Figure 7.5. For more details, refer to Ref. [9].

### 7.3.3.3    Fast Base Station Switching

The fast base station handoff is also a form of soft handoff which is supported by both the BS and the MS. The MS maintains a list of BSs referred to as the active set and then continuously monitors it. As in the case of MDHO, the MS may perform ranging and maintain a valid CID with the BSs of the active set. Nevertheless, the MS is allowed to communicate with only one BS called the anchor BS as depicted in Figure 7.6. In fact, the MS is registered and synchronized with the anchor BS; both entities exchange UL and DL traffic including management messages. The anchor BS may be changed from frame to frame with respect to the BS selection scheme. In that case, the connection is switched to the new anchor BS without performing explicit handoff signaling as the MS simply reports the ID of the newly selected BS on the CQICH. Note that every frame can be sent via a different BS belonging to the active set.

The anchor BS may be updated by implementing two mechanisms: the handover MAC management method and the fast anchor BS selection mechanism. The first updating mechanism is based on the exchange of five types of MAC management messages while the second updating mechanism transmit anchor BS selection information on the fast feedback channel. The new anchor BS should belong to the current diversity set; its selection is based on the signal strength reported by the MS. Adding BSs to the diversity set and removing others is done on the basis of the comparison of their

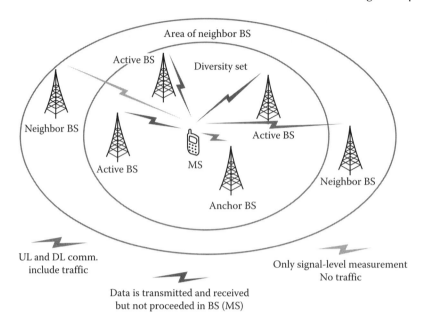

**FIGURE 7.6**   Fast base station switching.

long-term CINR to H_Add_Threshold and H_Delete_Threshold. Note that FBSS and MDHO have many similarities and achieve better performance compared to HHO. Nevertheless, they are more complex as they require the BSs of the active set or the diversity set to be synchronized, use the same carrier frequency, and share the network entry information.

### 7.3.3.4   Seamless Handoff

Seamless mobility intends to achieve a seamless handoff between different data networks and access technologies. Examples of such handoff occur between WiMAX and WiFi networks, WiMAX and UMTS networks, WiFi and 3G mobile networks including CDMA 2000 and UMTS, etc [10]. Subscribers, who are becoming more and more demanding, wish to have voice, video, and data connections anywhere and anytime and expect to roam within and between networks at low cost while preserving the initial security level and QoS guarantees. To take up such challenges, the used handsets should integrate multiple radio interfaces and support a wide range of applications while implementing power management. Such mixed-network devices must be able to automatically detect and select the best wireless network. On the other hand, networks should provide high bandwidth while implementing intelligent networking mechanisms including seamless roaming and cross-network identity and authentication. For instance, an efficient and usage-model appropriate means of establishing identity should be provided. When considering seamless mobility between WiFi and WiMAX, it is evident not to initiate handoff when the WiFi coverage is available as WiFi provides high bandwidth and achieves good performance. The received signal strength level, provided by the physical layer and the bandwidth of network layer, may be the guiding parameters that determine when to initialize handover and how to choose the best WiFi access point. Link-layer triggers should be defined to help IP handoff preparation and execution while cross-layer information exchange should fasten the handoff process. To guarantee an interrupted user connection during handoff between different networks, IEEE proposes a new standard referred to as IEEE 802.21. A more detailed description of vertical handoff and IEEE 802.21 specifications will be found in Section 7.5.1.

## 7.4   WiMAX HANDOFF CONTROL

### 7.4.1   HANDOFF INITIATION

IEEE 802.16e specifications have defined MS-initiated handoff and BS-initiated handoff to cover the majority of the possible handoff scenarios. Other network entities may also trigger handoff to enhance the overall performances.

#### 7.4.1.1   Mobile-Initiated Handoff

An MS may decide to change its serving BS after losing signal quality or after detecting that a higher QoS can be disserved by another BS. In such situations, the MS will initiate the handoff. Nevertheless, IEEE 802.16e specifications did not specify the methodology of deciding whether or not to perform the handoff; they only focused on the mechanisms that should be implemented to collect information about the neighboring BSs for taking the handoff decision. In fact, each BS should transmit on the broadcast connection a mobile neighbor advertisement message MOB_NBR-ADV informing the listening MSs of the characteristics of any neighboring BS. Such a message includes the identifier of each neighboring BS, its frequency, the supported services, and its available radio resources such as its available channels. Upon getting such information, each MS should be able to take the handoff decision in the light of scanning of possible target BSs. The scanning procedure begins when the MS sends its serving BS a MOB_SCN-REQ message to inform the serving BS that it wishes to scan the neighboring BSs. The message indicates a length of time in frames for this interval and the type of association that will be used for scanning. The aim of association is to enable the MS acquiring and recording ranging parameters and service availability information to select the proper BS target [11]. The specifications define three levels of possible associations, which are the association without coordination, the association with coordination, and the network-assisted association reporting. With the association without coordination, the target BS has no knowledge of the MS. Association with coordination means that the serving BS will coordinate association with the requested target BS and then respond to the requesting MS. With the network-assisted association reporting, the serving BS coordinates association with the requested target BS, a RNG-RSP message is sent back over the backbone to the serving BS. The latter collects all received RNG-RSP messages from all scanned BSs and then sends them to the MS in the form of a MOB_ASC_REPORT message. Next, the serving BS responds with a MOB_SCN-RSP message specifying the length of the approved scan and the association type that will be used. Upon receiving that message, the MS may scan its neighbors by synchronizing with a given BSs DL transmissions and estimating the quality of the physical channel. After performing scanning, the MS sends a MOB_MSHO-REQ message to its serving BS on the basic connection. That message includes a list of BSs recommended by the MS as targets. Upon receiving such message, the serving BS sends HO-prenotification messages to all BSs specified in the MOB_MSHO-REQ message and waits for the corresponding HO-prenotificaion-response messages to analyze them. Next, the serving BS generates the MOB_BSHO-RSP message indicating a list of target stations and sends it back to the MS on the basic connection. The MS may now perform or cancel the handoff; it informs its serving BS via a MOB_HO-IND message sent on the basic connection. If it performs handoff, the MS will inform its serving BS that it is leaving it while providing the parameters of the target BS and then it registers with the target BS. The target BS may, upon receiving the HO-prenotification message, include a fast ranging information element (IE) that provides the MS with a noncontention-based initial ranging opportunity to minimize the handoff process latency.

#### 7.4.1.2   Base Station Initiated Handoff

The serving BS may decide to no longer manage an MS and initiate handoff for it. This occurs generally when the serving BS can no longer provide the required QoS or when it detects that the MS is moving out of its coverage area. Although the causes of a BS-initiated handoff are similar to the causes

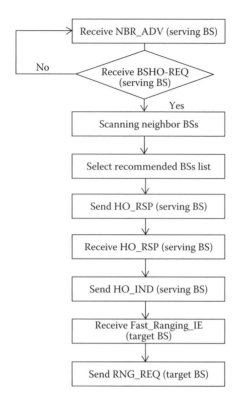

**FIGURE 7.7**   The BS initiated handoff at the MS level.

of an MS-initiated handoff, it is useful to let the BS decide to centralize the handoff procedure. In fact, the MSs are generally tiny equipments with limited power and computing resources; therefore, it is important to implement the handoff process at the BS level. The serving BS continues broadcasting the MOB_NBR-ADV message for the served MSs, but it orders the MS that needs to perform handoff via a MOB_BSHO-REQ message to start scanning the neighboring BSs. The MOB_BSHO-REQ message transmitted on the basic connection defines a list of recommended target BSs along with service level predictions and channel details. Upon receiving that message, the BS starts the scanning procedure and sends back a MOB_BSHO-RSP message to the serving BS indicating a list of recommended BSs. The rest of the handoff process is similar to the MS-initiated handoff case. In fact, the MS waits for the list of the target BSs and then sends a HO-IND message to its serving BS. Upon receiving the fast ranging IE, the MS sends the RNG-REQ ranging request message to the target BS to register with it. The BS-initiated handoff process, described earlier, is depicted by the flow chart in Figure 7.7.

### 7.4.1.3   Network-Initiated Handoff

The network can initiate handover depending on its current status. Such a decision can be made after the evaluation of the payload of different BSs or the data throughput at the reference points. In profile A, ASN GWs may initiate handover of MSs under their control [8]. The network-initiated handoff procedure depicted by Figure 7.8 begins by a prehandover operation, during which the serving ASN GW collects status information from the BSs and the MSs to decide whether a network-initiated handover is required. If it is the case, the ASN GW sends a HO_Directive message to the serving BS to order it to handoff some MSs to other BSs while providing it with a list of recommended BSs and starting a timer. The ASN GW may also specify how many payload should be migrated to other BSs to achieve load balancing as it may indicate the list of the recommended MSs that

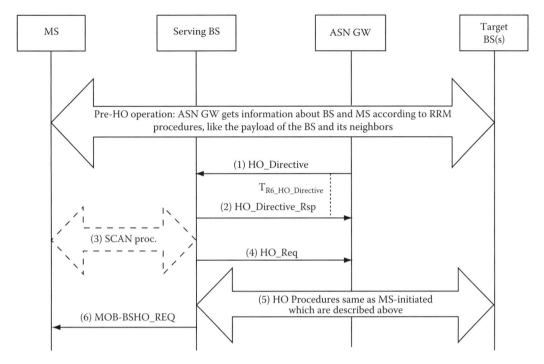

**FIGURE 7.8** The preparation phase of a network-initiated handoff.

need to be handed over. The serving BS should respond by a HO_Directive_Rsp message to make the ASN GW stop the timer. The serving BS selects some candidate MSs based on the information maintained by it and the list given by the HO_Directive message and then it may order some candidates to achieve scanning to get their neighbors' information. The serving BS will then select some suitable MSs for handover and send separately a HO_Req message relative to each MS to the serving ASN-GW. The following procedure is the same as the process of MS-initiated handoff described earlier. When the process of handover preparation is finalized by the network, the serving BS will send a MOB-BSHO_REQ message to each MS to order it to hand over to the target BS.

## 7.4.2 CELL SELECTION

Cell selection/resection enables a correct network topology acquisition and guides the handover process in selecting the best target BS.

### 7.4.2.1 Neighbor Advertisement from Serving Base Station

Each BS in the network should broadcast information describing the network topology via MOB_NBR-ADV messages. As stated earlier, these messages carry channel information for neighboring BSs normally provided by each BS's own DCD/UCD messages [11]. The serving BS may obtain such neighbors-related information over the backbone before broadcasting it to the managed MSs. Thanks to MOB_NBR-ADV messages, MSs are able to synchronize with the neighboring BSs without being obliged to monitor transmission for individual DCD/UCD broadcasts. The standard specifications fix the maximum period of sending the MOB_NBR-ADV message to 1 s so that an MS moving at high speed through the coverage area of each BS may get the message and perform handoff. It is valuable to note that the period of sending the MOB_NBR-ADV message determines the maximum speed, with which an MS is allowed to move through the network; therefore, it must be carefully chosen. Optimizing the handoff process requires a selection of the most suitable target BS that fits mobility path and application needs. To achieve that goal, MSs have to scan multiple channels to discover neighboring BSs and then select the best target. That selection may be based on different parameters such as the measured signal strength, the packet delay, the error ratio, the throughput, and the security level [12]. Cell reselection is achieved when the MS scans and/or associates with more than one BS to evaluate their suitability as handover target. The MS may integrate information obtained from a MOB_NBR_ADV message to give insight into available neighboring BSs for cell reselection. Note that the serving BS may allocate scanning intervals or sleep intervals for the cell reselection activity. However, such activity does not require terminating the ongoing connection with the serving BS.

#### 7.4.2.2 Periodic Intervals for Scanning Neighbor

Usually, the serving BS allocates time intervals known as scanning intervals to the MSs. Unfortunately, channel scanning can be a relatively time-consuming activity; therefore, MSs should process it and obtain the neighboring BSs list before performing handoff. The duration and frequency of scanning should be carefully determined to interleave scanning period and normal operations without affecting the network performances and the provided QoS. It is clear that a long scanning period increases the packets jitter and the end-to-end delay while imposing large buffer sizes. Contrarily, a short scanning period requires multiple iterations and increases the overall scanning duration. The scanning procedure depicted in Figure 7.9 begins when an MS sends a MOB-SCN_REQ message to its serving BS to request the allocation of a group of scanning intervals while indicating the estimated duration of time required for the scan [11]. The serving BS replies by a MOB-SCN_RSP message denying the request or stating the scanning interval duration that should be at least as long as requested by the MS. If no MOB-SCN_RSP message is received within a timer, the MS may retransmit the MOB-SCN_REQ message. The serving BS may also send an unsolicited MOB-SCN_RSP message with a value of zero associated with the scan duration to trigger the MS to report scanning result. Upon the receipt of a positive MOB-SCN_RSP message, the MS may begin scanning for one or more BSs during the time interval stated in that message. The MS may attempt to synchronize with the DL transmission of the scanned BS and estimate the quality of the PHY channel. IEEE 802.16e specifies a default scanning strategy requiring that each MS keeps a nonvolatile storage where it saves the

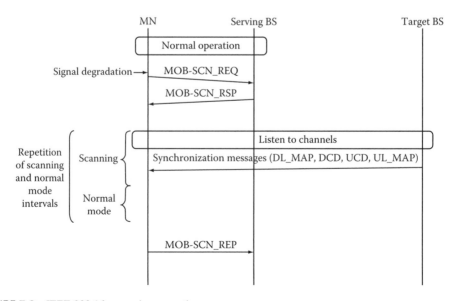

**FIGURE 7.9** IEEE 802.16e scanning operation.

last set of operational parameters. When the MS intends to acquire a DL channel, it should use its stored information. However, if that MS fails to obtain the DL channel, it will continuously scan the possible channels of the DL frequencies until it finds a DL signal. IEEE 802.16e specifications support temporarily suspending the communication between the BS and the MS during the scanning period. The exchange of MOB-SCN_REQ and MOB-SCN_RSP messages enables each entity to buffer packets while the normal communication is temporarily suspended [11,12]. An MS may end scanning and return to the normal operation mode anytime during any scanning interval; this is achieved by sending a MAC PDU message such as a bandwidth request to the target BS. At the end of scanning, the MS should report the scan status to its serving BS via a MOB-SCN_REP message.

### 7.4.3 Measurements for Handoff Optimization

Handoff optimization is a key challenge for the network management as it results in the enhancement of the network performances by optimizing throughput, routing, delay profiles, delivered QoS, and communication costs. Therefore, mobile WiMAX should take into consideration numerous parameters that implement a particular handoff policy to optimize the handoff performance, especially in case of interaction with heterogeneous networking technologies. It is valuable to note that IEEE 802.16e specifications consider the handover decision algorithms beyond the scope of the standard [11].

#### 7.4.3.1 Monitoring Parameters

Radio resource management (RRM) specifications should guarantee efficient resource utilization in WiMAX networks by assisting functions such as QoS provision, service flow admission control, and mobility management. RRM mechanisms shall be implemented in ASNs either with BSs that directly communicate between them or with BSs having no direct communication between them or at a centralized RRM entity that does not reside in a BS but that collects and monitors radio resource indicators from several BSs. Each BS should collect data about the neighboring BSs either through static configuration data or through a different RRM entity aware of the dynamic load of neighboring BSs. RRM mechanisms should support network functions in taking the required decisions. For instance, RRM may assist handoff preparation and control to improve the overall performances. For example, RRM mechanisms may optimize the system load control by selecting the most suitable target BS during handoff or updating the list of recommended BSs to handoff. The radio resources available at a BS where a BS-ID defines a sector with a single frequency assignment may be a good hint for BS selection during network entry or handoff. More specifically, the available radio resource indicator, which gives the percentage of reported average available subchannels and symbols resources per frame, may be used to select recommended target BSs that achieve approximately equal load. Averaging is calculated over a configurable time interval with a default value of 200 frames [8]. Choosing the more suitable BS as a target BS is done through scanning and association activities. The criteria that enlighten the choice may include the link quality in the UL direction. Such handoff monitoring parameters are measured by the MSs and then sent in a form of report to the managed BS, which may deliver them to the radio resource controller (RRC). The RRC, which is generally implemented at the ASN GW or at a BS, controls multiple radio resource agents (RRAs) located at the BS level. RRAs maintain a database of collected radio resource indicators received by the managed MSs or by the respective BSs. For instance, the PHY reports for DL and UL per MS include a set of parameters specified in Ref. [13]. In fact, IEEE 802.16d amendments define the receive signal strength indicator (RSSI) and the CINR as two main signal quality indicators that help in assigning BSs and selecting adaptive burst profiles. The specifications also define the mean and the standard deviation because the channel behavior varies in time. These indicators enable the channel quality monitoring and may be augmented to include measurements related to QoS parameters such as the burst error rate [8]. It is valuable to note that the RSSI measurements do not require receiver demodulation lock; therefore, they provide reliable channel-strength estimation even at low signal levels. Each BS has to collect the RSSI measurements while each MS shall obtain an RSSI measurement in an implementation-specific fashion. After performing successive RSSI measurements, the MS should derive and update estimates of the mean and the standard deviation of the RSSI and then report them in units of dBm via a REP-RSP message [13]. To obtain such a report, statistics are quantized in 1 dB increments ranging from $-40$ to $-123$ dBm while the values outside that range are assigned the closest extreme value within the scale. IEEE 802.16d specifications do not impose how to estimate the RSSI of a single message, but they claim the relative accuracy of a single-signal-strength measurement taken from a single message to be 2 dB with an absolute accuracy of 4 dB. Nevertheless, the standard specifications propose one possible method to estimate the RSSI of a signal of interest at the antenna connector by equation: $\text{RSSI} = 10^{-G_{rt}/10} * [(1.2567 * 10^4 V_c^2)/(2^{2B})R] * (1/N \sum_{n=0}^{N-1} |Y_{\text{I or Q}}[k,n]|)^2 mW$ where $B$ is ADC precision (number of bits of ADC), $R$ is ADC input resistance Ohm, $V_c$ is ADC input clip level (Volts), $G_{rt}$ is analog gain from antenna connector to ADC input, $Y_{\text{I or Q}}[k,n]$ is $n$th sample at the ADC output of I- or Q-branch within signal $k$, and $N$ is number of samples. IEEE 802.16d specifications also define how to update the (linear) mean RSSI statistics (mW) derived from a multiplicity of single messages, how to calculate the mean estimate in dBm, and how to update the expectation-squared statistic to solve for the standard deviation in dB (for more details, refer to Ref. [13]). On the other hand, CINR measurements require receiver lock but they enlighten us on the current operating condition of the receiver including the signal strength and the interference and noise levels [13]. Each BS has to collect the CINR measurements while each MS shall obtain a CINR measurement in an implementation-specific fashion. After performing successive CINR measurements, the MS should derive and update estimates of the mean and the standard deviation of the CINR and then report them in units of dB REP-RSP messages [13]. IEEE 802.16d specifications do not impose how to estimate the CINR of a single message, but they claim the relative accuracy of a single signal strength measurement taken from a single message to be 1 dB with an absolute accuracy of 2 dB.

### 7.4.3.2 Optimization Functions

The handoff process should not decrease the overall performance; therefore, it should implement mechanisms that optimize the delay to support real-time applications such as VoIP and video streaming. Connection dropping should also be reduced. IEEE 802.16e and mobile WiMAX specifications define prescan mechanism to measure the radio connection and select the target BS before handoff execution. Nevertheless, the handoff procedure should include not only layer 2 handoff but also the IP layer handoff for IP-based services. Figure 7.10 below shows that layer 3 handoff, which includes movement detection, IP configuration, and location registration subphases, highly affects the overall handoff delay and requires much more time to be executed than layer 2 handoff. Optimizing such delay is a hot research issue. For instance, the fast handoff procedure proposed in Ref. [14] proposes to process the configuration of CoA, duplicated address detection (DAD), etc. in advance to reduce the handoff latency. However, the mobile IP procedure always begins after ending layer two

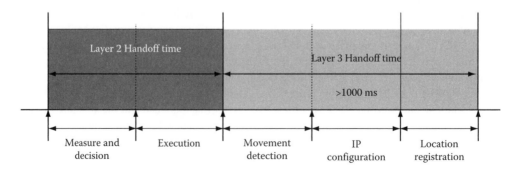

**FIGURE 7.10**  The handoff process for IP based services.

handoff so that the handoff delay is the summation of the layer 2 handoff delay and layer 3 handoff delay. Consequently, it is imperative to consider the correlation between layer 2 and layer 3 within a cross-layer scheme [15]. Authors in Ref. [15] analyzed the signaling message flow sequence and the format of both IEEE 802.16e and fast MIPv6 to correlate both handoff procedures and minimize the required signaling overhead. For instance, authors in Ref. [15] propose to integrate some layer 3 handoff information with the MOB_HO_IND message and the RNG_REQ as they share semantic characteristics while performing the handoff.

Regarding QoS, mobile WiMAX specifications define different classes of services having different requirements in the quality of the traffic delivered to MSs. This quality is generally measured in terms of data integrity, latency, and jitter. The handoff process should be optimized to not highly affect these parameters. For instance, maintaining data integrity during handoff means that the packet loss, duplication, or reordering rates will not be considerably increased while the impact on the DP setup latency/jitter should be minimized [8]. From the QoS point of view, there are controlled handoffs and uncontrolled handoffs. A controlled handoff should respect the following conditions:

- If the handoff is initiated by the MS, the latter should send to its serving BS a list of potential targets.
- Network should base its target selection on the list of potential targets provided by the MS.
- Network should inform the MS with the list of available targets for handoff. If that list is empty, the network will refuse to accept MS handoff. The available targets list should be a subset of the one requested by the MS or reported by it.
- MS should process handoff by moving to one of the provided targets or it should cancel the handoff. The decision is sent via the MOB_HO-IND message.

When any of these conditions is not respected, the handoff is considered uncontrolled and does not provide any QoS guarantees. In the worst case, the MS can connect to the target BS without any indications given to that target BS inducing an unpredictive handoff [8]. To provide data integrity and optimize the handoff process, several mechanisms are provided and classified into two main groups: DP setup mechanisms and DP synchronization mechanisms. DP setup mechanisms refer mainly to buffering and bi-multicasting. Buffering consists in saving the traffic of the services for which data integrity is required at the DP originator or the DP terminator level. Nevertheless, the buffering point may change during handoff based on the data integrity mechanism selection. Moreover, buffering may be done only during handoff or for simplicity within the lifetime of a session. On the other hand, multicasting refers to multicasting downstream traffic at the originator endpoint of the DP while bicasting consists in bicasting traffic to the serving element and to only one target. It is worth noticing that bicasting achieves better performances when it is combined with buffering. DP synchronization mechanisms aim at guaranteeing data delivery in different data functions, which buffered the different DPs (serving and target) used to deliver the data during handoff. It is achieved by using either sequence numbers, data retrieving, or Ack window with sequence number disablement. A sequence number is attached to each SDU in the ASN DP and then incremented by one every time it is forwarded in the DP. Data retrieving does not require the definition of sequence numbers; the anchor DP function buffers or copies the data during handoff preparation, and when a final target BS is identified, the serving BS will push back all the nonsent packets to anchor/target DF. When Ack window with sequence number disablement is implemented, data storage buffers in anchor DP are released by full or partial ACKs from the serving BS without requiring sequence numbers. Vertical handoff between mobile WiMAX and other wireless networks represents a hot research topic as it requires the implementation of optimization algorithms that are able to decide when to initiate handoff and which access network to choose while minimizing the handoff latency and preserving the security context and the required QoS. Authors in Ref. [16] have developed a vertical handoff decision algorithm that guarantees load balancing among attachments points such as APs and BSs while maximizing the collective battery lifetime of MSs. They also propose a route selection algorithm that selects the most suitable AP that minimizes the collective battery lifetime while achieving load balancing [16].

## 7.5 HETEROGENEOUS HANDOFF

Subscribers are becoming more and more demanding regarding roaming across different networking technologies such as WiFi, WiMAX, CDMA, and xDSL as they claim service continuity with satisfying QoS and security levels. On the other hand, achieving seamless handoff between different access technologies is a challenging issue as it obeys to different performance, QoS, and security constraints. More specifically, heterogeneous handoff requires collecting network information from different entities that are not necessarily interoperable and then applying optimized policies to take the wisest handoff decisions and processing them.

### 7.5.1 IEEE 802.21 HANDOFF STANDARD

IEEE 802.21 specifications aim at defining a framework that supports information exchange between entities belonging to different access networks and helps in taking handoff decisions; they also define a set of functional components that have to execute those decisions [17]. The proposed framework is built around the media independent handover function (MIHF), which lies between layer 2 and layer 3 and provides higher layers with abstracted services through a unified interface. Three services can be distinguished: the event service, the command service, and the information service.

*Media independent event services (MIES):* In general, handover is triggered by multiple events that may occur at the terminal or the network level. The MIES provides event classification, event filtering, and event reporting corresponding to dynamic changes in link characteristics, link status, and link quality [18]. Figure 7.11 shows that events can be classified into link events and MIH events. Besides, the event model follows the subscription/notification procedure. More specifically, the MIHF registers link event notifications with the interface while any upper-layers entity may register for an MIHF event notification. Link events are generated by both the physical and the MAC layers and then sent to the MIHF which has to report them to any entity that has registered an MIH event or a remote MIH event. MIH events are generated by the MIHF and may be local or remote. Local events occur at the client level while remote events take place in the network entities. Both link and MIH events are classified into six categories, which are administrative, state change, link

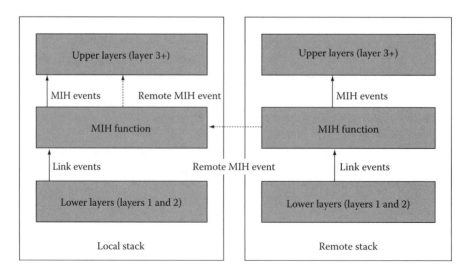

**FIGURE 7.11**   Media-independent event services.

parameter, predictive, link synchronous, and link transmission. The most common events include link up, link down, link parameters change, link going down, link handover imminent, etc. Upon being notified about certain events, the upper layer uses the command service to control the links to switch to a new PoA when needed [37,38].

*Media independent command service:* The media independent command service (MICS) enables MIH users to manage and control link behavior relevant to handovers and mobility [18]. More specifically, MICS commands aim at gathering information about the status of the connected links along with transferring higher-layer mobility and connectivity decisions to lower layers. As shown in Figure 7.11, upper layers generate MIH commands which may be local or become remote depending on the destination entity. These MIH commands are translated by the MIHF to link commands which differ depending on the access network being used. Examples of MIH commands include MIH poll, MIH scan, MIH configure, and MIH switch. These intend to poll connected links to get informed about their most recent status, to scan for newly discovered links, to configure the new links and then to switch between the available links [17,18].

*Media independent information service:* Mobile terminals need to discover the heterogeneous neighboring networks and communicate with their entities to optimize the handoff process and facilitate seamless handoff when roaming across these networks. To address this issue, media independent information service (MIIS) provides the capability for obtaining the necessary information for handovers [18]. Such information includes neighbor maps, link-layer information, and availability of services related to a particular geographic area. MIIS enables all layers to share IEs required for making handoff decisions; it also specifies the information structure and its representation while implementing a query/response type mechanism. Information may be either static such as the names and providers of the mobile terminals neighboring networks or dynamic such as channel information, MAC addresses, security information, etc. MIIS fulfils quick data transfer understood by different access technologies with very little decoding complexity. In fact, reports can be transferred either by means of type-length value (TLV) or using a schema referred to as resource description framework (RDF) which is represented in eXtensible Markup Language (XML). As layer 2 information may not be available or may not be rich enough to make intelligent handoff decisions, higher-layer services may be consulted to assist in the mobility decision making process.

*Service access points:* service access points (SAPs) represent the API through which the MIHF can communicate with the upper-layer and the lower-layer entities using IEEE 802.1 primitives. More specifically, we distinguish three types of SAPs: lower-layer SAPs, upper-layer SAPs, and management plane SAPs. Lower-layer SAPs vary according to the network access type as defined by each MAC and PHY relative to the corresponding access technology. On the other hand, upper-layer SAPs define the interface between the MIHF and upper-layer mobility management protocols or policy engines implemented within a client or a network entity. Finally, management SAPs describe the interfaces between the MIHF and the management plane of various networks.

## 7.5.2 Handoff with UMTS and WiFi

Mobile WiMAX intends to combine the advantages of both WiFi networks and third generation networks such as UMTS while intercommunicating with both.

### 7.5.2.1 Handoff with UMTS

If we carefully examine the reference models of both WiMAX and UMTS networks, we will notice that they share many similarities. For instance, the WiMAX ASN may be directly mapped to the UMTS terrestrial radio access network (UTRAN) while the CSN may be mapped to the UMTS core network as their functionalities are distributed in the same fashion [7]. Moreover, the QoS classes of service provided by both technologies enable a direct transfer of traffic flows from one network to another. In fact, the UMTS conversational class may be mapped to the UGSs class, the UMTS streaming class to the nrtPS, the UMTS interactive service class to the (extended) rtPS, and the UMTS background class to the BE class. Regarding security, both UMTS and WiMAX networks support mutual authentication and provide integrity, replay protection, confidentiality, and nonrepudiation guarantees. Nevertheless, moving from WiMAX to UMTS will induce degradation in the provided data rates. Network planning authorities may choose between deploying loose or tight coupling between WiMAX and UMTS networks depending on the degree of integration of both networks [7]. When loose coupling is implemented, access to the 3G AAA services is guaranteed by the packet data gateway (PDG) edge routers connected to the WiMAX network or routed through the Internet. PDG routers provide a tunnel termination gateway (TTG) between the WiMAX network and the mobile core network while providing charging gateway interfaces, IP address allocation, authentication in external networks, and single access to mobile core network packet data domain services [19]. Meanwhile, WiMAX and UMTS traffics are separated so that WiMAX providers can implement their own mobility, authentication, and billing mechanisms while signing roaming agreements with 3G providers. 3G providers will benefit from extending the coverage of their networks while reducing the deployment costs and serving rural or inaccessible areas. However, loose coupling requires the implementation of higher level mobility management protocols that need to be media independent; thus adding complexity and latency when roaming between different access networks. To address this issue, each network should share its admission control and resource management information with the neighboring networks.

When tight coupling is implemented, WiMAX and UMTS networks will use the same core network components including gateways, AAA entities, and infrastructure. More specifically, the WiMAX network will be connected to the UMTS gateway GPRS support node (GGSN) and the packet-switching domain of the UMTS core network. The UMTS considers the WiMAX network as a radio network controller (RNC); therefore, mobile terminals need to implement the UMTS protocol stacks. Compared to loose coupling, tight coupling facilitates the management procedures as the same billing and authentication, and mobility protocols can be used for both networks. Besides, vertical handoff may be easily initialized and optimized as the core network possesses a clear view of resource status in both networks. However, introducing WiMAX traffic into the UMTS network should be preceded by adapting some core network entities to handle new types of load and traffic patterns to avoid capacity conflicts. Last but not least, the tight-coupling approach is less flexible in coverage extensions than loose coupling; consequently, it is generally practical to adopt it when WiAMX and UMTS networks are owned by the same operator and the integration is implemented by means of adding patches to the existing components.

### 7.5.2.2 Handoff with WiFi

IEEE 802.11e standard aims at providing QoS guarantees at the LAN scale by specifying differentiations mechanisms at the WiFi MAC layer. However, WiFi networks-limited coverage prevents them from providing continuous Internet-related services and real-time applications support anywhere and at any time. On the other hand, IEEE 802.16e, from which inherits mobile WiMAX, faces some problems related with energy saving and quality providing in some indoor areas [20]. Consequently, integrating WiMAX and WiFi is a hot topic that is currently attracting an increasing number of researchers. Authors in Ref. [20] propose a multilayer network protocol architecture that aims at addressing QoS and mobility issues for integrating WiFi and WiMAX. In fact, they designed a two-tier network that considers a 802.16e cell as overlay cell that overlays a few WiFi cells and 802.11e cells as underlay cell cluster. When the MS stills under the coverage of WiFi cells, handoff will be performed horizontally as WiFi can provide high bandwidth and good performance. The AP, which offers the highest bandwidth value and reduces the unnecessary handover probability due to signal strength dropping down, will be selected as target AP. Authors in Ref. [20] propose to analyze multiple parameters in a cross-layer fashion to optimize the handoff decision. Examples of such parameters, which are adopted during the simulation, include residential time, WiMAX-cell capacity, and blocking and dropping probability.

The proposed network stack supports three IEEE 802.11 physical layers which are 802.11b, 802.11g, and 802.11n, providing data rates of 11, 54, and 250 Mbps. The network stack also includes IEEE 802.16 and IEEE 802.16e physical and MAC layers. A handover layer based on the IEEE 802.21 standard lies between the lower layers and the network layer implementing the Fast Mobile IPv6 protocol. Management entities are present at each layer and implement all the required provisioning, maintenance, operation, and administration functions. They are also in charge of communicating with servers processing the QoS policies, the mobility decisions, and the profiles storage. Authors in Ref. [20] propose to define unified layer 2 abstractions in the IEEE 802.21 to support layer 3 fast handoff. Moreover, WiFi/WiMAX and Fast Mobile IP interaction is achieved by the IEEE 802.21 primitives. They also base the handoff decision on the analysis of different link parameters such as signal strength, velocity of MNs, delay, service level prediction, etc. For instance, the target cell selection is speed sensitive, which means that the MNs are directed to the appropriate cell layer according to their velocity to decrease the blocking and dropping probabilities. Authors in Ref. [20] have also added a handoff protection mechanism consisting in reserving two free guard channels for handoff usage in advance to minimize the dropping of handoff calls. The proposed mobility management scheme depends on three types of arrival hosts in the WiFi cell [20]. If a filtered MN arrives in the WiFi cell from neighboring WiMAX cells and overlaid WiMAX cell, the scheme does not change its state. However, if a nonfiltered MN arrives in the WiFi cell from overlaid WiMAX cell and initial session itself, and if the residential time is longer than the residential time threshold and the WiMAX cell has enough capacity, then it will be overflowed to the WiMAX cell to reduce the handoff probability. Otherwise, it will be assigned to the WiFi cell. Authors have also defined overflow thresholds to reduce the dropping of a handoff call and the blocking of a new call. When a nonfiltered MN arrives in the WiFi cell and the blocking and dropping probabilities of the WiMAX cell are less than the overflow threshold, it will be overflowed into the WiMAX cell. Otherwise, it will be assigned to the WiFi cell.

## 7.6 HANDOFF SECURITY

Managing subscribers mobility while guaranteeing security implies managing handoff security which includes maintaining authentication, service flows, and key distribution between the BSs involved in the handoff process.

### 7.6.1 KEY MANAGEMENT PROTOCOL

IEEE 802.16e specifications guarantee authentication by using a public-key interchange protocol which establishes the keys and then uses them for authenticating the communicating entities. These keys are then used to produce other keys that will protect management messages integrity and transport the traffic encryption keys (TEKs). Two privacy key management protocols are defined: the PKMv1 and the PKMv2. PKMv2 is an enhancement of the PKMv1 as it enables both device and user authentication and defines a new key hierarchy while supporting advanced encryption standard (AES)-block cipher-based message authentication code (CMAC), AES keys wraps, and multicast and broadcast services (MBS). WiMAX 802.16e defines the Privacy Key Management Protocol (PKMv2) which enables an authentication scheme based on Extensible Authentication Protocol (EAP). The EAP-based authentication begins at the initial network entry when the MS sends an EAP Start message to its managing BS containing uniquely a message header. Next, the MS and the BS start an EAP conversation by exchanging an EAP Transfer message without HMAC/CMAC digest. If the BS has to send an EAP Success message during the EAP conversation, it should send an EAP payload to the MS with PKMv2 EAP Complete message signed by a newly generated EAP integrated key (EIK). The MS can then validate the message and possess EIK and pairwise master key (PMK). Nevertheless, if the MS and the BS negotiate the double EAP mode consisting in the authenticated EAP after EAP, the BS will wait for the EAP Start message before beginning the second round of the EAP authorization. After ending the authorization process, both the MS and the BS will generate the authorization key (AK) according to the primary authentication key (PAK), PMK, and PMK2 in their possession. After that, a security association-TEK three-way handshake will be performed [21]. Note that the authenticated-EAP authorization procedure protects the EAP payload by a HMAC or a CMAC digest using the EIK generated at the first EAP round.

### 7.6.2 AUTHENTICATION

The mobile WiMAX specifications stated in Ref. [8] define the AAA framework as the provider of the authentication services, the authorization services, and the accounting services. The authentication services consist in achieving the device, the user or the combined device, and user authentication along with the mutual authentication while the authorization services include the delivery of information to configure the session for access, mobility, QoS, and other applications. Accounting services manage the information required for both prepaid and postpaid billing and any information that may be used for auditing session activity by both the home NSP and visited NSP. The ASN defines one or more network access server(s) (NASs) which are viewed as the first AAA client where AAA messages originate and authentication and authorization attributes are delivered to AAA applications including the authenticator, the mobility applications (Proxy-Mobile IP, PMIP; foreign agent, FA), and the QoS applications. Mobile WiMAX specifications state that authentication and authorization have to be based on EAP. Different EAP methods can be supported but they need to support the provisioned credential types [8].

User authentication is based on PKMv2 which is used to transfer EAP over the IEEE 802.16 air interface between the MS and the BS. The authenticator may not be located at the BS level; therefore, the BS may forward the EAP messages using the authentication relay protocol to the authenticator. The latter will encapsulate the EAP messages with conformance to the AAA protocol packets format and then forward them via one or more AAA proxies to the AAA server within the CSN of the home NSP. The AAA server keeps the matching between the subscription and the correspondent subscriber. When the MS is roaming, multiple AAA brokers with AAA proxies may be involved in the authentication process. Mobile WiMAX specifications impose the use of simple or double EAP authentication schemes for achieving device authentication and reject the RSA-based authentication in this case. The EAP methods used for authenticating mobile WiMAX devices have to generate the master session key (MSK) and the extended master session key (EMSK). The network access identifier (NAI) is used as identifier within EAP-based user and device network access authentication. Device credentials are either a device cert or a device PSK while the EAP device identifier should be an AMC address or an NAI in the form of MAC_address@NSP_domain depending on where the device authentication terminates. The PKMv2 procedure occurring at the network entry of the MS is depicted in Figure 7.12.

First, the initiation of the network entry according to the IEEE 802.16e specifications begins by a successful ranging that is followed by the exchange of SBC messages between the MS and the ASN. That SBC negotiation consists in negotiating the PKM version, PKMv2 security capabilities, and authorization policy. When the 802.16 air link is established between the BS and the MS, a link activation notification is sent to the authenticator that will immediately begin the EAP sequence. The EAP exchange may then begin; it results in the attribution of a particular NAI realm. The shared master session key (MSK) and the EMSK are then established in both the MS and the AAA server. The MSK will be transferred to the authenticator in the ASN within a RADIUS accept message while the EMSK will be retained at the home AAA server. The MS and the authenticator will generate the pairwise master key (PMK) from the MSK. Besides, the MS and the AAA server will generate the mobility keys from the EMSK, thus marking the end of the authentication part of the authorization flow. The authenticator and the MS will generate the AK from the PMK and then the key distributor entity at the authenticator level will deliver the AK and its context to the key receiver entity in the

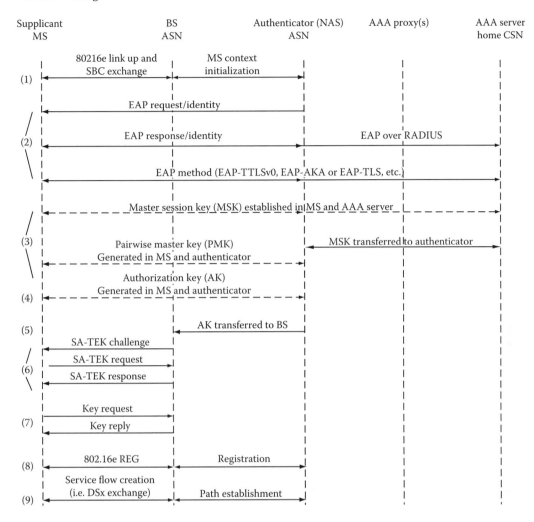

**FIGURE 7.12**　PKMv2 Authorization and authentication procedure.

serving BS. The key receiver entity will cash the received information and generate the subsequent subordinate IEEE 802.16e keys from the AK and its context. The MS and the BS will also process the PKMv2 three-way handshake procedure which provides the MS with the list of SA descriptors identifying the primary and the static SAs. At this step, the MS and the BS may use the newly acquired AK for MAC management messages protection. For each SA, the MS will request two TEKs from the BS. The managing BS will randomly generate the keys, encrypt them using the symmetric KEK key and then transfer them to the BS. Next, the MS will process a network registration and the authenticator will be informed of its competition. Finally, the authenticator will trigger the service flow and the DP establishment process. During handoff, the PKMv2 procedure needs to be optimized. For instance, when the mobile moves within the same mobility domain, the AK will be validated by signing and verifying a frame via the CMAC using the AK that is newly derived from the same PMK as long as the PMK remains valid. The AK validation also needs to be combined with ranging. Last but not least, it should be possible to share the TEK within the same mobility domain when handover procedure between two BSs can transfer TEK context information. If the TEK is shared between a set of BSs, those BSs are considered as the same security entities within the same trusted domain.

### 7.6.3 Data and Control Traffic Protection

IEEE 802.16e designers tried to address the security flaws of the IEEE 802.16d related to the protection of the control and the data traffic. For instance, IEEE 802.16e specifications introduce the new data link cipher Advanced Encryption Standard Counter with cipher-block-chaining MAC referred to as AES-CCM (counter with cipher-block chaining message authentication code). CCM mixes the counter mode encryption for data confidentiality with the CBC-MAC for data authenticity. This new cipher addresses the most fundamental deficiency of the original data protection scheme which is the lack-of-a-data-authenticity mechanism. The AES-CCM has been selected for many reasons. First, the U.S. National Institute of Standards and Technology has stated that CCM will become an approved mode for AES [22]. Second, this mode is used in IEEE 802.11i. CCM can protect authenticated but unencrypted data so that the encryption scheme protects the GMH. Third, no intellectual property claims have been made against CCM. When using the AES-CCM cipher, the transmitter has to define a unique nonce that is a per-packet encryption randomizer. As specified by IEEE 802.16i standard, IEEE 802.16e adds a packet number to each MPDU to ensure each nonce's uniqueness while the receiver should verify that the received packets correctly decrypt under AES-CCM and have a monotonically increasing packet number. A traffic encryption state machine with periodic TEK refresh improves the data traffic protection.

The control message's protection is achieved using the AES based CMAC where CMAC refers to the cipher-based MAC or the MD5-based HMAC schemes. When adopting AES-CMAC, the computation of the keyed hash value defined in the CMAC Tuple should use the CMAC Algorithm with AES. The DL authentication key CMAC_KEY_D will be used to authenticate messages in the DL direction while the UL authentication key CMAC_KEY_U authenticates messages in the UL direction. Theses keys are generated from the AK [11]. Compared to the checksum mechanism and the error-detecting codes, AES-CMAC guarantees stronger data integrity as it is able to detect intentional and unauthorized data modifications as well as accidental nonmalicious modifications [23]. AES-CMAC achieves the security goals of the HMAC; nevertheless, it is more appropriate to use it for information systems in which the symmetric key block cipher AES is more available than a hash function.

### 7.6.4 Security of Heterogeneous Handoff

Handover between different access networks is a hot research issue because it involves numerous challenges such as security, QoS provision, and billing. If we consider the handoff between IEEE 802.16e or mobile WiMAX and 3G, we will notice that the provided security services should be readdressed. For instance, an MS that accesses Internet services while switching from IEEE 802.16e to the 3G network or vice versa needs to properly process the authorization, authentication, and accounting steps to be able to maintain the ongoing session. If that MS has to execute AAA procedures all over again, the AAA server will be excessively requested while the handoff delay will increase as the MS will perform complicated processes. For all these reasons, the efficiency of the network will be reduced while the ongoing session may be interrupted. Authors in Ref. [24] have particularly addressed the authentication issue in case of vertical handover between IEEE 802.16e and 3G networks. The exemplary network environment they defined is made up of an IEEE 802.16e network, a 3G network, an AAA server, and an interbase station protocol (IBSP) mobility agent server. The 3G network includes an SGSN, a RNC, and multiple 3G BSs. The IEEE 802.16e network is made up of multiple BSs and an Access Router (AR) responsible for managing the wireless areas of several BSs.

When the MS tries to access the Internet from the IEEE 802.16e network via the BS, the managing BS of that MS will access the AAA server via the AR and then execute the AAA processes. After achieving AAA, the MS will be able to access the Internet while other IEEE 802.16e BSs will share the information on that MS by receiving the IBSP message multicasted from the managing BS. The IBSP message is first unicasted using UDP by the managing BS to the IBSP mobility agent server which is in charge of forwarding it. If the MS changes its managing BS but stills in the same IEEE 802.16e network during the session, the new managing BS will multicast the IBSP message throughout the entire network and update the location information of the MS without consulting the IBSP mobility agent server. The IBSP should guarantee the secure exchange of the MSs security context occurring between the current BS and the new BS during handoff. The 3G BS encompasses a virtual BS, a Node-B (NB), and a virtual BS-to-NB (VBS2NB) communication controller. The virtual BS is the representative BS of the 3G network which may be connected to Internet via the SGSN. When it receives the UDP packet with the IBSP message, the 3G BS treats it so that the IEEE 802.16e and the 3G networks can share the same information. The VBS2NB enables the NB to communicate with the virtual BS. The IBSP mobility agent server plays the role of intermediate between the IEEE 802.16e and the 3G networks. It is composed of an IBSP processing module, a representative BS list, and an MS database. The IBSP processing module treats the IBSP messages so that they may be shared between BSs in both access networks connected by the ISP. The representative BS list contains IP addresses of representative BSs in all access IEEE 802.16e and 3G networks that may be connected to each other by the ISP. The MS database contains all information related to the MS and found in the IBSP message such as the authentication key, the BSID, and the accounting information. That database is updated every time the MS enters a new access network. When the MS roams from the IEEE 802.16e network to the 3G network, the MS should issue a request to the 3G BS while the NB in the BS should generate a request for authentication information on the MS to the VBS2NB controller by using the MS-related information. The VBS2NB issues a request for authentication to the virtual BS. The latter should verify whether it possesses the requested authentication information. In that case, the BS will provide that information to the VBS2NB; otherwise, the BS should indicate that it has failed in providing the required data. The VBS2NB will transmit the information to the NB which will decide whether to perform an authentication process on the MS based on the received information. Next, the NB issues a request for PPP connecting the MS to the SGSN via the RNC without requiring generating additional authentication request for authenticating the MS to the SGSN. When the MS returns to the IEEE 802.16e network, the same procedures will be executed in the reverse order.

## REFERENCES

[1] R. Persaud, *Core Network Mobility*: *Active MPLS*, ISBN-3832255648, Shaker Verlag, GmbH, Germany, 2006.

[2] M. Handley, E. Schooler, H. Schulzrinne, and J. Rosenberg, Session Initiation Protocol, RFC 2543.

[3] Mobile WiMAX Part I: A technical overview and performance evaluation, Prepared on behalf of the WiMAX Forum, February 21, 2006. Available at www.wimaxforum.org

[4] S. Fili, Fixed, nomadic, portable and mobile applications for 802.16-2004 and 802.16e WiMAX Networks, WiMAX Forum, November 2005.

[5] J. Wolnicki, The IEEE 802.16 WiMAX broadband wireless access; Physical layer (PHY), medium access control layer (MAC), radio resource management (RRM), January 14, 2005.

[6] A.V. Aikaterini, Security of IEEE 802.16, Master thesis, Department of Computer and Systems Science, Royal Institute of Technology, 2006, available at dsv.su.se/en/seclab/pages/pdf-files/2006-x-332.pdf, last access on April 20, 2008.

[7] M. Carlberg Lax and A. Dammander, WiMAX—A study of mobility and a MAC-layer implementation in GloMoSim, Master thesis in Computing Science, April 6, 2006.

[8] WiMAX Forum Network Architecture, (Stage 2: Architecture tenets, reference model and reference points), Part 2, release 1.1.0, July 2007, available at www.wimaxforum.org

[9] Z. Becvar and J. Zelenka, Handovers in the Mobile WiMAX, In *Research in Telecommunication Technology 2006 - Proceedings [CD-ROM]*, 2006, vol. I, pp. 147–150. ISBN 80-214-3243-8.

[10] D. Pareek, *The Business of WiMAX*, ISBN-13 978-0-470-02691-5 (HB), John Wiley & Sons Ltd, Chichester, UK, 2006.

[11] IEEE standard for local and metropolitan area networks Part 16: Air interface for fixed and mobile broadband wireless access systems amendment 2: Physical and medium access control layers for combined

fixed and mobile operation in licensed bands and corrigendum 1, *IEEE Computer Society and the IEEE Microwave Theory and Techniques Society*, February 28, 2006.

[12] R. Rouil and N. Golmie, Adaptive Channel Scanning for IEEE 802.16e, Military Communication Conference (MILCOM), Etats-Unis, 2006.

[13] IEEE Std 802.16-2004. Part 16: Air interface for fixed broadband wireless access systems. Technical Report, June 2004. IEEE Standard for Local and metropolitan area networks.

[14] R. Koodli, Fast Handovers for Mobile IPv6, IETF RFC-4068, July 2005.

[15] F. Hsieh, Y. Chen, and P. Wu, Cross layer design of handoffs in IEEE 802.16e Network, *2006 International Computer Symposium Conference*, P703–P708, January 29, 2007.

[16] S. Lee, K. Sriramy, K. Kim, J. Lee, Y. Kim, and N. Golmie, Vertical handoff decision algorithms for providing optimized performance in heterogeneous wireless networks, to be presented in part at the IEEE GLOBECOM 2007, Washington, D.C., November 2007.

[17] A. Dutta, S. Das, D. Famolari, Y. Ohba, K. Taniuchi, T. Kodama, and H. Schulzrinne, Seamless handover across heterogeneous networks—an IEEE 802.21 centric approach, WPMC 2005.

[18] Draft IEEE standard for local and metropolitan area networks: Media independent handover services. IEEE P802.21/D01.00. March 2006.

[19] Mobile WiMAX Security, airspan white paper, wimax forum certified.

[20] S. Yang and J. Wu, Mobility management schemes with fast handover in integrated WiFi and WiMAX networks, Workshop ICS 2006.

[21] D. Pang, L. Tian, J. Hu, J. Zhou, and J. Shi, Overview and Analysis of IEEE 802.16e Security, Auswireless 2006 Conference, available at http://epress.lib.uts.edu.au/dspace/handle/2100/172, last access on 20 2008 April.

[22] D. Johnston and J. Walker, Overview of IEEE 802.16 Security, *Security and Privacy*, 2(3), 40–48, May–June, 2004.

[23] JH. Song, J. Lee, T. Iwata, The AES-CMAC Algorithm, IETF RFC-4493.

[24] P. Kim and Y. Kim, New authentication mechanism for vertical handovers between IEEE 802.16e and 3G wireless networks, *International Journal of Computer Science and Network Security*, 6(9B), 138–143, September 2006.

# 8 Power Management in Mobile WiMAX

*Georgios S. Paschos, Ioannis Papapanagiotou, and Thomas Michael Bohnert*

## CONTENTS

Mobile devices tend to incorporate more and more processing units and functionalities which has a negative effect on battery lifetime. For this reason IEEE 802.16e workgroup has standardized those mechanisms that would augment battery lifetime in a worldwide interoperability for microwave access (WiMAX) network without affecting the quality-of-service (QoS) performance. This chapter gives an overview of those mechanisms, includes the methods to analyze them, and identifies those elements that could decrease the waiting time when in sleep mode.

## 8.1 INTRODUCTION

By definition, designed for mobile devices, whose utility is inherently constrained by limited energy supply, efficient power saving mechanisms are decisive criteria for IEEE 802.16e to become a compelling alternative for mobile broadband wireless access (BWA). This applies to any mobile worldwide interoperability for microwave access (WiMAX) device, regardless of its size and application and therefore ranging from a tiny personal digital assistant (PDA) to mobile phones to table PCs and notebooks.

Power saving becomes important as all these devices continue to incorporate more and more functionalities per integration unit, a development that has been predicted already years ago and which is known as *Moore's law*. By doing so, manufacturers serve customer's demand for consolidation; the ideal device is small, light, but feature rich. Hence, while early mobile phones were solely designed for voice services, today's generation integrate MP3 player, camera, positioning service and many other energy-hungry features. The combined energy demand of such peripheral features, for instance and according to Ref. [1, Fig. 2.1 in Chapter 2], make up for almost 30 percent of the total for a Nokia 6630. Clearly, this also entails a whole set of new thermal requirements on mobile device design.

This development poses a considerable challenge to battery performance just as with computing power increasing energy accumulation is expected. Yet advances in battery capacity lacks considerably behind, like stated in Ref. [2] which claims that battery capacity has only improved by 80 percent in the last ten years while computing power doubles every 18 months according to Moore's law. Nevertheless and on top of that, a recent Taylor Nelson Sofres (TNS) research project reveals that "Two-thirds of mobile phone and PDA users rate 'two-days of battery life during active use' as the most important feature of the ideal converged device of the future" [3]. Similar to this finding, it is generally agreed that the operational time for a notebook shall be in the range of 2–5 hours. As of today, the only efficient means to cater to these user expectations is to counter this development by ever more efficient power saving mechanisms.

Motivating research on power consumption to overcome usability and performance constraints can be considered as the standard argumentation. But in fact, there is a much more important, frequently neglected reason: Saving energy means safeguarding our environment. According to Ref. [4], in 2008, 70 million notebooks will be sold, roughly double the number than back in 2001. That means 70 million new power consumers and equal number of new batteries, yet one of the most expensive energy source in terms of production, which usually involves many valuable resources and hazardous chemicals, but also in terms of deployment. Disposing unusable batteries is a complicated business and the implicative environmental consequences do not require any further elucidation. Hence, saving power—and amid the total lifetime of a battery—can only be of utmost importance.

In conclusion, effective power saving mechanisms for IEEE 802.16e, which is supposed to seize the lead in BWA technology, are important to improve the usability of mobile devices and to accomplish this in the most environmentally friendly manner. The chapter is divided into two main parts. In the first part, we analyze how the power management mechanism functions in both point-to-point and relay architecture, and set some important deductions for the sleeping time. In the second part, we identify the tools for the performance analysis of power saving mechanisms and set the initial pace for some major, open research issues.

## 8.2 OVERVIEW OF THE STANDARDS REGARDING POWER MANAGEMENT

### 8.2.1 INTRODUCTION AND DEFINITIONS

The basic principle of any power saving mechanism, including those adopted by the IEEE 802.16 standard family [5,6], is to have some prenegotiated intervals, in which the mobile device is absent from the serving base station (BS) air interface. The state of the mobile station (MS) in such intervals is called *sleep state*. In this state, the MS is considered by the BS unavailable for either Downlink or Uplink traffic. Between sleep states, the MS wakes up and exchanges messages with the BS in an interval called *listening window*. This procedure forms the *sleep mode operation* of an MS. To ensure backward compatibility and minimize the cost of portable devices, the IEEE 802.16e standard defines sleep mode as an optional feature for a MS. For the BS, as expected, it is mandatory.

In WiMAX networks, the location of a MS is managed by a two way process: *Paging* and *Location Update* [7]. Paging is invoked by the BS and informs the MS to update its position. Then the MS responds with a location update to inform the BS. When there is no traffic addressed to the MS, the MS is allowed to enter an optional idle state and reduce the location update rate. In any case, when the MS goes to sleep mode, it becomes unavailable and locationing becomes challenging.

WiMAX networks support strict QoS levels to cater error and delay prone applications (e.g., VoIP, video etc.). This is done by air interface access differentiation according to predefined, particular classes. This implies that IEEE 802.16e energy conservation mechanisms have to take these features into account. For each MS, the serving BS maintains one or several contexts where each one is related to a certain power saving class (PSC). A PSC is defined as a group of connections with common demand properties. Such PSCs differ by their parameter sets, procedures of activation/deactivation, and policies with respect to a MS availability for data transmission. If a MS has several different connections with a BS, then it may also maintain several PSCs. For example, each unsolicited grant service (UGS) type of connection requires a separate saving class instance, whereas several best-effort (BE) or nonreal-time variable rate (NRT-VR) connections can function in a single saving class.

To accommodate for multiple saving classes, there are two intervals. *Unavailability interval*, which is the time interval that does not overlap with any listening window of any active PCS, and *availability interval* which does not overlap with any unavailability interval. During the first, the MS may power down one or more physical operation components or perform other activities which do not require communication with the serving BS. The BS on the other side buffers the service data units (SDUs) destined to the MS for a predetermined period. If this period is expired while the MS is still unavailable, the packets are dropped. During an availability interval, the MS is capable of receiving all downlink transmissions and performs all the normal operations as defined in the legacy IEEE 802.16 standard [6].

## 8.2.2  MESSAGES AND TRANSACTIONS

The 802.16 standards can support 255 different management type of messages. According to IEEE 802.16e [1], 66 messages are defined from which 50–66 are used explicitly for handover and sleep mode operation. In the following, we describe the most important messages for sleep mode operation. Each message has three parts. The first part is casually MOB (implying Mobility) while the second part carries the scope of the message (SLP, TRF, HO, BSHO, MSHO, PAG, etc.). For the case of sleep mode operation, we are mainly concerned with SLP (implying Sleep) and TRF (implying Traffic). The final part of each message is usually REQ (implying Request), RSP (implying Response), or IND (implying Indication).

**MOB_SLP-REQ**: Sleep Request Message, type: 50, Connection: Basic.
A MS sends this message to request definition/activation of several PSCs. The actual definition occurs when a MS suggests a PSC for an incoming connection. The message contains a *Power Saving Class ID*, which is a unique identifier for a group of PSCs associated with a MS. This ID is also used for the exchange of many similar message types and response messages. Other parameters are also defined like the *Initial Sleep Window*, *Final Sleep Window Base*, and *Exponent* quantities measured in IEEE 802.16e frames. Moreover if the MS handles more than one class, the *Number_of_CIDs* field carries this number. There are several other fields beyond the scope of this chapter.

**MOB_SLP-RSP**: Sleep Response Message, type: 51, Connection: Basic.
This message is sent in response to a request for the definition/activation of a PSC from the BS to a MS using a broadcast CID (Connection ID) or the MS's basic CID. If a new definition was requested, then a new PSC is defined and the assigned ID is returned. When the MS receives this message, it

activates the defined PSC. The MOB_SLP-RSP message contains fields such as the *Length of Data*, for the number of bytes per PSC, the start frame number for the first sleep window, and the relative intervals. Additionally, CIDs and windows (base and exponent) are described in MOB_SLP-REQ.

**MOB_TRF-IND**: Traffic Indication Message, type 52, Connection: Broadcast.
By the time the MS wakes up, it starts listening for a possible MOB_TRF-IND message, sent from the BS on broadcast CID or sleep mode multicast CID. This message is sent from a BS to a MS in the listening interval to indicate if there has been traffic addressed to the MS while it remained in sleep mode and functions only when there are one or more PSC IDs defined for PSC type I (as described below). Any other MS ignores this message. An explicit occasion in which the BS may arbitrarily include a positive indication for a MS is if the MS's periodic ranging operation is scheduled to start in the next sleep window.

### 8.2.3 CLASSES AND SPECIFICATIONS

In this section, the functionality of each PSC is explained in detail. Keep in mind that PSCs are designed to deal with QoS and energy conservation at the same time and they apply in correlation with HO (handover) operation and relay systems.

#### 8.2.3.1 Power Saving Class of Type I

PSC type I is based on the exponential sleep mode algorithm. The sleeping window is doubled periodically until it reaches a maximum value (called Final Sleep Window). This type of PSC applies for best effort and nonreal-time variable rate (NRT-VR) services (HTTP requests, e-mail, and FTP). The following parameters must be set and sent with the request (MOB_SLP-REQ) to activate this PSC.

- Initial sleep window ($T_{min}$)
- Final sleep window base
- Listening window ($L$)
- Final sleep window exponent
- Start frame number for first sleep window
- Traffic triggered wakening flag

Sleeping mode can only be entered after successful parameter negotiation between the BS and the MS. The first of the above parameters defines the duration of the initial sleep window ($T_{min}$). After the MS wakes up from the first sleep window, and starts a fixed-interval listening window, it listens for positive MOB_TRF-IND messages. In case of a negative message the initial sleep window is doubled, as mentioned above. The procedure goes on until the sleep window is equal to the final sleep window, $T_{max}$, which is calculated using the final sleep window base and exponent. When the algorithm reaches this final state, the sleep window remains constant as long as there is no traffic arriving in the associated connection. Note here that other kind of traffic may well be active but the sleep mode state depends only on the associated one. The above process can be expressed as

$$T_j = \min\left(2^{j-1}T_{min}, T_{max}\right) \tag{8.1}$$

where $T_{max}$ = Final Sleep Window Base $* 2^{\text{Final Sleep Window Exponent}}$.

During a sleep window, packets addressed to PSC type I are temporarily buffered at the BS until the beginning of the next listening window. The MS returns from sleep mode as soon as it receives a positive indication embodied by a MOB_TRF-IND message. If the MS aims to return to sleep mode, after the completion of the transmission, it has to initiate a new parameter negotiation transaction with the BS. This is done with a new MOB_SLP-REQ message.

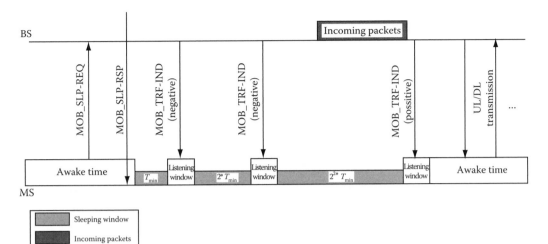

**FIGURE 8.1**  Power saving class: Type 1.

Finally, some remarks on this specific PSC. The MS is not expected to send or receive any MAC SDU during the listening interval. The PSC can, however, be deactivated not only by a positive MOB‑TRF-IND message, but also when the MS has to transmit a bandwidth request with respect to a connection belonging to this PSC.

### 8.2.3.2   Power Saving Class of Type II

Power saving class type II is recommended for connections carrying real time constant or constant bit rate traffic. Such connections usually require strict delay guarantees since they support applications such as voice over IP (VoIP) or real-time video streaming. This PCS is similar to type I with the difference that sleep intervals are now fixed (Figure 8.1). The parameters are

- Initial sleep window ($T_{min}$)
- Listening window ($L$)
- Start frame number (for first sleep window)

Here the sleep windows have the same length as the initial window ($T_{min}$) and thus only three fields are used in total. Similarly, the listening window is also constant. The PSC becomes active after the negotiation of the "start frame number for the first sleep window" and during the listening interval, the BS does not notify the MS by transmitting a traffic indication message, MOB‑TRF-IND, instead both parties, MS and BS, exchange packets directly.

The structure of this class is based on the isochronous behavior of the expected flows. By turning the transceiver off in the interval that the flow is inactive, important energy savings can be achieved. To minimize the delay, the sleeping window shall be equal to the interarrival time and no extra overhead is required to reenter sleep mode. Opposite to the PSC type I, the MS may send or receive any MAC SDU at connections comprising the PSC as well as acknowledgments to them during a listening window. Moreover each MS has as many PSCs of type II instances as connections belonging to this category (Figure 8.2).

### 8.2.3.3   Power Saving Class of Type III

Multicast connections supported by legacy IEEE 802.16 as well as management operations (e.g., periodic ranging, advertisement message broadcasting, etc.) are covered by this PSC. The parameters are

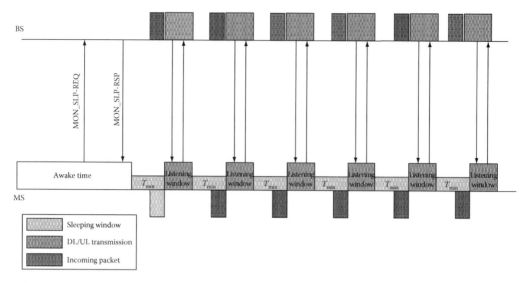

**FIGURE 8.2**   Power saving class: Type 2.

- Final window base
- Final sleep window exponent
- Start frame number (for first sleep window)

Power saving classes of this type are activated/deactivated for MOB_SLP-REQ/MOB_SLP-RSP, bandwidth request, and uplink sleep control header or downlink sleep control extended subheader transactions. Sleep operation is initiated after the start frame number and lasts for the duration of the final sleep window specified in "final sleep window exponent." After the completion of the sleep interval the MS exits the PSC type III, which means that each PSC type III sleep mode lasts only one cycle. To extend the MS availability by using this PSC, the BS or MS has to add another sleep request at the time the sleep window finishes so a new PSC type III instance will appear. Moreover, the PSC type II may also be activated by the TLV (type/length/value) encoding included in an RNG-RSP (Ranging Response) message. In this case, the sleep window of the class starts with the frame after the RNG-RSP message is transmitted and ends in the frame indicated by next periodic TLV.

The operation for the multicast case is more straightforward. A BS can guess the next portion of data and allocates the sleep window in between multicast traffic arrivals. For the ranging case, the sleep window is set equal to the time interval for a periodic ranging transaction. This PSC can also be used for locationing reasons.

### 8.2.4   Hybrid Case and Important Issues

As described in the standard, Figure 8.3 shows two PSCs: One for type I, in which the sleeping window is doubled as long as no traffic is addressed to it, and one for type II where the sleeping intervals are of constant length. Finally, availability and unavailability intervals are shown too. During unavailability interval the MS is said to be "sleeping."

Usually most of NRT-VR and BE connections are handled by PSC type I, whereas each UGS connection requires a single type II PSC. This is because bandwidth allocation for the first two scheduling classes is fully under control by the BS and does not necessarily strictly follow the traffic arrival pattern. On the other hand, for the case of UGS, each traffic source might have an individual, periodic interarrival time (ON/OFF pattern). Assume two VoIP calls, for example, one with 20 ms and another one with 30 ms interarrival time. If the larger interarrival time is used as a sleep window

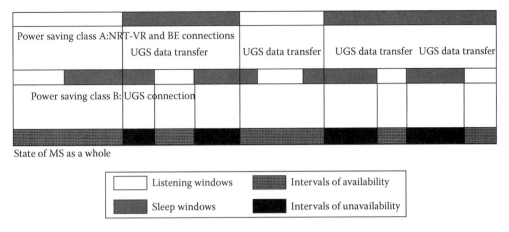

| | | | | |
|---|---|---|---|---|
| Power saving class A:NRT-VR and BE connections | | | | |
| UGS data transfer | | UGS data transfer | UGS data transfer | UGS data transfer |

Power saving class B: UGS connection

State of MS as a whole

| | |
|---|---|
| ☐ Listening windows | ▨ Intervals of availability |
| ▨ Sleep windows | ■ Intervals of unavailability |

**FIGURE 8.3**  Hybrid case for IEEE 802.16e.

then the first one would suffer a multiple of 10 ms additional delays for each transmission (without adding any other network delay). But if a sleeping window less than 30 ms would be used, then the ON interval for the second call could not fit into, which also means extra delay. If, however, both applications had the same interarrival time, then the sleep window is specified as the OFF periods between these calls. Nevertheless, management reasons of these two flows dictates that a seperate intance of PSC type II is used.

Finally, it has to be mentioned that as long as the number of UGS connections increases, the MS's total ON time does increase as well. This relationship would be directly proportional if the MS would only hold PSCs of type II. In WiMAX networks, however, ranging and other management operations require the use of PSC type III, and follow a decoupled operation. Deriving the final unavailability periods becomes more complex in this real hybrid case. The above analysis yields some deductions for the PSC type (e.g., VoIP calls) and the sleeping time:

**Deduction 1**. *Each* UGS *and* RTPS-CR/VR *connection needs to be handled by a separate power saving class type II with individual parameters.*
**Deduction 2**. *The energy efficiency is inversely proportional to incoming load.*
**Deduction 3**. *The access delay is improved when load is high, but not in a straightforward manner.*

### 8.2.5  PERIODIC RANGING IN SLEEP MODE

The standard defines periodic ranging in sleep mode functions in great detail. Ranging is a very important mechanism in IEEE 802.16 networks since the MS is using it to adjust the power and sychronize with the OFDM symbol. When the MS is in sleep mode, the ranging uplink transmission can be allocated in three different ways.

1. During the listening window, a BS may allocate an uplink transmission opportunity for periodic ranging.
2. A BS can activate a PSC type III to keep the MS in active state until the assignment of an uplink transmission opportunity for periodic ranging.
3. As mentioned above, the RNG-RSP (*or* MOB_SLP-RSP) can include the next periodic ranging TLV. From this the MS may know when the next periodic ranging opportunity shall occur. From the next periodic ranging TLV a MS may decode all consequent UL-MAP (uplink MAP) messages waiting for an uplink unicast transmission opportunity. When its own opportunity takes place the ranging procedure can be executed as normal.

It is important to mention that successful ranging does not deactivate the PSCs. After successful ranging, the BS announces the next ranging time in an RNG-RSP message. Additionally, if there is downlink traffic in the BS's queue then a DL traffic indication is addressed to the MS. This way, the MS can exit the sleep mode algorithm earlier and reduce the imminent response delay.

### 8.2.6 SLEEP MODE OPERATION IN WiMAX RELAYING NETWORKS

In the forthcoming IEEE 80216j standard [8], sleep mode operation has also been incorporated. In such relay networks, a BS coordinates all communications to and from subscriber stations. Although the initial standard provides a large coverage distance of up to 50 km under line-of-sight (LOS) conditions and typical cell radii of up to 8 km under non-line-of-sight (NLOS) conditions, there was still a huge demand for enhancing the network coverage without compromising system throughput. One of the possible options to achieve this was to introduce relay stations (RS) which bridge the communications between an SS out of coverage and the BS [9]. The great advantage of this multihop communication concept is that coverage is extended with cheap devices at a small cost of throughput degradation. RSs are divided into three categories, namely fixed RS (FRS) which are installed at fixed locations, nomadic RS (NRS) installed for a specific time duration, and mobile RS (MRS) which are mobile units operating with battery cells (Figure 8.4).

Useful deployment of MRS with a battery power source or low-power fixed/NRS powered by solar power or battery require high-power efficiency. Moreover, this feature reduces the interference generated by RSs. Here, we briefly describe how sleep mode can function in relay scenarios.

In a centralized scheduling system, where the network is coordinated by the BS (named in Relaying Networks as MR-BS), the MR-BS informs all the stations of the network including both the RSs and the MSs, about listening and sleeping windows. In a second phase, the RS coordinates with the MSs the finding of common availability and unavailability intervals. If all the MSs connected to a single RS are sleeping then the RS can enter the sleep mode too. Similar to the initial IEEE 802.16e standard, the RS can request activation of RS sleep mode by sending RS_SLP-REQ message to the MR-BS. Note that now respective message types begin not with MOB as before but with *RS*.

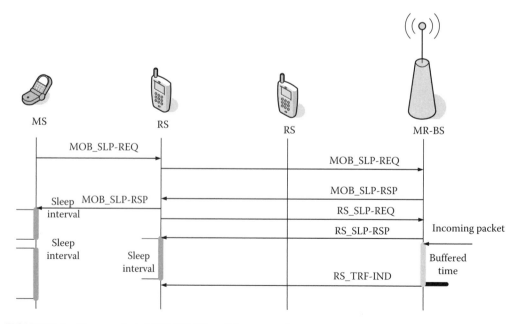

**FIGURE 8.4**   Sleep mode in IEEE 802.16j multihop networks.

In order for the RS to generate the listening and sleep windows, via a *RS_SLP-REQ*, the RS shall keep record of the information sleeping patterns of associated MSs. When an RS enters sleep mode it can be awakened by the serving MR-BS or by itself. The MR-BS can use the existing *MOB_TRF-IND* to awaken a sleeping RS.

Sleep Mode can be divided into

- Full RS Sleep Mode: This mode is entered if there is traffic at any relay and access link while the RS stays in sleep mode. All associated RSs and MSs connections are suspended RS sleeping.
- Partial RS Sleep Mode: This mode is similar to the previous case. The difference is that certain management messages are sent at predefined intervals to support MS network entry, re-entry, and handover.

A question remains whether these mobile relay stations can withstand the heavy usage. To what extent can such a terminal survive only on battery, and how energy efficient can IEEE 802.16e be in case of distributing information to many nodes?

## 8.3 PERFORMANCE ANALYSIS

The target of performance analysis is to quantify the behavior of the system and predict whether the performance of a new protocol or algorithm will meet the expectations. When analyzing a power management algorithm, the prime characteristic under examination is naturally the consumed energy during a period or lifecycle or equivalently the average power consumption. However, other measures demonstrating the trade-offs from these algorithms can be of great interest, too, like access delay, throughput, and blocking probability. Access delay is defined as the extra delay imposed by the power management algorithm because of the terminal unavailability. Throughput is the amount of data sent to a destination over a period and blocking probability is casually the probability that a connection will not be established successfully. This can happen if the connection is dropped after being unattended for a given time threshold because of the access delay.

Focussing on the IEEE standards for mobility and power management, it is evident that the proposed sleep mode algorithm imposes an access delay to every incoming flow that requires the attention of a terminal that is running the algorithm. The basic idea behind all models for performance analysis in the literature is to identify the time spent in intervals sleep and listening, given a traffic model which is usually Poisson arrivals. The energy consumption can be defined in terms of known values of power consumption per possible terminal state ($P_S$ for sleep state and $P_L$ for listening state). Access delay on the other hand is defined by the interarrival time statistics and the state of the sleep algorithm.

### 8.3.1 FIRST APPROACH

The first attempt to analyze the performance of IEEE 802.16e sleep mode algorithm was published in 2005 [10]. This paper, despite the many assumptions it was based on, served as the leading and inspiring paper in the field. In this section, we derive the basic approach to analyze the performance of these sleep algorithms. This analysis is similar to Ref. [10].

Assume an IEEE 802.16e terminal being part of a WiMAX network and located at a stationary point. Incoming packets arrive at this terminal according to a Poisson process with rate $\lambda$. Whenever there are no incoming packets, the terminal initiates immediately the sleep mode algorithm. Recall from Equation 8.1 that the sleep interval $T_j$ depends on the state $j$ of the sleep mode algorithm. The probability of $e_j$ {at least one packet arrived during the sleep interval $j$} will be then

$$P(e_j) = 1 - e^{-\lambda(T_j+L)}, \tag{8.2}$$

where $L$ is the listening interval. The probability that the first arrival happened on $j$ state will be

$$P(n = j) = e^{-\lambda \sum_{i=1}^{j-1}(T_i+L)}(1 - e^{-\lambda(T_j+L)}). \tag{8.3}$$

In the above, the first term corresponds to conditioning that no arrival has happened in the first $j - 1$ states. The second term is simply $P(e_j)$. In continuation, average delay $D$ and average power consumption $P$ can be computed by

$$E[D] = \sum_{j=1}^{\infty} P(n = j)(T_j + L)/2, \tag{8.4}$$

and

$$E[P] = \sum_{j=1}^{\infty} P(n = j)\frac{\sum_{k=1}^{j}(T_k P_S + LP_L)}{\sum_{k=1}^{j}(T_k + L)}. \tag{8.5}$$

In Equation 8.4, we assume that all possible arrivals in state $j$ are not ordered and distributed identically and uniformly in $T_j$. In Equation 8.5, we deviate from Ref. [10] by scaling the computed energy over time. This way the trade-off between power consumption and access delay is obvious. Moreover, it is clear why both short and long sleep states are required in case of a wide range of expected arrival rates.

Using Equations 8.4 and 8.5, we can numerically show the effect that the arrival rate has on average access delay and average power consumption. These relations are depicted in Figures 8.5 through 8.8 for several values of $T_{min}$ and $T_{max}$.

As expected, high arrival rate decreases the unavailability periods, and therefore, there is not much room for energy conservation. On the other hand, access delay is improved (decreased) since the small interarrival values do not allow the exponent of the sleep period to increase significantly. When the load is high enough, access delay depends on the $T_{min}$ value while when it is small enough it mostly depends on $T_{max}$. The trade-off between power consumption and access delay is obvious

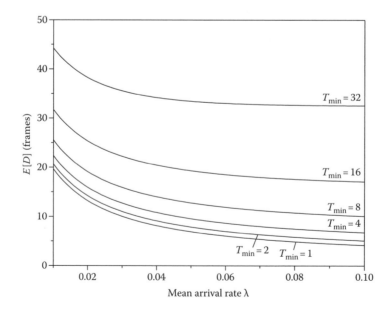

**FIGURE 8.5**  Average access delay for several values of $T_{min}$. In this case $T_{max} = 1024$ and $L = 1$ were used.

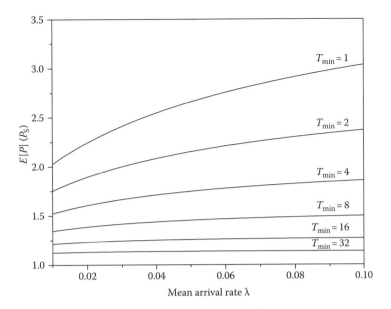

**FIGURE 8.6** Average power consumption for several values of $T_{min}$. In this case $T_{max} = 1024$, $P_S = 1$, $P_L = 10$, and $L = 1$ were used.

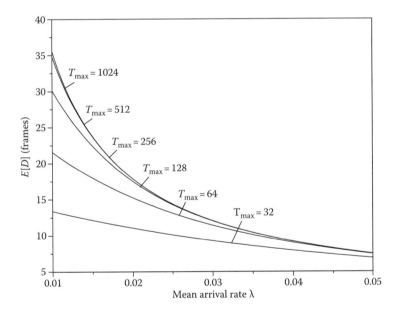

**FIGURE 8.7** Average access delay for several values of $T_{max}$. In this case $T_{min} = 1$ and $L = 1$ were used.

in these figures. The standardization of the exponential algorithm for sleep state targets exactly this trade-off.

Note that average power consumption in Figures 8.6 and 8.8 is different from Ref. [10] because we are using average power instead of energy.

Unfortunately, the above analysis is limited in many ways because of the several assumptions made. To delve into several features of the standard, some assumptions must be relaxed. In the following section, we discuss these assumptions.

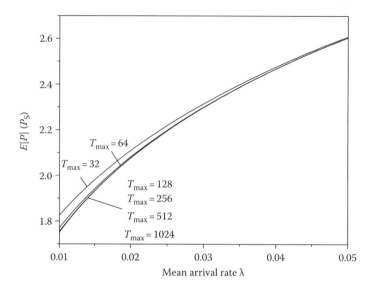

**FIGURE 8.8** Average power consumption for several values of $T_{max}$. In this case $T_{min} = 1$, $P_S = 1$, $P_L = 10$, and $L = 1$ were used.

### 8.3.2 MAIN ASSUMPTIONS

It is worthy to stress on the assumptions made in Ref. [10], since the list will lead us to the series of following papers that simply try to relax on these assumptions and therefore achieve higher accuracy.

#### 8.3.2.1 Arrival Rate

The arrivals are considered in Ref. [10] to follow Poisson process with rate $\lambda$. This assumption is crucial for the mathematical analysis since it simplifies the problem. Nevertheless in any case, this arrival process describes the way the packets are arriving at a distant network host, and therefore it is not straightforward to model without realizing a possible undergoing application. However, the type-1 class of sleep mode is also working when there is no application going on.

Some authors assume different arrival models without, but nevertheless, succeeding in reasoning these models. In Ref. [11], an Erlang distribution is used as a highly customizable distribution. The authors finally claim that the traffic modeling plays an important role in the results of performance analysis. In Ref. [12], the author uses hyper-Erlang distribution resulting in the same conclusion.

In Ref. [16], a different model is proposed using multiple interrupted Poisson processes (IPPs). It would be a challenge to model the arrival distribution like in Ref. [16], because this would give insight into the hierarchical case of packet level sleep mode and flow level sleep mode. It is foreseen that the operation of the sleep mode algorithm can take place in two different states, the one between toggling on and off an application in the terminal (flow level) and the other under an ongoing application where packets have long interarrival durations (packet level). It is expected that this kind of modeling will unveil a balance between the gain of the algorithm in these two different layers.

#### 8.3.2.2 Outgoing Traffic

Another reasonable assumption of Ref. [10] is that only the incoming traffic is considered. However, Zhang and Fujise [13] prove that taking into account the outgoing traffic improves significantly the access delay. This improvement actually stems from suppression of unavailability periods which in turn raises the power consumption and lowers the delay. The same result is found by Xiao in a parallel work [14].

### 8.3.2.3 Sleep Mode Setup Time

As explained in the previous section, the terminal must exchange MOB_SLP-REQ messages before initiating the sleep mode algorithm. This message transaction lasts for a period durating a small number of frames. It is proposed in Ref. [12] that this period must be subtracted from the interarrival time of the incoming packets to correctly compute the access delay. In this paper, the resulting interevent times are called packet residual interarrival times. This effect is also modeled in Refs. [15,17].

This time offset is usually small enough to neglect in case of interarrival times between two application streams. However, the effect could be dominating in case of using the sleep mode algorithm during an ongoing flow. In this case, the interarrival times would be generally smaller and comparable to this time difference. In this perspective, the sleep mode setup time remains an important issue.

### 8.3.2.4 Energy Cost of State Switching

Switching on and off the transceiver is known to consume a fixed amount of extra energy. In Ref. [18], the authors model this switching cost to compute the final energy consumption. However, modern specification sheets for WiMAX transceiver provide low consumption idle states where the switching to active state is immediate and consumes negligible amounts of energy [19]. Therefore, it seems that modeling the cost of state switching is not worthwhile.

### 8.3.2.5 Multiclass Scenario

Most of the papers in the literature so far focused on PSC type-1 class sleep mode. As explained above, this class refers to terminals that have only low QoS ongoing flows or no flows at all. However, considering a realistic scenario, one would have to take into account multiple connections of several QoS levels and include possible network management operations and therefore, combine many different sleep mode algorithms (one of type-1 and possibly many of type-2 and type-3) for only one terminal. These several algorithms, running in the same time, define the common availability periods of the terminal which in turn define the final unavailability periods. The terminal is allowed to go to sleep only on that commonly accepted unavailability periods (periods that the attention of the terminal is not required by the BS).

Kong and Tsang, in Refs. [20,21], take into account two different type of sleep mode classes. However, their work is focussed on selecting one of the two possible classes depending on delay and energy consumption optimality. Therefore, the task to analyze the performance of multiclass scenario remains open.

What is more important is the type-3 class. This class is used by the BS itself to organize multicast communications, locate the MS, and perform periodic ranging. Locationing is especially needed in mobile IEEE 802.16e networks to keep track of the mobile distance from the BS and perform handover operations. Including mobility into sleep mode modeling seems to require the addition of at least one extra sleep algorithm of type-3. Mobile scenarios are always multiclass. This assumption seems to be very important for the final results since high speed mobility tends to impose very frequent location updates. Then, periodic ranging is also very important and sometimes requires an extra PSC instance. It is interesting to investigate the effect of management operations in power consumptions of WiMAX terminals through a hybrid multiclass scenario.

### 8.3.2.6 Full Model Analysis

Most of the published work focuses on estimating the expectation of access delay and energy consumption. The mean value is indeed an intuitive characteristic and clearly valuable for comparisons and testing. Only Lee and Cho [15], show the possibility of computing the variance of these measures. On the other hand, derivation of high-order moments or even of the distribution of these measures can be possible for simple interarrival models and could prove to be insightful.

### 8.3.2.7 Queueing

The analysis described above as well as most of the literature assumes that once the terminal shifts to awake mode it immediately serves the packet. However, by assuming nonzero service time, and for high-rate arrivals it is expected that queueing phenomena will occur. Specifically, it is shown in Ref. [18], that the sleep mode algorithm can be modeled by a server with vacations and a $M/G/1/K$ queue. In Ref. [22], the authors give a generic promising model that can be used in many cases and applications to analyze and optimize the several parameters.

## 8.4 OPEN ISSUES IN PERFORMANCE ANALYSIS

This section summarizes the current challenges in performance analysis of sleep mode algorithms in IEEE 802.16e. The most important issue that remains open is the performance of these algorithms under a multiclass scenario. This problem is associated with the question: How efficient can power saving be in a realistic multiclass case? To tackle such a problem, one must take into account multiple power saving classes of different types operating simultaneously at the same terminal. Another issue in the same context is the possibility of canceling type-1 class in case of a highly populated scenario. The terminal will be able to acquire the MOB-TRF-IND information through the availability periods of other class types algorithms.

Another open issue is the definition of a proper statistical model for the arrivals. This model ideally would have the property of describing both arrivals at the flow level and the packet level. By achieving this, the gain in these two different cases can be investigated. On the other hand, using the current Poisson modeling, deriving the full stochastic model of energy consumption and access delay is also an open problem.

Finally, it is still an open issue to combine the above approaches to improve fine tuning of the performance modeling. This includes, the modeling of outgoing traffic effect and the sleep mode setup time. The final question to be answered is: How much can IEEE 802.16e sleep mode algorithms actually help in a realistic mobile scenario?

## 8.5 CONCLUSION

This chapter presented technical details regarding the IEEE 802.16e standards and particularly the sleep mode operation in WiMAX networks. Several important aspects of the standard were explained and insights to arising problems were given. The second part of the chapter introduced the performance analysis of power saving class type I and summed up the existing literature on the field. The critical assumptions in the analysis were pointed out while insights to analysis of the other two power saving classes were presented. In the previous paragraph open issues were summarized. Determining the efficiency of these algorithms in real operation is not a fully covered matter yet.

## REFERENCES

[1] F. H. P. Fitzek and M. D. Katz, *Cognitive Wireless Networks*, Springer, June 2007.

[2] C. Andersson, D. Freeman, I. Jnes, A. Johnston, and S. Ljung. *Mobile Media and Applications, from Concept to Cash: Successful Service Creation and Launch*, Wiley, 2006.

[3] Taylor Nelson Sofres (TNS), Two day battery life tops wish list for future, *Mobile Market Research Report*, September 2005.

[4] Allied Business Intelligence, Inc Q403, *Pyramid Research Report Nov 03, Stat/MDR*, January 04.

[5] IEEE 802.16e Workgroup, Part 16: Air interface for mobile broadband wireless access systems— Amendment for physical and medium access control layers for combined fixed and mobile operation in licensed bands, February 2005.

[6] IEEE 802.16 Workgroup, Part 16: Air interface for fixed broadband wireless access systems—standard for local and metropolitan area networks, June 2004.

[7] Hyun-Ho Choi and Dong-Ho Cho, Mobility support for IEEE 802.16e system, *WiMAX Handbook, Vol. Standards and Security*, pp. 103–128, CRC Press Taylor and Francis Group.

[8] IEEE 802.16j Mobile Multihop Relay Workgroup, Part 16: Air interface for relay broadband wireless access systems—Standard for local and metropolitan area networks, IEEE 802.16j/D1, 2007.

[9] G. S. Paschos, P. Mannersalo, and T. M. Bohnert, Cell capacity of IEEE 802.16 coverage extension, in *Proceedings of IEEE Consumer Communications and Networking Conference (CCNC '08)*, Las Vegas, Nevada, January 2008.

[10] Y. Xiao, Energy saving mechanism in the IEEE 802.16e wireless MAN, *IEEE Communications Letters*, 9(7), 595–597, July 2005.

[11] N. M. P. Nejatian and M. M. Nayebi, Evaluating the effect of non-Poisson traffic patterns on power consumption of sleep mode in the IEEE 802.16e MAC, in *Proceedings of IEEE Wireless and Optical Communication Networks (WOCN '07)*, pp. 1–5, Montreal, Quebec, July 2007.

[12] Y. Zhang, Performance modeling of energy management mechanism in IEEE 802.16e mobile WiMAX, in *Proceedings of IEEE Wireless Communications and Networking Conference (WCNC '07)*, pp. 3205–3209, Hong Kong, China, March 2007.

[13] Y. Zhang and M. Fujise, Energy management in the IEEE 802.16e MAC, *IEEE Communications Letters*, 10(4), 311–313, April 2006.

[14] Y. Xiao, Performance analysis of an energy saving mechanism in the IEEE 802.16e wireless MAN, in *Proceedings of IEEE Consumer Communications and Networking Conference (CCNC '06)*, pp. 406–410, Las Vegas, Nevada, January 2006.

[15] J.-R. Lee and D.-H. Cho, Performance evaluation of energy-saving mechanism based on probabilistic sleep interval decision algorithm in IEEE 802.16e, *IEEE Transactions on Vehicular Technology*, 56(4), 1773–1780, July 2007.

[16] C. R. Baugh, 4IPP traffic model for IEEE 802.16.3, IEEE 802.16.3c-00/51, Meeting #10, Tampa, Florida, October 2000.

[17] K. Han and S. Choi, Performance analysis of sleep mode operation in IEEE 802.16e mobile broadband wireless access systems, in *Proceedings of IEEE Vehicular Technology Conference (VTC '06 Spring)*, Vol. 3, pp. 1141–1145, Melbourne, Australia, May 2006.

[18] Y. Park and G. U. Hwang, Performance modelling and analysis of the sleep-mode in IEEE 802.16e WMAN, in *Proceedings of IEEE Vehicular Technology Conference (VTC '07 Spring)*, pp. 2801–2806, Dublin, Ireland, April 2007.

[19] WiMAX Transceiver AT86RF535A Specifications sheet, Atmel, Feb. 2006.

[20] L. Kong and D. H. K. Tsang, Performance study of power saving classes of type I and type II in IEEE 802.16e, in *Proceedings of 31st IEEE Conference on Local Computer Networks*, pp. 20–27, Tampa, Florida, November 2006.

[21] L. Kong and D. H. K. Tsang, Optimal selection of power saving classes in IEEE 802.16e, in *Proceedings of IEEE Wireless Communications and Networking Conference (WCNC '07)*, pp. 1836–1841, Hong Kong, China, March 2007.

[22] S. Alouf, E. Altman, and A. P. Azad, Analysis of an M/G/1 queue with repeated inhomogeneous vacations—Application to IEEE 802.16e power saving, Research Report RR-6488, INRIA, 2008, http://hal.inria.fr/inria-00266552.

# 9 Multimedia over Mobile WiMAX

*Wei Wang, Dongming Peng, Honggang Wang,
Hamid Sharif, and Hsiao-Hwa Chen*

## CONTENTS

In this chapter, we review the recent advancements of multimedia applications over mobile worldwide interoperability for microwave access (WiMAX), and study the multimedia delivery challenges and transmission strategies over mobile WiMAX networks. The generic multimedia transmission problem over mobile WiMAX networks is abstractly formulated as a cross-layer resource allocation problem. We further consider image transmission as a particular example, discussing an 802.16e protocol compliant resource allocation scheme to optimize reconstructed image quality with strict delay constraints. The focus of this chapter is in threefolds: (1) we provide a comprehensive review of state-of-the-art multimedia delivery optimization techniques over WiMAX networks, and summarize them in a cross-layer based multimedia resource allocation fashion; (2) we further discuss a new position-value oriented resource allocation paradigm for image transmission, which allows more efficient unequal error protection; (3) we also discuss WiMAX format compliant resource allocation schemes, identifying fragmentation and selective repeat automatic repeat request (SR-ARQ) as optimization strategies without violating the standardized components in 802.16e. Extensive simulation studies are also provided to show the comparative analysis of quality-latency performance between the position-based resource allocation approach and layer-based approaches.

## 9.1   INTRODUCTION

Mobile worldwide interoperability for microwave access (WiMAX) technology based on recent IEEE 802.16e specifications and Orthogonal Frequency Division Multiple Access (OFDMA) physical layer air interface [1,2] has become one of the most promising broadband wireless protocols to support high throughput and mobility over large coverage areas. This technology can provide fast and inexpensive broadband access to markets that lack infrastructure such as rural areas and unwired

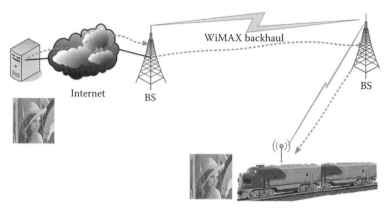

Mobile WiMAX client access

**FIGURE 9.1**  Multimedia applications over mobile WiMAX networks.

countries [3]. WiMAX can also serve as backhaul networks for client accesses to hot spots using different technologies including 802.11a/b/g as well as 802.16d/e. The most prominent characteristics of mobile WiMAX is that it can provide large distance data services for up to 31 mi [3] with high data rate transmissions. With rapid growth of online multimedia services, supporting sensitive multimedia streaming with low latency over wireless networks especially mobile WiMAX becomes a focus of many research and development activities [4]. Figure 9.1 illustrates an example of multimedia delivery applications over mobile WiMAX networks for railroads. The high-speed 802.16d protocol is applied to backhaul network, and mobile WiMAX (802.16e with SOFDMA) is applied to client access networks.

Multimedia streaming is bandwidth intensive, delay sensitive but loss tolerant, and bit errors or packet losses are inevitable in WiMAX networks due to the error-prone characteristic of wireless channel. Another important characteristic of multimedia streaming is unequal importance, where different packets in the stream have different perceptional values in terms of reducing distortion (i.e., some packets in the stream may be much more important than other packets). Fortunately, the connection oriented medium access control (MAC) layer of mobile WiMAX is designed to provide flexible quality of service (QoS) to different applications, which lays foundations to support multimedia streaming over WiMAX networks in two aspects: traffic differentiation among connections and resource allocation adaptation inside each connection. Each application traffic flow (e.g., video, voice, data) can be mapped to one or multiple service flows, and each service flow is further mapped into a logical connection with a unique 16-bit connection identifier (CID). The Service Data Unit (SDU, e.g., an H.264 video frame or a JPEG2000 image packet) from upper layer is dispatched with proper CID by SDU classifier, and MAC common part sublayer performs fragmentation and retransmission as well as QoS control. There are five QoS classes defined in mobile WiMAX: Unsolicited Grant Service (UGS), extended real-time polling service (ertPS), real-time polling service (rtPS), non real-time polling service (nrtPS), and best effort (BE) [1,2]. Such WiMAX flow scheduling architecture is illustrated in Figure 9.2, and the attributes of each service class in terms of traffic differentiation are described as follows. Example applications are also shown for each type of service class.

*UGS:* The UGS service class is especially designed to support real-time service flows which generate fixed-size data packets on a periodical basis, and the UGS service class offers real-time periodic bandwidth grants, which eliminates the bandwidth request overhead and latency [1]. Typical applications for UGS are T1/E1 data flows, G.711 based voice-over-IP (VoIP) traffic without silence repression, etc. [1].

*ertPS:* IEEE 802.16e standard introduces extended real-time polling service, which allows 802.16e to manage traffic rates and transmission policies, as well as improves latency and jitter

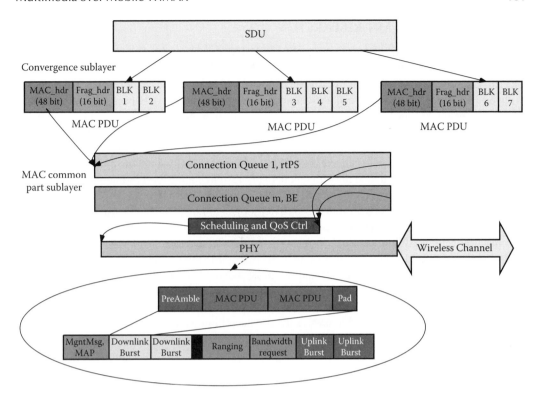

**FIGURE 9.2**   Packet delivery process in mobile WiMAX.

performance, where the ertPS service class is built on the efficiency of both UGS and rtPS [2]. The base station (BS) provides unicast bandwidth grants in an unsolicited manner similar to UGS. The difference between UGS and ertPS is that UGS bandwidth allocations are fixed while ertPS allocations are dynamic. The advantages afforded by ertPS are especially important in support of VoIP applications without silence repression, which generate variable size data packets on a periodic basis [2].

*rtPS:* The rtPS service class is designed to support real-time service flows with variable data size packets on a periodical basis, and it offers unicast and periodical request opportunities real-timely, which allows the subscriber station (SS) to specify the size of desirable bandwidth grant [1]. rtPS incurs more bandwidth request-grant overhead than UGS, but it improves data transmission efficiency and bandwidth resource utilization. Typical applications for rtPS include Moving Pictures Expert Group (MPEG) video streaming, video conferences, and IPTV [5].

*nrtPS:* The nrtPS service class is designed to assure service flows receiving bandwidth request opportunities even during network congestions, where BS offers unicast polling service to SS on a regular basis [1]. It is especially suitable for delay tolerant data streams such as HTTP-based Internet Web browsing, FTP-based file transferring, etc. [2].

*BE:* The intent of BE service class is to support data streams without minimum bandwidth allocation requirement [1]. The BE service (for instance, e-mail service) is on a resource available basis, where no throughput or delay guarantees are provided [6].

Overall, the versatile QoS framework in WiMAX has significant flexibility to differentiate application streams (e.g., voice, video/image, data) and to provide different services to these streams. More important, the flexible scheduling architecture provides considerable advantages to the resource allocation adaptation and optimization in each media stream.

## 9.2   STATE-OF-THE-ART RESEARCH

Generally, real-time multimedia streaming poses significant challenges in wireless networks due to the time-varying nature of wireless channels, limited bandwidth, channel state fluctuation, inevitable bit error and packet loss, ambient noise, and interferences. Many solutions have been proposed to deal with the challenges for real-time media streaming over generic wireless networks. Research focusing on network resource allocation such as [7–9] proposed effective solutions for improving performance of delay sensitive multimedia streaming over wireless local area networks (WLANs). These solutions used different error resilient protection techniques such as packetization and retransmissions to different media quality layers to achieve best effort multimedia quality with rate or delay constraints. Research in Ref. [10] formulated the efficient network resource allocation problem as joint optimal selection of transmission strategies across PHY, MAC, and APP layers, which maximized multimedia quality or perceived peak signal noise ratio (PSNR) subject to rate and delay constraint. Research in Ref. [11] furthered this approach to determine optimal cross-layer strategies based on classification and machine learning. Optimal MAC layer retry limits were predicted for various video packets transmitted over 802.11a WLANs, according to the perception importance of each video packet and current channel conditions. The unicast and multicast video streaming optimization problems over WLANs were addressed in Ref. [12], where hybrid automatic repeat request (ARQ) combining PHY layer forward error correction (FEC) and link layer retransmission were described for unicast flows, and the multicast optimization problem was solved via combining progressive source coding and low layer FEC. Another hybrid ARQ scheme combing Reed–Solomon (RS) coding and rate compatible punctured convolution (RCPC) coding for H.263 coded wireless video streaming was proposed in Ref. [13].

Unfortunately, cross-layer optimization with specific WiMAX consideration for multimedia streaming was not extensively discussed in most of the researches. In Ref. [14] a queuing-theoretic and optimization-based model for radio resource management in IEEE 802.16-based multiservice broadband wireless access (BWA) networks was proposed. Joint bandwidth allocation (BA) and connection admission control (CAC) were performed with packet level and connection level QoS consideration. In Ref. [15], they further presented the architecture for integrating hot spot 802.11 WLANs with 802.16 based multihop wireless mesh infrastructure to relay WLAN traffic to Internet. In that approach, the bandwidth allocation was presented with a bargaining game formulation for fair resource allocation, and an admission control policy was proposed to maximize the utilities for different types of connections. The research in Ref. [16] verified, via simulation, the effectiveness of rtPS, nrtPS, and BE in managing traffic sources, and the results highlighted that the rtPS scheduling service was a very robust scheduling service for meeting the delay requirements of multimedia applications. Research in Ref. [17] addressed resource allocation problems regarding dynamic subcarrier allocation, adaptive power allocation, CAC, and capacity planning in OFDMA wireless metropolitan areas networks (WMAN). Research in Ref. [18] proposed an adaptive bandwidth allocation and admission control scheme for polling service (PS) in an IEEE 802.16-based WMAN. A noncooperative game was proposed, admission control policy was described, and the solution was determined by the Nash equilibrium for the amount of bandwidth allocated to a new connection, ensuring QoS for all connections in the system. The research in Ref. [19] focused on scheduling and resource allocation in a cross-layer fashion. The principles of joint scheduling and resource allocation for IEEE 802.16 operating in adaptive modulation coding (AMC) mode were described, and the critical roles played by physical layer considerations, especially inter-cell interference estimation and channel state awareness were discussed. However, those mentioned researches focused on binary data transmission in mobile WiMAX, and transmission strategy optimization for multimedia content was not fully considered. The research in Ref. [20] studied the performance of voice packet transmissions and BS resource utilization using the three types of scheduling services in IEEE 802.16-based backhaul networks. They demonstrated that while the UGS achieves the best latency performance, the rtPS service could utilize the BS resource more efficiently and flexibly, trading-off between packet transmission performance and BS resource allocation efficiency. According to their studies, the appropriate choice of the frame size was important in both the rtPS and ertPS services to reduce delay and packet loss. Similar research regarding VoIP over WiMAX was found in Ref. [21]. However, they specifically considered the characteristics of VoIP. Content-based unequal error protection (UEP) for video/image streaming was not considered. The research in Ref. [6] showed that a good scheduling control was the key field to support coexisting real-time and nonreal-time traffic flows in mobile WiMAX. They especially suggested that for H.264/AVC-based scalable video coding, it was crucial to separate different video layer packets into different connections with different treatment of protection and retransmission. However, this was based on the assumption that no adaptive resource allocation exists in each connection, which may cause significant overhead on multiple connections' management.

Since IEEE 802.16e-based mobile WiMAX is a relatively new development, very few protocol compliant resource allocation strategies with respect to multimedia streaming have been proposed in literature. The challenges for multimedia streaming over WiMAX networks entail the definition of a MAC effectively supporting multimedia streaming while efficiently exploring limited radio resources [22]. The IEEE 802.16e standard already has build-in QoS features to support different classes of services, therefore, the radio resource allocation strategies and the scheduling algorithms for multimedia streaming between BS and SS are left open to specific vendor implementations. On the other hand, most of the previous works regarding wireless multimedia focus on traditional layer based UEP, and inequality between Position and Value (P–V) information [23] has largely been ignored. Detailed description of position and value diversity in multimedia streaming can be found in Refs. [23,24]. To sum up, the joint consideration of multimedia content and resource allocation with protocol compliance will provide significant potentials for improving delay sensitive media quality performance over standardized mobile WiMAX.

## 9.3 GENERALIZED MATHEMATICAL FORMULATION

Real-time multimedia streaming over mobile WiMAX network is often delay-sensitive but loss tolerant. On the other hand, different segments of the code stream (video or image media stream) have significantly different perception importance, and these different segments in the code stream require significant different error protections. Thus, multimedia streaming optimization over WiMAX networks can be mathematically generalized as a delay bounded quality optimization problem. This problem is cross layer in nature involving application layer distortion analysis in image/video codec as well as network/radio resource allocation strategy design in WiMAX. Let $E[\cdot]$ denote the mathematical expectation, $D$ denote the total quality indicator or distortion reduction [9,11] (the reduced amount of media distortion or the improved media quality brought by all the segments in the code stream), and $\sum \overline{T}$ denote the expected total delay for delivering the whole image code stream or group pictures in video. The problem for multimedia delivery over mobile WiMAX networks can be abstractly formulated as follows: Find the cross-layer transmission strategies $r$ (e.g., rate-distortion truncation point, WiMAX MAC fragmentation threshold or selective repeat ARQ block size, selective repeat ARQ retry limit, etc.) that maximize the expectation of the reconstructed media quality $E[D]$, while meeting the total delay constraint $T_{\max}$.

$$r = \arg \max \left( E[D] \right) \tag{9.1}$$

s.t.

$$\sum \overline{T} \leq T_{\max} \tag{9.2}$$

## 9.4 MULTIMEDIA DISTORTION ANALYSIS

Because images/video packets or frames are dominated by a mixture of stationary low frequency backgrounds and transient high-frequency edges, a wavelet transform is very efficient in capturing

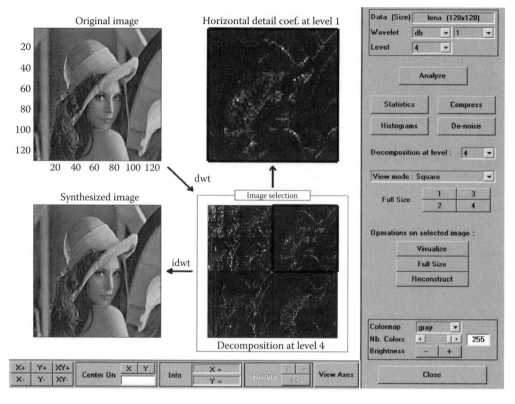

**FIGURE 9.3** Original lena image (128*128 pixels, 8 bpp), wavelet decomposition and reconstruction.

the bulk of image energy in a fraction of coefficients to facilitate compression [25]. Wavelet-based image compression techniques such as zerotree [26,27] or EBCOT [28] produce excellent scalability features and rate-distortion (R-D) performance for robust multimedia transmission over wireless channels. The wavelet decomposition is illustrated in Figure 9.3, where energy concentration in the low-frequency bands facilitates the construction of embedded code streams. The embedded nature of the compressed code stream provides the basis for scalable video/image coding by fine-tuning the R-D trade-off. The source encoding can be stopped as soon as a target bit-rate is met, or the decoding process can be stopped at any low desirable bit-rate by truncating the code stream. Typically, the code stream is composed of different quality layers in descending order with base layer providing the rough image and enhancement layers providing quality refinement. Different layers in the code streams have significant different perception importance to the end users. Losing the base layer may cause serious distortion for reconstructed pictures, while losing quality enhancement layers can still achieve acceptable picture qualities.

### 9.4.1 Layered Image/Video Coding

MPEG-4 introduced in 1998 was designated by the ISO/IEC MPEG under the formal standard ISO/IEC 14496, which was primarily aimed to low bit-rate video applications over networks [29]. The layer-based quality enhancement concept has been widely applied to scalable video coding (SVC) in MPEG-4 Part 10 H.264/advanced video coding (AVC) and JPEG2000 progressive image coding standards. The source coding bit-rate variety advantage has laid foundations of a number of emerging multimedia applications over bandwidth limited wireless networks such as IPTV, video on demand, online video gaming, etc. Both SVC in MPEG-4 and quality progression in JPEG2000 provide considerable advantage for error-robust multimedia streaming over time-varying wireless channels especially in mobile environments (e.g., mobile WiMAX networks). Without losing generality we use MPEG-4 video coding in this section and JPEG2000 in Section 9.4.2 as multimedia coding examples.

The video sources are coded into a couple of quality layers via SVC, starting with the rough pictures in low bit-rates followed by higher layers refinement data for quality enhancement in higher bit-rates. The rough pictures in base layers are much more important in terms of perception than the refinement data in enhancement layers, which deserve more protection upon transmission in wireless channels; the refinement data in enhancement layers can be discarded during transmission when bandwidth is limited. For each wireless mobile terminal in the mobile WiMAX networks, for example a moving vehicle, the available transmission bandwidth resource is fluctuating due to different locations of the vehicle and the corresponding path losses as well as the channel errors [30]. With SVC applied to each video stream on each wireless terminal, the actual traffic pumped into WiMAX networks from application layer can be adaptively controlled while keeping a rate-distortion optimize manner: if the available bandwidth is low due to high channel error probability, only the base layers of the rough pictures will be transmitted; when the channel condition becomes better with higher available bandwidth, both base layers and refinement layers will be transmitted to improve the perception quality.

Another important factor in multimedia streaming is the inter-packet dependency. It is typical that the dependency graph of multimedia packets is composed of packetized group of pictures, and the multimedia packet coding is correlated involving complex dependency among those packets. If a set of image/video packets are received, only the packets whose ancestors have all been received can be decoded [31]. Figure 9.4 illustrates the typical code stream dependency for layer-based embedded media stream. The inter-packet dependency provides opportunities for resource allocation and adaptation for multimedia streaming over WiMAX, where the packets with more descendents are much more important than those descendents. For example, for the layered-dependent media packets

```
Layer Based Multimedia Decoding:
Step 1:
       CumulativeSuccess=TRUE; iteration=0;
Step 2:
       While ( iteration < number of layers ) {
               Decode the layer (denoted by iteration).
               if (decoding successfully)
                       Then CumulativeSuccess=TRUE;
                       Else CumulativeSuccess=FALSE; break;
       }
Step 3:
       Output the decoded stream up to layer iteration.

IBP Based Video Decoding:
Step 1:
       Decode I frame. If fail, return;
Step 2:
       Find and decode the next P frame.
       If fail, go to Step 4; Otherwise pPFrame = the found P frame.
Step 3:
       While ( pPFrame != NULL ) {
               Decode the B frames ahead of pPFrame;
               Find and decode the next P frame.
               If decoding fail, break;
               if ( the current P frame == the last P frame ) pPFrame = NULL; Else pPFrame =pPFrame ->Next;
       }
Step 4:
       Output the decoded stream.
```

**FIGURE 9.4**  Typical packet dependency of layer based embedded media stream structure.

in the Figure 9.4, each packet is associated with a distortion reduction value denoting the quality gain if this packet is successfully received and decoded. If the packets in layer 2 can make contribution to the decoded media, all the packets in layer 0 and layer 1 must be received and decoded successfully; otherwise the packets in layer 2 are useless in terms of decoding even if they are transmitted without a single bit error.

Based on the analysis of unequal importance and inter-packet dependency, the UEP-based resource allocation strategies can be generalized for image/video streaming over wireless channels: network resource allocation and adaptation are applied to the media streaming according to the distortion reduction (importance) of each packet and the inter-packet dependency. The ancestor packets with more dependent children packets are protected with more network resources including stronger FEC capability, robust modulation schemes, and higher ARQ retry limits, etc. the descendent packets with less dependent children packets are less protected to save communication resources.

### 9.4.2 POSITION-VALUE ENHANCEMENT

Besides the layer-based quality scalability, wavelet-based image compressions also produce shape and position information of the regions or the objects in the picture, as well as the lighting magnitude value information describing those regions or objects [23,24]. Without losing generality, we use wavelet-based image coding as an example for multimedia content. The shapes or regions of the objects in the picture are much more important than the lighting value magnitudes of these objects. Errors in shape and region information lead to high distortion of reconstructed images, while errors in pixel magnitudes are more tolerable during transmission and decoding. This is because the shape and region information impacts the magnitude value information associated with those regions when the image is percept by end users. Furthermore, the shape and region information can be desirably translated to position information segments (e.g., *p-segment*) and the lightening magnitude information can be translated into value information segments (e.g., *v-segment*) by wavelet-based progressive compression in each quality layer. The p-segments denote how small-magnitude wavelet coefficients or insignificant wavelet coefficients are clustered, while the v-segments denote how large-magnitude wavelet coefficients are valued. Layers in the code stream represent the quality improvement manner, while the p-segments and v-segments in each layer represent the data dependency. The p-segments and v-segments can be easily identified from zerotree-based or EBCOT-based code-streams. The final code stream structure is composed of p-segments and v-segments in decreasing importance order as shown in Figure 9.5.

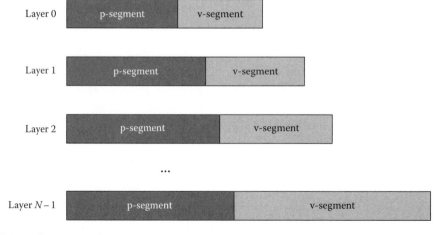

**FIGURE 9.5**   Code stream format for scalable quality layers and position-value separation in each layer.

Now we see how to separate p-segments and v-segments in each quality layer. The zerotree-based compression techniques generally involve dominant coding pass to extract the tree structures, as well as the subdominant pass to refine the leaves on the tree. These coding passes are invoked layer by layer in a bit-plane progressive way. In significant pass of each zerotree bit-plane coding loop, a half decreasing threshold $\delta$ is specified. A wavelet coefficient is encoded as positive or negative significant pixel if its magnitude is over $\delta$. The positive or negative nature is determined according to the sign of that coefficient. A coefficient may be encoded as a zerotree root if its magnitude and all the descendents' magnitudes are all below $\delta$, and itself is not a descendent of previous tree root. Otherwise this wavelet coefficient is encoded as an isolated zero. Because all of these positive or negative significant symbols, isolated zero and tree root symbols contain tree structure information, they are put to the p-segment of the current bit-plane layer. Then subdominant pass is invoked for magnitude refinement. The magnitude bit of each positive or negative significant symbol is determined according to the threshold $\delta$, and is put to the v-segment of that bit-plane layer. Thus, p-segment and v-segment are formed layer by layer with the half decreasing threshold $\delta$. Because p-segments contain zerotree structures and v-segments contain magnitude values, incorrect symbols in p-segments cause future bits to be mis-interpreted in decoding process while incorrect bits in v-segments are tolerable for errors.

The EBCOT-based JPEG2000 is a two-tiered wavelet coder, where embedded block coding resides in tier-1 and rate-distortion optimization resides in tier-2 [32]. Without losing generality, we only discuss the coding process in one code block (tier-1), because the p-segments and v-segments separation interacting with context formation (CF) and arithmetic coding (AC) resides in tier-1 intra code block coding, and the p-segments and v-segments in all other code blocks can be separated in the same way. Unlike zerotree compressions' coefficient by coefficient coding, JPEG2000 tier-1 coder processes each code block bit-plane by bit-plane from the most significant bit (MSB) to the least significant bit (LSB) after quantization. The p-segments and v-segments are also formed bit-plane by bit-plane in an embedded manner. In each bit-plane coding loop, each quantization sample bit is scanned and encoded in one of the significant propagation pass, magnitude refinement pass, and cleanup pass. In significant propagation pass, if a sample bit is insignificant ("0" bit in the current bit-plane) but has at least one immediate significant context neighbor (at lease a "1" bit occurs in the current/previous bit-plane), the zero coding (ZC) and sign coding (SC) coding primitives are invoked according to one of the 19 contexts defined in JPEG2000. The output codeword of this sample after ZC and SC are put to p-segment in this bit-plane, because the coded sample determines the positions of neighboring significant coefficients. In other words, it determines the structure of the code stream. The p-segment in this bit-plane is partly formed after significant propagation pass. In the following magnitude refinement pass, the significant sample bits are scanned and processed. If a sample is already significant, magnitude refinement (MR) primitive is invoked according to the context of eight immediate neighbors' significant states. The codeword after MR is put to v-segment in this bit-plane because it denotes magnitude information only, containing no position information of the significant samples. After magnitude refinement pass, v-segment in this bit-plane is completely formed. In the final cleanup pass, all the uncoded samples in the first two passes are coded by invoking ZC and run-length coding (RLC) primitives according to each sample's context. The code words after cleanup pass are put to the p-segment in this bit-plane, because the positions of significant samples are determined by how insignificant wavelet coefficients are clustered. Till now p-segment in this bit-plane is also formed. Then the following bit-planes are scanned, and all the p-segments and v-segments are formed bit-plane by bit-plane.

The UEP-based resource allocation with position-value enhancement is similar to layer based resource allocation where packets in base layers are more reliably protected than packets in quality refinement layers. Different from layer-based UEP, position packets are protected more reliably than value packets in each quality layer. Then the network resource allocation strategy is optimized according to the distortion reduction of each packet and the calculated dependency graph among these packets. Details of these strategies can be referred to Refs. [23,33]. Some position packets may have low values of distortion reduction but a lot of descendents, and these packets will be protected more effectively; the value packets with low distortion reduction and few descendents will be less reliably protected to save communication resources.

## 9.5  NETWORK RESOURCE ALLOCATION

Network resources can be dynamically adjusted and adapted in MAC or PHY layers for mobile WiMAX, for example, physical layer FEC strategy, modulation scheme, transmission power control, link layer scheduling strategy, fragmentation threshold, and ARQ retry limit. Here, we focus our discussion on the resource management strategies that can be easily applied to mobile WiMAX networks. In WiMAX networks, each higher layer SDU usually consists of multiple link layer Protocol Data Units (PDUs). Each SDU of a traffic flow, for example, a JPEG2000 coded image stream, is dispatched to a specific 802.16e connection by SDU classifier in convergence sublayer. The connection is associated with a set of QoS requirement parameters, and the delay budget $T_{max}$ for transmitting the whole JPEG2000 image stream is the one we discuss in this chapter. Also in this chapter, we specifically consider the transmission strategy optimization within an IEEE 802.16e connection, where the multiple connection management overhead is effectively obviated. We specifically consider the MAC layer delay performance of different fragmentation and retransmission strategies, which can be seamlessly applied to mobile WiMAX without violating what has already been defined in the IEEE 802.16e standard. In mobile WiMAX and the IEEE 802.16e standard, Selective Repeat based Automatic Repeat reQuest (SR-ARQ) is defined as the default ARQ strategy for optional performance enhancement, where the characteristics of SR-ARQ for mobile WiMAX are summarized as follows [1,2]:

1. The SR-ARQ in mobile WiMAX is enabled per connection basis.
2. A WiMAX connection must have SR-ARQ enabled or not, but it cannot have a mixture mode of both SR-ARQ and non-SR-ARQ.
3. During connection establishment process, SR-ARQ is negotiated using dynamic service addition (DSA) and dynamic service change (DSC) messages. The fragmentation threshold ARQ_BLOCK_SIZE is negotiated and the smaller one provided by BS and SS is chosen for the SR-ARQ enabled connection between BS and SS.
4. The SR-ARQ feedback bitmap is sent in the MAC management message via basic management connection between BS and SS, or in the piggyback message via the reverse link of data connection.
5. SR-ARQ feedback bitmap cannot be re-fragmented.

The SR-ARQ operation in mobile WiMAX is described in Figure 9.6. Without losing generality, we use downlink transmission in TDD mode as the example to describe the SR-ARQ process. The bandwidth resource is divided into fix-sized time frames with duration $T$, and the frame duration $T$ is further divided into downlink and uplink subframes with an adaptive boundary separated by a transmit/receive transition gap (TTG). The time frames are separated by a receive/transmit transition gap (RTG). The downlink subframe is composed of preambles, DL_MAP, UL_MAP, DCD, UCD control messages as well as burst transmission opportunities allocated for each SS. The Uplink subframe is further composed of ranging and bandwidth request slots, as well as transmission opportunity grants for each SS. In each upper layer SDU transmission, the SDU is fragmented into fix-sized SR-ARQ blocks and these blocks are dispatched into a specific connection queue. During the downlink transmission opportunity to the destination SS, these SR-ARQ blocks are transmitted in the time-varying

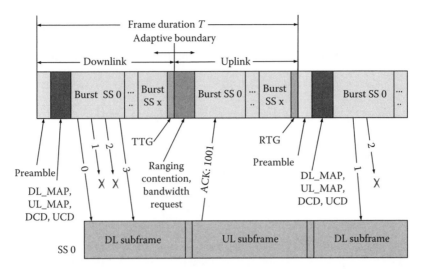

**FIGURE 9.6** SR-ARQ operations for mobile WiMAX. The detailed concept is explained in the 802.16(e) standards.

and error-prone wireless channel and some of them may be lost due to bit errors. It is worth noting that the chance of collision is minimal in the time slot scheduling based WiMAX networks, and the major packet loss is due to physical layer bit or symbol errors. The receiver SS responds with an SR-ARQ ACK bitmap to provide the receiving status of the SDU during the uplink transmission opportunity in the same frame duration, and those erroneous or lost blocks are negatively acknowledged. In the next frame duration $T$, the BS retransmits only negatively acknowledged blocks as well as the new data blocks, until the SDU is successfully delivered to the SS.

To support delay sensitive multimedia streaming, the SDU delivery delay and loss ratio are more meaningful than those of PDU when selective repeat is considered. It is easy to generically express the delay expectation $\overline{T}$ and average packet loss ratio $\xi$ for a specific SDU in one SR-ARQ enabled WiMAX connection. The close form expressions of these factors are typically required for resource allocation optimization in Equations 9.1 and 9.2. Let $e_b$ denote the measured bit error rate (BER) in the wireless channel, $H$ denote the MAC header of each fragment, and assume $K$ is the number of SR-ARQ blocks for an upper layer SDU with length $L$, and $F$ be the link layer fragmentation threshold (or the SR-ARQ block size), then $K = \lceil L/F \rceil$, where $\lceil \bullet \rceil$ denotes the ceiling function. Also let $M_{max}$ denote the link layer retry limit. The error probability $e_p$ of each PDU (SR-ARQ block) can be expressed in the following equation:

$$e_p = 1 - (1 - e_b)^{F+H} \tag{9.3}$$

Let $M(K)$ denote the round of transmissions with $K$ SR-ARQ blocks. The probability of successfully delivering a SDU with $i$ times retry is expressed as follows, which is similar to the discrete probability density function (PDF) in Ref. [34]:

$$\text{prob.}[M(K) = i] = \left(1 - e_p^i\right)^K - \left(1 - e_p^{i-1}\right)^K \tag{9.4}$$

Thus, the average loss rate $\xi$ of each SDU can be easily approximated as follows by getting the complementary-cumulative distribution function (CCDF) of the PDF.

$$\xi = 1 - \left(1 - e_p^{M_{max}}\right)^K \tag{9.5}$$

The expected round of transmissions of each SDU can be expressed as $\overline{M}$ given the retry limit $M_{max}$. Intuitively, it is a weighted summation of the successfully delivery probability multiplying the corresponding transmission rounds, and the probability of delivery failure multiplying the retry limit.

$$\overline{M} = \sum_{i=1}^{M_{max}} i \cdot \left( \left( 1 - e_p^i \right)^K - \left( 1 - e_p^{i-1} \right)^K \right) + M_{max} \cdot \left( 1 - \left( 1 - e_p^{M_{max}} \right)^K \right) \tag{9.6}$$

Given the known time frame duration $T$, which dominates the round trip time (RTT) in mobile WiMAX networks, the expected delay of delivering a SDU can be approximated as $\overline{T}$:

$$\overline{T} = \overline{M} \cdot T \tag{9.7}$$

So far the expected loss ratio and delay of delivering an upper layer SDU can both be expressed in close form functions of link layer fragmentation threshold $F$ and SR-ARQ retry limit $M_{max}$. Thus, by optimally assigning proper fragmentation threshold and SR-ARQ retry limits to different p-segments and v-segments, as well as selecting the rate-distortion truncation point of the whole multimedia code stream, the eventual media quality can be maximized within the overall delay budget.

To generically solve the delay constrained distortion reduction optimization problem, it is necessary to find the optimal transmission strategies including rate-distortion truncation point, fragmentation threshold, and retry limit for each code stream segment. However, this approach will incur too many optimization parameters leading to high optimization complexity. For example, for the code stream with $N$ quality layers and each segment to be assigned with a fragmentation threshold and retry limit, totally 32 parameters need to be optimized. To reduce the optimization complexity, p-segments and v-segments can be grouped with a pair of fragmentation threshold and assigned retry limit in a way similar to Refs. [23,33]. This is because the segment size is almost increasing monotonously which leads to layer-based UEP in terms of packet loss ratio: the segments in coarse quality layers have low packet loss ratio and the segments in fine quality layers have high packet loss ratio. This kind of grouping also preserves the unequal error protection between p-segments and v-segments because they are assigned with different transmission strategies. Thus, only five parameters including the rate-distortion truncation point $N$, fragmentation threshold for p-segments $F(p)$, fragmentation threshold for v-segments $F(v)$, retry limit for p-segments $M_{max}(p)$, and retry limit for v-segments $M_{max}(v)$ need to be optimized. With such reasonable simplification, the optimization problem can be solved with a specific designed genetic algorithm shown as follows. Details of the algorithm and analysis can be referred from Ref. [33].

1. Read the input delay constraint $T_{max}$, and measure the instantaneous channel BER $e_b$ via a feed back channel. A slow changing channel environment is typically assumed for multimedia cross-layer optimization over wireless networks, and the channel information can be fed back in a way similar to Ref. [4].
2. Encode the parameter vector $\{N, F(p), F(v), M_{max}(p), M_{max}(v)\}$ as a chromosome. Initialize the population size $\alpha$ and desirable number of generations $\chi$. Create the first generation randomly.
3. Evaluate the genetic fitness $\phi$ for each chromosome in the current population $\pi$, and sort the chromosome in descending fitness order. The fitness is evaluation according to the distortion reduction expectation $\phi = E[D]$ calculated using the coded chromosome $E[D] \leftarrow \{N, F(p), F(v), M_{max}(p), M_{max}(v)\}$ according to the measured distortion reduction of each packet and the calculated packet dependency graph according to decoding correlation.
4. Cross over of the chromosomes according to their fitness values $\phi$. The crossing over probability for chromosome $i$ is proportional to the fitness value: $p(i) = \phi(i) / \sum_{j=0}^{\alpha-1} \phi(j)$. This

approach ensures the better solutions with higher fitness values will have more opportunities to produce elite children in the next generation. Go to step 3 if $\pi \leq \chi$. Otherwise go to step 5.

5. In the last generation, output the chromosome with the highest fitness value. This is the solution of resource allocation algorithm.

## 9.6  QUANTITATIVE PERFORMANCE STUDY

The delay-distortion performance of the position-value based resource allocation is compared in this section with traditional layer-based resource allocation in mobile WiMAX. The parameters of the simulation study are listed as follows. Default time frame duration $T = 0.004$ second, default channel BER is 0.0001, MAC header is 6 bytes, and fragmentation subheader is 2 bytes. Frequency bandwidth is 20 MHz, and 64-QAM with $3/4$ coding rate and $1/4$ cyclic prefix are used.

Figures 9.7 and 9.8 indicate the average loss ratio and expected delay trade-off for delivering a typical SDU with 1400 bytes using different fragmentation thresholds and SR-ARQ retry limit strategies. From these figures it is clear to see, with larger fragmentation number (lower fragmentation threshold and shorter PDU length accordingly) and higher SR-ARQ retry limit, the SDU packet loss ratio is decreased considerably. However, the penalty of such packet loss ratio decreasing is the prolonged delay of successful SDU delivery, mainly because of the retransmission latency. This is because mobile WiMAX is basically TDMA based scheduling, retransmission has to be reissued in the next frame duration. Thus, quality and latency form the trade-off that can be fine-tuned in mobile WiMAX transmission strategies optimization.

Figure 9.9 depict the delay-distortion performance comparison of the position-value based approach and layer-based approaches. For both of these approaches, image qualities with loose delay constraints are better than those with strict delay constraints. This is because with loose delay constraints, more network resource especially the SR-ARQ retransmissions can be allocated to the code stream, which improves the packet delivery ratio and thus the picture quality considerably.

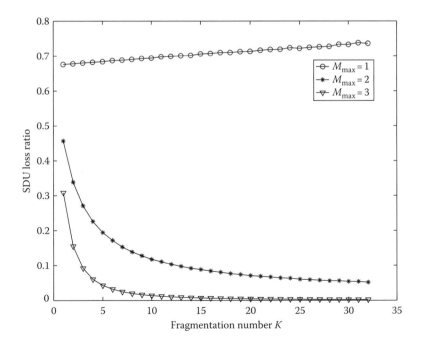

**FIGURE 9.7**    SDU loss ratio for an application layer packet with 1400-byte length, at channel BER 1e-4.

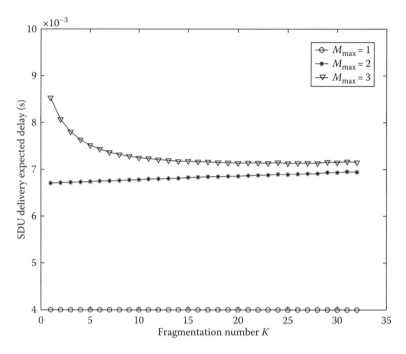

**FIGURE 9.8** SDU delay expectation for an application layer packet with 1400-byte length, at channel BER 1e-4.

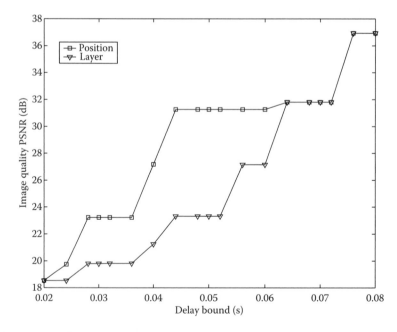

**FIGURE 9.9** Image quality and delay-bound for different resource allocation schemes at BER 1e-4.

The position-value based approach achieves better delay-distortion performance than layer-based approach with the same latency budget constraint. Layer-based UEP approaches allocate resource according to the importance of different layers in code stream, and important layers containing coarse image information are more effectively protected while unimportant layers containing image fine details are less protected. The position oriented approach allocates resource more efficiently by considering not only different layers' unequal importance, but also the unequal importance of position and value information in each layer. With the position-value based resource allocation, the p-segments especially those in the coarse image quality layers are more effectively protected to improve image quality; and the v-segments especially those in fine details enhancement layers are less protected to reduce delay penalty. From this figure we can see, with 1e-4 channel BER the position-value based approach shows quality improvement up to 7–8 dB in terms of PSNR over traditional layer based approach with the same delay constraint. In the worst cases, i.e., with ultra tight or ultra loose delay constraints, the position-based approach has similar performance as layer based approach. This is because it has either not enough network resource or over excessive network resource for allocation. In those situations, the performance is not confined by efficiency of resource allocation, but the amount of network resource itself.

## 9.7   CONCLUSION

In this chapter, we have discussed the multimedia delivery challenges in mobile WiMAX networks, and reviewed state-of-the-art multimedia transmission optimization strategies over mobile WiMAX networks. The problem of multimedia delivery over mobile WiMAX is generically formulated as a resource allocation optimization problem. It can be solved by efficient resource allocation within the standard scope. Furthermore, we have discussed the concepts of a new position-value oriented protocol compliant resource allocation scheme and traditional layer-based resource allocation approach to optimize quality for delay sensitive multimedia streaming over mobile WiMAX networks. The discussion has also included joint fragmentation and SR-ARQ optimization strategies without violating what have already been standardized in mobile WiMAX. The multimedia delivery performance of the position-value based approach was then compared with layer-based resource allocation schemes in a typical mobile WiMAX environment, and the quantitative analysis of the quality-latency performance was also presented.

## 9.8   OPEN ISSUES

In this chapter, we have been focusing on the fragmentation and SR-ARQ-based WiMAX complaint resource allocation scheme in MAC layer, and physical AMC strategy optimization is not discussed. It is desirable that AMC is incorporated into the cross-layer optimization for multimedia streaming optimization over mobile WiMAX. However, incorporating AMC brings further challenges: the fragmentation and SR-ARQ can be optimized within each 802.16e connection, because SR-ARQ is enable per-connection basis in mobile WiMAX and each connection may have its own fragmentation and SR-ARQ strategy. On the other hand, AMC is typically adjusted or optimized per-station basis, which may involve multiple connections conveying different media streams with different QoS requirements and different fragmentation and retransmission strategies. Some of the researches in literature such as Refs. [39,40] discuss the joint optimization of multiple image bitstreams, but standard compliant resource allocation for mobile WiMAX is not considered in their researches. How to optimize multiple media streams jointly within the standard defined AMC and SR-ARQ resource allocation scopes still deserves in-depth study.

Another challenging issue is multimedia security over mobile WiMAX. Multimedia content is bulk size in nature, and encryption/decryption is painstakingly slow due to high computational overheads. It is impossible to encrypt all the media content in the code stream in real time, thus selective encryption must be issued for each media stream in each WiMAX connection. However, the security sublayer in IEEE 802.16e standard is not designed for multimedia streams but for general binary data, thus it cannot provide flexible selective encryption scheme adapting to media content and wireless channel information. Some of the previous researches such as Refs. [32,34–37] talked about multimedia selective encryption scheme, but none of them have specifically considered the multimedia selective encryption support in mobile WiMAX. Further more, the position-value based code stream partition is naturally compatible to selective encryption because it can desirably extract the skeleton of code stream structure for selective encryption and UEP. But how to formulate security oriented cross-layer optimization, and how to provide WiMAX protocol complaint security support for multimedia streaming will still be very challenging.

## REFERENCES

[1] IEEE Std 802.16-2004, Air interface for fixed broadband wireless access systems, October 2004.

[2] IEEE Std 802.16e, Air interface for fixed and mobile broadband wireless access systems, February 2005.

[3] Z. Abichar, Y. Peng, and J. Chang, WiMax: The emergence of wireless broadband, *IEEE IT Professional,* 8(4), 44–48, July 2006.

[4] M. Chatterje, S. Sengupta, and S. Ganguly, Feedback-based real-time streaming over WiMAX, *IEEE Wireless Communications*, 14(1), 64–71, February 2007.

[5] J. She, F. Hou, P. Ho, and L. Xie, IPTV over WiMAX: Key success factors, challenges, and solutions, *IEEE Communications Magazine,* 45(8), 87–93, August 2007.

[6] C. Huang, H. Juan, M. Lin, and C. Chang, Radio resource management of heterogeneous services in mobile WiMAX systems, *IEEE Wireless Communications,* 14(1), 20–26, February 2007.

[7] M. van Der Schaar and N. Sai Shankar, Cross-layer wireless multimedia transmission: Challenges, principles, and new paradigms, *IEEE Wireless Communications,* 12(4), 50–58, August 2005.

[8] M. van der Schaar and D. Turaga, Cross-layer packetization and retransmission strategies for delay-sensitive wireless multimedia transmission, *IEEE Transactions on Multimedia,* 9(1), 185–197, January 2007.

[9] L. Qiong and M. van der Schaar, Providing adaptive QoS to layered video over wireless local area networks through real-time retry limit adaptation, *IEEE Transactions on Multimedia,* 6(2), 278–290, April 2004.

[10] M. van Der Schaar and N. Sai Shankar, Cross-layer wireless multimedia transmission: Challenges, principles, and new paradigms, *IEEE Wireless Communications*, 12(4), 50–58, August 2005.

[11] M. van de Schaar, D. Turaga, and R. Wong, Classification based system for cross layer optimized wireless video transmission, *IEEE Transactions on Multimedia*, 8(5), 1082–1095, October 2006.

[12] A. Majumdar, D. Sachs, I. Kozintsev, K. Ramchandran, and M. Yeung, Multicast and unicast real-time video streaming over wireless LANs, *IEEE Transactions on Circuits and Systems for Video Technology*, 12(6), 524–534, June 2002.

[13] L. Hang and M. Zarki, Performance of H.263 video transmission over wireless channels using hybrid ARQ, *IEEE Journal on Selected Areas in Communications*, 15(9), 1775–1786, December 1997.

[14] D. Niyato and E. Hossain, A queuing-theoretic and optimization-based model for radio resource management in IEEE 802.16 broadband wireless networks, *IEEE Transactions on Mobile Computing*, 55(11), 1472–1488, November 2006.

[15] D. Niyato and E. Hossian, Integration of IEEE 802.11 WLANs with IEEE 802.16-based multihop infrastructure mesh/relay networks: A game theoretic approach to radio resource management, *IEEE Network Magazine,* 21(3), 6–14, May–June 2007.

[16] C. Cicconetti, A. Erta, L. Lenzini, and E. Mingozzi, Performance evaluation of the IEEE 802.16 MAC for QoS support, *IEEE Transactions on Mobile Computing*, 6(1), January 2007.

[17] S. Ali, K. Lee, and V. Leung, Dynamic resource allocation in OFDMA wireless metropolitan area networks, *IEEE Wireless Communications*, 14(1), 6–13, February 2007.

[18] D. Niyato and E. Hossain, An approach to bandwidth allocation and admission control for polling service in IEEE 802.16, *IEEE Wireless Communications*, 14(1), 27–35, February 2007.

[19] L. Badia, A. Baiocchi, A. Todini, S. Merlin, S. Pupolin, A. Zanella, and M. Zorzi, On the impact of physical layer awareness on scheduling and resource allocation in broadband multicellular IEEE 802.16 systems, *IEEE Wireless Communications*, 14(1), 36–44, February 2007.

[20] D. Zhao and X. Shen, Performance of packet voice transmission using IEEE 802.16 protocol, *IEEE Wireless Communications*, 14(1), 44–51, February 2007.

[21] S. Sengupta, M. Chatterjee, S. Ganguly, and R. Izmailov, Improving R-score of VoIP streams over WiMax, *IEEE International Conference on Communications,* vol. 2, pp. 866–871, June 2006.

[22] F. De Pellegrini, D. Miorandi, E. Salvadori, and N. Scalabrino, Quality-of-service support in WiMAX networks: Issues and experimental results, *CREATE-NET Technical Report,* TR-20060009. June 2006.

[23] W. Wang, D. Peng, H. Wang, H. Sharif, and H. H. Chen, Optimal image component transmissions in multirate wireless sensor networks, in *Proceedings of IEEE Global Communications Conference (GLOBECOM)*, Washington, DC, November 2007.

[24] W. Wang, D. Peng, H. Wang, and H. Sharif, A cross layer resource allocation scheme for secure image delivery in wireless sensor networks, in *Proceedings of ACM International Wireless Communications and Mobile Computing Conference (IWCMC)*, Honolulu, HI, August 2007.

[25] R. Hamzaoui, V. Stankovic, and Z. Xiong, Optimized error protection of scalable image bitstreams, *IEEE Signal Processing Magazine*, 22(6), 91–107, November 2005.

[26] J. M. Shapiro, Embedded image coding using zerotrees of wavelet coefficients, *IEEE Transactions on Signal Processing*, 41(12), 3445–3462, December 1993.

[27] A. Said and W. A. Pearlman, A new, fast, and efficient image codec based on set partitioning in hierarchical trees, *IEEE Transactions on Circuits and Systems for Video Technology*, 6(3), 243–250, June 1996.

[28] D. Taubman, High performance scalable image compression with EBCOT, *IEEE Transactions on Image Processing*, 9(7), 1158–1170, July 2000.

[29] http://en.wikipedia.org/wiki/MPEG-4

[30] H. Juan, H. Huang, C. Huang, and T. Chiang, Scalable video streaming over mobile WiMAX, in *Proceedings of IEEE International Symposium on Circuits and Systems*, New Orleans, LA, pp. 3463–3466, May 2007.

[31] P. Chou and Z. Miao, Rate-distortion optimized streaming of packetized media, *IEEE Transactions on Multimedia*, 8(2), 390–404, April 2006.

[32] C. Lian, K. Chen, H. Chen, and L. Chen, Analysis and architecture design of block-coding engine for EBCOT in JPEG 2000, *IEEE Transactions on Circuits Systems Video Technology*, 13(3), 219–230, March 2003.

[33] W. Wang, D. Peng, H. Wang, H. Sharif, and H. H. Chen, Energy-constrained quality optimization for secure image transmission in wireless sensor networks, *Advances in Multimedia*, 2007(2), 1–9, 2007.

[34] W. Luo, K. Balachandran, S. Nanda, and K. Chang, Delay analysis of selective-repeat ARQ with applications to link adaptation in wireless packet data systems, *IEEE Transactions Wireless Communications*, 4(3), 1017–1029, May 2005.

[35] Z. Wu, A. Bilgin, and M. Marcellin, Joint source/channel coding for multiple images, *IEEE Transactions on Communications*, 53(10), 1648–1654, October 2005.

[36] Z. Wu, R. Jandhyala, A. Bilgin, and M. W. Marcellin, Joint source/channel coding for multiple video sequences with JPEG2000, in *Proceedings of Data Compression Conference*, Snowbird, UT, pp. 489–489, 2005.

[37] T. Lookabaugh and D. C. Sicker, Selective encryption for consumer applications, *IEEE Communications Magazine*, 42(5), 124–129, May 2004.

[38] H. Cheng and X. Li, Partial encryption of compressed images and videos, *IEEE Transactions on Signal Processing*, 48(8), 2439–2451, August 2000.

[39] W. Zeng and S. Lei, Efficient frequency domain selective scrambling of digital video, *IEEE Transactions on Multimedia*, 5(1), 118–129, March 2003.

[40] M. Grangetto, E. Magli, and G. Olmo, Multimedia selective encryption by means of randomized arithmetic coding, *IEEE Transactions on Multimedia*, 8(5), 905–917, October 2006.

# 10 Relay-Assisted Mobile WiMAX

*Yan Q. Bian, Yong Sun, Matthew W. Webb, and Andrew R. Nix*

## CONTENTS

Mobile worldwide interoperability for microwave access (WiMAX) promises to deliver high data rates over extensive areas to high user densities. The latest development of WiMAX provides both multiple-input multiple-output (MIMO) and multihop relay. However, there are major concerns including leveraging high data rates and large cellular coverage in network planning to compete with other beyond 3G networks. For mobile WiMAX to be successful, it must achieve high spectral efficiency and large coverage using limited available spectrum for system optimization. This chapter gives a technical overview and presents design strategies to achieve highly efficient mobile WiMAX. Directional distributed relaying is introduced for effective multiuser transmission. Furthermore, channel measurements are performed and analyzed for a multihop relay in mobile WiMAX. Practical applications are also investigated within a realistic urban environment through ray-tracing techniques.

## 10.1   INTRODUCTION

There are two major technological and social trends significantly changing people's lives: wireless communications and the Internet. Leveraging these two trends, worldwide interoperability for microwave access (WiMAX) creates a new utility enabling the development of new services and new Internet business models. In particular, the full potential of WiMAX will be realized when it is used for innovative nomadic and mobile broadband applications [1]. With the finalization of the IEEE 802.16e standard [2] and upcoming test and certification of WiMAX products [3], mobile broadband services are becoming a reality. The 802.16e standard provides broadband wireless Internet Protocol (IP) access to support a variety of services (such as voice, data, and multimedia) on virtually any device. The operation of WiMAX is currently limited to a number of licensed frequency bands below 6 GHz for reliably supporting non-line-of-sight (NLoS) operations. The 802.16e standard has also become a part of the IMT-2000 family.

WiMAX is often quoted as combining long transmission ranges (e.g., in a macrocell) with high data capacities (multi megabit per second throughput to end users). Power and spectral efficiency is key to a successful WiMAX deployment. The mobile WiMAX physical (PHY) layer is based on scalable orthogonal frequency division multiple access (SOFDMA) technology, which enables flexible channelization. The new technologies employed by mobile WiMAX result in higher data transfer rates, simpler mobility management, and lower infrastructure costs compared to current 3G systems. The underlying scenario for mobile WiMAX is an outdoor environment with multiple users within a cell. Hence scheduling (allowing a fair and efficient distribution of resources) and interference (from intracell and intercell) become important issues.

Radio relaying can address many of the challenges faced in the deployment of mobile WiMAX [4,5] and its potential benefits were studied in Refs. [6,7]. The relay system targets one of the biggest challenges in next generation mobile wireless access (MWA), namely the provision of high data rate coverage in a cost-effective and ubiquitous manner. Within a multihop relay network, a multihop link can be formed between the base station (BS) and a distant mobile station (MS) using a number of intermediate relay stations (RS). To avoid interference between the relay links, the simplest approach is to assign unique radio resources to each link. Using this approach, the multihop users will rapidly drain the system of valuable radio resource. The wireless medium is a precious infrastructure commodity and the situation is especially acute at lower frequencies (where the radio signal propagation characteristics are more favorable) when a significant amount of radio spectrum is needed to provide ubiquitous wireless broadband connectivity. Spectrum predictions for future cellular networks indicate large shortfalls by 2010, if not before [8]. Hence the above approach can only be used for a very small number of very high-value MSs, or in applications where spectral efficiency is not vital, such as military or disaster relief communication networks. Given its commercial applications, relaying in the context of WiMAX must conserve radio spectrum and emphasize the need for high spectral efficiency. Enhancing radio resource efficiency is a key challenge in a competitive business development.

This chapter focuses on the efficiency of relay transmission. We present leading edge techniques, and merge both theoretical analysis and practical application for a mobile WiMAX system with multihop relay. Directional distributed relaying is then proposed to achieve high data throughput with reduced demands on radio resource.

## 10.2   CURRENT DEVELOPMENTS IN WiMAX TECHNOLOGY

WiMAX technology encompasses broadband wireless equipment which is designed in compliance with the IEEE 802.16 standard and certified by the WiMAX Forum. The IEEE 802.16e standard leverages several differences and enhancements over the 802.16-2004 standard [9] to support mobile subscribers.

*Scalable orthogonal frequency division multiple access (SOFDMA)*: Introduced in the 802.16e amendment over fixed WiMAX's OFDM, SOFDMA supports scalable channel bandwidths from

1.25 to 20 MHz, using quadrature amplitude modulation (QAM; 16QAM or 64QAM) or quaternary phase shift keying (QPSK) modulation. SOFDMA enables additional resource allocation flexibility and adaptively optimized multiuser performance.

*Advanced antenna technologies*: MIMO PHY layer techniques have the potential to significantly increase bandwidth efficiency based on the premise that operation occurs in a rich scattering multipath environment [10]. 802.16e defines optional support for such advanced antenna technologies. Major advantages of MIMO include diversity gains, multiplexing gains, interference suppression, and array gains. The inclusion of MIMO techniques alongside flexible subchannelization and adaptive modulation and coding (AMC) enables mobile WiMAX technology to improve system coverage and capacity.

In addition, 802.16e presents many advanced features for performance enhancements, such as handover support, quality-of-service (QoS) support, and energy savings mechanisms for handheld support, etc.

*Multihop relay*: Another milestone in the development of WiMAX was the introduction of multihop relay running as the 802.16j multihop relay project [4], which targets on OFDMA PHY layer and Medium Access Control (MAC) layer enhancements for licensed bands to enable the operation of RSs. The objectives of 802.16j are to enhance coverage, throughput, and system capacity by specifying 802.16 multihop relay capabilities and functionalities of interoperable RSs and BSs. Several technical topics were focused on, mainly including relay concepts, frame structure, network entry, bandwidth request, security, mobility management, routing, path management, interference control and radio resource management, etc.

The concept of multihop relaying is already well developed in the fixed telecommunications world. Microwave radio relays have been widely used to transmit digital and analog signals over long distances, with examples including telephony and broadcast television. With the evolution of mobile networks, wireless relays have been further developed for cellular transmission [11]. Analog repeaters are sometimes used in cellular systems to extend coverage into regions that are uncovered by the standard network [12]. Digital relaying for cellular applications was initially investigated in Ref. [13] to enhance coverage for delay-insensitive traffic. More recently, Streaming21 of the United States announced the availability of a 3G relay server in mid-2006, a carrier-grade mobile streaming solution that allows mobile operators and content providers to deliver multimedia contents to mobile phone subscribers over GPRS and 3G networks [14]. Now, the relay concept is being further developed in 802.16j to supporting both digital repeater and decode-forward (DF) relaying with various techniques, such as cooperative relaying, intelligent radio resource management (RRM) for radio resource reuse, smart antenna on RSs for direction-controlled transmission, relay grouping, etc.

While the IEEE and the WiMAX Forum strive to address the technological challenges of high mobile, NLOS WiMAX services, large throughput and coverage, etc., commercial service providers face additional operational challenges including spectrum limitations, security vulnerabilities, and QoS implementations. Many researchers have been working toward developing mechanisms that provide highly efficient mobile WiMAX, which is the core topic in this chapter. Further development is expected in a newly approved project from the IEEE, namely 802.16m—Advanced Air Interface to meet IMT-advance requirements. It is targeting data rates of 100 Mbps for mobile applications and 1 Gbps for fixed applications, cellular, macro- and microcell coverage, with currently no restrictions on the RF bandwidth.

## 10.3   MULTIHOP TOPOLOGY AND CAPACITY ANALYSIS

Figure 10.1 illustrates the multihop relay topology with some usage scenarios, including fixed, nomadic, and mobile relaying. Different from mesh topology with multiple connections routing by subscriber equipment, multihop relay is a dedicated carrier-owned infrastructure with a tree-based topology and one end of the path is the BS. With the tree structure, we can derive several capacity formulas with regard to different applications.

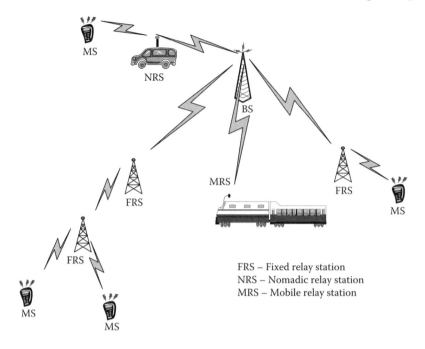

**FIGURE 10.1**  Illustration of multihop relay topology with typical usage scenarios.

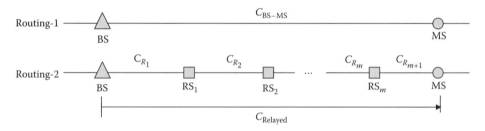

**FIGURE 10.2**  A topology with two routings for transmission between a BS and an MS.

## 10.3.1  MULTIHOP RELAY CAPACITY AND BOTTLENECK PROBLEM

Without loss of generality, we utilize a simple topology with only two routings for the transmission between a BS and an MS as shown in Figure 10.2. Routing-1 is the direct access path from the BS with a capacity of $C_{BS-MS}$. Routing-2, is a relayed transmission path between the BS and the MS with $m$ RSs. We call Routing-1 an access-routing and Routing-2 a relayed routing. The single user capacity through the relayed routing, $C_{Relayed}$, can be derived with the following different relay transmissions in time division duplex (TDD) mode.

*Scheme 1: Amplify-and-forward (AF) relay*

With this relay transmission, each RS can only forward its received signal to the next RS until it reaches MS. The relayed capacity is highly limited by the worst link quality and can be expressed as

$$C_{\text{Relayed\_AF}} = \frac{\arg\min_{m+1}\left(C_{R_1}, \ldots, C_{R_m}, C_{R_{m+1}}\right)}{m+1} \qquad (10.1)$$

where $C_{R_i}$ denotes the relayed capacity through the $j$th hop. It is shown clearly that the relayed link capacity is the minimum value of the capacity vector, which also indicates the bottleneck problem, especially if the minimum capacity occurs on a relay link.

*Scheme 2: Decode-and-forward (DF) relay*

For this scheme, there are two typical conditions. The first one is that the radio resources for each access station (including both BS and RSs) are fixed. In this case, the relay link capacity is fundamentally the same as that in Equation 10.1. In contrast, if radio resources are fully flexible and assigned according to each link quality, then the capacity of the relayed routing can be derived as

$$C_{\text{Relay\_DF}} = \frac{C_{R\max}}{\sum_{j=1}^{m+1} \frac{C_{R\max}}{C_{R_j}}} \tag{10.2}$$

where $C_{R\max} = \arg\max_{m+1} (C_{R_1}, \ldots, C_{R_m}, C_{R_{m+1}})$ represents the largest capacity link. From Equations 10.1 and 10.2, we can see that with the DF scheme with dynamic resource assignment it is possible to smooth the bottleneck problem. For a clearer demonstration, we adopt 2-hop relay (with one BS–RS relay link and one RS–MS access link) as an example. The bottleneck problem is mainly due to the lower capacity on the BS–RS link rather than the RS–MS link. Figure 10.3 presents the capacity comparison between AF and DF. In this study, the relay link (BS–RS) has a constant capacity (10 Mbps) and the access link (RS–MS) capacity varies from 1 to 50 Mbps. With the AF scheme, the relayed capacity increased linearly with the increase of access link capacity until reaching the capacity of relay link. In contrast, with the DF scheme, the relayed capacity continues increasing with the increase of access link. Without consideration of noise enhancement in the AF scheme, the relayed capacity is the same at the point that the two links have the same capacity.

The relayed routing might require more radio resources. It implies a challenge of efficiency for practical relay deployment which requires advanced techniques to develop high efficient relay system. The rest of this chapter will concentrate on relay efficiency and its developed techniques. The bottleneck problem in this section can be normally resolved by user access control linked with QoS control scheme especially for the system which employs AF relay scheme. The idea case without losing any radio resources is to allocate a user on the relayed routing if both access link and relay link achieve the same QoS.

## 10.3.2 Formulation of Relay Efficiency

To measure the efficiency of a relay deployment, we now introduce an effective system capacity metric $(C_{\text{eff}})$. This is intended to leverage link level capacity gain and system capacity gain. Assuming there

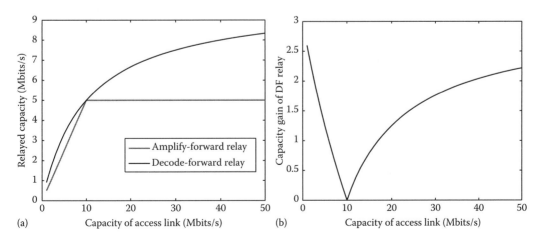

**FIGURE 10.3** Analysis of relay SNR-gain required for high relay efficiency.

are a total of $s$ MSs to be allocated, and $p$ of these users are served via multihop relays, $C_{\text{eff}}$ can be expressed as

$$C_{\text{eff}} = \sum_{i=1}^{s-p} C_i^{\text{BS}} + \frac{\sum_{j=1}^{p} C_j^{\text{RS}}}{N_{\text{rc}}} \tag{10.3}$$

where $C^{\text{BS}}$ and $C^{\text{RS}}$ denote the capacity without relaying (BS access capacity) and the capacity with relaying (relayed capacity). $N_{\text{rc}} \in \{1, 2, \ldots, m+1\}$ is the number of radio resources employed in the relays. While the BS–RS and/or RS–RS links are ideal (no bottleneck problems), the effective relay efficiency is derived as

$$\xi_{\text{c}} = \frac{\sum_{i=1}^{s-p} C_i^{\text{BS}} + \left(\sum_{j=1}^{p} C_j^{\text{RS}}\right)/N_{\text{rc}}}{\sum_{k=1}^{s} C_k^{\text{BS}}} \tag{10.4}$$

From a system-level point of view, the effective relay efficiency must be greater than one. In the case where $\xi_{\text{c}} < 1$, radio resource sharing should be considered. However, resource sharing may introduce interference.

### 10.3.3 SISO MULTIHOP RELAY WITHOUT RADIO RESOURCE SHARING

We start with a simple relay deployment for a SISO system with a single user. Without radio resource sharing, the system relies on the use of unique radio resources for each link, including the BS–RS, BS–MS, RS–RS, and RS–MS. We define relay SNR-gain as $G_{\text{SNR}} = \text{SNR}_{\text{relay}} - \text{SNR}_{\text{access}}$, where the $\text{SNR}_{\text{relay}}$ and $\text{SNR}_{\text{access}}$ represent the signal-to-noise ratio (SNR) of the last hop between the RS and MS (shortened to relay-SNR) and the BS-access link (directly between the BS and MS, shortened to access-SNR) respectively. The effective relay efficiency is derived from Ref. [4], and the capacities of BS access and relay access can be obtained from the Shannon capacity equation [15] based on the $\text{SNR}_{\text{access}}$ and $\text{SNR}_{\text{relay}}$ value.

Figure 10.4 shows the relay efficiency for a 2-hop relaying with two radio resources. It indicates that the relay efficiency does not linearly increase with $G_{\text{SNR}}$, and for a certain required efficiency

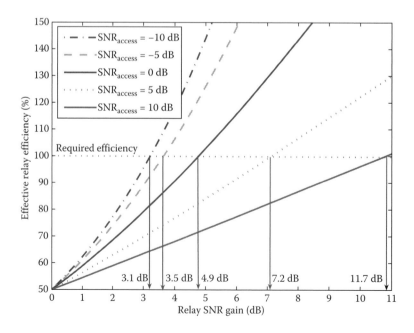

**FIGURE 10.4** Analysis of relay SNR-gain required for high relay efficiency (2-hop with two radio resources).

the value of $G_{\mathrm{SNR}}$ varies according to $\mathrm{SNR}_{\mathrm{access}}$. Also, the requirement of the relay SNR-gain is much lower if the value of $\mathrm{SNR}_{\mathrm{access}}$ is low, e.g., when $\mathrm{SNR}_{\mathrm{access}}$ is $-10\,\mathrm{dB}$ the relay SNR-gain only needs to be around 3.1 dB for 100% relay efficiency. However, to maintain a high level of relay efficiency, the relay SNR-gain must exceed $10\,\mathrm{dB}$ when $\mathrm{SNR}_{\mathrm{access}}$ is $10\,\mathrm{dB}$. If a relay system has more than 2-hop relays, the required relay SNR-gain could be even higher [7].

## 10.3.4 INTERFERENCE IMPACTS ON MIMO MULTIHOP RELAY

Given the scarcity of radio resources, to achieve a highly efficient relay deployment it is necessary to implement radio resource sharing. With resource sharing, the critical issue is the interference introduced into the system. Fundamentally, there are two basic factors that must be taken into account when quantifying a MIMO relaying application: (1) the value of the relay SNR-gain and (2) the condition of the MIMO channel matrix.

If no prior channel knowledge is available at transmitter, the effective relay efficiency for a MIMO multihop relay system, with $M$ transmit and $N$ receiver antennas can be rewritten for the downlink (DL) case, as

$$
\xi_c = \left[ 1 \Big/ \sum_{k=1}^{s} \log_2 \det \left( \mathbf{I}_N + \frac{1}{M} \cdot \frac{P_{s,k}^{\mathrm{BS}}}{P_{1,k} + P_{n,k}} \cdot \mathbf{H}_k \mathbf{H}_k^H \right) \right]
$$
$$
\times \left[ \begin{array}{l} \displaystyle\sum_{i=1}^{s-p} \log_2 \det \left( \mathbf{I}_N + \frac{1}{M} \cdot \frac{P_{s,i}^{\mathrm{BS}}}{P_{1,i} + \sum_{ii=1}^{q_j^{\mathrm{RS}}} P_{1,i,ii}^{\mathrm{RS}} + P_{n,i}} \cdot \mathbf{H}_i \mathbf{H}_i^H \right) \\[4mm] + \dfrac{1}{N_{\mathrm{rc}}} \displaystyle\sum_{j=1}^{p} \log_2 \det \left( \mathbf{I}_N + \frac{1}{M} \cdot \frac{P_{s,j}^{RS}}{P_{1,j} + \sum_{jj=1, jj \neq j}^{q_j^{\mathrm{RS}}} P_{1,j,jj}^{\mathrm{RS}} + P_{n,j}} \cdot \mathbf{H}_j \mathbf{H}_j^H \right) \end{array} \right] \tag{10.5}
$$

where $P_s^{\mathrm{BS}}$ and $P_s^{\mathrm{RS}}$ represent the signal power received from the BS and RS. $P_I^{\mathrm{BS}}$, $P_I^{\mathrm{RS}}$, and $P_I$ denote the cochannel interference (CCI) power received from the BS, RS, and any other resources (e.g., from other cells), respectively. $l_i^{\mathrm{RS}}$ denotes the number of RSs that use the same radio resource as the $i$th user (BS–MS), and $q_j^{\mathrm{RS}}$ represents the number of RSs that use the same radio resource as the $j$th user (RS–MS). From Equation 10.5 we deduce the impact of interference and radio resource usage. Increasing any of the interference terms or the value of $N_{\mathrm{rc}}$ could reduce the relay gain. Reducing $N_{\mathrm{rc}}$ could also increase the CCI. Furthermore, the channel correlation is an important factor for MIMO transmissions.

Figure 10.5 presents the impact of interference on MIMO systems with 2-hop relays. We have defined two thresholds: one is at 100 percent efficiency (which is required to compute the minimum SIR value) and the other is at 200 percent efficiency, which implies the relay deployment has doubled the capacity of the direct BS–MS link. Results indicate that the high level MIMO configuration is more tolerant to interference. For example, the SISO system requires 4.5 dB more SIR than an $8 \times 8$ MIMO approach at 200 percent relay efficiency. Note that in this study the MIMO channels are uncorrelated.

MIMO channel correlation is detrimental to MIMO systems, especially in high-order MIMO transceivers with small antenna separations. The effective relay efficiency versus channel correlation value is shown graphically in Figure 10.6. For a $2 \times 2$ MIMO system we deliberately set both the $\mathrm{SNR}_{\mathrm{access}}$ and the relay SNR-gain to 6 dB (since this produces 100 percent effective relay efficiency for a SISO system when the BS–MS link and the RS–MS link are uncorrelated). Three extreme scenarios are now considered: (1) the BS–MS link is correlated but the RS–MS link is uncorrelated, (2) both the BS–MS and the RS–MS links are correlated with identical correlation values, and (3) the RS–MS link is correlated but the BS–MS link is uncorrelated. For the first scenario, the relay improves the link channel properties and consequently improves the relay efficiency. In contrast, for

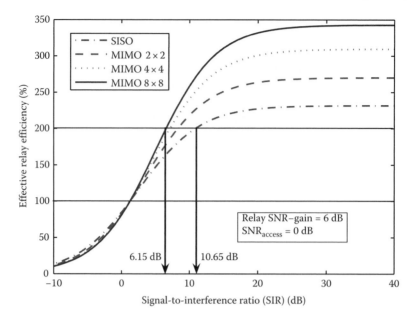

**FIGURE 10.5**   Analysis of relay SIR required for high relay efficiency.

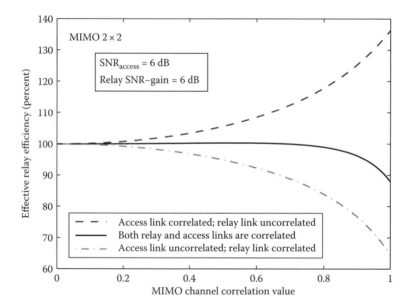

**FIGURE 10.6**   Impact of channel correlation on a relayed MIMO system.

the third scenario the correlated RS–MS link severely degrades the relay efficiency. However, it is interesting to see that the second scenario shows that a high channel correlation value (greater than 0.6) can destroy the relay gain and dramatically decrease the relay efficiency.

### 10.3.5   EFFECTS OF MULTIUSER TRANSMISSION

To investigate the relay capacity gain with multiuser transmission, we consider equivalently equal relay efficiency, which is defined as that each user's relay in multiuser transmission achieves equal efficiency for both with and without radio resource sharing. With resource sharing deployment, single user relay efficiency can be expressed as

$$\xi_{ws} = \frac{C_{i\_ws}^{RS}}{C_{i\_ws}^{BS}} \tag{10.6}$$

where $C_{i\_ws}^{RS}$, $C_{i\_ws}^{BS}$ represent the relayed capacity and BS access capacity of the $i$th user respectively, within the radio resource sharing deployment. For the relay without resource sharing but to achieve the same efficiency as that in Equation 10.6, it has to meet either $C_{i\_wo}^{RS} = n_{i\_rc} \cdot C_{i\_ws}^{RS}$ or $C_{i\_wo}^{BS} = C_{i\_ws}^{BS}/n_{i\_rc}$ where $n_{i\_rc}$ is the number of resources for the case of radio resource sharing, $C_{i\_wo}^{RS}$, $C_{i\_wo}^{BS}$ denote relayed capacity and BS access capacity of the $i$th user, without sharing. It can be seen that, the efficiency loss by employing more resources could be recovered by relay capacity gain. These represent two realistic application scenarios and considerations. First it is dependent on relay link capacity with $n_{i\_rc}$ times the capacity of relay with resource sharing. Alternatively, the equal system gain could be achieved when the relay employed in a system where the BS-access link has much low capacity.

The cases presented above are based on single user relay deployment. For multiuser application, it would be interesting to show the different gain achieved from each single user to the system capacity gain. Figure 10.7a presents the achievable effective relay efficiency versus number of users for a 2-hop relaying for two group results (as seen in Figure 10.7b), where the "number of users" is defined as the number of users through relaying. For the results presented in Figure 10.7, we assume 15 users to form an OFDMA transmission (802.16e defines up to 15 users can be supported by 512 FFT DL PUSC OFDMA profile). It also assumes that all the BS-access capacities are equal. The single user relay capacity gain without resource sharing is 3 dB higher than the one with resource sharing because of a 2-hop relaying. Figure 10.7a illustrates the efficiency gain (defined as the difference between the relay with and without resource sharing). In general, if all users in multiuser transmission are through relays, all cases can achieve the same system efficiency gain, as shown in Figure 10.7a. However, with some users through relaying, multiuser achieves capacity gain which is not linear with number of users. Figure 10.7b indicates that the relay with resource sharing achieves higher multiuser relay efficiency.

Based on previous assumptions, both achieve the highest relay efficiency when all users are through relaying. While evaluating the system efficiency with a few users through relaying, it is shown clearly that each single-relayed transmission contributes more to system efficiency with radio resource sharing. From this study it is shown that, in any case, the relay with radio

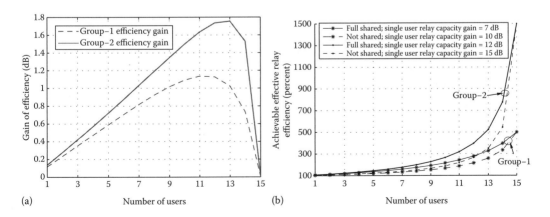

**FIGURE 10.7**  Relay efficiency with multiple users.

resource sharing has the potential to achieve optimal multiuser relay gain. Hence, with an OFDMA multiuser application, it is ideal to adopt flexible channelization, AMC on each user, and efficient radio resource management.

## 10.4 HIGHLY EFFICIENT MULTIHOP RELAY TOPOLOGIES

The big challenge for broadband wireless system design comes up with the right balance between capacity and coverage that offers good quality and reliability at a reasonable cost. It is important to look at system spectral efficiency more broadly to include the notion of coverage area. Results presented in previous sections have demonstrated the high potential benefits for relay deployment with radio resource sharing, in terms of interference, MIMO combination, and multiuser transmission. To implement the radio resource reuse and achieve highly efficient relay deployments, appropriate frequency reuse and multiuser access strategies are required. Relay systems must be based on a topology that fully exploits effective resource assignment based on the spatial separation of nodes. In this section, we propose directional distributed relay for highly efficient multiuser transmission with reduced demands on radio resource.

Figure 10.8 depicts the directional distributed relaying architecture. This is based on a paired radio resource transmission scheme, and it is possible to achieve one radio resource to one user (or one group of users) in average, even with multihop relay. The radio resource can be defined as either frequency (e.g., subcarriers in an OFDMA symbol) or time (e.g., OFDMA time slots). Transmissions in the BS coverage are the same as the IEEE 802.16e standard. For relay links, paired transmissions are applied, where the BS forms two directional beams, or uses two sector antennas to communicate with RS1 and RS2 simultaneously. A paired radio resources are required: $f_1$ and $f_2$. The first radio

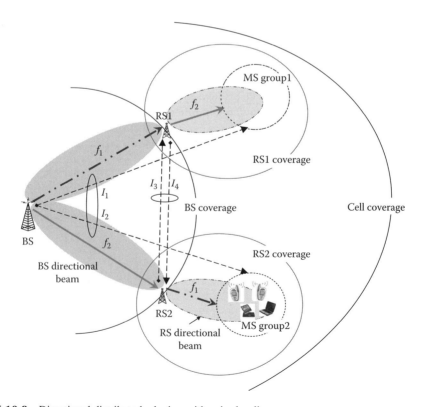

**FIGURE 10.8**   Directional distributed relaying with paired radio resource.

resource ($f_1$) is applied to the RS–BS1 link and also to the RS2–MS links (in the RS2 coverage); while the second resource ($f_2$) is applied to the BS–RS2 link and also to the RS1–MS links (in the RS1 coverage). Radio resources are shared between the RSs and MSs. Each end-user employs a single pair of radio resources, on average.

Using the sharing scheme outlined above the interference can be controlled at the BS and RS nodes. In this relay configuration there are only two sets of interference, as also illustrated in Figure 10.8. The interference between the BS and MS groups ($I_1$ and $I_2$) can be detected and controlled by the BS. First, the BS could employ an adaptive array to exploit the spatial separation of the groups. Second, since the received power by each MS in each MS group is known to the BS, the BS can apply interference avoidance [16,17] between the two groups based on measured signal to interference plus noise ratio (SINR) and power control, where the transmit power of the two RSs are controlled for balancing the SINR according to the service requirement. Furthermore, in this scenario the expected level of interference is small because the BS connects to the MSs through a relay, which means the relay SNR-gain will be much higher than the $SNR_{access}$ level. Interference between RSs ($I_3$ and $I_4$) can be reduced by array processing (including the use of sector antennas) at the RSs. Interference measurement for the efficient resource assignment can be achieved during the neighborhood discovery procedure, which is being developed in Refs. [18,19]. To achieve high levels of SINR (e.g., 10–25 dB), array processing, including the use of sector antennas at the RS, is desirable.

This proposed topology is fully compatible with the existing 802.16e standard and no modifications are required at MSs. Alternative deployments topologies are also possible based on the same concept, such as a single RS to cover a coverage hole (Section 10.6). In such cases, the radio resource sharing is performed between the RS and its BS. It could be complicated for statistical studies as the performance is fully dependent on the deployment scenario. However, it is much more feasible in a realistic application environment by employing real channel measurements and ray tracers.

## 10.5  MULTIHOP RELAY CHANNEL MEASUREMENTS

Recently, several channel models have been developed [20–23] for various environments and system topologies such as the COST 231 [20], Stanford University Interim (SUI) [23], and WINNER [21] models. Unfortunately, statistical models of this type make a number of general assumptions that are not always met in practical WiMAX scenarios. For example, the SUI models taken from Ref. [9] are suitable for FWA (rather than MWA) and they are based only on cellular-type measurements in the 1.9 GHz band [22]. The COST 231 models do not formally apply beyond 2 GHz. The WINNER models do reach to frequencies at 5 GHz [24], but does not include frequency correction factors to enable their use in the 3.5 GHz band envisaged for WiMAX. Furthermore, there is little by way of empirical data collected in specifically multihop scenarios in the 3.5 GHz making an evaluation of these models in their intended deployments difficult. This part presents the results of a measurement campaign conducted in outdoor city environment to subject these models to empirical scrutiny.

### 10.5.1  Measurement Campaign

Measurements were carried out along seven routes in Bristol city centre, United Kingdom, comprised of five RS locations as shown in Figure 10.9. These routes were identified from a ray-tracing analysis as being likely to exhibit deep shadowing from the BS (which was mounted on the roof of a building approximately 30 m above the top of a hill). The system under consideration is a 2-hop DL, as shown schematically in Figure 10.10. A commercial 3.5 GHz panel antenna was used with a beamwidth of 90° in azimuth and 10° in elevation with 17 dBi gain. Since the BS antenna was not omnidirectional, its orientation and downtilt are important. Prior to commencing the main campaign, power measurements were taken at each location for downtilts between 0° and 10° following which during measurement the antenna was set to the downtilt which maximized the received power.

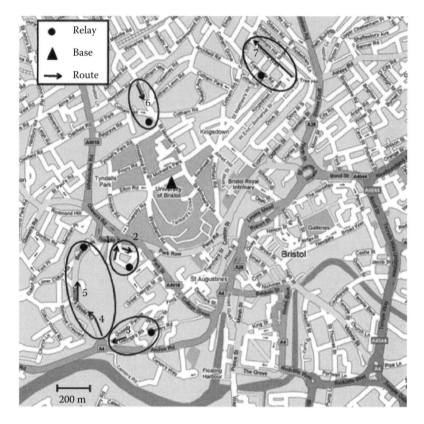

**FIGURE 10.9**   Map of Bristol, United Kingdom, showing the BS, the 7 MS routes, and the 5 RS locations.

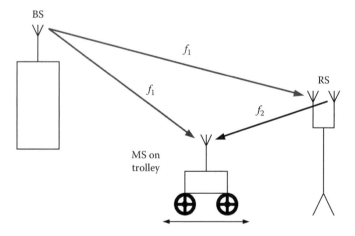

**FIGURE 10.10**   Schematic of multihop measurement system. $f_1 = 2.58$ GHz, $f_2 = 3.467$ GHz.

The antenna was oriented approximately southwesterly for routes 1–5 and northeasterly for routes 6 and 7.

The RS was on a portable pump-up mast that was left fixed in place once on location. Two dipoles were mounted on top of the mast at either end of a beam that could be rotated in a plane. One end was fixed to act as the RS–MS transmitter, and the other end rotated around it to act as the BS–RS receiver; this rotation was to permit measurements of the local variation of the BS–RS signal.

The BS transmitted 20 W before antenna gain and after cable loss, etc., on 3.59 GHz and the RS 3.67 W on 3.467 GHz.

The MS was constructed on a trolley that was pulled or pushed along the selected routes. A laptop on the trolley provided "command and control." The two wheels on one axle of the trolley were equipped with electronic pulse counters that incremented the counts on the laptop approximately every 4 mm. Also connected to the laptop was a spectrum analyzer which received the signal from the antenna. The antenna (mounted at 1.65 m above ground) was the same type of dipole as used at the RS. The mean length of a route was about 90 m, and each route was repeated with the RS at a selection of heights of {2,3,4,5} m.

Path lengths were determined by using survey grade GPS equipment to obtain positional fixes for the start and end of each route and the trolley's distance pulses to estimate the route in between. Elevation was assumed to change linearly along a route. Path loss at a particular point was determined as the difference between the transmitted and received power after taking account of the system gains and losses in cables, amplifiers, etc. The effect of antenna patterns was approximated as being the gain in the LoS direction between the transmitter and receiver.

## 10.5.2  CALCULATION OF MEAN PATH LOSS AND SHADOWING

With one measurement approximately every 2 cm at a frequency of 3.5 GHz, the fast fading is captured. To extract the slow fading for calculation of mean path loss and shadowing, the fast fading was averaged through a sliding window of 40 wavelengths with successive windows overlapping by 30 wavelengths. An example of this process is shown in Figure 10.11 for the BS–MS link of route 6.

The shadowing on a wireless link is the signal fluctuations around the local mean level. For the purposes of this part, the local mean level will be obtained by fitting a least-squares regression curve to the data at each location and taking this as the mean path loss around which the shadowing fluctuations occur. For the BS–MS link, the data comprising all of the runs together will be treated as a single variable, but for the RS–MS link, the data collected at each different RS height will be treated separately. An example of the resulting shadowing distribution is shown in Figure 10.11b, on normal probability axes, so that a normal distribution with the same mean and variance as the actual data would be exactly a straight line. This is evidently a normally distributed variable, and calculation shows that it has zero mean and standard deviation of 3.5 dB. Similar distributions were found for the other routes and for the RS–MS link.

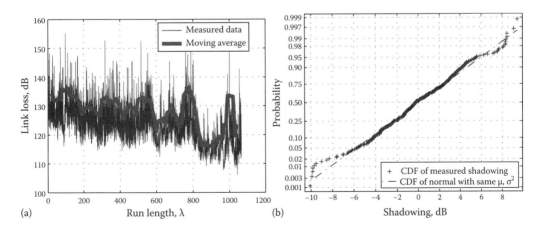

**FIGURE 10.11**   Example of BS–MS fast- and slow-fading along route 6, with RS at 5 m.

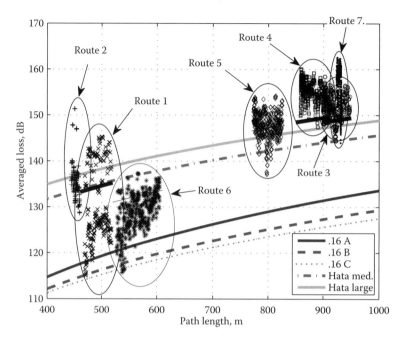

**FIGURE 10.12** Overview of BS–MS link for all routes. COST 231 WI NLOS medium city model shown with thick black lines.

### 10.5.3 ANALYSIS OF RELAY AND ACCESS LINKS

#### 10.5.3.1 BS–MS Links

Figure 10.12 summarizes the results from the BS–MS links. Although there was no change to the BS–MS link itself from run to run at a given location, environmental factors inevitably caused some random variation in the results. Overall, however, the various routes show distinct behaviors, with a general trend of loss increasing with path length. Although there was no change to the BS–MS link from. The 802.16d SUI models (.16d "A" to .16d "C") are shown along with the COST 231 Hata model for the relevant range of path lengths. Clearly, the 802.16d models are substantially underestimating the mean path loss at all the locations considered here, but the mismatch is especially pronounced at longer path lengths where the measured data lie about 25 dB above the modeled path loss. For the Hata model, however, there is a much smaller mismatch in general. This is perhaps unexpected since the Hata model is not specified above 2 GHz whereas the 802.16d models are specifically corrected to 3.59 GHz and based on cellular environments not unlike those used here (but with a receive antenna mounted on a 2 m-high vehicle). The Hata model for a large city comes within 10 dB of the mid-range link losses for all the locations, and closer than this for some, particularly along routes 3 and 5.

The COST 231 Walfisch-Ikegami (WI) models depend on location-specific features, and so curves for these are shown over the applicable ranges only and for clarity for the medium city NLoS only; the large city adds about 7 dB at 3.5 GHz. It can be seen that the WI model in general gives better predictions of the measured path loss on this link than the 802.16d or Hata models. The medium city and large city NLoS WI models spanned the measured range at most locations, or passed through some part of the measured data even though the WI model is not formally specified at frequencies beyond 2 GHz. This analysis agrees with the recommendation in Ref. [24] to use the NLoS WI model for this type of link (urban, from above- to below-rooftop).

The values of the shadowing standard deviation $\sigma_s$ calculated for the various routes are given in Table 10.1, along with those used by the models in Ref. [24]. The 802.16d models and WINNER

**TABLE 10.1**

**Summary of Shadowing Standard Deviations**

| | | | | Shadowing Standard Deviation, dB | | | |
|---|---|---|---|---|---|---|---|
| | BS → MS | RS → MS | IEEE | WINNER | | | |
| Route | (Measured) | (Measured) | 802.16d | B5a | C2 | B1 LoS | B1 NLoS |
| 1 | 3.7 | 3.8–7.2 | | | | | |
| 2 | 4.2 | 3.1–3.7 | | | | | |
| 3 | 2.4 | 3.8–5.3 | | | | | |
| 4 | 2.0 | 1.5–3.0 | 8.2–10.6 | 3.4 | 8 | 2.3 | 3.1 |
| 5 | 2.6 | 2.7–4.2 | | | | | |
| 6 | 3.5 | 3.0–3.4 | | | | | |
| 7 | 3.7 | 0.6–1.4 | | | | | |

Range of RS–MS values relates to changing RS antenna height. Range of IEEE 802.16d values spans models "A" to "D."

B5a are designed for path lengths on the order of several kilometers whereas WINNER C2 and B1 are designed for hundreds of meters to a few kilometers. The WINNER C2 shadowing value of 8 dB is suggested in Ref. [24] for cases where the transmitter is above roof-top and the receiver below in an urban macrocell. Although this is a good description of the BS–MS link, the measurements have found much lower values of shadowing—within the narrow range 2–4 dB. This is in part because of the run length (about 90 m on average) that these values were measured along; longer runs are likely to result in greater variability around the mean and thus increased estimates of SS. The other WINNER models, B5a and B1, predict values of 2.3–3.5 dB that are close to those found by measurement, suggesting that these are good shadowing models for the BS–MS link. The 802.16d models suggest shadowing values much higher than have been found here but, like the WINNER C2, are intended to apply to path lengths an order of magnitude longer than that used in the measurement campaign and must therefore encompass greater environmental variation.

### 10.5.3.2 RS–MS Links

The key differences between the BS–MS and RS–MS links are that the RS link is in some locations LoS (especially routes 1, 2, and 3) and much closer to the MS than the BS is. Figure 10.13 shows an overview of the averaged link losses for all seven routes. The 802.16d models are shown along with the COST 231 models for the relevant range of path lengths and a transmit antenna height of 3.5 m, the mean of {2,3,4,5}m. For this link the Hata models are consistently overestimating the path loss by about 10 dB, while the 802.16d models do pass broadly through the region of the measurements.

Whereas for the BS–MS link, the WI model matched rather better than the 802.16d and Hata models, here it is found to be usually further from the measured data than the Hata or 802.16d models, whether in LoS or NLoS mode. The NLoS mode typically over-predicts by 10–20 dB and the LoS mode under-predicts, as shown in Figure 10.13. Route 3 had the clearest physical LoS for the RS–MS link.

The shadowing statistics for RS–MS link are also summarized in Table 10.1, with the range of shadowing values given by the various RS heights listed. Similar to the earlier results, they are significantly smaller than the 802.16d models and much closer to the WINNER models. Routes 2, 5, and 6 (RS–MS links along those routes was NLoS) are particularly good matches to the NLoS WINNER B1 model, which is described in Ref. [24] as intended to match urban microcells with both transmit and receive below roof level. The most LoS route is route 3, but this reports rather higher shadowing than other routes because of the unusually large loss variations which are caused by a transient localized effect.

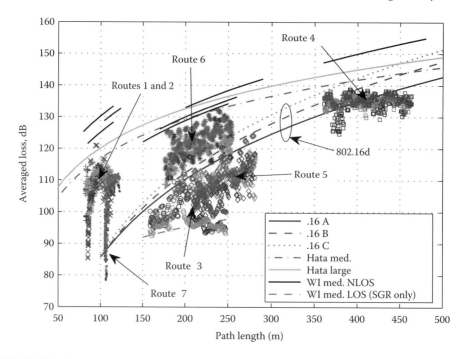

**FIGURE 10.13**  Overview of RS–MS link for all routes. Shades indicate different RS heights [5, 4, 3, 2] m.

#### 10.5.3.3  BS–RS Links

Measurements of the BS signal power at the RS were taken by attaching a dipole to the end of a rotating arm at the top of the RS (Section 10.5), and taking measurements at approximately 45° steps. This operation was carried out each time the height of the RS was changed at a location. The results are shown in Figure 10.14. Clearly the best fit is 802.16d "A" but the fit of all three 802.16d models is much better than for the other two links. The other models are less useful, with Hata especially overpredicting quite substantially.

The antenna-height correction factor (AHCF) is a component of the link-loss models that accounts for changes in the loss caused by changes in the height of the receiving antenna, and it can be derived as

$$\text{AHCF} = \begin{cases} ah_{RS} + b & \text{Hata} \\ a \log_{10}(h_{RS}/b) & \text{802.16d} \end{cases} \tag{10.7}$$

where $h_{RS}$ is the RS height. Values for the parameters $a$ and $b$ are given in the models, and here the fit of curves using those values will be compared to taking data-derived (DD) values from regression analysis of the measured data. After removing the local mean path loss by regression, only the shadowing and AHCF are left since path length is essentially fixed as RS height changes and there is no frequency correction to be applied. The variability shown in Figure 10.15 is caused by the statistical shadowing, the mean level of which is the AHCF as shown.

Parameter values for the different approaches are in Table 10.2. The DD Hata form has a gradient that is somewhat smaller than the specified form, but the most significant difference is the vertical offset. The gradient, $a$, of the DD 802.16d model is between the two categories of the specified model, suggesting that the specified models are a reasonably good match to this environment. As with the Hata models, there is a different vertical offset. In this case, however, that is attributable to the different base RS height—in the specified model, it is based around offsets from a 2 m high RS hence $b = 2$; here the mean height was 3.5 m hence the DD value for $b$ of 3.5. By using $b = 3.5$,

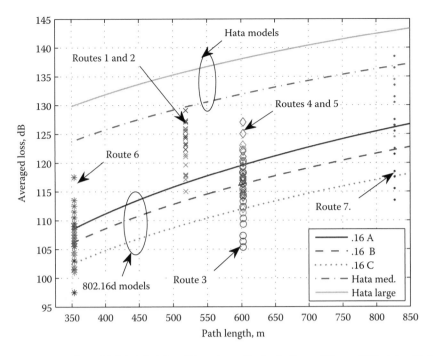

**FIGURE 10.14**  Overview of BS–RS link for all locations. Shades indicate different RS heights [5, 4, 3, 2] m.

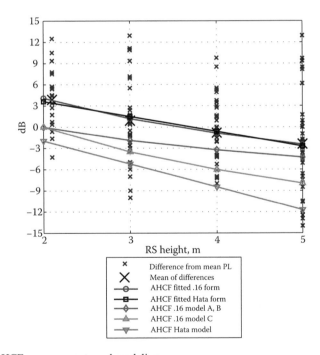

**FIGURE 10.15**  AHCF measurements and modeling.

the key differences between the model and the measurements are resolved. This suggests that the form of the 802.16d model is a good one, based on this measurement campaign, but that for the best match, one must incorporate the actual RS height rather than rely on the useful, but less-accurate

**TABLE 10.2**
**AHCF Parameters**

| Model<br>$[ah_{RS} + b]$ | a | b | $\epsilon$, dB |
|---|---|---|---|
| *(a) Hata model AHCF* | | | |
| DD Hata | −2.12 | 7.85 | 0.4 |
| Model Hata | −3.25 | 4.53 | 7.4 |
| **Model**<br>$[a\log_{10}(h_{RS}/b)]$ | a | b | $\epsilon$, dB |
| *(b) 802.16d model AHCF* | | | |
| DD .16d | −16.5 | 3.5 | 0.2 |
| Model .16d A, B | −10.8 | 2 | 2.9 |
| Model .16d C | −20 | 2 | 4.9 |

value in the standardized model. The much smaller RMS errors $\epsilon$ for the standardized 802.16d than Hata affirm this, although refitting Hata can eliminate the RMS error difference.

## 10.6 MULTIHOP STUDY BASED ON RAY-TRACING

As mentioned before, a major issue for radio resource sharing is the interference. Greater efficiencies in spectrum use can be achieved by coupling channel-quality information in the resource-allocation process. Ray-tracing is an effective tool for network setup, evaluation, and optimization. To demonstrate the potential of the proposed resource sharing framework, a site specific ray-tracing propagation model is used in this section to provide realistic environment specific propagation data for the BS–MS, BS–RS, and RS–MS links.

The ray-tracing tool used in this work was verified with measurement data, and has been used in many previous WLAN and WiMAX system evaluations, e.g., Refs. [25–27]. The ray-tracing model takes individual buildings, trees, and terrain contours into account and determines specific multipaths based on scattering and diffraction. Mobile transitions from LoS to NLoS are naturally handled by the algorithms. In this study, we can easily utilize the ray tracer to establish the typical multihop relay application scenarios, such as coverage hole and cell edge. Also, it can directly demonstrate interference paths and strengths which is more suitable for studies presented in previous sections.

### 10.6.1 SIMULATION SCENARIO

Because mobile terminals are allowed to move freely at street level in the MWA scenario, local surrounding large and small obstructions and terrain contours result in path loss, shadowing, and multipath fading. Interferences from other signals also distort the transmitted signal in an unpredictable and time-varying fashion.

A realistic MWA scenario is now analyzed based on a region of Central Bristol, United Kingdom (Figure 10.16a). A single BS location was chosen on the roof top of a tall central building (30 m above ground level). Eighty-five MS units were distributed over this geographic area at street level with heights of 1.5 m. The BS with 3-sectors was assigned an EIRP of 57.3 dBm (based on a 15 dBi 1200 sector BS antenna). First, the raw multipath components (MPCs) are created using the above ray-tracing tool. Based on isotropic ray-traced channel data, the ETSI specific antenna beam patterns [28,29] are then incorporated via spatial convolution. Figure 10.16b presents the distribution of received power from the BS to the surrounding region at street level. A severe coverage hole is clearly visible, and this is mainly caused by variations in the terrain height. Two methods could be

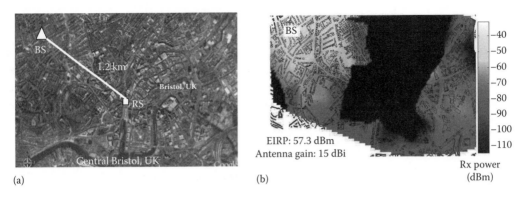

(a)                                                                              (b)

**FIGURE 10.16**   Simulated macrocell in Bristol, United Kingdom.

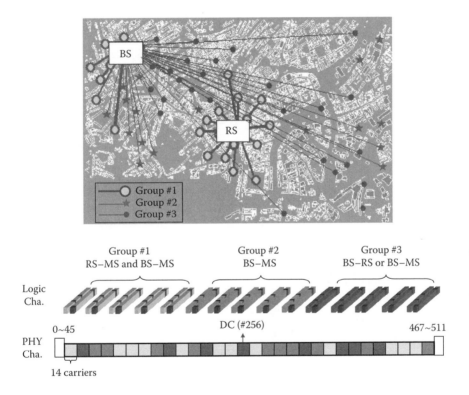

**FIGURE 10.17**   BS and RS radio resource location.

used to achieve acceptable WiMAX coverage in this macrocell. First, a second BS could be deployed in the coverage hole. However, this would add to the infrastructure costs and hence it may be more effective to deploy an RS node.

Figure 10.17 shows the locations of MSs and their affiliations to either the BS or RS. The locations of BS, RS and MSs are overlaid on a terrain map of Bristol city-centre. For each MS a line is drawn to indicate whether communication occurs via the BS or RS. The choice of connection type depends on the received SINR. At a given location, signal power is obtained by sum of all received MPCs, which are transmitted by an assigned BS sector. Interferences are caused by MPCs which comes from either RS (for BS access) or BS (for RS access) if interfering users occupy same frequency resource as the detect user. It should be noted that in NLoS conditions the MS may connect to an adjacent sector since this is determined by the direction of the strongest path.

Figure 10.17 also demonstrates subchannel allocation for a 3-users OFDMA operating with a 5 MHz channel bandwidth, based on 512-FFT DL PUSC (Partial Usage of Sub-Channels) OFDMA TDD profile as specified in 802.16e. For the 512-FFT DL PUSC OFDMA, a total of 15 subchannels are mapped (after renumbering and permuting) in one OFDMA symbol. Each sector has access to one-third of the total number of subcarriers. There are 3 groups (one per sector), with 5 subchannels in each group. Each subchannel comprises 2 clusters (14 physical subcarriers in each cluster).

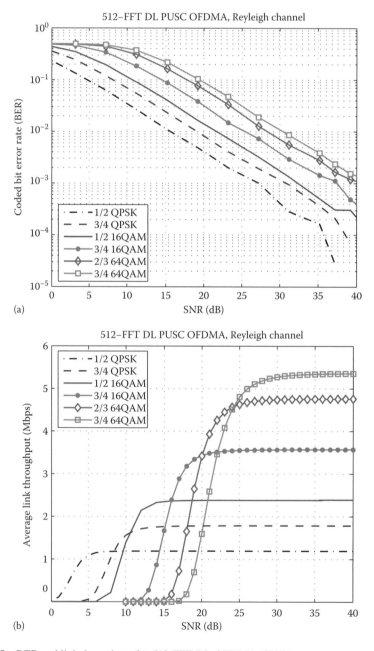

**FIGURE 10.18**   BER and link throughput for 512-FFT DL OFDMA (SISO).

This results in a total of 420 subcarriers in each OFDMA symbol (360 data bearing carriers and 60 pilot carriers). When the RS is used to connect to an MS, a certain number of timeslots (or alternatively subchannels) must be assigned in the covering sector to support the BS–RS link. Here we assume that BS–RS link and BS–MS link (taken from the sector covering the RS) are in group #3. When radio resource sharing is applied, a number of RS–MS and BS–MS links are supported simultaneously using the same resources. For all the RS–MS links, and also the BS–MS links where the antenna beam is steered away from the RS, it is possible to share group #1. Group #2 is used for those MSs that connect to one of the BS sectors (i.e., but not the sector coving the RS); these can be located near to the RS if required.

## 10.6.2 NUMERICAL RESULTS

A WiMAX PHY simulator was developed within a Monte Carlo simulator to evaluate system performance in terms of expected throughput and outage capacity [6]. Figure 10.18 shows BER and link throughput performance versus SNR. The throughput is defined as the maximum possible transmission rate such that the probability of error is arbitrarily small. The link throughput for each user can be calculated from packet error rate (PER) by $C_{link} = \frac{N_D N_b R_{FEC} R_{STC}}{T_s} \times (1 - PER)$, where $T_s$, $N_D$, $N_b$, $R_{FEC}$, and $R_{STC}$ denote the OFDMA symbol duration, the number of assigned data subcarriers, the number of bits per subcarrier, and the FEC coding rate for the user. Results indicate that higher models increase the link throughput but require higher SNR to achieve a low PER.

With assumptions that the system has (1) fair scheduling such that all users have an equal opportunity to access the BS (users are uniformly selected without any other constraints, e.g., SNR), (2) ideal channel estimation, and (3) optimal MIMO link adaptation, such as AMC mode with the highest throughput (assuming PER < 10%) should be selected for each receive location.

Table 10.3 compares 2 × 1 STBC system performance in terms of coverage and throughput for three types of deployment, named as without RS, with RS, no sharing, and with RS and sharing. Within our simulation scenario, antenna separation at the BS was set to 10 wavelengths to improve the transmit diversity gain. Transmit power of 23 dBm and an identical three sector antenna is used at the RS. The MS units are assumed to employ omni directional antennas. As expected, the RS improves the coverage, although users now suffer CCI from the BS because of radio resource sharing. Compared to the case without using RS, using RS improves coverage and capacity, e.g., all MS can be communicated within the 2-hop relay network without radio resource sharing. With resource

## TABLE 10.3
### Performance of Resource Sharing for 2-Hop Relay System (2 x 1 STBC)

|  | Without RS | With RS, No Sharing | | With RS and Sharing | |
|---|---|---|---|---|---|
|  |  | BS Zone | RS Zone | BS Zone | RS Zone |
| Filed | 9 | 0 | 0 | 1 | 0 |
| 1/2 QPSK | 0 | 0 | 0 | 1 | 0 |
| 3/4 QPSK | 1 | 1 | 0 | 1 | 0 |
| 1/2 16QAM | 0 | 0 | 0 | 1 | 1 |
| 3/4 16QAM | 1 | 1 | 0 | 2 | 1 |
| 2/3 64QAM | 1 | 1 | 0 | 2 | 1 |
| 3/4 64QAM | 73 | 67 | 15 | 62 | 12 |
| Average throughput | 4.72 Mbps | 5.29 Mbps | | 5.05 Mbps | |
| Number of recourses | 1 | 2 | | 1 | |
| Spectral efficiency (Mbps/Hz) | 0.94 | 0.53 | | 1.01 | |

sharing, capacity can be less than with RS, no sharing and this is due to fewer users operating at the highest AMC mode because of interference. However, with the concept of a directional distributed relay, there is only one radio resource required in this area. Together with above link adaptation and dynamic channel allocation, the achievable spectral efficiency is 1.01 Mbps/Hz, which is much higher than the second case (0.53 Mbps/Hz). Importantly, this leads to a high data throughput with reduced demands on radio resources.

## 10.7   CONCLUSIONS AND RECOMMENDATIONS

Achieving high throughputs with low outage probability in a large cell is a challenge, but this drives effective deployment with MIMO and multihop relay techniques, which have been intensively explored in this chapter. We have analyzed the effective relay efficiency. For relay systems without radio resource sharing, a higher relay SNR gain was required. This implies that the system requires a higher transmit power at the RS. In contrast, radio resource sharing offers high potential for a mobile WiMAX network when relays are deployed. Radio resource sharing is very applicable to MIMO relaying. In addition to multiuser transmission, the relay with radio resource sharing has the potential to achieve optimal multiuser relay gain.

To achieve the practical application of radio resource sharing, a relay system must be based on a topology that fully exploits effective resource assignment based on the spatial separation of nodes. Directional distributed relay topologies were introduced for highly efficient relay deployment. This scheme is fully backward compatible with the current mobile WiMAX standard.

Greater efficiencies in spectrum use can be achieved by coupling channel-quality information in the resource-allocation process. However, this study is heavily dependent upon realistic application environments. Channel modeling becomes critical to reflect application environments. With this purpose, a multihop relay measurement campaign was performed and is presented in this chapter. The measurement results have been analyzed and compared with several statistical and standardized models. It was found that the 802.16d model only matched well to the BS–RS link in terms of path loss and antenna-height correction factor. For the BS–MS link, the WINNER models were a good match to the shadowing, and the COST 231 Hata and WI models were useful for predicting path loss. The RS–MS link was also well matched by the WINNER shadowing values, but none of the path-loss models considered were a good match to any of individual measurement locations.

To demonstrate the potential of the directional distributed relay architecture in a realistic outdoor environment, a site-specific ray-tracing propagation model is used. Results show that, compared to a relay system without resource sharing, the implementation of resource sharing improves capacity significantly. For the OFDMA multiuser transmission, it is ideal to adopt flexible channelization and MIMO LA on different users, and multihop relay with efficient radio resource management.

## REFERENCES

[1] J. G. Andrews, A. Ghosh, and R. Muhamed, *Fundamentals of WiMAX*, Prentice Hall, 2007.
[2] IEEE Std 802.16e-2005, Part 16: Air interface for fixed and mobile broadband wireless access systems, February 2006.
[3] WiMAX ForumTM Mobile system profile, Release 1.0 Approved specification, November 2006.
[4] http://ieee802.org/16/relay
[5] R. Pabst, B. H. Walke, D. C. Schultz, P. Herhold, H. Yanikomeroglu, S. Mukherjee, H. Viswanathan, M. Lott, W. Zirwas, M. Dohler, H. Aghvami, D. D. Falconer, and G. P. Fettweis, Relay-based deployment concepts for wireless and mobile broadband radio, *IEEE Communications Magazine*, 42(9): 80–89, September 2004.
[6] Y. Bian, Y. Sun, A. Nix, and J. McGeehan, High efficient mobile WiMAX with MIMO and multihop relay, *Journal on Communications*, Academy Publisher, 2(5): 7–15, August 2007.
[7] Y. Sun, Y. Bian, A. Nix, and P. Strauch, Study of radio resource sharing for future mobile WiMAX with relay, *IEEE Mobile WiMAX'07*, March 2007.

[8] Radio Communications Agency, Strategy for the future use of the radio spectrum in the UK (2002), 2002.

[9] IEEE Std 802.16TM-2004, Part 16: Air interface for fixed broadband wireless access systems, October 2004.

[10] A. Paulraj, R. Nabar, and D. Gore, *Introduction to Space-Time Wireless Communications*. Cambridge University Press, 2003.

[11] G. N. Aggelou and R. Tafazolli, On the relaying capability of next generation GSM cellular networks, *IEEE Personal Communications*, pp. 40–47, February 2001.

[12] E. Drucker, Development and application of a cellular repeater, *IEEE Vehicular Technology Conference*, pp. 321-325, June 1988.

[13] V. Sreng, H. Yanikomeroglu, and D. Falconer, Coverage enhancement through two-hop relaying in cellular radio systems, *IEEE Wireless Communications and Networking Conference, 2002, (WCNC2002)*, 2: 881–885, March 2002.

[14] www.3g.co.uk/PR/July2006/3332.htm

[15] C. Shannon, A mathematical theory of communication, *Bell Labs Technical Journal*, 27, 379–423, 623–656, July and October 1948.

[16] C. Rose, S. Ulukus, and R. D. Yates, Wireless systems and interference avoidance, *IEEE Transactions on Wireless Communications*, 1(3): 415–428, July 2002.

[17] Shin Horng Wong and Ian J. Wassell, Channel allocation for broadband fixed Wireless access, *IEEE Wireless Personal Multimedia Communications*, 2002, 2: 626–636, October 2002.

[18] I. Fu, Y. Sun, W. Chen, et al., Neighborhood discovery and measurement for fixed/nomadic RS in IEEE 802.16j multi-hop relay network IEEE, IEEE C802.16j-07/171r2, March 2007.

[19] W. Chen, D. Viorel, Y. Sun, et al., Interference detection and measurement in OFDMA relay networks, IEEE C802.16j-07/229r5, March 2007.

[20] E. Damosso, Ed., Digital mobile radio towards future generation systems—COST 231 Final Report, Luxembourg: European Commission, 1999.

[21] D. S. Baum, H. El-Sallabi, T. Jämsä, et al., Final report on link level and system level channel models, IST-2003-507581 WINNER D5.4 v1.4, November 2005.

[22] V. Erceg, L. J. Greenstein, S. Y. Tjandra et al., An empirically based path loss model for wireless channels in suburban environments, *IEEE Journal on Selected Areas in Communication*, 17(7): 1205–1211, July 1999.

[23] IEEE 802.16a-03/01, Channel models for fixed wireless applications, June 2003.

[24] G. Senarath, W. Tong, P. Zhu, et al., Multi-hop relay system evaluation methodology (channel model and performance metric), IEEE 802.16j-06/013r3, February 2007.

[25] Y. Bian, A. Nix, E. Tameh, and J. McGeehan, MIMO-OFDM WLAN architectures, area coverage and link adaptation for urban hotspots, *IEEE Transactions on Vehicular Tech.*, 57(4): 2364–2374, 2008.

[26] Y. Bian and A. Nix, Throughput and coverage analysis of a multi-element broadband fixed wireless access (BFWA) system in the presence of co-channel interference *IEEE VTC2005-Fall*, September 2006.

[27] C. Williams, Y. Bian, M. A. Beach, and A. R. Nix, An assessment of interference cancellation applied to BWA, *IEEE Mobile WiMAX 07*, March 2007.

[28] 2004/421/UK, UK interface requirement 2015; Public fixed wireless access radio systems operating within the 3 to 11 GHz frequency bands administered by Ofcom, V6.2, February 2005.

[29] ETSI EN302.326-3 V1.1.1, Fixed radio systems multipoint equipment and antennas; Part 3: Harmonized EN covering the essential requirements of article 3.2 of the R&TTE directive for multipoint radio antennas, October 2005.

# 11 Routing and Scheduling for WiMAX Mesh Networks

*Jianhua He, Xiaoming Fu, Jie Xiang, Yan Zhang, and Zuoyin Tang*

## CONTENTS

In addition to the mandatory implementation of point-to-multipoint (PMP) mode, WiMAX networks can be optionally configured to work in Mesh mode, to achieve increased reliability, coverage, and reduced network costs. Although IEEE 802.16 standards specify several quality of service (QoS) schemes and related message formats for WiMAX networks, the problems of scheduling algorithms for both PMP and Mesh mode are left unsolved. Routing algorithms for WiMAX networks are outside the scope of the standard work as well. In this chapter, we investigate the issues of routing and scheduling for WiMAX mesh networks. We overview the mesh mechanisms specified in the IEEE 802.16 standard and survey the existing research on scheduling and routing for WiMAX mesh networks. Then both distributed and centralized routing algorithms are studied, and their effectiveness on alleviating potential network congestion is compared. The scheduling problem is mathematically modeled by taking into account the interference constraints. Solutions are developed which can maximize the utilization of network capacity subject to fairness constraints on the allocation of scarce wireless bandwidth.

## 11.1   INTRODUCTION

In today's telecommunications, networking and services are changing in a rapid way to support next generation Internet (NGI) user environment. Wireless networks will play an important role in NGI. Wireless broadband networks are being increasingly deployed and used in the last mile for extending or enhancing Internet connectivity for fixed or mobile clients located on the edge of the wired network [1].

With high data rate, large network coverage, strong QoS capabilities, and cheap network deployment and maintenance costs, WiMAX is regarded as a disruptive wireless technology and has many potential applications [1,2]. It is expected to support business applications, for which QoS support will be a necessity. Depending on the applications and network investment, WiMAX network can be configured to work in different modes, point-to-multipoint (PMP) or Mesh mode. An illustration of PMP and mesh mode is presented in Figure 11.1. For example, the network can have a simple base station (BS) working in PMP mode and serving multiple subscriber stations (SSs) if the potential SSs can be covered by the BS. Mesh topology is an optional configuration for WiMAX networks. In the Mesh mode, traffic demands are aggregated at a set of SS nodes which are equipped with 802.16 interfaces. Subsequently, the traffic demands at SS nodes are delivered to a set of BSs nodes which functions in the PMP mode. These BS stations can be connected by a backhaul and connected to Internet access point (IAP) nodes. An amendment to the IEEE 802.16 specifications (where WiMAX is based) is IEEE 802.16j, Multihop Relay Specification. IEEE 802.16j is being developed by IEEE 802.16's Relay Task Group. It is expected to extend reach/coverage through relaying [3]. In this chapter, we will focus on the Mesh mode.

Wireless mesh network offers increased reliability, coverage, and reduced network costs [4,5]. There are extensive research, standardization, and commercial development activities on mesh networks [4–7]. For example, several IEEE special task groups have been established to define the requirements for mesh networking in wireless personal area networks (WPANs), wireless local area networks (WLANs), and wireless metropolitan area networks (WMANs). A brief description of the standardization activities can be found in Ref. [4]. Wireless mesh networks can be a prospective solution for broadband wireless Internet access in a flexible and cost-effective manner. However, wireless mesh networks also raise a number of research challenges, e.g., network routing, scheduling, QoS support, network management, etc. [4,8]. Those challenges are faced by WiMAX mesh networks without exception. WiMAX Mesh mode is defined with OFDM for frequency between 2 and 11 GHz and time division multiple access (TDMA) is used in the Medium Access Control (MAC) layer to support multiple users. Unlike the single hop wireless networks, routing algorithms are required to determine routes for the connections between a SS and a BS. As WiMAX networks operate synchronously in a time-slotted mode, it is also necessary to allocate time slots without collision over the network to achieve assigned bandwidth for each connection. More challenging is that the routing and scheduling for WiMAX networks are tightly coupled. The routing and scheduling problem for WiMAX networks is different from 802.11-based mesh networks. In the 802.11-based mesh networks, the MAC layer is contention-based; routing algorithms and MAC layer protocols can be designed and operated separately.

Although IEEE 802.16 standards specify several QoS schemes and related message formats, the problems of scheduling algorithms for both PMP and Mesh mode are left unsolved. Routing algorithms for WiMAX networks are outside the scope of the standard work as well. In this chapter, we will investigate the issues of routing and scheduling in WiMAX mesh networks. Both distributed and centralized routing algorithms will be studied, and their effectiveness on alleviating potential network congestion will be compared. The scheduling problem will also be mathematically modeled by

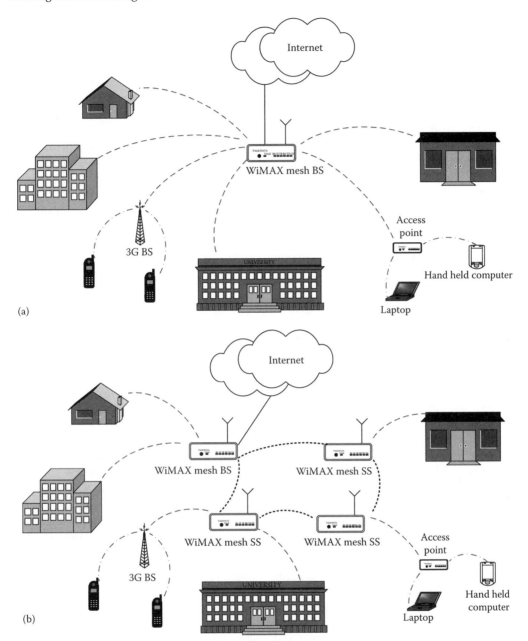

**FIGURE 11.1** WiMAX (a) PMP network and (b) mesh network architectures.

taking into account the interference constraints. Solutions are developed to maximize the utilization of network capacity subject to fairness constraints on allocation of scarce wireless resource among the SSs.

The chapter is organized as follows. The WiMAX mechanisms defined for Mesh mode is overviewed in Section 11.2. Existing research on scheduling and routing for WiMAX mesh networks is presented in Section 11.3. Both distributed and centralized routing algorithms will be presented in Section 11.4. A scheduling problem for WiMAX mesh networks is mathematically modeled and solved in Section 11.5. Typical numerical results are presented in Section 11.6. Finally, we conclude the chapter and discuss the open research issues in Section 11.7.

## 11.2   OVERVIEW OF 802.16 MECHANISMS FOR MESH MODE

Unlike the PMP mode that only allows communication between the BS and SS, each station is able to create direct communication links to a number of other stations in the network instead of communicating only with a BS. However, in typical network deployments, there will still be certain nodes that provide the BS function of connecting the mesh network to the backbone networks. When using Mesh centralized scheduling to be described below, these BS nodes perform much of the same basic functions as the BSs do in PMP mode. Communication in all these links in the network are controlled by a centralized algorithm (either by the BS or decentralized by all nodes periodically), scheduled in a distributed manner within each node's extended neighborhood, or scheduled using a combination of these. The stations that have direct links are called neighbors and forms a neighborhood. A node's neighbors are considered to be one hop away from the node. A two-hop extended neighborhood contains, additionally, all the neighbors of the neighborhood.

In this section, we will briefly introduce the frame structure, network entry procedures, bandwidth request, and grant mechanisms defined for WiMAX Mesh mode, which will be the base requirement for the design of routing and scheduling algorithms to be presented later.

### 11.2.1   IDENTIFICATIONS

There are several types of identifications that an SS will use for different purposes.

Each SS has a 48-bit universal MAC address, which uniquely defines the SS from other SSs. The MAC address is used during the network entry process and as part of the authorization process by which the candidate SS and the network verify the identity of each other.

After authorized to the network, a candidate SS will receive a 16-bit node identifier (node ID) upon a request to the Mesh BS. Node ID is the basis for identifying SSs during normal operation. The node ID is transferred in the Mesh subheader, which follows the generic MAC header, in both unicast and broadcast messages [2].

To facilitate communications with local neighboring SSs, an SS will use 8-bit link identifiers (link IDs). Each SS shall assign an ID for each link it has established to its neighbors. The link IDs are communicated during the Link Establishment process as neighboring SSs establish new links. The link ID is transmitted as part of the Connection ID in the generic MAC header in unicast messages. The link IDs are used in distributed scheduling to identify resource requests and grants. Because these messages are broadcast, the receiver nodes can determine the schedule using the transmitter node ID in the Mesh subheader, and the link ID in the payload of the MSH-DSCH (Mesh mode schedule with distributed scheduling) message. The MSH-DSCH message will be introduce later.

### 11.2.2   FRAME STRUCTURE

Unlike in PMP mode, there are no clearly separate downlink and uplink subframes in Mesh mode. A Mesh frame consists of a control and a data subframe [2]. There are two types of control subframes, which serve different functions, which will be discussed later. All transmissions in the control subframe are sent using QPSK-1/2 with the mandatory coding scheme. The data subframe is divided into minislots [2]. A scheduled allocation consists of one or more minislots.

#### 11.2.2.1   Network Control Subframe

The first type of control subframe is termed network control subframe, used to create and maintain cohesion between the different systems. The network control subframe is illustrated in Figure 11.2. Frames with a network control subframe occur periodically. The period of occurrence is indicated in the network descriptor. The length of the control subframe is fixed and of length OFDM symbols, which is also indicated in the network descriptor. During a network control subframe, the first seven symbols are allocated for network entry. The following symbols in the network control subframe are allocated to a number of network configuration (MSH-NCFG) messages and each MSH-NCFG takes seven symbols.

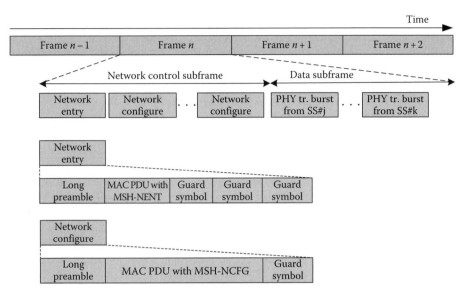

**FIGURE 11.2** WiMAX mesh network control subframe.

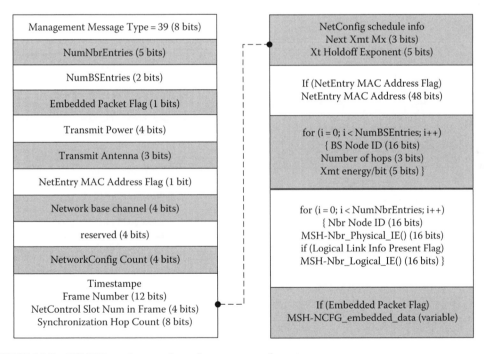

**FIGURE 11.3** WiMAX mesh network configure message format.

MSH-NCFG messages provide a basic level of communication between nodes in different nearby networks. All the nodes (BS and SS) in the mesh network will transmit MSH-NCFG [2]. The MSH-NCFG message format is shown in Figure 11.3. Through the MSH-NCFG message, a BS or an SS will report a number of its neighbors. The number of neighbors reported on may be a fraction of the whole set of neighbors known to this SS. A node will also report the Mesh BSs that its neighbors report and report the distances in hops to the BSs. The Embedded Packet Flag is used to indicate if an embedded data information element (IE) is included in the message. A network descriptor is one of the five defined embedded data IE that can be included in MSH-NCFG. Transmit Antenna indicates the logical antenna used for transmission of this message. Up to eight antenna, directions can be supported. NetworkConfig Count is the counter of MSH-NCFG messages transmitted by this node, which is used by neighbors to detect missed transmissions. Xmt Holdoff Exponent and Next Xmt Mx are the variables used by this node to indicate its next MSH-NCFG eligibility interval. The Xmt Holdoff Exponent is used to calculate XmtHoldoffTime, which is the number of MSH-NCFG transmit opportunities after Next Xmt Time that this node is not eligible to transmit MSH-NCFG packets. There are MSH-CTRL-LEN¨C1 opportunities per network control subframe, as indicated in network descriptor. Next Xmt Time is the next MSH-NCFG eligibility interval for this node, and is calculated by Xmt Holdoff Exponent and Next Xmt Mx.

Network descriptor is an important embedded data IE that can be included in MSH-NCFG. The parameters included in a network descriptor is shown in Figure 11.4. Similar to the downlink and uplink network descriptor in PMP mode, the network descriptor in Mesh mode defines the parameters associated to burst profiles, i.e., FEC code type, Mandatory Exit Threshold, and Mandatory Entry Threshold. It also defines the important parameters such as the length of control subframe (MSH-CTRL-LEN), and the number of DSCH opportunities in schedule control subframe (MSH-DSCH-NUM). The parameter Scheduling Frames defines how frequent a schedule control subframe will appear.

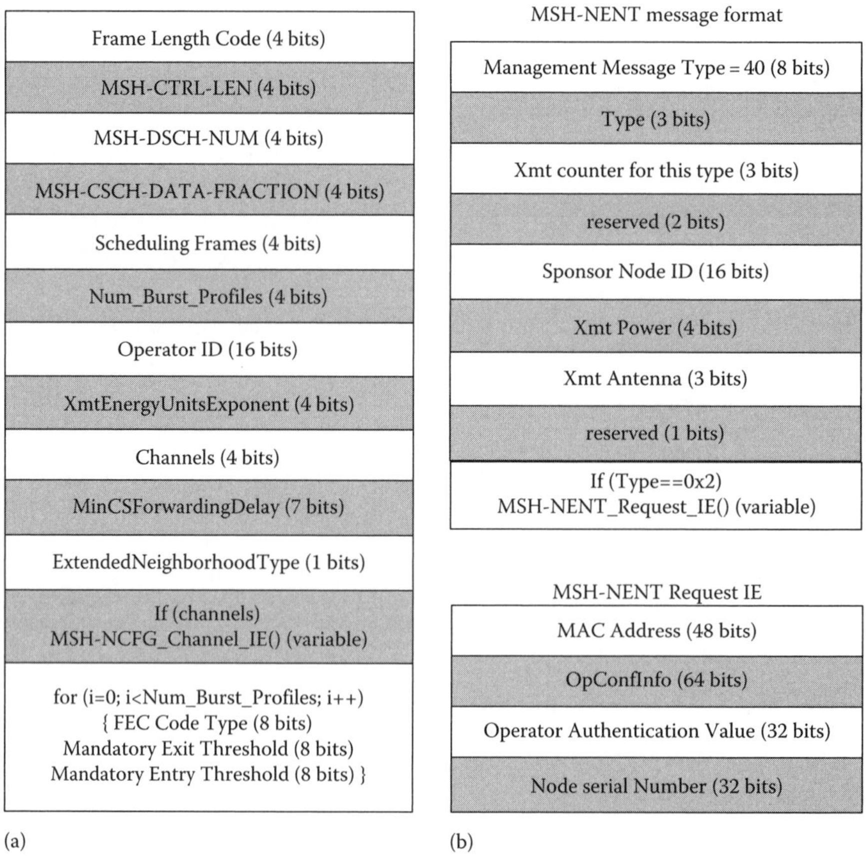

**FIGURE 11.4** WiMAX (a) Mesh network discriptor and (b) Mesh Network Entry message format.

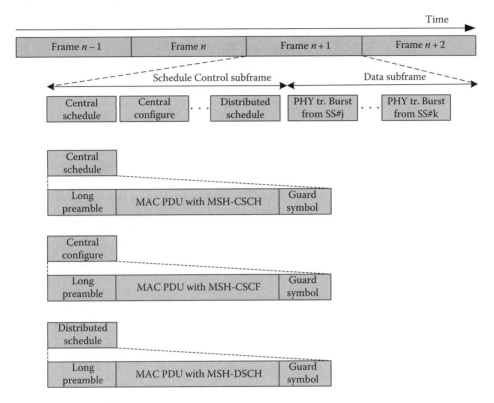

**FIGURE 11.5** WiMAX Mesh schedule control subframe.

## 11.2.2.2 Schedule Control Subframe

The second type of control subframe is termed schedule control, which is used to coordinate scheduling of data-transfers between systems. The schedule control subframe is shown in Figure 11.5 [2]. The first symbols are allocated to transmission bursts containing Mesh centralized scheduling message (MSH-CSCH) and Mesh centralized configuration message (MSH-CSCF), and the remainder of the schedule control subframe is allocated to transmission bursts containing Mesh distributed scheduling messages (MSH-DSCH) [2]. The message formats of MSH-CSCS, MSH-CSCF, and MSH-DSCH are shown in Figure 11.6.

- MSH-CSCH message: An MSH-CSCH message is created by a Mesh BS for the purpose of centralized scheduling, including collecting bandwidth request from the SSs and delivering the transmission schedule to the SSs. To deliver a transmission schedule to the SSs, an MSH-CSCH message is generated and broadcasted by the BS to all its neighbors. Then all the nodes with hop count lower than a threshold, which is configured by the BS, will forward the MSH-CSCH message to their neighbors that have a higher hop count. For the forwarding direction starting from the BS, the Grant/Request Flag in the MSH-CSCH message is set to 0. In addition, SSs can use MSH-CSCH messages to request bandwidth from the Mesh BS, by setting the Grant/Request Flag in the MSH-CSCH message to 1. Each SS reports the individual traffic demand requests of each child node in its subtree from the BS. The SSs in the subtree are those in the current scheduling tree to and from the Mesh BS. The scheduling tree is known to all nodes in the network and is ordered by node ID. The parameter Flow Scale Exponent is used to determine scale of the granted bandwidth. Its value typically depends on the number of nodes in the network, the achievable PHY bit rate, the traffic demand, and the provided service. For the downlink, Flow Scale Exponent gives the absolute

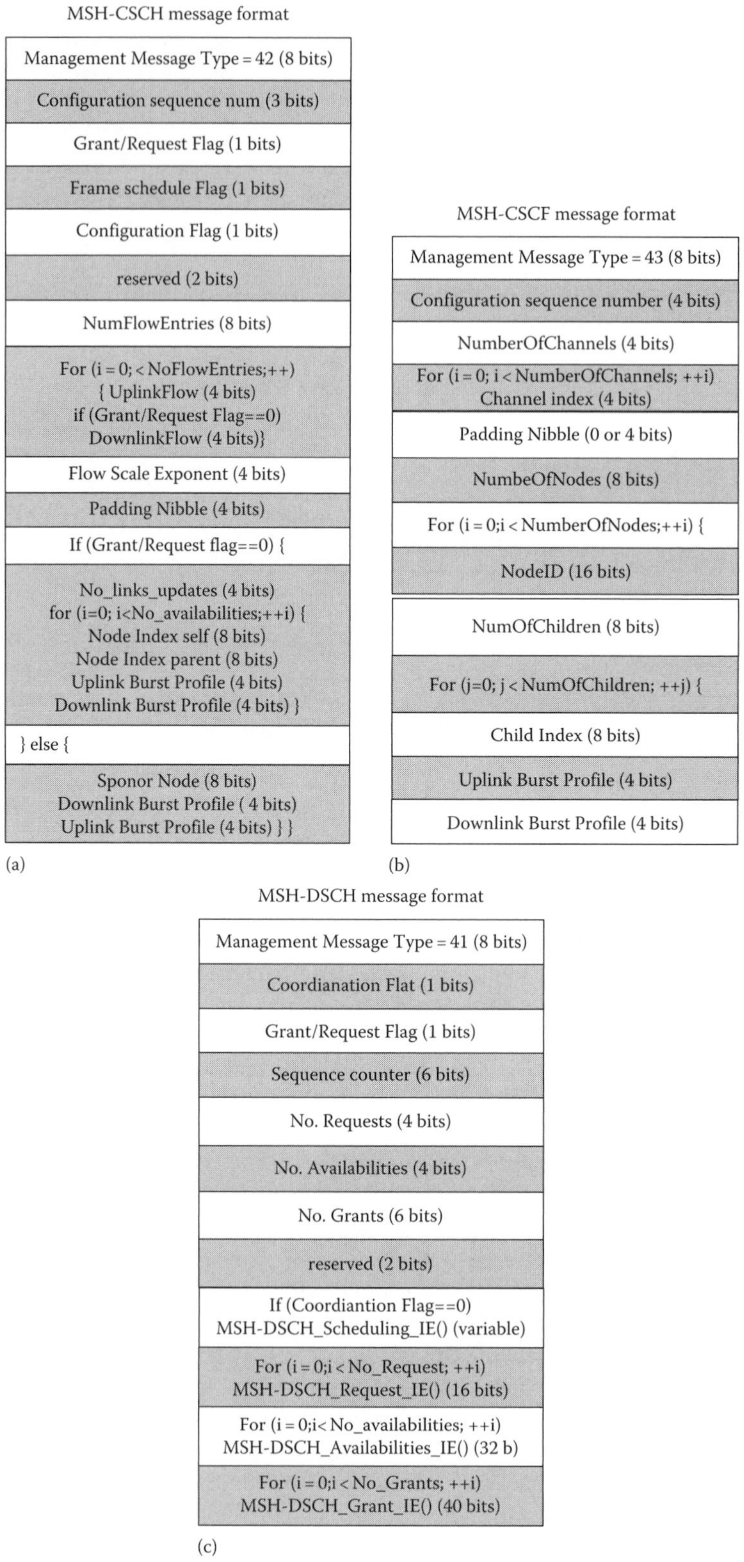

MSH-CSCH message format

| |
|---|
| Management Message Type = 42 (8 bits) |
| Configuration sequence num (3 bits) |
| Grant/Request Flag (1 bits) |
| Frame schedule Flag (1 bits) |
| Configuration Flag (1 bits) |
| reserved (2 bits) |
| NumFlowEntries (8 bits) |
| For (i = 0; < NoFlowEntries;++) { UplinkFlow (4 bits) if (Grant/Request Flag==0) DownlinkFlow (4 bits)} |
| Flow Scale Exponent (4 bits) |
| Padding Nibble (4 bits) |
| If (Grant/Request flag==0) { |
| No_links_updates (4 bits) for (i=0; i<No_availabilities;++i) { Node Index self (8 bits) Node Index parent (8 bits) Uplink Burst Profile (4 bits) Downlink Burst Profile (4 bits) } |
| } else { |
| Sponor Node (8 bits) Downlink Burst Profile ( 4 bits) Uplink Burst Profile (4 bits) } } |

(a)

MSH-CSCF message format

| |
|---|
| Management Message Type = 43 (8 bits) |
| Configuration sequence number (4 bits) |
| NumberOfChannels (4 bits) |
| For (i = 0; i < NumberOfChannels; ++i) Channel index (4 bits) |
| Padding Nibble (0 or 4 bits) |
| NumbeOfNodes (8 bits) |
| For (i = 0;i < NumberOfNodes;++i) { |
| NodeID (16 bits) |
| NumOfChildren (8 bits) |
| For (j=0; j < NumOfChildren; ++j) { |
| Child Index (8 bits) |
| Uplink Burst Profile (4 bits) |
| Downlink Burst Profile (4 bits) |

(b)

MSH-DSCH message format

| |
|---|
| Management Message Type = 41 (8 bits) |
| Coordianation Flat (1 bits) |
| Grant/Request Flag (1 bits) |
| Sequence counter (6 bits) |
| No. Requests (4 bits) |
| No. Availabilities (4 bits) |
| No. Grants (6 bits) |
| reserved (2 bits) |
| If (Coordiantion Flag==0) MSH-DSCH_Scheduling_IE() (variable) |
| For (i = 0;i < No_Request; ++i) MSH-DSCH_Request_IE() (16 bits) |
| For (i = 0;i< No_availabilities; ++i) MSH-DSCH_Availabilities_IE() (32 b) |
| For (i = 0;i < No_Grants; ++i) MSH-DSCH_Grant_IE() (40 bits) |

(c)

**FIGURE 11.6** Message formats for schedule control (a) coordinated scheduling message, (b) coordinated configuration message, and (c) distributed scheduling message.

values of flow granted, so the total minislot range allowed for centralized scheduling need not be used if not needed, with the remainder set aside for distributed scheduling. For the uplink, the lowest exponent possibility is used at each hop, with quantization of forwarded requests rounded up to avoid reducing any requests to zero.

- MSH-CSCF message: An MSH-CSCF message is also used in Mesh mode for the purpose of centralized scheduling, and specifically for configuration. The Mesh BS generates and broadcasts the MSH-CSCF message to all its neighbors, and all nodes forward (rebroadcast) the message according to its index number specified in the message. With each new configuration message, the number of Configuration sequence number in the message is incremented by 1. The parameter Number of Channels in the message indicates the number of channels available for centralized scheduling. And the parameter number of nodes determine the number of nodes in scheduling tree. For each node in the scheduling tree, the node ID of its one-hop neighbors will be given as well as the the Uplink/Downlink Burst Profile used for the link from/to the neighbors.
- MSH-DSCH message: During a schedule control subframe, MSH-DSCH-NUM distributed scheduling messages will occur. The number MSH-DSCH-NUM is indicated in the network descriptor included in an MSH-NCFG message. Distributed scheduling messages may also occur in the data subframe if not in conflict with the scheduling dictated in the control subframe. In the MSH-DSCH message, the parameter Coordination Flag indicates the type of the distributed scheduling, being 0 for coordinated and 1 for uncoordinated. Both type of distributed scheduling will require a threeway handshake (Request, Grant, and Grant confirmation) to establish a valid schedule. The parameter grant/request Flag indicate the type of the message, being 0 for Request message and 1 for Grant message. A number of request IEs, availability IEs, and grant IEs can be included in the message.

### 11.2.3  MESH NETWORK ENTRY MECHANISM

The standard specifies network entry mechanism to find sponsor nodes and establish links with neighbors [2]. In this chapter, the network entry mechanism is used as one of the alternatives to establish traffic routes for WiMAX mesh networks. Upon entering the mesh network, a new SS searches for MSH-NCFG to acquire coarse synchronization with the network. Once the physical layer has achieved synchronization, the MAC layer can acquire network parameters for the MSH-NCFG message. Meanwhile, the SS can builds a physical neighbor list, from which the SS can select a potential sponsor node out of the eligible sponsor nodes. How to select the sponsor node is not specified in the standard. The selection method will be discussed in more details later in this chapter. The SS then synchronizes its time to the potential sponsor node and sends a network entry request message to the potential sponsor node. If the candidate sponsor node accepts the request and opens a sponsor channel, the channel is ready for use to register with the BSs. After the SS is authorized to enter the network by the BS, it can request bandwidth from the BS via the sponsor node and can also establish links with the SSs other than the sponsor node [2].

### 11.2.4  BANDWIDTH ALLOCATION AND GRANT MECHANISMS

In the 802.16 standard, flexible bandwidth allocation and grant mechanisms have been defined for PMP mode to guarantee the QoS of various service flows. Those mechanisms include periodically polling, real-time polling, nonreal-time polling, contention-based bandwidth request scheme, poll-me bit, bandwidth stealing, and piggyback [2]. However, the bandwidth request and grant mechanisms can not be used in Mesh mode due to the multihop networking in Mesh mode. In WiMAX Mesh mode, all the communications in the links in the network are controlled by three ways, i.e., using a centralized scheduling algorithm, using a distributed scheduling algorithm within each node's extended neighborhood, or using a combination of these two types of algorithms. All the scheduling algorithms will be implemented based on the scheduling messages presented in Section 11.2.2.

In the coordinated distributed scheduling algorithm, all stations (BS and SSs) coordinate their transmissions in their extended two-hop neighborhood. Coordinated distributed scheduling does not rely on the operation of a BS and transmissions are not necessarily directed to or from the BS. Within the constraints of the coordinated schedules, uncoordinated distributed scheduling can be used for fast and ad-hoc setup of schedules on a link-by-link basis with directed requests. Grants of the uncoordinated schedules need to ensure that the resulting data transmissions do not cause collisions with the data and control traffic scheduled by the coordinated scheduling algorithms.

Although distributed scheduling algorithms are more scalable, they are inefficient in QoS guarantee. On the other hand, centralized scheduling ensures collision-free scheduling over the links in the network, typically in a more optimal manner than the distributed scheduling method. Therefore better QoS support and network bandwidth utilization can be achieved. In this chapter centralized scheduling will be the only scheduling algorithm studied.

### 11.2.4.1   Centralized Scheduling

With the centralized scheduling algorithm, transmission schedule for the SSs is defined by the BS. The BS determines the flow assignments from the resource requests from the SSs. Subsequently, the SSs validate the MSH-CSCH schedule and determine the actual schedule from these flow assignments. The assignments determined by the BS extends to those SSs not directly connected to the BS. Intermediate SSs are responsible for forwarding bandwidth requests for SSs listed in the routing tree that are further from the BS (i.e., more hops from the BS) and the MSH-CSCH message from the BS to their neighbors as required. The SS resource requests and the BS assignments are both transmitted during the schedule control subframe. Centralized scheduling ensures that transmissions are coordinated to ensure collision-free scheduling over the links in the routing tree to and from the BS. The centralized scheduling will persist over a duration that is greater than the cycle time to relay the new resource requests and distribute the updated schedule.

Determination of the flow assignment and routing tree is outside the scope of the standard, and will be addressed in later of this chapter.

### 11.2.4.2   Distributed Scheduling

Coordinated distributed scheduling ensures that transmissions are scheduled in a manner that does not rely on the operation of a BS, and that is not necessarily directed to or from the BS. In the coordinated distributed scheduling mode, all the stations (BS and SSs) need coordinate their transmissions in their extended two-hop neighborhood. The coordinated distributed scheduling mode uses some or the entire control portion of each frame to regularly transmit its own schedule and proposed schedule changes on a PMP basis to all its neighbors. Within a given channel all neighbor stations receive the same schedule transmissions. All the stations in a network have to use the same channel to transmit schedule information in a format of specific resource requests and grants.

Within the constraints of the coordinated schedules (distributed or centralized), uncoordinated distributed scheduling can be used for fast, ad-hoc setup of schedules on a link-by-link basis. Uncoordinated distributed schedules are established by directed requests and grants between two nodes, and shall be scheduled to ensure that the resulting data transmissions (and the request and grant packets themselves) do not cause collisions with the data and control traffic scheduled by the coordinated distributed nor the centralized scheduling methods.

The major differences between coordinated and uncoordinated distributed scheduling are in the portions where the MSH-DSCH messages are transmitted. In the coordinated case, the MSH-DSCH messages are scheduled in the control subframe in a collision-free manner. In the uncoordinated case, MSH-DSCH messages are transmitted in data subframes and may collide.

## 11.3 LITERATURE SURVEY

An in-depth survey on the general wireless mesh networks has been presented in Refs. [4,5], which covers the research activities and open issues from physical layer to applications, and from mobility management to power management, from designs with single channel single radio device to those with multiple channel multiple radio devices. However, most of the existing researches about wireless mesh networks are based on IEEE 802.11 standard. The routing and MAC layer protocols designed for 802.11 standard-based wireless mesh network cannot be used directly or efficiently for WiMAX mesh networks. And not much work has been done on WiMAX mesh networks. In this section, we will introduce the routing and scheduling related work on WiMAX mesh networks.

### 11.3.1 COORDINATED DISTRIBUTED SCHEDULING

The main idea of the coordinated distributed scheduling is to coordinate the transmission of MSH-DSCH messages over transmission opportunities in a collision-free manner. Through the exchanges of collision-free MSH-DSCH over control subframes, collision-free data slot reservations in the data subframes can be achieved.

To achieve the goal of collision-free MSH-DSCH transmission, nodes will exchange 2-hop or 3-hop neighborhood scheduling information with each other. Because nodes shall run the scheduling algorithm independently, a common algorithm has been specified in the standard for each node in the neighborhood to calculate the same schedule. The algorithm is random and predictable by dynamically constructing the seeds of a random number generator for each node according to a common rule. In particular, the seed for a given node is constructed based on its unique node ID and the index of the candidate transmission opportunity. The most important parameters that have significant impacts on the performance of coordinated distributed scheduling are Xmt Holdoff Exponet (3 bits) and Next Xmt Xm (5 bits), which have been introduced in Section 11.2.2. They can be used to control the contention on the transmission opportunities and improve bandwidth utilization.

Due to the importance of the coordinated distributed scheduling on the network performances, an analytical framework is needed to assess the performance of the scheduling scheme. Cao et al. analytically investigated how the channel contention is correlated with the total node number, exponent value, and network topology [9,10]. With the assumption that the transmit time sequences of all the nodes in the control subframe form statistically independent renewal processes, they developed methods for estimating the distributions of the node transmission interval and connection setup delay. The analytical method will be helpful for evaluating upper layer performance like throughput and delay. They implemented the coordinated distributed scheduling module in NS-2 and showed that their analytical model is quite accurate under various scenarios, including both single hop and multihop networks.

Based on Cao's analytical model, Bayer et al. presented an enhancement of the model [11]. In particularly, they evaluated the scalability of the coordinated distributed scheduling. A scalability problem was observed that leads to poor performance in dense networks and aggravates QoS provisioning. The problem may result from the election-based transmission timing mechanism for scheduling the transmission of MSH-DSCH messages. They propose a dynamic adaptation mechanism to counteract the scalability problem, in which the parameter Xmt Holdoff Exponet is dynamically and locally adjusted according to the network contention and the status of a node. The Next Xmt Xm is used as a contention indicator. If Next Xmt Xm used by the node or its neighbors exceeds a specified threshold, the Xmt Holdoff Exponet is increased. The status of a node is defined according to its transmission activity, if it is BS or if it is a sponsor node. Significant UDP throughput increase is observed with the application of the adaptation mechanism for both single and multiple hop network scenarios.

In the 802.16 standard and the above analytical work, it is assumed that the control messages can be transmitted without collision in the extended neighborhood (2-hop or 3-hop). However, such kind of interference model may not hold in practice. Zhu and Lu investigate the performance of coordinated distributed scheduling under a realistic interference model [12]. Extensive simulations were conducted to evaluate the reception collision performance of the scheduling mechanism. It was reported that the collision ratio of control messages can be as high as 20 percent for 2-hop extended neighborhood. They studied how to deal with the collision problem by appropriate configuration of parameters such as Xmt Holdoff Exponent.

## 11.3.2   COORDINATED CENTRALIZED SCHEDULING

An early work on multihop scheduling for WiMAX mesh networks is presented by Kim and Ganz in Ref. [13]. They proposed a fair multihop centralized scheduling algorithm. The algorithm consists of two phases, namely node ordering and link allocation, which is introduced as below.

In the beginning of the algorithm, each node sends bandwidth request messages to the BS. The BS uses a node ordering algorithm to determine an order list for the nodes and broadcast the ordering to the nodes. The nodes will forward the broadcast message according to the ordering. In the node ordering algorithm, a measure satisfaction index is defined as the ratio of average bandwidth allocated in a given number of frames to a node's weight ($W_i$). The weight can reflect the priority of the node. The total weight value $W_{i,\text{total}}$ is defined as its weight plus the sum of all the weights of its children nodes [13]:

$$W_{i,\text{total}} = W_i + \sum_{j \in C_i} W_j, \tag{11.1}$$

where $C_i$ denotes the set of child nodes of node $i$. The satisfaction index $S_i(f)$ of each node $i$ in a frame $f$ is defined as

$$S_i(f) = \frac{\sum_{fp=f-T}^{f-1} B_i(f_p)}{TW_{i,\text{total}}}, \tag{11.2}$$

where $B_i(f)$ denotes the link bandwidth allocated to node $i$ in frame $f$. After calculating the satisfaction index of all the nodes, the BS sorts bandwidth requests in increasing order of the satisfaction index of each node on a per-hop basis. The nodes closer to the BS have higher orders.

After the node ordering phase, each node computes its own schedule based on the MSC-CSCF. The computation uses two types of matrices: schedule matrix and collision matrix. Each node takes the order list broadcast by the BS, bandwidth requirement of each node, scheduled time slots, and network topology as the input. Its schedule is determined from the time slots in which the node is not identified in the collision matrix.

The proposed scheduling algorithm is stated to achieve maximum throughput in a heavily loaded wireless network while also guarantee fairness of channel access among different nodes. However, there are still three major problems in the scheduling algorithm. First, each node is required to compute its own schedule based on the order list in the link allocation phase. In the link allocation algorithm, each node will require the information of scheduled time slots by other nodes as input. However, such information is not easy to obtain in a distributed way. Second, there is not an efficient method to control the time slots that a node can request or reserved in a frame. Third, the forwarding tree is not constructed. The traffic from the children nodes of a node is also not taken into account in the link allocation algorithm.

A part of the problems in the scheduling algorithm proposed in Ref. [13] are addressed by Han et al. [14]. They proposed a collision-free centralized scheduling algorithm for WiMAX mesh networks. The traffic from children nodes is taken into account in the algorithm by means of designing a relay strategy. The scheduling algorithm is based on a simple routing tree. The routing tree is constructed by selecting the nearest neighbor with the minimum hop count to the BS as the sponsor node of this node. In the scheduling algorithm, service token is used to determine the eligibility of a node to be scheduled in a time slot. Initially service token of a link is assigned based on its traffic demand, with the purpose of guaranteeing fairness. A link can be scheduled only if the service token number of the transmitter of the link is nonzero. Each time after a link is assigned a time slot, the service token of the transmitter is decreased by one and that of the receiver is increased by one. Through this method of service token adjustment, the hop-by-hop relay model is integrated into the scheduling algorithm. During the scheduling process, there can be more than one links having nonzero service token. A link selection algorithm is further designed for the BS to determine the order of the links to schedule. The selection can be based on the four criteria: random, min interference, nearest to BS, and farthest to BS. The scheduling algorithm is evaluated in terms of scheduling length, channel utilization, and transmission delay, with comparison on the four selection criteria. However, optimization by the joint routing and scheduling is not investigated in their work.

### 11.3.3  ROUTING AND CENTRALIZED SCHEDULING

As the problems of routing and scheduling are tightly coupled for WiMAX mesh networks, joint routing and scheduling design have been an interesting research topic.

An early investigation on disjoint routing and scheduling for WiMAX mesh networks is reported by Haas et al. [15]. They developed an interference-aware framework for WiMAX mesh networks with the goal of achieving high utilization. The proposed framework includes an interference-aware route construction algorithm and an enhanced centralized scheduling scheme. Both traffic load demand and interference conditions are taken into account in the design. The interference-aware routing algorithm is designed by considering interference conditions in the network. A blocking metric $B(k)$ is defined for a given route from the Mesh BS toward an SS node $k$, which is used to model the interference level of routes in the networks. Let blocking value $b(j)$ of a node $j$ be the number of nodes that could be interfered in message reception if node $j$ transmits. Then blocking metric $B(k)$ of a route to the node $k$ will be the summation of the blocking values of nodes that transmit or forward packets along the route. Let $S(k)$ denote the set of nodes on the considered route from BS to node $k$.

$$B(k) = \sum_{j \in S(k)} b(j) \qquad (11.3)$$

According to a predefined sequence of nodes joining the mesh network, each time the node is processed to select a sponsor node with minimum blocking metric. A routing tree can be constructed consequently.

With the interference-aware routing tree, an interference-aware scheduling is proposed which exploit concurrent transmission opportunity to achieve high spectral utilization. The order of the links selected to reserve a time slot at each allocation iteration is based on the traffic demand. For each selected link, after it reserves a time slot at an allocation iteration, the traffic demand for this link decreases by one. At the same time, the scheduling algorithm find the maximum number of concurrent transmissions that are allowed in this slot. The iterative allocation continues until there is no unallocated traffic demand.

The above interference-aware routing and scheduling algorithms are simulated and compared to a basic scheduling algorithm and an optimal solution. In the basic scheduling algorithm, only one transmission is scheduled at each time slot, therefore spectral reuse is not utilized. The optimal solution is optimized with objective function of maximal network throughput. The proposed routing and scheduling schemes are reported to achieve near-optimal performance and significantly outperforms the basic algorithm. However, only small size networks are investigated and full fairness is ensured in the optimal solution.

In Ref. [16] the neighbor degree metric is used to construct routing tree. The neighbor degree metric of a node is set as a function of its transmission power and the neighbor number of this node. A cross-layer design architecture is proposed, in which transmit power control, routing construction, and centralized scheduling are taken into account. However, it is observed there is only a minor (3–7 percent) improvement on network throughput by employing the cross-layer design.

Tsai and Wang investigated the routing and admission control in WiMAX mesh networks [17]. They proposed a new routing method with shortest-widest efficient bandwidth (SWEB) as metric. The SWEB for a path is defined as follows. Let $p_{i,j}$ and $C_{i,j}$ denoted the packet error rate and capacity over a link $(i,j)$, respectively. Let $h$ denote the hop account of the path. Then the SWEB metric for a path $P$ can be calculated by [17]:

$$\text{SWEB} = \frac{\min\{C_{i,j}(1 - p_{i,j})|(i,j) \in P)\}}{2h}. \tag{11.4}$$

A node will select the path from itself to BS with the largest path SWEB. They also proposed a token bucket-based admission control method. However, centralized scheduling is not exploited. Another problem on the routing algorithm is that a node may be required to have several sponsor nodes due to the path selections of its children nodes.

Cao et al. proposed a joint routing and scheduling algorithm for WiMAX network under a new fairness constraints [18]. Shetiya and Sharma investigated QoS support in 802.16 mesh networks with centralized scheduling and shortest path routing algorithm [19]. However, only centralized routing algorithms are considered and spectral reuse is not allowed.

Due to the capabilities of direction antenna in increasing network throughput, applications of direction antenna to the general wireless mesh networks have been widely studied. Cao et al. investigated the cross-layer design for 802.16 multihop wireless backhaul networks [20]. They considered the joint optimal design of routing, MAC scheduling, and physical layer resource allocation. Beamforming antenna arrays are equipped at the physical layer. They introduced the notion of transmission set (TS) to separate the physical layer operations from those at the upper layers. TS is defined as a set of links in the network that can be simultaneously active at any given time. Due to the computation complexity of finding an optimal solution, a column generation approach is employed to identify TSs. Near-optimal performance in terms of scheduling cost is reported. However, the cross-layer design is observed to be complex and may require a long time to converge.

## 11.4 ROUTING ALGORITHMS

In this section and Section 11.5, we will investigate routing and scheduling algorithms for WiMAX mesh networks, with the target application of using WiMAX mesh networks to network cellular BSs and gateways to the Internet. The gateways of general cellular networks can be mobile switching center (MSC) in GSM networks or MSC and serving GPRS support node (SGSN) in WCDMA networks. In this network architecture, the gateways act as 802.16 mesh BSs in the 802.16 backhaul network. They manage the backhaul network and allocate bandwidth to the SSs. The cellular BSs act as 802.16 SSs in the backhaul network. In the remaining part of the chapter, the cellular BSs will be called SSs for simplicity. The SSs forward aggregated traffic from the mobile users to Internet in the uplink direction and deliver traffic to the mobile users from the Internet in the downlink direction via the 802.16 BSs. For simplicity, only uplink traffic with QoS requirement is considered in this chapter. However, the approaches of routing and scheduling can be applied to bidirection traffic. It is assumed that the backhaul topology and traffic demands from the SSs will not change frequently. Therefore, the frequency of updating traffic routes and scheduling will be low. It is feasible and beneficial to use high performance algorithms such as the optimal centralize scheduling and routing algorithms.

### 11.4.1 Classification

As IEEE 802.16 standards specify the MAC and PHY layer protocols, routing protocols are outside the scope of the standard work. We can identify the following ways of solving routing and scheduling (bandwidth allocation) problems for WiMAX mesh networks:

- Centralized routing and centralized scheduling (CRCS): All the traffic, position information of the stations are sent to the BSs. The BSs jointly solve the routes and schedule problems, and send the routes/schedule information back to the SSs.
- Distributed routing and centralized scheduling (DRCS): SSs distributedly find their routes to the BSs. After the routes are found, the SSs send their information to the BSs. The BSs determine collision-free transmission schedule.
- Distributed routing and distributed scheduling (DRDS): Stations distributively find their routes to the BSs. After SSs find their routes to BSs, they work with their next-hop stations toward the BSs to determine collision-free transmission schedules in a distributed way.
- Hybrid routing and scheduling (HRS): Both routing and scheduling algorithms can be implemented in either a distributed or a centralized way.

In WiMAX mesh networks, distributed scheduling can be used by SSs with their neighbor SSs to reserve time slots. However, the success of reserving time slots will depend on the availability of free time slots not reserved by the centralized scheduling and other neighbor stations. As centralized scheduling has priority over distributed scheduling in WiMAX, it will be easier to provide time slots reservation for end-to-end QoS support. Therefore, we will consider only centralized scheduling in this chapter.

Centralized scheduling algorithm will work together with either centralized or distributed routing algorithms. For the centralized routing algorithm, routing, bandwidth allocation, and scheduling problem can be jointly investigated. We will formulate the joint routing, allocation, and scheduling problem with an optimal mathematical model, which will be described in Section 11.5. The input to the optimal model from the routing point of view will be network connectivity matrix CM, $CM_{ij} = 1$ if and only if station $i$ is in the communication range of station $j$. After the optimal problem is solved, the route from an SS to a BS will be determined.

For the centralized scheduling and distributed routing algorithms, routes from SSs to BSs will be determined first. Then we can also derive network connectivity matrix CM for distributed routing. However, in this case, the station connectivity matrix CM will not only be determined by the communication ranges but also the traffic routes. We have $CM_{ij} = 1$ if and only if the link $e_{ij}$ between station $i$ and station $j$ is in any route of SSs to BSs. Then the connectivity matric CM can be input to the optimization model for centralized scheduling. In Section 11.4.2, we will introduce several distributed routing algorithms.

### 11.4.2 Distributed Routing

In the WiMAX mesh networks, the BSs will periodically broadcast network configuration messages, which are further forwarded by the neighbor stations. The forwarding process continues toward network edge until the predefined maximum number of hops is reached. For each forwarding neighbor stations, it will add the information of hops from itself to the BSs. Although routing algorithm is not specified in the standards, the distance information together with other local information can be utilized by a new SS to find a route to a BS.

For a new SS to join the WiMAX mesh network, it is required to listen to the network configuration/synchronization from its neighbor stations. After it hears network configuration message at least twice from a station, it can send request to join the network through this station. With the available local information, such as the signal strength, distance between, the new station can select which station to be its sponsor station to a BS. Therefore the distributed routing algorithm can be reduced to find a sponsor station for a route toward a BS. We simply give four methods.

- Random selection (RS): In this method, a new station will choose a neighbor station as its sponsor station randomly among the candidate stations with the same minimum number of hops to a BS.
- Minimum node ID (MID): Among all the neighbor stations with the same minimum number of hops to any BSs, the station with minimum node ID will be selected as the sponsor station.
- Maximum signal strength (MSS): The station with maximum average signal strength will be selected as the sponsor station among the candidate stations, which can achieve higher transmission data rate.
- Minimum aggregate traffic (MAT): To balance traffic routed over the neighbor stations, a new station can also choose the station with minimum aggregated traffic load among the candidate stations.

## 11.5  OPTIMAL BANDWIDTH ALLOCATION AND SCHEDULING ALGORITHM

In Section 11.4, we described the distributed routing algorithms. Both centralized and distributed routing algorithms can work with bandwidth allocation and scheduling scheme to provide optimal network performance. The required input from both centralized and distributed routing algorithms are the connectivity matrix defined in Section 11.4.1. The traffic routing and scheduling problem can be solved separately. Next we will formulate the bandwidth allocation model to allocate bandwidth to wireless links. Then time slots will be scheduled for the wireless links to achieve interference-free transmissions. In the currently considered problem, we assume that all stations operate in one operation frequency band with one omni-antenna. In practice, directional antennas and multiple operation frequency bands can be employed to maximize network capacity.

In the network, all of the WiMAX stations are connected by wireless links to the stations in their communication ranges and BSs have wireline links to fixed networks providing Internet access. The wireline links have finite capacities. The problems are then how to efficiently and economically route the traffic aggregated in the individual SSs to the fixed networks and allocate required bandwidth in each station on the traffic routes.

As network performance can be optimized in different ways, we choose two kinds of performance metrics for optimization. The first is the maximum delivered traffic to Internet, denoted by $T_s$. $T_s$ is the sum of the delivered traffic from individual SSs $T_i$, $T_s = \sum T_i$. As maximizing delivered traffic may starve the traffic from the SSs located far away from the BSs, we also consider the second performance metric $\lambda$ to ensure fairness. To provide fair delivery of traffic to fixed network, we request the ratios of successfully delivered traffic from the SSs to fixed networks to their traffic demands are equal to $\lambda$. Then the optimization objective is to maximize proportion $\lambda$ to achieve delivery fairness. If the constraints on the capacities of wireline links to fixed networks is released, the bottleneck will be inside the WiMAX backhaul. We set the optimization objective to maximize the overall traffic delivered to fixed networks. The variables used in the optimization model are described in Table.11.1.

With constrained wireline link capacity, the delivered traffic will not only be limited by the capacity of wireless networks, but also these of wireline links. For the wireless links, to achieve interference-free transmissions, if a station (say SS $t$) is transmitting to a neighbor station (say SS $r$) in one slot, there should be no other transmissions on the interfering links. An interfering link $e_{uv}$ of a link $e_{ij}$ means that either station $u$ or $v$ is in the interference range of station $i$ or $j$. We denote the set of interfering links of a link $e_{ij}$ by $I_{ij}$. Combining the constraints on the capacities wireline and wireless link and the interference-free wireless transmissions, we can formulate the optimization problem in the following equation:

**TABLE 11.1**

**Variables Description**

| | |
|---|---|
| $V$ | The set of all the WiMAX stations |
| $N$ | Number of WiMAX stations including SSs and BSs |
| $E$ | The set of all the wireless links |
| $R_{ij}$ | Capacity of wireless link from station $i$ to station $j$ |
| $Q_i$ | Traffic demands from station $i$ |
| $U_i$ | Capacity of wireline link from station $i$ to the fixed networks |
| $CM_{ij}$ | Station connectivity matrix, |
| | 1 if a station $i$ is in transmission range of station $j$, otherwise 0 |
| $I_{ij}$ | The set of interference links to the link $e_{ij}$, |
| | $i_{ij} \in I_{ij}$ if $i_{ij}$ interferes with $e_{ij}$ |
| $X_{ij}$ | Traffic routed from station $i$ to station $j$ |
| $F_i$ | Traffic routed from BS $i$ to fixed networks |

Maximize: $\lambda$
Subject to:

$$
\begin{cases}
\displaystyle\sum_{j \in V}^{CM_{ij}=1} (X_{ji} - X_{ij}) + \lambda Q_i = 0, & \forall i \in V,\ U_i = 0 \\[2ex]
\displaystyle\sum_{j \in V}^{CM_{ij}=1} (X_{ji} - X_{ij}) + \lambda Q_i - F_i = 0, & \forall i \in V,\ U_i > 0 \\[2ex]
\dfrac{X_{ij}}{R_{ij}} + \displaystyle\sum_{k,l \in V}^{e_{kl} \in I_{ij}} \dfrac{X_{kl}}{R_{kl}} \leq 1, & \forall i, j \in V,\ CM_{ij} = 1 \\[2ex]
X_{ij} \leq R_{ij}, & \forall i, j \in V \\[1ex]
X_{ii} = 0, & \forall i \in V \\[1ex]
F_i \leq U_i, & \forall i \in V
\end{cases}
\tag{11.5}
$$

If we assume that the wireline link capacity is not constrained compared to the wireless link data rate, the bottleneck will be inside the wireless networks. We can consider the optimization objective to be the delivered traffic without fairness consideration. Similar to the optimization model under fairness constraint, we can formulate the problem in the following equation:

Maximize: $\displaystyle\sum_{i \in V} T_i$
Subject to:

$$
\begin{cases}
\displaystyle\sum_{j \in V}^{CM_{i,j}=1} (X_{ji} - X_{ij}) + T_i = 0, & \forall i \in V,\ U_i = 0 \\[2ex]
\displaystyle\sum_{j \in V}^{CM_{i,j}=1} (X_{ji} - X_{ij}) + T_i - F_i = 0, & \forall i \in V,\ U_i > 0 \\[2ex]
\dfrac{X_{ij}}{R_{ij}} + \displaystyle\sum_{k,l \in V}^{e_{kl} \in I_{ij}} \dfrac{X_{kl}}{R_{kl}} \leq 1, & \forall i, j \in V,\ CM_{ij} = 1 \\[2ex]
0 \leq X_{ij} \leq R_{ij}, & \forall i, j \in V \\[1ex]
X_{ii} = 0, & \forall i \in V \\[1ex]
F_i \geq 0, & \forall i \in V
\end{cases}
\tag{11.6}
$$

As we have imposed the interference constraint in the optimization model, it is easy to derive collision-free scheduling to achieve the allocated bandwidth. A simple algorithm has been presented in Ref. [21].

## 11.6  SIMULATION RESULTS

In this section we will present some typical simulation and analytical results. The WiMAX mesh network is configured with 20 and 40 stations. For both network size, we have 2 BSs configurations, 1 BS (station 1), and 2 BSs (station 1 and 2), and the left stations are set as SSs. The network area is square with size of 3000 m × 3000 m for 20 stations and 5000 m × 5000 m for 40 stations, respectively. All the stations are randomly located in the network area. They have the same transmission range of 1500 m and interference range of 3000 m. The network topology used in the simulations with station IDs and station connectivities are shown in Figures 11.7 and 11.11. The data rates of all the wireless links are distance dependent. The relationship of distance $d$ and link data rate $R$ used in the simulation is shown in Table 11.2.

Typical simulation and analytical results for network with 20 stations are shown in Figures 11.8 and 11.9. In the figures, the routing algorithms represented by algorithm 1 to 5 are explained in Table 11.3. Each node has random traffic in the range of [0,22] Mbps to be delivered to the Internet through BSs. Figure 11.8 presents the normalized maximal λ versus different routing algorithms. The normalized maximal λ are obtained by dividing maximal λ from optimization model (1) for the different routing algorithms by the maximal λ value. It can be observed that for both BS configurations, centralized routing algorithm can achieve more than 30 percent performance improvement. The four distributed routing algorithms achieves similar performance while the MSS algorithm is a little better. As an example, Figure 11.10 shows the final traffic routes from the SSs to the BS (station 1) with centralized routing algorithm.

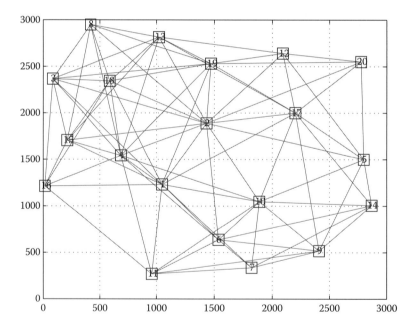

**FIGURE 11.7**   Network topology with 20 stations.

**TABLE 11.2**
**Station Distance and Link Data Rate**

| $d$ (km) | <0.25 | <0.5 | <0.75 | <1 | <1.25 | <1.5 |
|----------|-------|------|-------|----|-------|------|
| $R$ (Mbps) | 75 | 60 | 45 | 30 | 15 | 10 |

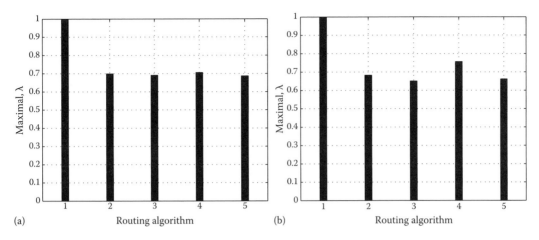

**FIGURE 11.8** Normalized maximal λ versus different routing and allocation algorithms with network size of 20 stations: (a) one BS (station 1) and (b) two BS (station 1 and 2).

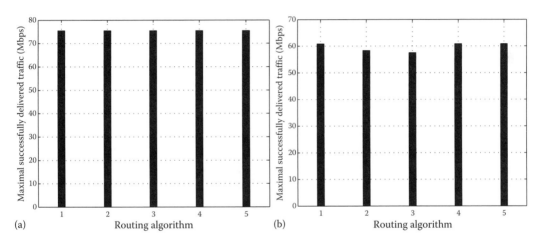

**FIGURE 11.9** Maximal successfully delivered traffic versus different routing and allocation algorithms with network size of 20 stations: (a) one BS (station 1) and (b) two BS (station 1 and 2).

**TABLE 11.3**
**Routing Algorithms**

| | |
|---|---|
| Algorithm 1 | Centralized routing algorithm |
| Algorithm 2 | Random routing |
| Algorithm 3 | Minimal station ID (MID) algorithm |
| Algorithm 4 | Maximal signal strength (MSS) algorithm |
| Algorithm 5 | Minimal aggregate traffic (MAT) algorithm |

Figure 11.9 presents the results of maximal successfully delivered traffic versus different routing algorithms for 20 stations network, which are obtained through optimization model (2). It is observed that joint centralized routing and scheduling does not achieve any performance advantages. This is because that the network bottleneck is in the wireless links. Only SSs close to BSs can successfully deliver their traffic to Internet, while SSs far away from BSs will be starved. It is noted that no

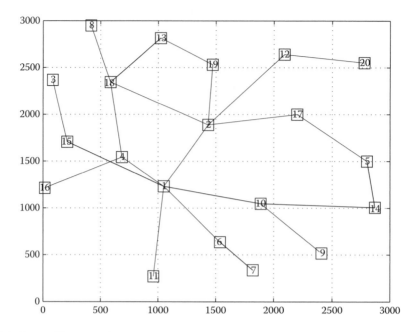

**FIGURE 11.10**   Traffic routes determined by centralized routing algorithm with two BSs.

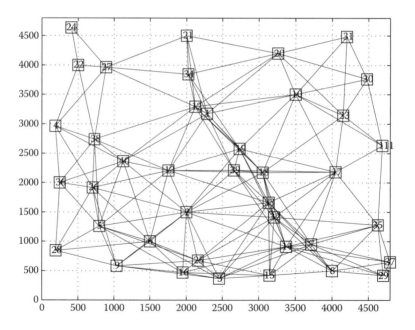

**FIGURE 11.11**   Network topology with 40 stations.

fairness constraints are put in the optimization model (2). As a result, the maximal $\lambda$ obtained from (2) is zero for all the investigated routing algorithms.

Similarly, the results on the maximal $\lambda$ and the maximal successfully delivered traffic for network size of 40 stations are presented in Figures 11.12 and 11.13. Similar performance improvements can also be observed for both BS configurations.

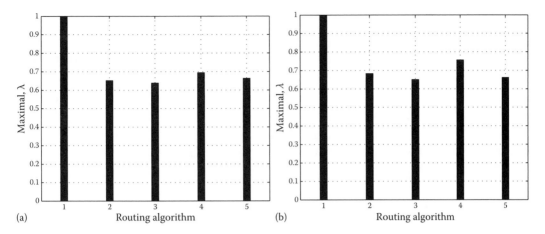

**FIGURE 11.12** Normalized maximal λ versus different routing and allocation algorithms with network size of 40 stations: (a) one BS (station 1) and (b) two BS (station 1 and 2).

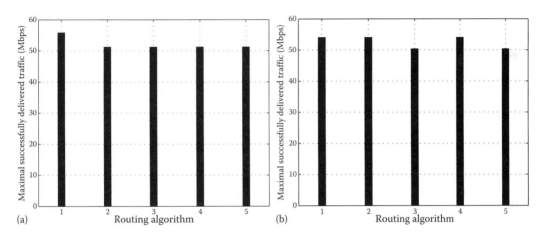

**FIGURE 11.13** Maximal successfully delivered traffic versus different routing and allocation algorithms with network size of 40 stations: (a) one BS (station 1) and (b) two BS (station 1 and 2).

## 11.7  CONCLUSION

Routing and bandwidth allocation are critical for QoS support over WiMAX networks. In this chapter, we investigate different routing and bandwidth allocation algorithms in WiMAX mesh networks. We proposed several distributed routing algorithms, which are compared with a centralized routing algorithm. The connection matrix of the stations are used to solve a bandwidth allocation problem. The problem is formulated by a mathematical optimization model. The solution developed from the optimization model can be used to schedule the time slots for stations transmission without collision. Simulation and analytical results with medium size networks show that joint centralized routing and scheduling can achieve the best performance with over 30 percent improvement. The improvement is achieved at the cost of increased network management complexity. In the future work, the overhead brought by centralized routing algorithm will be evaulated, and the benefit of the performance improvement will be justified. In addition, it is interesting to consider the problem of increasing network capacity by using multiple radio interface over multiple frequency channels in WiMAX mesh networks.

## ACKNOWLEDGMENTS

The work reported in this chapter has been partially funded by EPSRC through the HIPNet project, United Kingdom, and by the European Union through the Welsh Assembly Government, whose funding and support are gratefully acknowledged.

## REFERENCES

[1] C. Eklund, R. Marks, K.L. Stanwood, and S. Wang. IEEE standard 802.16: A technical overview of the WirelessMANTM air interface for broadband wireless access. *IEEE Communication Magazine*, 40(6), 98–107, June 2002.

[2] IEEE Std 802.16-2004. IEEE Standard for Local and metropolitan area networks Part 16: Air inteface for fixed broadband wireless access systems, 2004.

[3] IEEE 802.16's Relay Task Group. Webpage: http://wirelessman.org/relay/

[4] I.F. Akyildiz, X. Wang, and W. Wang. Wireless mesh networks: A survey. *Elsevier Journal of Computer Networks*, 47, 445–487, 2005.

[5] R. Bruno, M. Conti, and E. Gregori. Mesh networks: Commodity multihop ad hoc networks. *IEEE Communication Magazine*, 43(3), 123–131, 2005.

[6] P. Kyasanur and N.H. Vaidya. Routing and interface assignment in multi-channel multi-interface wireless networks. In: *Proc. IEEE WCNC 2005*, New Orleans, LA, pp. 2051–2056, March 2005.

[7] R. Draves, J. Padhye, and B. Zill. Routing in multi-radio, multi-hop wireless mesh networks. In: *Proc. of the 10th ACM Annual International Conference on Mobile Computing and Networking (MOBICOM 2004)*, pp. 114–128, September 2004.

[8] M. Ergen, S. Coleri, and P. Varaiya. QoS aware adaptive resource allocation techniques for fair scheduling in OFDMA based broadband wireless access system. *IEEE Transactions on Broadcasting*, 49(4), 362–370, December 2003.

[9] M. Cao, W. Ma, Q. Zhang, X. Wang, and W. Zhu. Modelling and performance analysis of the distributed scheduler in IEEE 802.16 Mesh mode. *ACM MobiHoc*, 2005.

[10] M. Cao, W. Ma, Q. Zhang, and X. Wang. Analysis of IEEE 802.16 mesh mode scheduler performance. *IEEE Transactions on Wireless Communications*, 4, 1455–1464, 2007.

[11] N. Bayer, X. Bangnan, V. Rakocevic, and J. Habermann. Improving the performance of the distributed scheduler in IEEE 802.16 mesh networks. *Vehicular Technology Conference 2007 Spring*, pp. 1193–1197, 2007.

[12] H. Zhu and K. Lu. On the interference modeling issues for coordinated distributed scheduling in IEEE 802.16 mesh networks. *IEEE Broadnets 2006*, pp. 1–10, 2006.

[13] D. Kim and A. Ganz. Fair and efficient multihop scheduling algorithm for IEEE 802.16 BWA Systems. *IEEE Broadnets 05*, Boston, MA, pp. 833–839, 2005.

[14] B. Han, W. Jia, and L. Lin. Performance evaluation of scheduling in IEEE 802.16 based wireless mesh networks. *Computer Communications*, 30, 782–792, 2007.

[15] H.Y. Wei, S. Ganguly, R. Izmailov, and Z.J. Haas. Interference-aware IEEE 802.16 WiMAX mesh networks. *IEEE VTC 2005 Spring*, Stockholm, Sweden, pp. 3102–3106, May 2005.

[16] M. Peng, Y. Wang, and W. Wang. Cross-layer design for tree-type routing and level-based centralised scheduling in IEEE 802.16 based wireless mesh networks. *IET Communications*, 1(5), 999–1006, 2007.

[17] T.-C. Tsai and C.-Y. Wang. Routing and admission control in IEEE 802.16 distributed mesh networks. *Wireless and Optical Communications Networks*, pp. 1–5, 2007.

[18] M. Cao, Vi. Raghunathan, and P. R. Kumar. A tractable algorithm for fair and efficient uplink scheduling of multi-hop WiMAX mesh networks. *IEEE WiMesh*, 2006.

[19] H. Shetiya and V. Sharma. Algorithms for routing and centralized scheduling to provide QoS in IEEE 802.16 mesh networks. *ACM workshop on Wireless Multimedia Networking and Performance Modelling*, October 2005.

[20] M. Cao, X. Wang, S.-J. Kim, and M. Madihian. Multi-hop wireless backhaul networks: A cross-layer design paradigm. *IEEE Journal on Selected Areas in Communications*, 25(4), 738–748, May 2007.

[21] M. Alicherry, R. Bhatia, and L. Li. Joint channel assignment and routing for throughput optimization in multi-radio wireless mesh networks. *ACM Mobicom05*, 2005.

# Part II

## WiMAX Planning and Optimization

# 12 WiMAX: Architecture, Planning, and Business Model

*Abdulrahman Yarali and Saifur Rahman*

## CONTENTS

The fixed/portable broadband wireless access is becoming a necessity for many residential and business subscribers worldwide. The demand is exploding as the pricing of broadband services is rapidly decreasing. The worldwide interoperability for microwave access (WiMAX) technology is an integral part of the portfolio by complementing 2G/3G mobile access, Digital Subscriber Line (DSL) broadband fixed access, and Wireless Fidelity (WiFi) hotspot access. In this chapter, an extended overview of WiMAX and its applications in higher generation wireless networks as a cost-effective solution to answering the challenges posed by the digital divide is presented. Technology behind WiMAX and its network architecture, design, and deployment are examined in addition to factors that impact WiMAX planning and performance. A WiMAX radio coverage simulation and analysis at different frequency bands for different demographic is presented. Furthermore, the WiMAX business models and a comparison with two enhanced third-generation (3G) technologies that are potential competitors to WiMAX are explored. The chapter concludes with identification of some of the applications, benefits, and drawbacks of WiMAX mobile networks.

## 12.1 INTRODUCTION

Telecommunications has grown at a tremendous rate in the last ten to twenty years. Improved semiconductor and electronics manufacturing technology, and the growth of the Internet and mobile telecommunications have been some of the factors which have fueled this growth in telecommunications. The deployment of state-of-the-art telecommunications infrastructure and services has, however, been restricted to the developed world. The least developed countries have been left in the technological dark ages with few or none of the next-generation networks installed. Developed countries now boast high-speed connections with a large percentage of homes having access to the Internet and broadband services at an affordable fee. The underdeveloped countries are yet to enjoy such facilities. This is referred to as the digital divide [1]. During the first World Summit on the Information Society (WSIS) held in Geneva in December 2003, the Digital Divide was defined as the unequal access to information and communication technologies (ICTs), where the least developed countries are separated from the developed countries because of a lack of technology particularly ICT [2].

The digital divide has persisted due to the relatively high cost of putting up modern telecommunications infrastructure. This is compounded by the fact that there are a number of different services available and each service requires its own technology and network [3]. Therefore, existing technologies such as Wireless Fidelity (WiFi), Digital Subscriber Line (DSL), Global System for Mobile communications (GSM), Integrated Services Digital Network (ISDN), and the relatively new 3G technologies have not been able to provide a total solution to closing the digital divide. Figure 12.1 illustrates the main network types and the prevalent technologies associated with each, mapped against usage models and access modes.

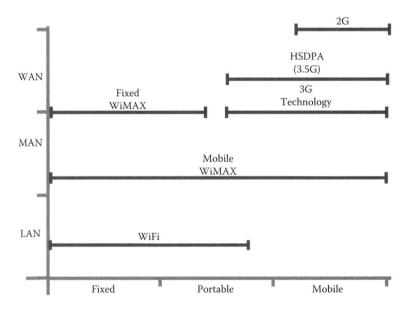

**FIGURE 12.1** Wireless network type and range. MAN—metropolitan area network (citywide, rural area), LAN—local area network (office, home, campus), and WAN—wide area network (countrywide, international).

WiMAX will boost today's fragmented broadband wireless access market and mobile WiMAX promises to offer a solution to closing the existing digital divide. WiMAX can address the fixed wireless access and portable Internet market, complementing other broadband wireless technologies. Government initiatives to reduce the digital divide are making gains for broadband wireless countries such as Australia, South Korea, Taiwan, and the United States have programs in place today, and there has been a push by the European Commission for more flexible spectrum policies.

WiMAX access can be easily integrated within both fixed and mobile architectures, enabling operators to integrate it within a single converged core network, thereby providing new capabilities for a user-centric broadband world.

WiMAX addresses the following needs which may answer the question of closing the digital divide [1]:

- Cost effective
- Offers high data rates
- Supports fixed, nomadic, and mobile applications thereby converging the fixed and mobile networks
- Easy to deploy and has flexible network architectures
- Supports interoperability with other networks
- Aimed at being the first truly a global wireless broadband network

WiMAX is a standard that is championed by the WiMAX forum which was formed in June 2001 to promote conformance to IEEE 802.16 standard. The WiMAX forum currently has more than 470 members comprising the majority of operators, component, and equipment companies in the communications ecosystem. The WiMAX forum promotes interoperability by working closely with IEEE and other standards groups such as the European Telecommunications Standards Institute (ETSI) which have their own versions of broadband wireless. Along these lines, the WiMAX forum works closely with service providers and regulators to ensure that WiMAX forum certified systems meet customer and government requirements.

The original WiMAX standard only catered for fixed and nomadic services. It was reviewed to address full mobility applications; hence, the mobile WiMAX standard, defined under the IEEE 802.16e specification was created. Mobile WiMAX supports full mobility, nomadic, and fixed systems to compete against DSL to cover isolated areas such as rural hot spots, private campus networks, and remote neighborhoods. Mobile WiMAX is more promising to be deployed as a cellular network that offers ubiquitous broadband services to mobile users to over large geographical areas. It can be deployed as a central office bypass to avoid using existing wired infrastructure for competitive local exchange carriers and wireless Internet service provider. Figure 12.2 shows the standard history for 802.16.

## 12.2  NETWORK ARCHITECTURE

The mobile WiMAX end-to-end network architecture is based on an All-Internet Protocol (IP) platform, all packet technology, and no circuit switch telephony. The end-to-end architecture makes the greatest possible use of IETF and IEEE standards and protocols along with the adoption of commonly available standard equipment.

The open IP architecture gives network operators great flexibility when selecting solutions that work with legacy networks or that use the most advanced technologies, and in determining what functionality they want their network to support. They can choose from a vertically integrated vendor that provides a turnkey solution or they can pick and choose from a dense ecosystem of best-of-breed players with a more narrow focus. The architecture allows modularity and flexibility to accommodate a broad range of deployment options such as small scale to large scale, urban, suburban, and rural

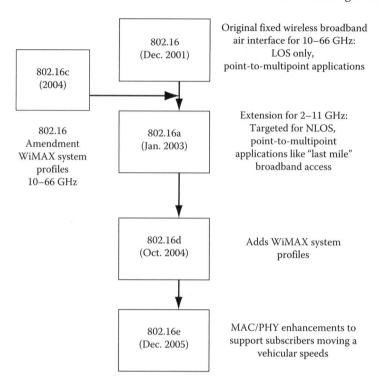

**FIGURE 12.2**  802.16 standard evolution.

coverage, mesh topologies, flat, hierarchical and their variant, and finally, coexistance of fixed, nomadic portable and mobile usage models [4].

Mobile WiMAX adds both the mobility and multiple-input multiple-output (MIMO) functionalities to the IEEE 802.16-2005 standard. It is one of two standards adopted by the WiMAX forum with the other one being the IEEE 802.16-2004. Mobile WiMAX network architecture mainly has three components. These include the access services network (ASN), the core services network (CSN), and the application services (AS) network. Figure 12.3 illustrates the interconnection of these networks. The WiMAX network supports the following key functions:

- All-IP access and core service networks
- Support for fixed, nomadic, and mobile access
- Interoperability with existing networks via internetworking functions
- Open interfaces between ASNs and between the ASN and the CSN
- Support for differential quality of service (QoS) depending on the application
- Unbundling of the access, core, and application service networks

## 12.2.1  ACCESS SERVICES NETWORK

The ASN is the access network of WiMAX and it provides the interface between the user and the core service network. Mandatory functions as defined by the WiMAX forum include the following:

- Handover
- Authentication through the proxy authentication, authorization, and accounting (AAA) server

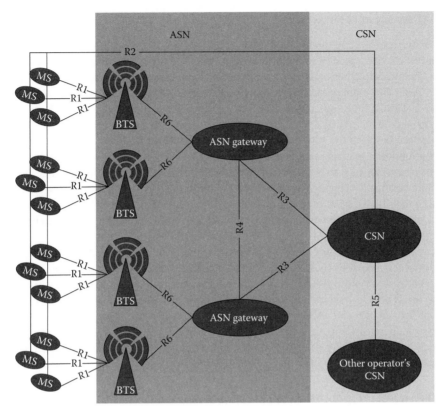

**FIGURE 12.3** WiMAX network architecture. (From Yegani, P., Cisco System white paper, WiMAX Overview, IETF-64, Vancouver, Canada, November 7–11, 2005. With permission.)

- Radio resource management
- Interoperability with other ASN's
- Relay of functionality between CSN and mobile station (MS), e.g., IP address allocation

Base station (BS): The cell equipment comprises the basic BS equipment, radio equipment, and BS link to the backbone network. The BS is what actually provides the interface between the mobile user and the WiMAX network. The coverage radius of a typical BS in urban areas is around 500–900 m [6]. In rural areas the operators are planning cells with a radius of 4 km. This is quite a realistic number now and quite similar to the coverage areas of GSM and UMTS high-speed downlink packet access (HSDPA) BSs today.

Deployment is driven either by the bandwidth required to meet demand, or by the geographic coverage required to cover the area. Based on the cell planning of other previous technologies, urban and suburban segments cell deployment will likely be driven by capacity. Rural segment deployment will likely be driven by the cell radius. For BTS systems, the emphasis is more on performance than on cost and size, although there still is an interest in low cost because WiMAX is a new deployment.

ASN gateway: The ASN gateway performs functions of connection and mobility management and interservice provider network boundaries through processing of subscriber control and bearer data traffic. It also serves as an Extensible Authentication Protocol (EAP) authenticator for subscriber identity and acts as a Remote Authentication Dial-In User Service (RADIUS) client to the operator's AAA servers.

## 12.2.2 Core Services Network

The CSN is the transport, authentication, and switching part of the network. It represents the core network in WiMAX. It consists of the home agent (HA) and the AAA system and also contains the IP servers, gateways to other networks, i.e., public switched telephone network (PSTN), and 3G.

WiMAX has five main open interfaces which include reference points R1, R2, R3, R4, and R5 interface [7]. The R1 interface interconnects the subscriber to the BS in the ASN and is the air interface defined on the physical layer and Medium Access Control (MAC) sublayer. The R2 is the logical interface between the mobile subscriber and the CSN. It is associated with authorization, IP host configuration management, services management, and mobility management. The R3 is the interface between the ASN and CSN and supports AAA, policy enforcement, and mobility management capabilities. The R4 is an interface between two ASNs. It is mainly concerned with coordinating mobility of MSs between different ASNs. The R5 is an interface between two CSNs and is concerned with internetworking between two CSNs. It is through this interface that activities such as roaming are carried out.

The unbundling of WiMAX divides the network based on functionality. The ASN falls under the network access provider (NAP). The NAP is a business entity that provides WiMAX network access to a network service provider (NSP). The NSP is a business entity that provides core network services to the WiMAX network and consists of the CSN. The Applications services fall under the applications service provider (ASP).

If network operator wants to reap the full benefits that WiMAX and its all-IP architecture can deliver, they need to carefully select the ASN and CSN solutions that best suit their requirements and provide all the functionality required while avoiding unnecessary complexity in their network.

## 12.3 TECHNOLOGIES EMPLOYED BY WiMAX

Mobile WiMAX operates in licensed frequency bands in the range of 2 to 6 MHz. The technologies employed by mobile WiMAX include the following:

- Scalable orthogonal frequency division multiple access (SOFDMA) on the physical layer
- MIMO
- IP
- Adaptive antenna systems (AAS)
- Adaptive modulation schemes (AMS)
- Advanced encryption standard (AES) encryption

## 12.3.1 Physical Layer

Mobile WiMAX will initially operate in the 2.3, 2.5, 3.3, and 3.4–3.8 GHz spectrum bands [8] using SOFDMA. OFDMA is perhaps the most important technology associated with WiMAX. SOFDMA is based on OFDMA which in turn is based on OFDM [9]. OFDM is a form of frequency division multiplexing, but it has higher spectral efficiency and resistance to multipath fading and path loss compared to other multiplexing methods. It divides the allocated frequency spectrum into subcarriers which are at right angles to each other. This reduces the possibility of cross-channel interference thereby allowing the subcarriers to overlap. This reduces the amount of frequency spectrum required, hence the high spectral efficiency. The reduced data rate of each stream reduces the possibility of intersymbol interference because there is more time between the arrival of symbols from different paths. This feature of OFDM makes it resistant to multipath fading and ideal for nonline of sight (NLOS) applications. In OFDMA each frequency subcarrier is divided into subchannels which can be accessed by multiple users hence increasing the capacity of OFDM [10].

Scalable OFDMA is a form of OFDMA which allows variable channel bandwidth allocation from 1.25 to 20 MHz. SOFDMA has capabilities which make it ideal for the implementation of

IP and hybrid automatic repeat request (HARQ). WiMAX also uses other features to enhance the performance of OFDMA. They include dynamic frequency shifting, MIMO, AAS, and software-defined radios. Dynamic frequency shifting monitors the signal and changes frequencies to avoid interference. Software-defined radios are controlled by changing software settings and this gives the equipment more flexibility when switching frequencies.

MIMO is a technology that has already found use in WiFi (IEEE 802.11n). MIMO multiplies the point-to-point spectral efficiency by using multiple antennas and RF chains at both the BS and the MS. MIMO achieves a multiplicative increase in throughput compared to single-input, single-output (SISO) architecture by carefully coding the transmitted signal across antennas, OFDM symbols, and frequency tones. These gains are achieved at no cost in bandwidth or transmit power [11].

AAS are spatial processing systems which combine antenna arrays with sophisticated signal processing. They reduce the effects of interference from multiple signal paths thereby also contributing to high capacity of the system and the use of mobile WiMAX in NLOS environments.

## 12.3.2   MAC SUBLAYER

The 802.16 MAC sublayer uses a scheduling algorithm for which the subscriber station only needs to compete for initial entry into the network. The scheduling algorithm also allows the BS to control QoS parameters by balancing the time-slot assignments among the application needs of the subscriber stations.

WiMAX supports QoS differentiation for different types of applications. The 802.16 standard defines the following types of services [12]:

- Unsolicited grant services (UGS): UGS is designed to support constant bit rate (CBR) services, such as T1/E1 emulation, and Voice-over-IP (VoIP) without silence suppression.
- Real-time polling services (rtPS): rtPS is designed to support real-time services that generate variable size data packets on a periodic basis, such as MPEG video or VoIP with silence suppression.
- Nonreal-time polling services (nrtPS): nrtPS is designed to support nonreal-time services that require variable size data grant burst types on a regular basis.
- Best effort (BE) services: BE services are typically provided by the Internet today for Web surfing.

## 12.4   NETWORK DIMENSIONING AND DESIGN

Designing, deploying, and managing any wireless cellular system requires clear objectives to be identified from the outset. These includes definition of the footprint coverage, the estimated number of users, the traffic load distribution, the penetration and growth rate, and internetwork access and roaming. Mobile WiMAX, which will be deployed like 2G and 3G cellular networks, supports fractional frequency. Fractional frequency reuse takes advantage of the fact that mobile WiMAX user transmit on subchannels and does not occupy an entire channel such as in 3G. The objective of the radio network dimensioning and design activity is to estimate the number of sites required to provide coverage and capacity for the targeted service areas and subscriber forecast. This process is based on many assumption such as uniform distribution of subscribers, homogenous morphology, and ideal site location. The main inputs required for network dimensioning are site equipment-specific parameters, marketing-specific parameters, and licenses regulation and propagation models [13]. Figure 12.4 shows the flow chart of activities performed in network design and planning, starting from data collection of marketing and design requirement input and achieving the business model to provide a nominal site plan using a network simulation software.

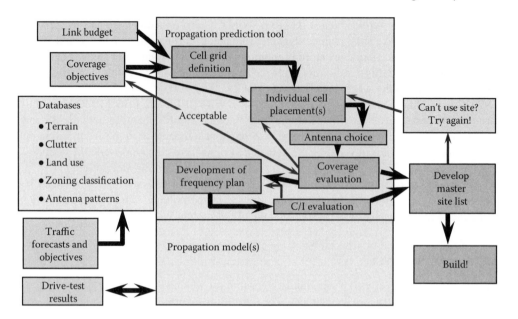

**FIGURE 12.4**   The cell planning process.

Mobile WiMAX is designed to complement existing 2G/3G access technologies with an "Always Best Connected" experience with voice and data connections. There is a large range of possible scenarios for the deployment of mobile WiMAX, but main four categories are [14]

- Fixed and mobile operator with enhanced data for GSM evolution (EDGE)/3G who uses mobile WiMAX as a complementary extension for data services
- Mobile only operator with EDGE/3G who uses mobile WiMAX as a complementary extension for data services
- Fixed operator who uses mobile WiMAX to compete with 3G operators for data and voice services
- New entrant who uses mobile WiMAX to move into mobile market—threat to incumbent mobile operator.

WiMAX operates in a mixture of licensed and unlicensed bands. The unlicensed bands are typically the 2.4- and 5.8-GHz bands. Licensed spectrum provides operators control over the usage of the band, allowing them to build a high-quality network. The unlicensed band, on the other hand, allows independence to provide backhaul services for hotspots. Typical area licensed WiMAX spectrum allocations are

- Lower 700 MHz (US) with 2 × 6 MHz channels
- 2.5 GHz Multichannel Multipoint Distribution Service with 15.5 MHz in US and 72 MHz in Canada
- 3.5 GHz Wireless Local Loop with 2 × 2 MHz channel blocks
- 5.8 GHz UNI (license exempt) with 80 MHz allocation

WiMAX access networks are often deployed in point-to-multipoint cellular fashion where a single BS provides wireless coverage to a set of end users stations within the coverage area. The technology behind WiMAX has been optimized to provide both large coverage distances of up to 30 km under line-of-sight (LOS) situations and typical cell range of up to 8 km under NLOS [15]. In an NLOS, a signal reaches the receiver through reflections, scattering, and diffractions. The signals arriving at the receiver consists of many components from direct and indirect paths with different delay spreads, attenuation, polarizations, and stability relative to the direct path. WiMAX technology solves or mitigates the problem resulting from NLOS conditions by using OFDMA, Subchannelization, directional antennas, transceiver diversity, adaptive modulation, error correction, and power control [16]. The NLOS technology also reduces installation expenses by making the under-the-eaves customer premise equipment (CPE) installation a reality and easing the difficulty of locating adequate CPE mounting locations.

Both LOS and NLOS coverage conditions are governed by propagation characteristics of their environment, radio link budget, and path loss. In both the cases, relays help to extend the range of the BS footprint coverage allowing for a cost-efficient deployment and service [17].

## 12.4.1  WiMAX Cell Site Design

One of the most important technical and business issues of any wireless technology is efficiently (cost and performance) providing coverage and capacity, while avoiding the build-out of a large number of new BSs. Cell design is performed with the help of a network planning tool using digital elevation and demographic maps. The first step in designing a wireless system is to develop a link budget. Link budget is the loss and gain sum of signal strength as it travels through different components in the path between a transmitter and receiver. The link budget determines the maximum cell radius of each BS for a given level of reliability and is comprised of two types of components: system related components and nonsystem related components [18].

These components are important factors when evaluating the complexity and speed in deploying at higher frequency bands, especially in unlicensed bands such as 5.8 GHz (licensed in some countries such as Russia). Other factors like interference from other surrounding networks will also impact network performance and QoS.

Path loss, shadow margin, environmental effects, and morphology are important factors when planning for an optimum coverage. The morphology and physical surroundings of a cell site play a very important role in determining the cell footprint. A cell site footprint can shrink from 7 km in a mostly flat area with light tree densities to 3 km in a hilly terrain with moderate-to-heavy tree densities [18]. With adaptation of Erceg model [19], the cell size for several carrier frequencies from 450 MHz to 3.5 GHz is estimated for WiMAX systems using path loss propagation models for flat rural, hilly rural, and urban environment. The following equation is used to estimate total path loss

$$L = A + 10\gamma \, \log\left(\frac{d}{d_0}\right) + S \tag{12.1}$$

$$A = 20 \, \log\left(\frac{4\pi d_0}{\lambda}\right) \tag{12.2}$$

$$\gamma = a - bh_b + \frac{c}{h_b} \tag{12.3}$$

In these equations: $L$ is the total path loss in dB, $d_0$ is the close-in reference distance and $d$ is the transmitter–receiver separation distance in km, $S$ is a random variable, $\lambda$ is the wavelength of the carrier, $A$ is free space path loss, $h_b$ is the BS height, $\gamma$ is path loss exponent and $a$, $b$, and $c$ are constant depending on morphology type.

Figure 12.5 illustrates a comparison of a path loss simulation for a WiMAX system for different frequency bands. In this study, a link budget of 142 dB which provides a 3 km cell coverage at 1900 MHz has been assumed [20]. To obtain the same cell radius of 3 km with 2.5 GHz frequency band an additional 4 dB for link budget is needed. In a coverage limited design scenario, this 4 dB corresponds to 22 percent reduction in cell coverage footprint and almost 70 percent increase in the cell

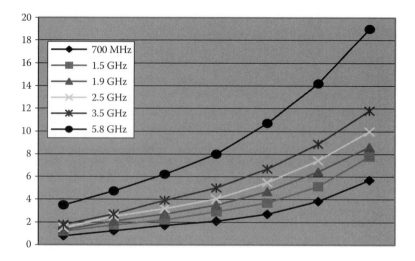

**FIGURE 12.5**   Path loss vs. WiMAX cell radius.

**TABLE 12.1**
**WiMAX Cell Count vs. Frequency**

| Frequency Band Operating Frequency | Cell Radius (km) | Link Budget (dB) | Cell Radius Reduction (percent) | Cell Count Increase (percent) |
|---|---|---|---|---|
| 1900 MHz | 3 | 142 | | |
| 2.5 GHz | 3 | 146 | 21–24 | 62–75 |
| 3.5 GHz | 3 | 151 | 42–46 | 200–250 |

count. Table 12.1 shows cell count calculation for 1900 MHz to 3.5 GHz to illustrate the impact that path loss can have, especially when deploying in higher frequency bands [21]. Figures 12.6 and 12.7 show the results of cell count estimation in a flat rural area for frequency operation of 450 MHz and 3.5 GHz. Assuming the equal distribution of the coding modulation schemes inside the cells and the probability of terrain coverage of 95 percent, the system capacity is lower for WiMAX systems at 450 MHz frequency, due to large cell size. Compared with existing cellular systems, WiMAX systems implement advanced radio features that compensate for the extra attenuation resulting from higher carrier frequency, larger transmission bandwidth, and deep indoor penetration.

The radio enhancement feature applicable to the fixed and mobile WiMAX is subchannelization. Other enhancement features that are only applicable to mobile WiMAX are convolutional turbo coding, repetition coding (3 dB gain), and HARQ.

Applying smart antennas or MIMO configuration in different topologies will enhance the cell site coverage footprint. Cell planning options and WiMAX technology features also allow interference and noise handling so that WiMAX can provide sufficient coverage [19]. Figure 12.8 shows global percentage of WiMAX deployment per frequency band.

## 12.5   THE WiMAX BUSINESS MODEL

The biggest challenges to deploying WiMAX-based services are business related. Carriers need financial capability to implement infrastructure. Each operator has to carefully identify its own

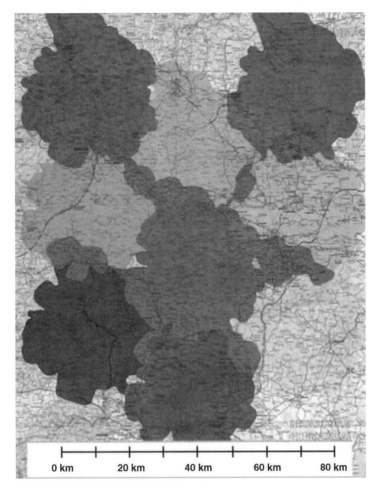

**FIGURE 12.6**   Cell radius for 450 MHz. (Courtesy of Jozef Stefan Institute.)

requirements, dictated by the type of services offered, the market segments targeted, the spectrum available, and the topography of the coverage area. There is no single solution that works for all, and operators need to make key choices about the management and core networks as they plan for their WiMAX networks.

An accurate business case analysis must take into account a wide variety of variables such as demographics, services, frequency band alternatives, capital expense items, operating expense items, and CPE equipment. The WiMAX business model can be looked from several perspectives. These include the equipment vendors, service providers and application providers, and customers [22]. WiMAX will have a larger impact long term than we have seen from cellular phones in the past two decades. Initial rollouts of WiMAX will begin mostly by competitive local phone service carriers and rural Internet service providers. Larger carriers will utilize fixed WiMAX to deliver services to residential customers many of whom are in underserved markets. WiMAX adoption in these underserved markets will be high due to lack of availability of high-speed data access. These deployments will generate capital to be reinvested for future deployments. Larger customer base will begin driving both the cost of carrier and customer equipment down. As the economy of scale makes deployment less expensive, mobile platforms will begin to appear. This development will be spread between high population centers and the rural markets that already have fixed platforms deployed. Fixed platform will act as a springboard for mobile deployment. Then interconnections

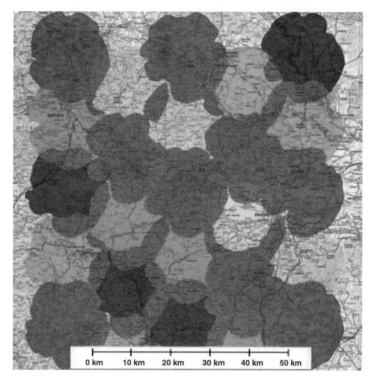

**FIGURE 12.7**    Cell radius for 3.5 GHz. (Courtesy of Jozef Stefan Institute.)

Percentage of WiMAX deployments per frequency band

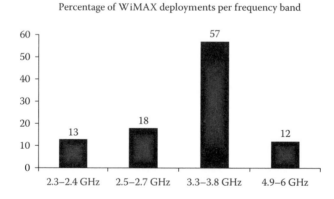

**FIGURE 12.8**    WiMAX deployment per band. (BWA Research, UK, Available at: www.wimaxcounts.com)

will begin to form between rural markets and metropolitan markets as carriers from cooperative agreements to share network resources. The economy of scale will increase exponentially at this point and we will notice a negative impact on traditional cellular, Internet, and voice services. Once the implementation of initial hot underserved rural markets and high-density metro areas is completed, springboard deployments will quickly take WiMAX coverage to the level of coverage offered by traditional wireless today. This process will move much faster than the deployment of cellular networks and devices for the following key reasons:

- Manufacturing process for WiMAX devices will be quite similar to that of wireless devices and mostly the changes will be in components and software.
- Readiness of the current wireless fixed and mobile market and waiting on new technology.
- As carriers built out wireless networks, most of the questions in this field have been answered and can now be applied to the development of a mirror network that provides WiMAX access.

### 12.5.1 EQUIPMENT VENDORS

WiMAX, as with many new technologies, is based on an open standard. Although standards increasingly play a crucial role in driving adoption, they are not sufficient to guarantee success. A standard-based technology will success only if a solid ecosystem of operators, vendors, and solution and content providers emerge to support it, as is in the case of WiMAX. WiMAX enables intervendor interoperability which brings lower costs, greater flexibility and freedom, and faster innovation to operators.

Within the WiMAX industry there is a strong commitment to ensuring full interoperability through certification and ad-hoc testing between vendors. It is important for network operators to realize how interoperability is established and what it covers so that they understand how different products, solutions, and applications from different vendors can coexist in the same WiMAX network. The advantages that interoperability brings are multiple. Some of these advantages are the ability to choose among vendors, flexibility when choosing the appropriate network elements and components, success to the latest cutting-edge technology, and an open architecture which makes it easier for operators to roll out new revenue-generation services and applications as they can rely on wider pool of suppliers.

The two categories of equipment vendors include the network equipment vendors and the terminal equipment vendors. Network equipment includes ASN and CSN equipment, and vendors include companies such as Motorola and ZTE of China. They will gain their profits through the sale of the equipment and through installation of the equipment. They may further have after sales agreements with the customers who are the service providers. Terminal equipment includes mobile phones, CPE, modems, laptops, smart phones, and PDAs and they are manufactured by companies like Nokia, Blackberry, Motorola, and Intel. They will gain their profits through the sale of the terminal equipment. Nokia, the world's top handset maker, expects to start selling cell phones using the WiMAX technology in 2008.

### 12.5.2 SERVICE PROVIDERS

As IP networks become faster (higher bandwidth) and more responsive (lower delay), the set of services implemented on IP-based networks has grown. This growth generates more revenue opportunities for service providers, and thus next-generation networks are all migrating toward IP technology. From an operator standpoint, services can be broken into four billable classes: (1) basic Internet services which are typically billed at a flat rate, (2) premium Internet services which are important not only to improve ARPU, but to add new services, (3) VPN services which can be billed by QoS level, and (4) operator premium services which are applications provided on the operator's network.

The service providers are expected to gain profits through the sale of the different services and applications that WiMAX is capable of carrying. The different services that can be offered on WiMAX networks include best effort VoIP, carrier class IP telephony through the IP multimedia core, music, video conferencing, streaming video, interactive gaming, mobile instant messaging (IM), IP television (IPTV), basic broadband wireless Internet, and other application-based services to corporate customers. The concept of unbundling the network reduces the barriers of entry into the mobile telecommunications industry because a provider does not need to own the whole network.

The business aspect of the service providers can also be looked at from two perspectives. The first one is where the service provider owns the whole system including the core network and the access

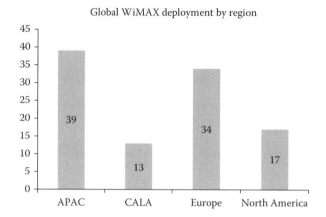

**FIGURE 12.9**   Global WiMAX networks. APAC = Asia Pacific, CALA = Caribbean and Latin America. (BWA Research, UK, Available at: www.wimaxcount.com)

network. The second option is the unbundled option where the access network and core networks exist as independent business entities.

In emerging markets such as Africa, and South Asia where telecom investment is still nascent and 3G yet to be launched, WiMAX makes complete business sense even at equal cost, better speeds, better spectrum utilization, and the promise of broadband to a much sparsely spread population. For developed economies, the United States for instance, the 2.3 spectrum band is believed to be more capex efficient and hence better than 3G and high-speed uplink packet access (HSUPA). More importantly, the phase in the capex cycle of a telecom operator will determine each operator strategy—whether to embrace WiMAX or stick to its existing technology. The WiMAX industry entered the year 2007 as a year for ecosystem buildup in the preparation for regional and nationwide deployments of WiMAX services. It appears that 2008 will be a make-or-break period for WiMAX. Figure 12.9 shows a global WiMAX deployment by region.

### 12.5.3  APPLICATION PROVIDERS

WiMAX has already revolutionized the broadband wireless market by standardizing broadband wireless access market, by opening up new service opportunities and by creating the environment for ubiquitous broadband services.

The aim is to provide the service that best fits the individual's needs. Applications can be developed in house by the service providers, outsourced from other companies or developed and sold directly to the end user by an independent applications development company.

Applications are based on IP, and IP applications are sent back or forth via WiMAX. This allows the users to develop applications independently from the underlying network infrastructure. Some applications will still be developed by operators but the vast majority will come from those working directly in the Internet crowd. For them and for the end users competing wireless technologies are very beneficial. Competition spurs network roll outs, offers possibility for new players in the market, and creates competition between device manufacturers. Also, new applications will be introduced more easily and much more quickly as they are no longer forced into a tight framework that takes long time to develop and from which it is difficult to get out again.

### 12.5.4  WiMAX CUSTOMERS

Prospective WiMAX customers can be grouped either geographically or by the level or volume of services. Geographical categories range from urban to rural customers, while categories according to size include individual customers and the corporate customers. Urban areas offer the highest density of customers with more business establishments. In such cases a higher number of cells which are small in size are required to meet the capacity requirements. These are the areas where more competition is expected. Rural areas are expected to have a lower penetration of customers, less corporate customers, and bigger cell sizes because emphasis is on coverage rather than capacity. Individual subscribers will use WiMAX for music downloads, interactive gaming, and personal broadband Internet, and will form a large percentage the total subscribers.

Corporate subscribers are also expected to contribute to revenues of WiMAX, and their interest will be in applications and services which will enhance their organizations apart from the basic telecommunications services.

Companies are poised to compete with each other in WiMAX network deployment, which will ensure that the prices will be competitive.

## 12.6 COMPARISON WITH COMPETING TECHNOLOGIES

At some point current 2G and 3G network operators will migrate to a 4G network technology. Mobile WiMAX is likely to face competition from 3G and 4G technology enhancements. They include the code division multiple access (CDMA) variants CDMA2000 and wideband-CDMA (WCDMA) and their enhancements which are 1x evolution data optimized (1xEVDO) and HSDPA, respectively. Unlike in the early days of the CDMA vs. GSM competition, this higher generation competition will be quite different and fruitful because for these new generations networks; the applications are separated and do not depend on each other. 4G networks will go far beyond 2G and 3G by mainly improving three parameters:

- Interface technology: 4G standards will make a radical change and will use OFDM [9]. The new modulation itself will not automatically bring an increase in speed but very much simplifies the following two enhancements:
- Channel bandwidth: 4G systems will use a bandwidth of up to 20 MHz, i.e., the channel offers four times more bandwidth than channels of current systems. As 20 MHz channels might not be available everywhere, most 4G systems will be scalable, e.g., in steps of 1.25 MHz. It can therefore be expected that 4G channel sizes will range from 5 to 20 MHz.
- MIMO: The idea of MIMO is to use the multipath phenomena. Although this behavior is often not desired, MIMO makes active use of it by using several antennas at the sender and receiver side, which allows the exchange of multiple data streams, each over a single individual wave front. Two or even four antennas are foreseen to be used in a device. How well this works is still to be determined in practice but it is likely that MIMO can increase throughput by a factor of two in urban environments.

Increasing channel size and using MIMO will increase throughput by about 8–10 times. Thus speeds of 40 Mbps per sector of a cell are thus possible. Using a commonly accepted evaluation methodology for 3G systems, mobile WiMAX has been simulated against the 3G enhancements [23]. These simulations have shown that

- Mobile WiMAX peak data rates are up to 5x better than 3G+ technologies.
- Mobile WiMAX spectral efficiency is 3x better than any 3G+ technology.
- Lower equipment cost for WiMAX due to certified products (compare with WiFi).
- WiMAX requires new infrastructure while high-speed packet access (HSPA) rides on UMTS.
- Roughly the same coverage (average ∼5 km).
- Roughly the same performance (average ∼2 Mbps per user).

- HSDPA launched in 2006 while HSUPA will come in 2008.
- WiMAX standard set end of 2005 and first products in 2006.
- HSPA has a higher acceptance with mobile operator.

## 12.6.1  1xEVDO

This standard is developed by the third generation partnership project 2 (3GPP2), the body responsible for CDMA and EVDO. 1xEVDO is an enhanced version of CDMA2000-1x. There are four versions that have been released, namely, Rev. 0, Rev. A, Rev. B, and Rev. C.

1xEVDO is a high-speed data only specification for 1.25 MHz frequency division duplex (FDD) channels with a peak downlink (DL) data rate of 2.4 Mbps.

Improvements to CDMA2000-1x in the 1xEVDO Rev. 0 specification include [9]:

- DL channel is changed from code division multiplexing (CDM) to time division multiplexing (TDM) to allow full transmission power to a single user.
- DL power control is replaced by closed-loop DL rate adaptation.
- Adaptive modulation and coding (AMC).
- HARQ.
- Fast DL scheduling.
- Soft handoff is replaced by a more bandwidth efficient "virtual" soft handoff.

1xEVDO Rev. 0, however, was designed to support only packet data services and not conversational services. In 1xEVDO Rev. A and EVDO Rev. C (also dubbed DORC), additional enhancements were added to the 1xEVDO specification. They include the following [8]:

- DL: Smaller packet sizes, higher DL peak data rate (up to 3.1 Mbps), and multiplexing packets from multiple users in the MAC layer.
- Uplink (UL): Support of HARQ, AMC, higher peak rates of 1.8 Mbps, and smaller frame size

These enhancements in both the UL and DL of 1xEVDO Rev. A allow it to support conversational services.

## 12.6.2  HSDPA/HSPA

The WCDMA specification was enhanced to create the high-speed downlink packet access (HSDPA) and then HSPA specifications. The enhancements in HSDPA include AMC, multicode operation, HARQ, higher DL peak rates (up to 14 Mbps), and decentralized architecture where scheduling functions are moved from the radio network controller (RNC) to Node-B, thus reducing latency and enabling fast scheduling.

HSPA adds enhancement to the UL of the WCDMA specifications. In Ref. [9] a quantitative comparison of mobile WiMAX, 1xEVDO, and HSPA system performance was conducted based on the commonly accepted 1xEVDV evaluation criterion. The mobile WiMAX system configuration was based on the WiMAX forum baseline minimum configuration. Table 12.2 illustrates a comparison of mobile WiMAX with 3G enhancements [24].

These technologies, i.e., EVDO, HSPDA, and mobile WiMAX have several performance enhancing features in common as follows [25]:

- AMC
- HARQ
- Fast scheduling
- Bandwidth efficient handoff

**TABLE 12.2**
**Comparison of Mobile WiMAX with 3G Enhancements**

| Parameter | | 1xEVDO Rev. A | 3xEVDO Rev. B | HSDPA | HSUPA | Mobile WiMAX |
|---|---|---|---|---|---|---|
| Duplex | | FDD | FDD | FDD | FDD | TDD |
| Occupied spectrum (MHz) | | 2.5 | 10 | 10 | 10 | 10 |
| Channel bandwidth (MHz) | DL | 1.25 | 5 | 5 | 5 | DL/UL = 3 |
| | UL | 1.25 | 5 | 5 | 5 | |
| Spectral efficiency | DL | 0.85 | 0.93 | 0.78 | 0.78 | 1.91 |
| | UL | 0.36 | 0.28 | 0.14 | 0.30 | 0.84 |
| Net information throughput | DL | 1.06 | 4.65 | 3.91 | 3.91 | 14.1 |
| per channel/sector (Mbps) | UL | 0.45 | 1.39 | 0.7 | 1.50 | 2.20 |

## 12.6.3 WiFi

WiMAX is different from WiFi in many respects. The WiFi MAC layer uses contention access. This causes users to compete for data throughput to the access point. WiFi also has problems with distance, interference, and throughput and that is why triple play (voice, data, video) technologies cannot be hosted on traditional WiFi. In contrast, 802.16 uses a scheduling algorithm. This algorithm allows the user to only compete once for the access point. This gives WiMAX inherent advantages in throughput, latency, spectral efficiency, and advanced antenna support.

Companies developing radical innovations may adopt different stances not only based on the strategic interests of the company but also by taking into other considerations such as the market and its needs and requirements, as well as other products it may carry.

When comparing WiFi and WiMAX, one is comparing their substitutability and complementary to existing technologies and how different companies have and will view them. WiMAX and WiFi can offer some potentially significant cost savings for mobile network operators by providing an alternate means to backhaul BS traffic from cell site to the BS controllers. Mobile network operators typically utilize some type of wired infrastructure that they must buy from an incumbent operator. A WiFi or WiMAX mesh can offer a much more cost-effective backhaul capability for BSs in metropolitan environments.

Using WiFi and WiMAX open broadband wireless standards and implementing mobile computing, governments and partners can quickly and cost-effectively deploy broadband to areas not currently served, with little or no disruption to existing infrastructures. Standards-compliant WLANs and proprietary WiFi mesh infrastructures are being installed rapidly and widely throughout the world. Standards-compliant WiMAX products can provide NLOS backhaul solutions for these local networks and WiMAX subscriber stations can currently provide Internet access to customers such as schools and other educational institutions and campuses.

The results of the comparison show that mobile WiMAX has better performance in all the areas listed above (where it shares performance enhancing features with EVDO and HSDPA/HSPA). Furthermore, the technologies on which mobile WiMAX is based result in lower equipment complexity and simpler mobility management due to the all-IP core network. They also provide mobile WiMAX systems with many other advantages over CDMA-based systems such as

- Tolerance to multipath and self-interference
- Scalable channel bandwidth
- Orthogonal UL multiple access
- Support for spectrally-efficient TDD
- Frequency-selective scheduling

**TABLE 12.3**
**Summary of WiMAX Applications**

| Class Description | Real Time | Application Type | Bandwidth |
|---|---|---|---|
| Interactive gaming | Yes | Interactive gaming | 50–85 Kbps |
| VoIP, video conferencing | Yes | VoIP | 4–64 Kbps |
| | | Videophone | 32–384 Kbps |
| Streaming media | Yes | Music/speech | 5–128 Kbps |
| | | Video clips | 20–384 Kbps |
| | | Movies streaming | >2 Mbps |
| Information technology | No | Instant messaging | <250 byte messages |
| | | Web browsing | >500 Kbps |
| | | Email (with attachments) | >500 Kbps |
| Media content download | No | Bulk data, Movie download | >1 Mbps |
| (store and forward) | | Peer to peer | >500 Kbps |

- Fractional frequency reuse
- Improved variable QoS
- Advanced antenna technology

## 12.7  APPLICATIONS

The WiMAX standard has been developed to address a wide range of applications. Based on its technical attributes and service classes, WiMAX is suited to supporting a large number of usage scenarios. Table 12.3 address a wide range of applications [26].

### 12.7.1  VoIP AND IP

Mobile WiMAX is an all-IP network. The use of OFDMA on the physical layer makes it capable of supporting IP applications. It is a wireless solution that not only offers competitive Internet access, but it can do the same for telephone service.

VoIP offers a wider range of voice services at reduced cost to subscribers and service providers alike. VoIP is expected to be one of the most popular WiMAX applications. Its value proposition is immediate to most users. Although WiMAX is not designed for switched cellular voice traffic as cellular technologies as are CDMA and WCDMA, it will provide full support for VoIP traffic because of QoS functionality and low latency. IPTV enables a WiMAX service provider to offer the same programming as cable or satellite TV service providers. IPTV, depending on compression algorithms [27], requires at least 1 Mbps of bandwidth between the WiMAX BS and the subscriber. In addition to IPTV programming, the service provider can also offer a variety of video on demand (VoD) services. IPTV over WiMAX also enables the service provider to offer local programming as well as revenue generating local advertising.

## 12.8  BENEFITS OF WiMAX

The WiMAX solution reflects the general trend in the communications industry toward unified packet-based voice and data networks. Fundamental benefits of this transition are reduced operation cost, improved network optimization, and better management of changes. The followings are some of the benefits of WiMAX.

*Wireless.* By using a WiMAX system, companies/residents no longer have to rip up buildings or streets or lay down expensive cables.

*High bandwidth.* WiMAX can provide shared data rates of up to 70 Mbps. This is enough bandwidth to support more than 60 businesses at once with T1-type connectivity. It can also support over a thousand homes at 1-Mbps DSL-level connectivity. Also, there will be a reduction in latency for all WiMAX communications.

*Long range.* The most significant benefit of WiMAX compared to existing wireless technologies is the range. WiMAX has a communication range of up to 40 km [28].

*Multi-application.* WiMAX uses the IP and is therefore capable of efficiently supporting all multimedia services from VoIP to high speed Internet and video transmission. It also supports a differentiated QoS enabling it to offer dynamic bandwidth allocation for different service types. WiMAX has the capacity to deliver services from households to small and medium enterprises, small office home office (SOHO), cybercafés, multimedia Tele-centers, schools and hospitals.

*Flexible architecture.* WiMAX supports several systems architectures, including point-to-point, point-to-multipoint, and ubiquitous coverage.

*High security.* The security of WiMAX is state of the art. WiMAX supports advanced encryption standard triple data encryption standard. WiMAX also has built-in VLAN support, which provides protection for data that is being transmitted by different users on the same BS. Both variants use privacy key management (PKM) for authentication between BS and SS station. WiMAX offers strong security measures to thwart a wide variety of security threats.

*QoS.* WiMAX can be dynamically optimized for a mix of traffic that is being carried.

*Multilevel service.* QoS is delivered generally based on the service-level agreement between the end user and the service provider.

*Interoperability.* WiMAX is based on international, vendor-neutral standard. This protects the early investment of an operator because it can select the equipments from different vendors.

*Low cost and quick deployment.* WiMAX requires little or no external plant construction compared with the deployment of wired solutions. BSs will cost under $20,000 but will still provide customers with T1-class connections [29].

*Worldwide standardization.* WiMAX is developed and supported by the WiMAX forum (more than 470 members). The WiMAX forum collaborates with different international standards organizations that are developing broadband wireless standards with the intent to provide interoperability among the standards. Some of the other broadband wireless standards include HiperMAN/HiperLAN (Europe) and WiBRO (South Korea). These standards are compatible with WiMAX at the physical layer. WiMAX will become a truly global technology-based standard for broadband and will guaranty interoperability, reliability, and evolving technology and will ensure equipment with very low cost.

## 12.9   DRAWBACKS OF WiMAX

The most significant challenge is that WiMAX is a new technology with emerging support.

*Hesitancy.* Companies are very hesitant of setting up WiMAX BSs today because it has not yet reached widespread use. Intel has made their Centrino laptop processors WiMAX enabled. All laptops are expected to have WiMAX by 2008 [30].

*Exclusion of start-up companies.* Even though cost provides a low barrier to entry, none of the start-up companies are projected to be major players in the development of WiMAX. Intel and Cisco seem to have an obvious advantage today, and by the time it reaches widespread use, large operators will find WiMAX to be a very attractive new way of raising revenues.

*Research and development.* For WiMAX to succeed, new products must be researched and developed to incorporate WiMAX. Without the help of major companies investing in this R&D, WiMAX could be gravely underutilized.

## 12.10   CONCLUSION

The combination of both advanced radio features and flexible end-to-end architecture makes WiMAX attractive solution for diverse operators. It provides many different services on one network, services which required different networks in the past. It also provides convergence of fixed and mobile networks. It provides high speed access to the subscriber at a reasonable cost, thereby enabling the service provider to make a profit from the technology, using economies of scale. It offers the advantage of reduced total cost of ownership during the lifetime of a network deployment. Standalone WiMAX networks are certainly feasible, but in most cases WiMAX access technology will be adopted by operators as an extension to their existing networks. This allows operators to make the most of their existing infrastructure such as BS sites, and IP service infrastructure for service and related AAA and billing systems.

Although it is clear that WCDMA has the advantage when referring to voice and soft handoff of voice, these advantages disappear for data-centric applications. There are some additional advantages of WCDMA in equipment performances; however, these advantages are not sufficient to overcome the advantages of OFDMA. As data traffic continues to grow, there will be an increasing need to off-load data from 3G to and OFDMA-based network optimized for data applications. Mobile WiMAX (802.16e) provides the only standards-based OFDMA WAN technology. WiMAX is an excellent complement to other wireless technologies that are designed to work in the LAN (WiFi) or that offer wider coverage but with more limited capacity (GSM, CDMA, UMTS, EVDO). Recent inclusion of WiMAX in IMT2000, and the ITU decision may push the CDMA giant further toward adopting 802.16e. WiMAX and future wireless networks that aspire to offer 4G services will attempt to become unified communications systems that fit diverse markets and have very different sets of customers and requirements. The common architecture is supposed to result in an overall advance in technology and a reduction in costs.

In regard to WiMAX planning and cell design, the radio enhancement feature applicable to fixed and mobile WiMAX compensate for the extra attenuation resulting from higher carrier frequency, larger transmission bandwidth, and deep indoor penetration.

WiMAX is expected to take prominence in about five years (2012). The strengths of WiMAX lie in its ability to address the requirements of modern telecommunications networks and the commitment that has been shown to its development and wide acceptance by a number of leading equipment vendors and service providers. The biggest challenges to deploying WiMAX-based services do not stem very much from the spectrum, but from business case issues.

## REFERENCES

[1] G. Smyth, Wireless Technologies and e-learning: Bridging the digital divide, Intel Corporation, December 2006.
[2] G. Cayla, S. Cohen, and D. Guigon, WiMAX an efficient tool to bridge the digital divide, WiMAX Forum, November 2005, p. 2.
[3] S. Rahman and M. Pipattanasomporn, Alternate technologies for telecommunications and internet access in remote locations. In *Proceedings of 2002 3rd Mediterranean Conference and Exhibition on Power Generation, Transmission, Distribution and Energy Conversion*, Greece, November 2002.
[4] L. Bai, Analysis of the market for WiMAX services, Thesis, Lyngby, Denmark, May 2007.
[5] P. Yegani, Cisco Systems white paper, WiMAX Overview, IETF-64 November 7–11, Vancouver, Canada, 2005, p. 4.
[6] B. Puzzolante, G. Redaelli, and G. Grazia, Nationwide implementation of a WiMAX mobile access network at CEFRIEL, STEM, 2006.

[7] L. Nuaymi, *WiMAX Technology for Broadband and Wireless Access*, John Wiley, New York, 2007.

[8] M. Paolini, Mobile WiMAX: The best personal broadband experience! WiMAX forum, June 2006. pp. 6.

[9] Flarion, OFDM for mobile data communication, August 2004.

[10] G. Parsaee and A. Yarali, OFDMA for the 4th generation cellular networks, (CCECE 2004), Niagara Falls, Canada, May 2–5, 2004.

[11] A. Salvekar, S. Sumeet, L. Qinghua, V. Minh-Anh, and Q. Xiaoshu, Multiple-antenna technology in WiMAX systems, *Intel Technology Journal*, 8(3), 235, August 20, 2004.

[12] G. Nair, J. Chou, T. Madejski, K. Perycz, D. Putzolu, and J. Sydir, IEEE 802.16 medium access control and service provisioning, *Intel Technology Journal*, 8(3), 217, August 20, 2004.

[13] B. Upase, M. Hunukumbure, and S. Vadagana, Radio network dimensioning and planning for WiMAX networks, *Fujitsu Sci. Tech. J.* 43(4), 435–450, October 2007.

[14] White paper, A comparative analysis of mobile WiMAX deployment alternatives in the access networks, WiMAX forum, May 2007.

[15] White Paper, IEEE 802.16a Standard and WiMAX igniting broadband wireless access, WiMAX forum, 2003.

[16] G. Parsaee, A. Yarali, and E. Ebrahimzad, MMSE-DFE equalizer design for OFDM systems with insufficient cyclic prefix, *Proceeding of IEEE VTC04 conference*, September 26–29, 2004, Los Angeles, CA.

[17] C. Hoymann, M. Dittrich, S. Goebbels, and B. Walke, Dimensioning cellular WiMAX Part II: Multihop WiMAX networks, *in Proceedings of IEEE mobile WiMAX*, March 2007.

[18] A. Yarali, J. Wheatley, J. Ponder, M. Sharifi, and H. Behrooz, Wireless cellular and PCS network cell count estimator, ASEE Conference at E Peoria, IL, March 26, 2004.

[19] V. Erceg, Channel models for broadband fixed wireless systems, IEEE 802.16.3c-2000/47.

[20] A. Fellah and R. Syputa, WiMAX and broadband wireless (Sub-11GHz) worldwide market analysis and trends 2006–2012, September 2006.

[21] White Paper, Consideration for deploying mobile WiMAX at various frequencies, Nortel.

[22] WiMAX: The business case, WiMAX Forum, 2004.

[23] White paper, Understanding WiMAX and 3G for portable/mobile broadband wireless, Intel, December 2004.

[24] A. Yarali, B. Mbula, and A. Tumula, WiMAX: A key to bridging the digital divide, *Proceedings of IEEE SoutheastCon 2007*, Richmond, VA.

[25] C. Hoymann and S. Goebbels, Dimensioning cellular WiMAX Part I: Single hop networks, in *Proceedings of 13th European Wireless Conference*, 2007.

[26] D. Gray, Mobile WiMAX–Part II: A comparative analysis, May 2006. pp. 14–41.

[27] White paper, WiMAX deployment considerations for fixed wireless access in the 2.5 GHz and 3.5 GHz licensed bands, WiMAX forum, June 2005.

[28] A. Yarali and A. Cherry, Internet Protocol Television (IPTV), *Proceeding of IEEE TENCON05*, Melbourne, Australia, November 21–24, 2005.

[29] M. Sauter, WiMAX base station prices and coverage ranges, March 31, 2007. http://mobilesociety. typepad.com/mobile_life/2007/03/wimax_base_stat.html

[30] WiMAX weekly news: Weekly clip report, April 26th–May 3rd 2007, WiMAX Forum.

# 13 WiMAX Networks Dimensioning

*Konstantinos Ntagkounakis and Bayan Sharif*

## CONTENTS

Network Dimensioning is a process that combines analytical and pragmatic design methodologies to achieve WiMAX deployment that are optimized for high performance, lower costs, and reduced time-to-market. Dimensioning is essentially the process where the WiMAX system capabilities are correlated with the network design objectives and performance indicators to define the most suitable deployment strategy and hence estimate the equipment size (list of materials). Successful dimensioning is based on simple analysis and design principles but most importantly past deployment experience, especially with regard to accuracy. The first part of this chapter provides a description of the necessary steps to perform dimensioning, while the second part presents advanced approaches and previous experience in the context of deployment strategy. The chapter concludes with open issues to enhance the process of dimensioning.

## 13.1  INTRODUCTION

The process of dimensioning is usually performed several times per network case. Initially, when an operator issues a request for proposal (RFP), a network design case study is the most common criterion for the selection of an equipment vendor/integrator. After the initial screening and during the negotiations, this exercise may be repeated with more accurate input data (i.e., business plan, performance indicators) for the purposes of the contract where a final budget should be agreed. However, the most important dimensioning process is performed after the award of the contract and during the preliminary design. At this point accuracy is crucial and hence every care should be taken to achieve it. The customer must finalize the business plan and the RF designer should carefully select the best deployment strategy. Dimensioning results are used as input in the preliminary network design, which includes the distribution of estimated equipment to service areas, derivation of base station (BS) nominal positions, and preliminary BS configuration (i.e., sectorization, air-interface parameterization).

The impact of dimensioning on the business perspective of a WiMAX network investment is quite significant, simply because it allows an investor (i.e., operator, Internet service provider) to select the vendor/integrator based on the best proposed strategy. The benefits are: a diversity of strategies that would increase the know-how, reduce capital expenses (CapEX)/operating expenses (OpEX) and network complexity, and improve time-to-market. From the vendor/integrator perspective, an accurate dimensioning would increase the chances for more contract awards, reduce the network design and implementation complexity and most importantly guarantee customer satisfaction. Furthermore, it would contribute to the overall success of WiMAX technology ensuring more investments and market growth.

## 13.2  DEPLOYMENT SCENARIOS

A major feature of WiMAX compared to other wireless access technologies is that it breaks the barrier of addressing a single customer profile. Global system for mobile communications (GSM)/universal mobile telecommunications system (UMTS) provide mainly voice and low speed internet to mobile subscribers, while local multipoint distribution service (LMDS)/wireless local loop (WLL) offer higher bandwidth services to fixed subscribers. WiMAX can offer broadband services to all fixed, nomadic, and eventually mobile subscribers, [1], according to the aims of the latest IEEE 802.16e standard. This major advantage for WiMAX technology offers greater flexibility and scalability; however it presents more design challenges. A conceptual presentation of deployment scenarios,

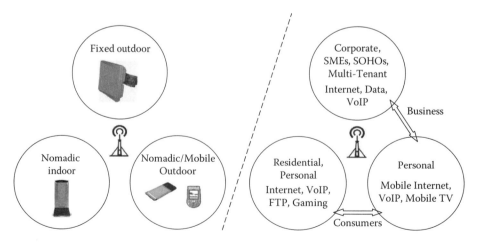

**FIGURE 13.1**  Abstract of WiMAX deployment scenarios.

based on equipment, services, and potential customer profiles is presented in Figure 13.1. Each "sector" represents a WiMAX terminal profile:

- *Fixed-outdoor* units (including antenna, RF subsystem, modem), which can be installed on the rooftop or outer building walls for maximizing link performance. A cable connects the unit to an indoor interface terminal that provides Ethernet and VoIP ports [2].
- *Fixed/portable indoor* units (intergraded antenna, RF baseband and interface in a single box), which are installed indoors close to a window or the outer wall. The unit is portable within the indoor space, however it requires power supply [3].
- *Nomadic/mobile* units (PCMCIA cards, handheld devices), which are truly portable (mobile in future versions) and can be used in outdoor and indoor spaces [4].

Each terminal profile is built with different performance capabilities and cost towards specific customer profiles. Fixed-outdoor terminals are capable of long range, robust links that can transfer high-bandwidth and delay sensitive services with low impact on network air-interface resources, hence they are more suitable for corporate, small-to-medium enterprises (SMEs), and small-offices-home-offices (SOHOs). The higher hardware and installation costs are balanced by higher revenues. Fixed-indoor terminals have considerably less cost and are self-installable, albeit with smaller link range. Such terminals address the mass market of residential access. Finally the nomadic and portable terminals require even greater network design margins and usually address individual customers at specific service areas (such as community/camp networks). Observing Section 13.1 it can be seen that business customers are likely to prefer a combination of fixed-outdoor and nomadic units, while residential and personal customers will probably select either a fixed-indoor or a nomadic unit. As WiMAX technology progresses, more system gain will be achieved in the air-interface thus resulting in higher cell ranges and increased percentage of nomadic terminals mainly at the expense of fixed-indoor units.

The continuous development of WiMAX technology from IEEE 802.16-2004 standard to the IEEE 802.16e amendment, [5,6], has led to significant improvements in the air-interface. Recent advances include higher BS transmit power, advanced antenna systems (MIMO, beamforming (BF)), improved radio resource management through the OFDMA profile, improved coding techniques which reduce the signal-to-interference and noise ratio (SINR) thresholds, efficient uplink (UL) subchannelization, and flexible frequency reuse. The current amendment of WiMAX offers more than 15 dB increase in the system gain over previous versions which drastically extends the radio coverage, and can therefore reach indoor customers even when using portable/mobile terminals. As the WiMAX system gain increases due to the continuous enhancement of the air-interface, in the context of dimensioning, the network size for a specific deployment is reduced, and so is the up-front investment.

## 13.3 DIMENSIONING PROCESS

The process of dimensioning involves a sequence of steps, which serve different requirements such as capacity or coverage estimations, to conclude to a final outcome as shown in Figure 13.2. The input consists of the business plan, the assets, and the key performance indicators (KPIs). The first action is then to verify that all necessary information is available and clarified, alternatively an interactive "questions and answers" session is necessary. During the input analysis, the designers utilize their theoretical and technological expertise to define the dimensioning strategy. In some cases this analysis identifies that potential business plan assumptions could affect the network design process, and a discussion is likely to be initiated with the customer. Such an approach is more appropriate after the award of a contract. The following step is the processing of service characteristics such as the Internet rates and Voice-over-Internet Protocol (VoIP) codecs. In most cases the services are provided in the business plan with their marketing description and it is necessary to identify the impact of each

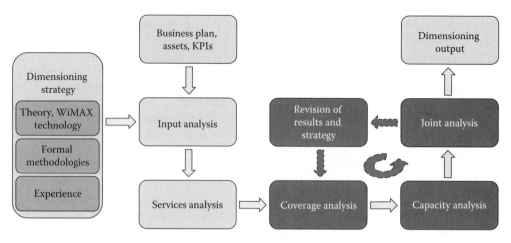

**FIGURE 13.2**  Network dimensioning process.

service on the WiMAX air-interface. In the next step, the coverage analysis is performed: the provided service areas are identified and the required number of points of presence (PoP) is estimated. A PoP refers to a WiMAX site, with at least three sectors, which is capable of providing a 360° footprint.

Further to the coverage analysis, the service areas should be also offered sufficient capacity as dictated by customer numbers and services profiles. Therefore the next step is to estimate the required number of sectors to achieve this capacity. The estimates of PoP and sectors are then used to determine the configuration of the BS in the final joint analysis step. It is common after the joint analysis for designers to evaluate the results and proceed with customizations to further improve the solution. In this case, coverage, capacity, and joint analysis may be revised several times before providing the final output. A description of the strategy, the final bill of material (BoM), and a discussion of the assumptions and methodology are mainly the outcome of a dimensioning study.

### 13.3.1  BUSINESS PLAN, ASSETS, AND KEY PERFORMANCE INDICATORS

A well-defined business plan can be the most significant driver behind a successful investment on WiMAX technology. From a cost point of view, a very ambitious plan in terms of subscribers/services will certainly impact CapEX and OpEX, whereas from a design point of view, the impact comes when the requirements are not extensively defined or they contradict the technology capabilities. A revision of the business plan during or after the network design may compromise the network performance or create additional design costs but most importantly it will result in implementation delays. A list with the most common information provided with a business plan is presented as follows:

1. *Service area(s)*: defined with geocoded polygons, including the size in km$^2$, and the terrain profile details (i.e., urban, suburban, rural, average building height, etc.).
2. *Coverage type*: such as fixed-outdoor, on rooftop, or on outer walls, fixed-indoor, nomadic outdoor/indoor, mobile outdoor or any combination thereof.
3. *Subscriber profile(s)*: such as residential, small business, corporate. Subscriber profiles may relate to a specific type of coverage and service.
4. *Subscriber distribution*: subscriber numbers per profile, per service area, and per deployment year, according to the scalability plan.
5. *Service profile(s)*: such as VoIP, broadband Internet, VPN along with their distinct characteristics (i.e., VoIP codecs, peak information rates, contention factors, etc.). Service profiles may relate to specific subscriber profiles and coverage types.
6. *Available spectrum*: defined as paired, along with local regulations concerning the allowed channelization and duplex schemes.

7. *Existing infrastructure*: such as sites that can be reused, available backhauling equipment with Ethernet interface, and core network PoPs.
8. *Cartographic data*: such as high-resolution digital maps with buildings.
9. *Key performance indicators*: such as coverage objective in terms of percentage of the service area, differentiated per terminal type, where a stable QPSK link can be achieved.
10. *Customer requirement*: such as duplex scheme, number of sectors/BS, channel bandwidth, reuse scheme, type of sites, deployment strategy.

### 13.3.2 STRATEGY

During request for information (RFI)/RFP stages, a dimensioning exercise may be requested by a customer, mainly for two reasons: either to acquire know-how by differentiated proposals or to identify the more cost-efficient solution. In the first case, the requirements are usually relaxed so that the participant vendors/integrators can design with flexibility, while the provided information (i.e., business plan, assets, service areas) is hypothetical. The submitted studies will probably be presented in various formats and most certainly based on diverse assumptions. In such case a direct comparison among the studies is complicated, and usually a more defined exercise is the next step. In the second approach, the case study is well defined so that the design assumptions are either implied or directly mentioned. The results are now directly comparable, hence a clear ranking list can be obtained. From RF network designer point of view a different strategy should be followed: showing flexibility in the network design and perhaps providing several alternatives for the first approach, while a more strict, cost-optimum solution is more appropriate for the second approach.

### 13.3.3 SERVICES DIMENSIONING

The scope behind this process is to translate the "marketing" description of services, for different customer profiles, and estimate their impact on the air-interface. To successfully offer broadband access over a wireless link, mainly two conditions should be met: capacity availability and a quality-of-service (QoS) mechanism. WiMAX technology is truly broadband in terms of capacity and applies a sophisticated QoS mechanism, a description of which can be found in Refs. [5,6]. In this section, the three most common categories of WiMAX services will be analyzed, namely VoIP, broadband data, and guaranteed bandwidth.

#### 13.3.3.1 VoIP Service

VoIP service is essentially the packet-based equivalent of existing circuit-based public switched telephone network (PSTN) voice service [7]. In this case, the VoIP dimensioning approach is based on existing methodologies, although there are differences on how the WiMAX air-interface internally handles the VoIP traffic. Analogous to telephony, the voice service characteristics such as total lines, active lines, grade of service (GoS), and traffic activity are estimated within the congestion point. In WiMAX networks the congestion point is a sector, which uses predefined air-interface resources out of which a dedicated portion should be available for the VoIP service. A failure to reserve these resources would result in further GoS degradation, which in the case of packet-based VoIP service could result in degradation of several or all ongoing sector calls. Moreover, it can safely be assumed that behind a sector (i.e., backhaul, core network) there is dedicated bandwidth to handle all VoIP and data incoming and outgoing traffic.

There are two paths to achieving VoIP service in a sector: to estimate the required reserved capacity for a given number of concurrent calls, or conversely, to estimate how many concurrent calls can be accommodated for a given available VoIP capacity. In both cases the "missing" variable is the required rate per call (duplex for time division duplex (TDD) systems and simplex for frequency division duplex (FDD) systems), which depends upon the selected voice codec [8]. The voice codec essentially transforms the voice stream into data packets, however on top of the data packet a number

**TABLE 13.1**

**Simplex Rate (Kbps) for ITU Voice Codecs**

| Voice Codec | Encoding Method | Data Rate | IP Rate | Ethernet Rate |
|---|---|---|---|---|
| G.729a | CS-ACELP | 8 | 40 | 48 |
| G.711 | PCM | 64 | 80 | 87 |
| G.723.1 | Multi-rate | 6.4 | 17 | 23 |

of headers are attached prior to the air-interface. These headers are related to the Ethernet and Internet protocols [9]. Indicative simplex rates are provided in Table 13.1, depending on the codec and necessary headers. The VoIP service results in symmetric downlink (DL) and UL traffic.

Table 13.1 indicates the data rate that should be considered in VoIP service calculations according to the network setup. Some WiMAX products apply header compression schemes and may slightly reduce the required data rates. Furthermore, in the near future the voice silence compression scheme will be applied, as soon as the Enhanced-rtPS traffic flow is developed in products, resulting in even lower rate requirement. Additional information concerning the header compression and more detailed codec characteristics (packet size, sample duration, and packets per second) can be found in Refs. [10,11].

The most common dimensioning requirement is to estimate the required VoIP capacity per sector, which is directed to the unsolicited grand service (UGS) or real-time polling service (rtPS) flow [5]. This estimation is based on the required number of concurrent calls in a sector, during busy hour. Given that VoIP service operates analogous to PSTN, the active calls $N$ can be estimated by utilizing the Erlang B Equation 13.1, and hence depend on the sector traffic activity $A$ and the target blocking probability (grade of service) $P_b$. The selection of Erlang B equation is based on three important assumptions: (1) Poisson call arrivals, (2) fixed or exponentially distributed call holding times, and (3) blocked calls are cleared.

$$P_b = \frac{\frac{A^N}{N!}}{\sum_{X=0}^{N} \frac{A^X}{X!}}$$

(13.1)

Consider a simple example, where each customer has two telephone lines with 20 mE, 40 mE, and 75 mE traffic activity per line and $P_b = 0.01$. Multiplying the subscribers, the lines and the traffic activity/line, the overall sector traffic activity is estimated. Using Equation 13.1, the number of active calls, $N$, is calculated. The graph of $N$ versus the number of subscribers in a sector is depicted in Figure 13.3. It can be observed that as the traffic activity per line increases, $N$ increases accordingly driving up the required sector throughput dedicated to VoIP. The through put is calculated by multiplying $N$ with the codec rate from to Table 13.1. However the true impact of passing VoIP packets through the air-interface is still not considered. VoIP packets are generally smaller (depending on codec data rate and sampling interval) compared to average data packet and hence increase the signaling overhead in the frame. A margin of up to 20 percent can be considered on top of the above estimated VoIP throughput to account for this effect. Some studies have investigated the performance of VoIP over WiMAX, [12,13], however since the scheduling methodologies are out of the scope of the IEEE 802.16e standard, each product manufacturer should provide the RF network designers with an insight into the scheduling methodology and VoIP service modeling recommendations.

### 13.3.3.2 Broadband Data Service

The impact of broadband data service on sector capacity can be interpreted in a straight forward manner. Equivalent to the ADSL service, wireless broadband Internet is offered to customers in

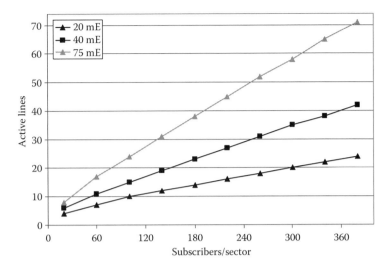

**FIGURE 13.3**  Active VoIP lines versus subscribers/sector.

terms of DL/UL peak information rate (PIR) in Mbps. This is the maximum subscriber throughput rate for a noncongested network. However, network design best-practice and investment costs impose that only a portion of the subscribers will use the resources concurrently, while broadband Internet is a non-delay-sensitive, best-effort (BE) service. It is common practice among operators to over-subscribe PIR, which means that during busy hour the subscribers will be restricted to lower rates. The committed information rate (CIR) which should be guaranteed to a subscriber during busy hour, as a performance indicator, is estimated by Equation 13.2 where $O$ is the data over-subscription ratio.

$$R_{cir} = \frac{R_{pir}}{O} \tag{13.2}$$

$$R_{total} = \sum (R_{cir,i} \times N_{s,i}) \tag{13.3}$$

It is common to have multiple data service profiles (i.e., for residential or business use) per service plan. The overall data throughput rate, $R_{total}$ is the summation of the per profile rates, $R_i$, each of which can be obtained by multiplying the profile's CIR, $R_{cir,i}$ with the corresponding number of subscribers, $N_{s,i}$. All the expressions for the service rates, $R$, denote the sum of DL and UL rates for TDD analysis and the higher rate between DL, UL, for FDD analysis.

### 13.3.3.3  Guaranteed Bandwidth Service

The guaranteed bandwidth service is a form of data service which exhibits strict QoS requirements, and is usually provided to corporate customers for establishing VPN or as leased line equivalent. This type of service is assigned to the UGS service flow to ensure priority among other services and constant availability and is usually not over-subscribed ($O = 1$). This type of service directly reduces the sector throughput (which is shared by other service flows) and hence can greatly impact the network design. Usually up to 20 percent of radio resources can be devoted to guaranteed bandwidth service.

### 13.3.3.4  QoS over the WiMAX Air-Interface

A main objective of dimensioning is to ensure that all services are reserved with sufficient throughput in a sector so that their QoS characteristics are preserved through the air-interface. The IEEE 802.16e standard, [6], has defined service flows with specific characteristics, such as the UGS, rtPS, nrtPS,

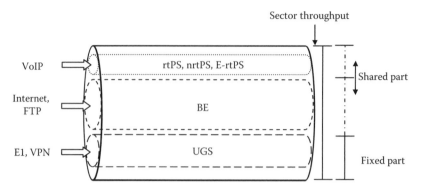

**FIGURE 13.4**  WiMAX air-interface QoS mechanism.

E-rtPS, and BE. All the services are mapped on one of the flows and accordingly have priority during the scheduling operation. The portion of throughput dedicated to each flow can be estimated by dividing the corresponding required rate (i.e., VoIP for rtPS) by the overall sector throughput. A mandatory rule is that the overall rate should be lower than the average sector throughput, as estimated under operating conditions. A safety capacity margin of 5 percent is usually included to avoid stretching the scheduling mechanism to its limits. An illustration of the WiMAX air-interface QoS mechanism is shown in Figure 13.4, where the sector throughput is shown as a big pipe which contains smaller pipes, which correspond to service flows. The portion of the sector pipe that corresponds to the UGS service is fixed and can be set during the WiMAX system configuration (i.e., 20 percent), while the remaining portion is shared among the other service flows dynamically. The rtPS, nrtPS, E-rtPS flows have priority over BE flow, althougth their rates can be limited by policy mechanisms such as call admission control (CAC).

### 13.3.4  COVERAGE DIMENSIONING

A primary objective when designing a WiMAX network is to provide radio coverage to a specified service area and type of subscribers. The purpose of coverage dimensioning is to ensure that a sufficient number of BS will be deployed and that the resulting coverage will satisfy the performance indicators. The process is simple: the service area ($km^2$) is divided by the cell footprint to produce the necessary points of presence (PoP) where a WiMAX BS will be deployed. The service area is defined in the business plan, however the cell footprint depends on the deployment scenario and product configuration/performance. To calculate the cell footprint, a very significant step is to estimate the maximum system range.

The maximum system range is defined as the range for which the system can achieve a performance threshold, usually in terms of received signal strength (RSS). RSS is estimated by Equation 13.4 and takes into consideration the system gains such as transmitter power, $P$, the antenna gains (per element, BF, MIMO), $G$, the signal processing gains (hybrid ARQ, repetition), $G_{sp}$, the system losses such as distance-dependent path loss with shadowing and fading, $L_d$, the penetration loss, $L_p$, and the the design margins (implementation, reliability, mobility, interference[*]), $M$.

$$S = P + G + G_{sp} - L_d - L_p - M \text{ (dBm)} \tag{13.4}$$

The RSS ($S$) threshold depends on the signal-to-noise ratio (SNR) threshold and the noise floor ($N_{th}$). For proper system operation there is an SNR value for which the decoding of the received signal results in lower than $10^{-6}$ bit error rate. Since the SNR thresholds depend on the modulation and

---

[*] Interference, for tight frequency reuse, reduces the maximum range by 1–3 dB.

coding scheme (PHY mode), the maximum system range that corresponds to the RSS of the lower scheme is considered i.e., QPSK. For Mobile WiMAX and considering a 5 MHz channel bandwidth a typical value would be around $S = -97$ dBm.

As mentioned above, the maximum system range depends on the deployment scenario and product configuration. For example the use of MIMO increases the antenna gains by several dB. In general for different terminals (fixed outdoor, indoor) the gains, losses, and margins can differ substantially. Another parameter, which is terminal independent, is the coverage certainty indicator. Such an indicator can be defined as "achieving a QPSK-link at 85 percent of locations in a cell and for 99 time availability," hence affecting the shadowing and fast fading components of the path loss.

Considering typical WiMAX deployment scenarios, the indicative DL maximum system range, based on Stanford University Interim (SUI) models, [14], is presented in Table 13.2. The DL power is $P = 30$ dBm, the element gain is 17 dBi and $SNR_{QPSK} = 4$. The terminal antenna gain is 16, 10, and 4 dBi, for outdoor, indoor, and mobile units, respectively. Column two shows the EIRP with MIMO 2x and BF 8x. It can be observed that the system range is highly dependent upon the deployment scenario and can range from several kilometers to just few hundred meters. A very important issue when defining the maximum range is to investigate both the DL and UL. In most cases the system gain is similar, however sometimes the UL achieves smaller range, and this should be taken into account.

The BS footprint, is estimated by the *operating* system range which, depending on the deployment scenario, can be the maximum or a percentage of it. For rural areas and outdoor terminals the maximum range can be used, however, for urban areas and mobile terminals a certain overlap among adjacent cells may be desirable for mobility and handover. Hence a margin of up to 10 percent can be applied. Another parameter that affects the footprint is the cell shape, the network layout and the number of sectors. A square cell shape is most common in fixed WiMAX networks, while the hexagonal cell is preferred for mobile WiMAX networks. The cell footprint can be derived by Equation 13.5 where $r$ is the operating range.

$$F_{sq} = 2r^2, \quad F_{hex} = \frac{3}{2}\sqrt{3}r^2 \qquad (13.5)$$

The result of Equation 13.5 provides nonadjacent cell overlapping footprint. To assume that the whole footprint is available, at least three sectors per cell are necessary. The relation between the operating range, the footprint, as well as an indication of the required PoP to cover a particular area of $100 \, km^2$ is highlighted in Table 13.3. It can be observed that the square cells have smaller footprint while, as the operating range increases even by few hundred meters the PoP are substantially

**TABLE 13.2**

**SUI-Based System Ranges for Various Scenarios**

| Scenario/Terminal | EIRP (dB) | $G_{sp}$ (dB) | $L_p$ (dB) | $M$ (dB) | Range (km) SUI-A | SUI-B | SUI-C |
|---|---|---|---|---|---|---|---|
| MIMO/outdoor | 50 | 5 | 0 | 10 | 3.8 | 5.2 | 7.9 |
| BF/outdoor | 56 | 9 | 0 | 10 | 6.2 | 8.8 | 13.5 |
| MIMO/indoor | 50 | 5 | 15 | 12 | 1.2 | 1.5 | 2.2 |
| BF/indoor | 56 | 9 | 15 | 12 | 1.9 | 2.5 | 3.7 |
| MIMO/mob./out. | 50 | 5 | 0 | 15 | 1.6 | 2.1 | 3.0 |
| BF/mob./out. | 56 | 9 | 0 | 15 | 2.6 | 3.4 | 5.2 |
| MIMO/mob./ind. | 50 | 5 | 18 | 15 | 0.7 | 0.8 | 1.1 |
| BF/mob./ind. | 56 | 9 | 18 | 15 | 0.9 | 1.3 | 1.9 |

**TABLE 13.3**

**Cell Footprint for Different Ranges and PoP for Covering a 100 km² Area**

| Operating Range (km) | Square Cells | | Hexagonal Cells | |
|---|---|---|---|---|
| | Footprint (km²) | PoP | Footprint (km²) | PoP |
| 0.7 | 0.98 | 103 | 1.27 | 79 |
| 0.8 | 1.28 | 79 | 1.66 | 61 |
| 1.0 | 2.00 | 50 | 2.60 | 39 |
| 1.4 | 3.92 | 26 | 5.09 | 20 |
| 2.0 | 8.00 | 13 | 10.39 | 10 |
| 2.8 | 15.68 | 7 | 20.37 | 5 |
| 4.0 | 32.00 | 4 | 41.57 | 3 |

reduced. Clearly during dimensioning and when the major objective is coverage for mobile terminals, the adoption of hexagonal cells results in fewer PoP, hence reducing the budgetary cost. On the other hand, when high capacity is needed, square cells have an advantage providing more capacity per km². Additionally, there are four candidate neighbor BS available to select the best server or perform capacity balancing.

A confusing situation is to select the operating range for hybrid coverage scenarios with more than one terminal profile or to adopt a scalability plan which dictates phase coverage upgrade. When hybrid coverage is required from network launch, then the operating range is determined by the worst case terminal profile/deployment scenario (refer to Table 13.2). Thus for hybrid outdoor/indoor, the indoor terminal can operate within a shorter range, which will be considered as the limiting factor when estimating the PoP footprint. When coverage is to be upgraded from fixed outdoor to indoor in the near future, a good approach is to dimension the network directly for indoor, however deploy a portion of the BS that would satisfy the initial outdoor coverage requirement.

### 13.3.5  CAPACITY DIMENSIONING

Further to providing adequate radio coverage to customers, the next equally important objective is to ensure sufficient air-interface capacity (throughput) to offer a wide range of services. Given a business plan, the network capacity is driven by the potential subscriber number and the committed rates for data and VoIP services, as discussed in services dimensioning. The purpose of capacity dimensioning is to translate the capacity into number of sectors, which then have to be distributed for the estimated number of PoP. The missing parameter is average operating sector throughput, which finally determines how many subscribers can be served in a sector. Defining the average sector throughput is quite complex as it involves interpretation of a wide range of parameters such as

- *Accurate equipment specifications*: PHY modes and SINR thresholds, advanced antenna system impact, interference mitigation mechanisms, scheduling intelligence, and terminal types.
- *Deployment considerations*: Available spectrum, reuse factor and channel assignment method, coverage type(s), propagation environment, and site suitability.
- *Designer capabilities*: Very good knowledge of the IEEE 802.16e standard, formal training or academic expertise in planning methodologies and their impact on practical networks, past deployment experience.

The sector throughput is usually provided as recommendation by the system vendor, however it can vary a lot depending on the deployment scenario. During dimensioning a more accurate

**TABLE 13.4**

**Ethernet Rates per PHY Mode (IEEE 802.16e System)**

| PHY Mode (CTC) | SINR (dB) | 5 MHz Bandwidth | | 10 MHz Bandwidth | |
|---|---|---|---|---|---|
| | | DL (Mbps) | UL( Mbps) | DL (Mbps) | UL (Mbps) |
| QPSK 1/2 | 2.9 | 1.4 | 0.9 | 2.7 | 1.7 |
| QPSK 3/4 | 6.3 | 2.2 | 1.3 | 4.3 | 2.7 |
| 16QAM 1/2 | 8.6 | 2.8 | 1.7 | 5.8 | 3.6 |
| 16QAM 3/4 | 12.7 | 4.3 | 2.6 | 8.6 | 5.4 |
| 64QAM 1/2 | 16.9 | 5.8 | 3.5 | 11.6 | 7.1 |
| 64QAM 3/4 | 18.0 | 6.5 | 3.9 | 12.9 | 8.2 |

estimation can be made through two approaches: (1) By extensive RF planning simulations and statistical processing that would result in a PHY mode regions map and give the average throughput. If clutter data are available this approach may provide a better view of the network performance. (2) By previous experience and interpretation of the above factors based on system performance observation. The former approach is time consuming, however it is much more accurate and is usually followed during actual network design. The latter is more appropriate during dimensioning.

Typically the equipment specifications include a table that matches the available PHY modes with their SINR requirement and the resulting Ethernet throughput (for different channel bandwidths) as in Table 13.4. The SINR values are given in the standard [6]. The provided Ethernet rates correspond to a 60/40 DL/UL asymmetry and hence the TDD rate is the sum of DL and UL rates. The upper throughput bound can be achieved for an interference-free sector where terminals are located only in the 64QAM, 3/4 region, achieving around 10 Mbps TDD for the 5 MHz bandwidth. During practical deployments, the terminals will be scattered across the whole cell footprint, hence operating in various modes. Furthermore, interference due to frequency reuse may further downgrade the PHY mode for a particular terminal, especially if the number of channels is limited. A default assumption is to consider, as average sector throughput, the one corresponding to 16QAM, 1/2. Thereafter if the deployment conditions are favorable, as in the case of fixed-outdoor terminals or when enough spectrum is available for relaxed reuse, higher throughput should be expected. The throughput can be also enhanced by means of MIMO techniques and this should also be taken into account. It should be noted that the values in Table 13.4 refer to full usage of subcarrier (FUSC) permutation scheme, where all subchannels are allocated to users, hence the whole channel is exploited. In case segmentation is considered, i.e., partial usage of subcarrier (PUSC), the users in a sector utilize a specific segment (1–6 subchannels), and therefore the throughput in this case is reduced accordingly. Based on the standard, there will be regions in the DL and UL subframes for both FUSC and PUSC and in this case an average throughput condition should be expected. In most products, during the sector configuration, an RF designer can select or exclude segmentation according to deployment conditions.

The process of selecting frequency reuse and channel allocation in sectors is very important for both capacity and coverage. In mobile WiMAX this can be done in a flexible manner, although frequency planning cannot be avoided. This is due to the possibility of nonuniform network layout, in most cases, where frequency planning may improve performance. According to mobile WiMAX terminology the reuse is denoted as 1.$x$.$y$, where $x$ denotes the cell sectors and $y$ the available channels. There are two main schemes under consideration: global reuse 1.$x$.1, where a single channel is used everywhere and cell/cluster reuse 1.$x$.$y$, where $y = nx$, $n = 1, 2, 3$. The most appropriate scheme is adopted, based on the systems's special capabilities to reject or tolerate interference (i.e., via BF). An indicative performance of the most common (Sections 13.3.1 and 13.3.3) schemes for nomadic/mobile terminals (most sensitive to interference) is presented in Figure 13.5. It can be

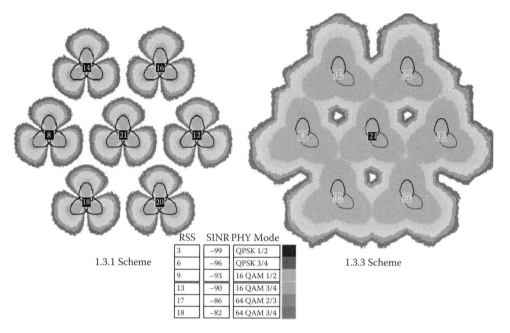

| RSS | SINR | PHY Mode |  |
|-----|------|----------|--|
| 3   | −99  | QPSK 1/2 |  |
| 6   | −96  | QPSK 3/4 |  |
| 9   | −93  | 16 QAM 1/2 |  |
| 13  | −90  | 16 QAM 3/4 |  |
| 17  | −86  | 64 QAM 2/3 |  |
| 18  | −82  | 64 QAM 3/4 |  |

1.3.1 Scheme                                                                 1.3.3 Scheme

**FIGURE 13.5**   SINR map for 1.3.1(FUSC) and 1.3.3(PUSC) schemes.

observed that for the 1.3.1 scheme the SINR drops well below the 3 dB threshold for the lowest modulation scheme, QPSK, hence a significant part of the cell footprint, especially among adjacent sectors, has no coverage. In the case of the 1.3.3 scheme, the interference appears closer to the cell edges and hence the coverage blanks spots are much smaller. It should be noted that in the 1.3.3 scheme the higher order PHY mode schemes extend to a larger region, hence indicating an improved sector throughput. Furthermore, when employing PUSC instead of FUSC, the 1.3.1 scheme behaves essentially as 1.3.3, while 1.3.3 as 1.3.9. It is typical for a sector to operate in 1.3.1 FUSC mode for terminals with good link quality and short link distance and in PUSC mode, which is equivalent to 1.3.3 for terminals that would otherwise achieve low SINR (due to low signal strength or interference).

Knowing the average sector throughput as described in previous paragraphs, capacity dimensioning can be completed as follows: Initially an analysis on the customers that can be accommodated in a sector is performed. This is done by analyzing the service plan and calculating the average data and VoIP CIR per service and customer. Then a graph that shows the throughput versus the subscribers is obtained, as in Figure 13.6. The average sector throughput is projected in the sector total graph indicating the corresponding subscriber number. Considering a 10 Mbps throughput, each sector can accommodate around 70 customers. Moreover a graph such as in Figure 13.6 is quite useful also during network operation as when the throughput takes a higher value, due to special circumstances, additional subscribers can be served. The second step is to divide the number of customers per area with the customers/sector to calculate the required number of sectors.

### 13.3.6   JOINT DIMENSIONING

In Sections 13.3.4 and 13.3.5, the number of PoP and sectors were estimated according to the requirements of a dimensioning project. The final step, as shown in Figure 13.2, is to combine these results into the optimum BS configuration. Clearly the estimated PoP and sectors are the absolute minimum according the needs of coverage and capacity, respectively. At this stage joint consideration may suggest that more PoP or sectors may be necessary. There are three possible conditions:

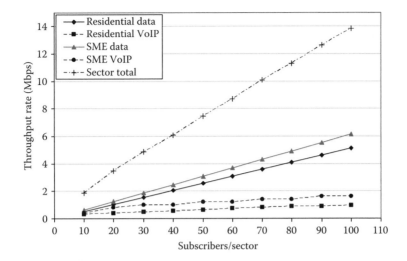

**FIGURE 13.6** Estimated sector and services throughput versus subscribers.

- *Balanced network*: The number of PoP approaches $1/x$ of the number of sectors, which means that in each PoP roughly $x$ sectors will be deployed. The number of sectors for blanket coverage should be $3 < x < 6$, where $x = 3$ for Mobile WiMAX. This condition ensures both the integrity of the footprint and satisfies the capacity requirement.
- *Coverage-limited network*: The number of PoP is quite higher than $1/3$ of sectors. The network is coverage limited and in this case the number of sectors should be increased until the previous condition is met. The fact that the original business plan leads to a coverage-limited network should be stated in the dimensioning study. CapEX is driven by coverage performance indicators, while the additional sectors will further increase the air-interface capacity. Operators may want to revise the size of the service area, or exploit the additional capacity.
- *Capacity-limited network*: The number of PoP is quite lower than $1/3$ of sectors, which indicates either additional PoP or higher sectorization (sectors/PoP). Increasing the number of PoP will trigger additional CapEX and OpEX in terms of site acquisition and preparation. Therefore, if a higher sectorization scheme is possible, such as when the terminals are fixed-outdoor, fixed-indoor, or nomadic where handover is not necessary, it should be preferred as a cost-optimum solution. When the network needs to accommodate mobile terminals and provide handover capability, the sectors should be $3 < x < 4$ and more PoP may be needed. An alternative approach would be to deploy a dual layer cell where 6 sectors of 120° are used, however each pair of sectors (i.e., 1 and 4) is assigned the same azimuth. For a dual layer cell at least 6 channels are necessary for the frequency reuse of 1.3.3.

The selected number of sectors per PoP defines the BS configuration in terms of frequency reuse/channel assignment, and antenna beamwidth/azimuth/tilt, while other air-interface parameters are not related to dimensioning. Capacity or coverage dimensioning should be revised based on the above-mentioned conditions, for coverage or capacity limited cases, respectively. A comparison between initial requirement and actual achievement should be included in the dimensioning study.

### 13.3.7 BACKHAUL DIMENSIONING INPUT

One result of the dimensioning of the access network is the required capacity by the backhaul network. A typical backhaul network consists of point-to-point or point-to-multipoint radio links or even fiber rings in recent networks. In most cases the information rate that is originated to and from the backhaul network should be calculated to evaluate an existing infrastructure or consider the cost to deploy a backhaul network along with WiMAX.

The total committed information rate of a WiMAX PoP is linear to the number of sectors and average sector throughput, which mainly depends on channel bandwidth and deployment scenario. Considering trisector cells, and average throughput of 9 Mbps (5 MHz channel bandwidth), the total TDD committed rate is 27 Mbps. To estimate the FDD equivalent, which is directed to and from the backhaul network, the total rate is multiplied by the DL/(DL + UL) ratio. Hence the traffic that should be reserved in the backhaul network for a PoP with 2/3 DL WiMAX traffic is 18 Mbps. For a dual layer cell or 10 MHz channel bandwidth the backhaul rate per PoP would be doubled, around 36 Mbps. It is acceptable to include a 5 percent margin as in practice it is common to observe small deviations in sector throughput. During the deployment and the operation the exact backhaul traffic can be monitored through the management system and design adjustments can be made accordingly.

In certain occasions the network designers are requested to propose a backhaul system along with the WiMAX network. Although this is not in the scope of this chapter, however there are some interesting observations that can be highlighted. Clearly the multipoint systems are more cost effective than point-to-point radio systems when the WiMAX PoP number is high, the PoP are quite close to each other and when their backhaul rate is such that more than 3–4 WiMAX PoP per multipoint sector can be served (around 10–12 per site). This indicates that the WiMAX site backhaul traffic should be from 15 to 25 Mbps, which mainly refers to trisector cells with 5 MHz channel. For higher bandwidth systems, such as dual layer cells or with 10 MHz bandwidth per site the point-to-point solution may offer the higher required capacity. Again in a dense urban environment where sites are usually deployed with separation of 0.3–0.7 km the use of fiber (if available) could be a better solution. Concluding, the best backhaul system strategy can be determined by evaluating the number of WiMAX PoP, their backhaul rate, and finally the PoP positioning. Note that although the WiMAX network dimensioning is done in such way to optimize the access network, this does not suggest that the backhaul network is also optimized and a separate study may be necessary.

## 13.4   TECHNOLOGY INTERACTION

Accurate interpretation of WiMAX technology is essential for the dimensioning process. By and large, WiMAX systems are required to conform to the IEEE 802.16e standard, however, parts of the system are intentionally unspecified to provide added-value and increase competition among product manufacturers. From the RF designer's point of view, the interpretation of WiMAX technology is already performed during the coverage (i.e., maximum and operating range estimation) and capacity dimensioning (i.e., average sector throughput estimation). From the operator's point of view the comparison is performed in terms of product capabilities. The interaction of evolving WiMAX technology in dimensioning is briefly described in this section.

### 13.4.1   EQUIPMENT SPECIFICATIONS

From a WiMAX access network design perspective, the most important parameters are related to the PHY and MAC characteristics of the air-interface. Typically all parameters that affect the link gain budget, such as transmit power, antenna gains, receiver sensitivities, advanced antenna systems are of great importance. Consider an example of a WiMAX access network that is intended to serve stationary indoor terminals in a typical suburban environment (i.e., SUI-C channel), hence a coverage-limited scenario. Comparing two systems with MIMO 2x and BF 8x configuration, it can be seen in Table 13.2 that the system range is around 2.2 and 3.7 km, respectively. This result is for the same amplifier output where the 6 dB system gain is due to the difference in antenna elements. Note that the diversity or BF gains are not considered for the range estimations since they are only applied in user traffic and not in the signaling part of the frame that usually limits the range. Applying

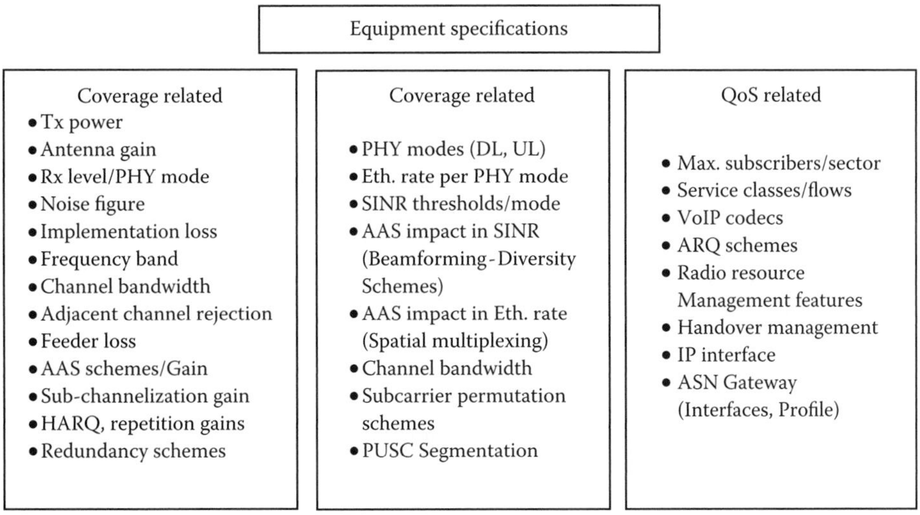

**FIGURE 13.7**   Important WiMAX air-interface parameters.

Equation 13.5, the footprint is estimated as 12.5 and 35.5 km², respectively. Results indicate that for coverage limited scenarios the higher system gain of a BF configuration can significantly reduce the network size. On the other hand, for mobility scenarios the use of MIMO is a more suitable approach. In general WiMAX has several air-interface profiles that may be best suited according to the deployment scenario.

In addition to the link gain budget, the sector capacity is equally significant for capacity-limited scenarios. A comparison of the spectral efficiency per modulation and the required SINR thresholds could indicate that some products may operate with higher efficiency than others. Capacity is further increased by advanced antenna systems, where spatial multiplexing could even double the spectral efficiency, while BF could reduce interference levels and upgrade the PHY mode.

A list of important parameters, with impact on coverage, capacity, and QoS are shown in Figure 13.7. These parameters should be provided for all combinations of operating frequency band-channel bandwidth, both for the DL and UL and for different terminal profiles. A vital parameter for dimensioning is the number of subscribers that can be supported in a sector due to availability of service flows. This number depends on the number of service types per user. Additionally, a description of the capabilities and performance of the radio resource management (scheduler) would be beneficial to the designers. It is common during dimensioning to assume that the system can preserve QoS but in many occasions a safety margin (i.e., 5 percent) is required. The scheduler's performance may downgrade close to full capacity load or for high number of subscribers per sector. Considering BE traffic, the impact is not as significant, however, this is not the case for VoIP and other delay-sensitive services. When handover is involved, it becomes more challenging to preserve QoS. It is evident from this section that WiMAX networks designers should have a thorough understanding of technology so as to foresee potential issues during network dimensioning and hence consider appropriate margins. A good insight into WiMAX air-interface performance issues, as well as explanation of basic concepts can be found in Ref. [15].

## 13.4.2   DIMENSIONING CERTAINTY AND MARGINS

As stated above, dimensioning accuracy is crucial. For vendors/integrators, it is essential that an RFP response should be financially prudent to increase chances for a contract award. However after the award, the network should be deployable, and considering that the majority of RFPs concern turn-key projects, a rough or underestimated offer can lead to miscalculation of the required network size and hence increase the implementation costs at the vendor's/integrator's expense. A balanced condition can be achieved by incorporating operating/performance and certainty margins. As discussed in Section 13.3.4 an operating margin is applied to coverage estimations, where the system range is selected smaller than the maximum. The same applies to capacity estimations where, for the average sector throughput, usually a small margin is considered. Considering the overlapping effect of various margins, a careless consideration can lead to overestimation. Specifically for coverage and capacity margins, as discussed in Section 13.3.6, the network will be either coverage-limited or capacity-limited. Therefore, in practice only one of the previously mentioned margins has impact on the financial offer. An important factor for defining margins is the quality of the provided business plan. The provision of extensive information would facilitate more accurate dimensioning.

It should be noted that in addition to the operating margins, another issue to consider is the network implementation margins. Although the RF designers take great caution to predict any possible causes of degradation, in many occasions problems may occur during the implementation. An engineering team, which is not well trained, or pays little attention to details, can make the difference between a successful and poor deployment. In cases where existing infrastructure is utilized, such as sharing of GSM sites, the condition of these sites or the restrictions posed by an operator usually cause problems. For example to save cost of antenna poles, operators may install WiMAX antennas below GSM antennas. This is contradictory to best practice since WiMAX operates in higher band and hence experiences higher propagation losses, and furthermore there is the issue of equipment RF isolation where a minimum separation distance should be maintained. Another example is with the installation of fixed-outdoor CPEs. Careless installation will result in suboptimum performance of such units, compromising their competitive advantage which is high capacity and robustness. In general, past deployment experience can make the difference in dimensioning, and it is preferred that the RF network designers which are involved in presales activities are the same that will have the responsibility of carrying out the final design and deployment supervision.

### 13.4.3  PRESENT-FUTURE TECHNOLOGIES

Over the last two years IEEE 802.16-2004 products have become more widely used and with several deployments in the field, basic experience with WiMAX technology has been increased. The new, upcoming IEEE 802.16e, which is an amendment to the IEEE 802.16-2004 standard, promises significant improvements. The major enhancements in WiMAX technology for the upcoming version of the standard are outlined in Table 13.5.

The most important enhancement concerns the system gain, which is roughly increased by 15–25 dB, and hence the cell range is also increased. For fixed/nomadic terminals it is evident that 802.16e is much more efficient and is also capable of catering for mobile terminals. Another decisive improvement is the introduction of AAS, which increases robustness through STC, MRC, and spectral efficiency through spatial multiplexing (SM) and BF. The transition to OFDMA clearly

**TABLE 13.5**
**Evolution of IEEE 802.16 Technology**

| Technology | IEEE 802.16d | IEEE 802.16e | IEEE 802.16j |
|---|---|---|---|
| Range (km) | 0.7 | 1.8 | 2 (per hop) |
| AAS | – | STC/MRC/SM/BF | Cooperation |
| Air-interface | OFDM | OFDMA | Relay-based |
| QoS | Basic | E-rtPS | Enhanced |
| Profile | IP Padios | ASN | ASN |
| Scenarios | Nomadic | Mobile | Mobile relaying |

boosts the radio resource management efficiency and improves QoS particularly in the presence of VoIP service. It is evident that the evolution of WiMAX targets three objectives: to increase the system gain and reach customers inside their homes/offices and on the move, to boost capacity so as to reduce service costs and be competitive with other access technologies, and finally to coexist in a seamless manner in the upcoming all IP networks.

This is clearly the case with the development IEEE 802.16j which introduces the concept of relays [16]. A relay-based network can in principle extend the range boundlessly, however, in practice 2-hop links are more likely to be implemented (for delay and throughput issues). From the designer's perspective if the first link (BS-relay) is line of sight then the system range can be several times higher than conventional systems, and hence for coverage-limited networks the dimensioning would result in much reduced costs. Furthermore, if the BS employs BF toward the relays, concurrent communication with several of them may be established in the form of spatial-division multiple access (SDMA), therefore boosting cell capacity.

While the all IP architecture is on the way (access service network (ASN) gateway), with major manufacturers of network products supporting this direction, the next WiMAX standard, IEEE 802.16m is also under consideration. IEEE 802.16m will revise the air-interface in the scope of international telecommunications union–radiocommunication sector (ITU-R) requirements for IMT-2000 and IMT-Advanced.

## 13.5 STRATEGY INTERACTION

The most significant part of dimensioning, which extensively affects the quality of the proposed solution, is the definition of the design approach or strategy. Defining dimensioning strategy is the most complex and challenging task since it involves an overall interpretation of various requirements and performance indicators during the network evolution phases, which may have little in common. Practice has shown that the network may evolve in time, not only in terms of size (expansion) but also in terms of terminal equipment and hence coverage type (upgrade to nomadic/mobile coverage) or a combination of these. Furthermore, it may evolve in terms of technology, where an upcoming standard may pose significant changes in the radio equipment. In this context, defining a strategy that ensures minimum additional investment and network upgrade complexity is the most important objective.

### 13.5.1 CUSTOMIZED NETWORK DESIGN

Customized network dimensioning provides a major advantage to vendors/integrators by highlighting their expertise and by indicating a cost-optimum solution. In many RFPs or project cases the developed business plan has extensive details which can be exploited in a very positive manner. A common occasion is that the overall service area is broken down into subareas with distinct characteristics, such as common type of customers and terminal profiles, common terrain, or service requirements. Such distinction allows a customized treatment of each subarea, where its characteristics are matched with an optimum solution, thus avoiding an overall rough approximation. The main customizations that can be applied in dimensioning, provided that the necessary information is available, are described below:

- *Nonuniform sectorization*: The use of different sectorization schemes, depending on the coverage type may result in reducing the required PoP. In some subareas there might be only SMEs, where a fixed-outdoor unit would be utilized. Therefore, in this case there is no restriction to use trisector cells. If the available spectrum is sufficient, up to eight sectors can be deployed in a PoP, depending on the capacity requirements.
- *Dual layer coverage*: In many occasions the subscribers will use both fixed-outdoor and mobile units hence in this case a more efficient approach is necessary. If extreme capacity is required, a high number of sectors can be used, building a trisector layer for the

mobiles and overlapping layer for the fixed, provided that frequency reuse can be applied. The neighboring list for mobile handovers will include only a sector in the first layer, while the demanding fixed links will be isolated in the second layer. This approach is very cost effective in cases where an extreme capacity demand would require a PoP every 0.2–0.5 km, if plain trisector cell layout was considered.

- *Nonuniform channel bandwidth*: Assuming that extreme capacity is still the performance indicator, by using higher channel bandwidth the capacity per PoP is increased and therefore the number of PoP is maintained at a reasonable level. While the overall network may be implemented with trisector cells, 5 MHz channel bandwidth and 1.3.3 reuse scheme (using 3 out of 4 available channels), a specific subarea with SMEs may be served by quad-sector cells, with 10 MHz bandwidth. In this case, the 10 MHz in the upper band will be assigned in sectors 1 and 3, while the 10 MHz in the lower band in sectors 2 and 4. This arrangement provides a 260 percent increase in capacity for the same footprint. It should be noted that the use of nonuniform channel bandwidth requires specialized sector configuration (transmit power, UL receive target level, and antenna arrangement) and extensive interference studies, applying the interference rejection filter methodology that takes into account the impact from trasmissions in all channels.
- *Hybrid coverage*: When both fixed-outdoor and nomadic/mobile coverage is required, the design approach is to select the operating range for the worst case condition. This usually leads to a huge upfront network size and investment. An alternative approach would be to consider that only a portion of the service area will be covered with the worst case terminal, hence the cell range can be selected higher. As the network evolves and additional PoP will be deployed for capacity upgrade, the percentage of the mobile coverage in the service area will also increase. This approach is particularly efficient for networks that extend to a large area, however with low subscriber density. The dimensioning is based on fixed-outdoor coverage that can reach higher ranges, however close to a BS there will be opportunity to use fixed-indoor/nomadic/mobile terminals also. This concept is highlighted in Ref. [17].
- *Divide and conquer principle*: To apply the previous customizations, the service area can be processed in pieces, where each subarea has a specific characteristic. If the segmentation of the service area is not available in the business plan, the designers can perform such action, by requesting or collecting more information (i.e., from site survey). This principle usually drives the network size up, and it is most useful for the actual design phase. The equipment quantities are increased in an exact manner because greater detail is taken into account during the actual design. This method approaches an optimum wireless network design.

It should be mentioned that the advances of the WiMAX air-interface allows greater flexibility and customizations during the wireless network design and hence balances the complexity of accommodating terminals with different profiles.

## 13.5.2 SCALABILITY

As mentioned previously, the most significant advantage of WiMAX is the flexibility and scalability of the air-interface. Therefore the concept "pay as you grow" is definitely applicable in WiMAX commercial networks. Scalability allows the reduction of initial investment and risks, and thereafter the network is expanded based on the market penetration and revenues. The majority of business plans provide the subscriber numbers and services per deployment year and in this context the dimensioning should be presented according to the scalability plan (network size/performance per year). The complexity in this case is not the increase of subscriber numbers or service area size, but when in parallel the coverage type is upgraded or new products (based on recently released standards) have to coexist with deployed equipment. The main types of scalability plan that have to be considered during dimensioning are presented as follows:

- *Network expansion*: Expansion may be in terms of subscriber numbers, service rates, or service areas or a combination and involves the deployment of new PoP or addition of sectors in existing PoP. A great challenge, in such scenario, is to optimize the positioning of PoP to achieve the best coverage and capacity outcome. It is more appropriate to determine the network size and PoP positions for the last deployment year, and then deploy only a subset of PoP that would satisfy the objectives of the initial phase.
- *Coverage upgrade*: It is common to allow the use of more demanding terminal profiles (i.e., nomadic/mobile) in the network after the first year so as to allow time to test the performance of the wireless network. In this case, not only the network is expanded from first to second year, but also the coverage should be upgraded too. The same approach "design for the future, deploy for present" as above should be applied, and the only difference is that a more dense network is probably required.
- *Technology upgrade*: The major change in terms of equipment is between the IEEE 801.16-2004 and 802.16e standards. The new products are based on software defined radio technology and therefore future standard releases will probably be implemented with minor changes. It is quite challenging to upgrade an existing fixed WiMAX network to coexist with mobile WiMAX, unless there is provision for additional spectrum. The major challenge is to replace subscriber equipment and restore access and it is likely that this transition phase will take a long time.

## 13.6   TECHNOECONOMICS OF DIMENSIONING

As already stated several times, the business plan and the dimensioning strategy are the main factors that affect the network size and overall investment. On top of the access network, a backhaul network should be implemented to connect the access with the core network, and there is also the core network infrastructure. The overall equipment depends on the number of PoP, almost in a linear manner. Further to the equipment costs, the deployment engineering costs should be also taken into consideration.

### 13.6.1   Cost Increasing Factors during Dimensioning

The dimensioning output may lead to an oversized/undersized network mainly for two reasons: either due to the business plan or due to a low quality study. In the first case it is the responsibility of the author of the business plan if it is not so realistic, while in the second case it is the responsibity of the designer if a study outcome is of low accuracy. The main parameters that impact the network size and hence the costs can be seen in Table 13.6. An ambitious business plan will most likely lead to a significant investment. However, in terms of network design and implementation, a huge

**TABLE 13.6**
**Dimensioning Size and Cost Increasing Factors**

| Factor | CapEX/OpEX | Complexity | Time-to-Market |
|---|---|---|---|
| Ambitious business plan | Very high | Very high | Very long |
| Unbalanced number of PoP | High | High | Long |
| Luck of scalability strategy | High | Very high | Long transition |
| Environmental changes | High | High | Long transition |

network increases design complexity which results in longer time-to-market. A scalable deployment is preferable since a higher design quality can be obtained. In contrast, lack of scalability would further extend the transition period until the network implementation is finalized. The costs and complexity also depend on the number of PoP. If the distance between PoP is very small (i.e., less than 0.5 km) then the design complexity is increased significantly. For capacity-limited networks that require too many sectors it would be cost effective to use higher sectorization per BS or a dual sector layer if possible, as described in Section 13.5.1. Finally it should be mentioned that for areas under heavy construction the expected terrain changes should be considered since they may result in a long design and implementation period. Any condition that would result in longer engineering times (design, implementation) would also further increase costs. Furthermore, terrain changes may alter the coverage and hence a revision may be necessary after a period.

## 13.7 CONCLUSION

Dimensioning is very significant process in designing WiMAX access networks. In this chapter, dimensioning was presented and analyzed in great detail. Although presented through a practical perspective, dimensioning can be very complex, particularly if specific details of the air-interface are taken into account to achieve maximum accuracy and hence optimum solutions. This chapter has aimed to inspire and guide the reader towards further enhancement of dimensioning, through analytical and pragmatic methodologies from academia and industry, respectively. Future work can be focused on three mains areas: (1) system capacity and impact of services, where a mathematical model for estimating average sector throughput could be developed. Furthermore, the coexistence of different services under realistic scheduling conditions could be investigated to estimate subscriber number that can be accommodated in a sector, for specific scenarios. (2) Impact of advanced antenna systems on coverage and capacity, which relate to propagation conditions and on frequency reuse. (3) Techno-economic analysis, where the spectrum versus equipment costs can be investigated. Finally, the presented methodology will have to be revised in the context of relay-based WiMAX networks which will probably be introduced in the next few years.

## REFERENCES

[1] WiMAX Forum white paper: Fixed, nomadic, portable and mobile applications for 802.16-2004 and 802.16e WiMAX networks, www.WiMAXforum.org. November 2005.
[2] WiMAX Forum white paper: Deployment considerations for fixed wireless access in the 2.5 GHz and 3.5 GHz Licensed Bands, www.WiMAXforum.org. June 2005.
[3] WiMAX Forum white paper: Deployments with self-installable indoor terminals, www.WiMAX forum.org. June 2005.
[4] WiMAX Forum white paper: Mobile WiMAX, Part 1: Overview and performance, Part 2: Comparative analysis, www.WiMAXforum.org. August 2006.
[5] IEEE 802.16-2004 Standard, www.ieee.org, 2004.
[6] IEEE 802.16e-2005 Standard, www.ieee.org, 2005.
[7] Goode B., Voice over Internet protocol, *Proceedings of the IEEE*, 90(9): 1495–1517, September 2002.
[8] ITU-T, G.7xx Voice Codecs, http://www.itu.int//rec/T-REC-g. March 2007.
[9] James Yu. Call admission control and traffic engineering of VoIP, in *Proceedings of 2nd International Conference on Digital Communications*, San Jose, CA, July 2007.
[10] Degermark M. et al., IP header compression, IETF RFC 2507, 1999.
[11] Casner S. et al., Compressing IP/UDP/RTP headers for low speed serial links, IETF 2508, 1999.
[12] Howon Lee et al., Performance analysis of scheduling algorithms for VoIP services in IEEE 802.16e systems, *IEEE VTC 2006*, 3: 1231–1235, May 2006.
[13] McBeath S. et al., Efficient signalling for VoIP in OFDMA, *IEEE WCNC 2007*, 2247–2252, March 2007.

[14] IEEE 802.16.3c-01/29r4, Channel models for fixed wireless applications, www.ieee.org, 2001.

[15] Andrews J. et al., *Fundamentals of WiMAX*, Prentice Hall, 2007.

[16] Pabst R., Relay-based Deployment concepts for wireless and mobile broadband radio, *IEEE Communications Magazine*, 42(9): 80–89, September 2004.

[17] K. Ntagkounakis, B. Sharif et al., Cost-efficient WiMAX network deployment: The hybrid outdoor/indoor dual-layer coverage approach, in *Proceedings of 17th IEEE PIMRC*, Greece, September 2007.

# 14 Network Planning for IEEE 802.16j Relay Networks

*Yang Yu, Vasken Genc, Seán Murphy,*
*and Liam Murphy*

## CONTENTS

In this chapter, a problem formulation for determining the optimal node location for base stations (BSs) and relay stations (RSs) in relay-based 802.16 networks is developed. A number of techniques are proposed to solve the resulting integer programming (IP) problem—these are compared in terms of the time taken to find a solution and the quality of the solution obtained. Finally, there is some analysis of the impact of the ratio of BS/RS costs on the solutions obtained.

Three techniques are studied to solve the IP problem: (1) a standard branch and bound mechanism, (2) an approach in which state space reduction techniques are applied in advance of the branch and bound algorithm, and (3) a clustering approach in which the problem is divided into a number of subproblems which are solved separately, followed by a final overall optimization step.

These different approaches were used to solve the problem. The results show that the more basic approach can be used to solve problems for small metropolitan areas; the state space reduction technique reduces the time taken to find a solution by about 50 percent. Finally, the clustering approach can be used to find solutions of approximately equivalent quality in about 30 percent of the time required in the first case.

After scalability tests were performed, some rudimentary experiments were performed in which the ratio of BS/RS cost was varied. The initial results show that for the scenarios studied, reducing the RS costs results in more RSs in the solution, while also decreasing the power required to communicate from the mobile device to its closest infrastructure node (BS or RS).

## 14.1 INTRODUCTION

IEEE 802.16 technologies are advancing rapidly. The base standards have been defined, industry is offering products to the marketplace, and network operators are deploying systems and offering service to subscribers. Mobile variants of the technology are now receiving a lot of attention: the WiBro system in Korea provides coverage for large population centers and has a substantial user base; the WiMax Forum is hosting a number of events designed to solve interoperability problems for 802.16e compliant systems, in advance of providing certification to products.

While all this momentum behind the technology bodes well, there are still significant issues that must be addressed before WiMax sees mass market adoption in many places throughout the world. In many countries, changes in spectrum allocation or regulations are necessary before WiMax systems can be rolled out. In some countries, the low cost of high-speed wired broadband connections make the business case for WiMax solutions less compelling. Subscriber terminals are still evolving, but the form factor of most current WiMax terminals is still quite large for mobility applications. So, while there are many reasons to believe that the future is bright for mobile WiMax technology, it will be some years before it reaches maturity.

One very important issue which is faced by network operators at initial network roll-out is how to provide maximum coverage at minimum cost. One approach which can be very useful in this context is to employ so-called relay network architectures. The essential idea is to use RSs, which are associated with BSs to effectively increase the coverage area of the BS at low cost. If the price point of the RSs is sufficiently low, this can result in a lower cost coverage solution than the traditional BS-based solution.

The 802.16 standards body has been developing standards for 802.16-based relay network architectures. More specifically, it is working on standards for BSs and RSs which enable them to work with legacy 802.16-2004 [1] and 802.16e [2] compatible devices—this will result in the 802.16j [3] standard. The 802.16j task group are focusing on issues associated with how to make minimal modifications to the signaling and frame structure such that operation with legacy systems is possible, while introducing a relay-based network architecture.

While the work of the 802.16j task group is still in the relatively early stages—the standard will most likely not be ratified until 2009 at the earliest—it is interesting to determine how the relay network architecture will impact network design and deployment. Because relays introduce a significant change to the network architecture, traditional network planning approaches are no longer applicable and new approaches are required.

In this chapter, new approaches to planning of 802.16j relay are described. The contributions of this chapter are twofold. First, there is a synthesis of previous proposals [4,5] to solve this problem—the basic branch and bound approach, the approach using state space reduction and the clustering approach. Second, there is an analysis of the impact of the ratio of BS/RS costs on the resulting solution.

The structure of the chapter is as follows. An overview of the main components of 802.16j is first provided. This is followed by a short discussion of radio propagation models suitable for this context. The main problem formulation is then presented, followed by a discussion of the different solution techniques. These different techniques are then compared in terms of scalability and solution quality and the results are presented. Finally, there is some analysis of the impact of the BS/RS costs.

## 14.2 OVERVIEW OF IEEE 802.16J

In IEEE 802.16j low cost RSs are introduced to provide enhanced coverage and capacity. Using such stations, an operator could deploy a network with wide coverage at a lower cost than using only (more) expensive BSs to provide good coverage, and increasing significantly the system throughput. As network utilization increases, these RSs could be replaced by BSs as required. The mesh architecture defined in WiMAX is already used to increase the coverage and the throughput of the system. However, this mode is not compatible with the point-to-multipoint (PMP) mode with no support of the OFDMA PHY, fast route change for mobile station (MS), etc. Hence, the standards organization has recognized this as an important area of development, and today a task group is charged with drafting a new standard: the IEEE 802.16j mobile multihop relay design to address these issues. The first draft of the IEEE 802.16j standard has just finished in August 2007.

### 14.2.1   IEEE 802.16J SCOPE

The IEEE 802.16j is aiming to develop a relay mode based on IEEE 802.16e by introducing RSs depending on the usage model:

- Coverage extension
- Capacity enhancement

In other words, the relay technology is first expected to improve the coverage reliability in geographic areas that are severely shadowed from the BS or to extend the range of a BS. In both cases, the RS enhances coverage by transmitting from an advantageous location closer to a disadvantaged SS than the BS. Second, it is expected to improve the throughput for users at the edges of an 802.16 cell. It has been recognized in previous 802.16 contributions that subscribers at the edges of a cell may be required to communicate at reduced rates. This is because received signal strength is lower at the cell edge. Finally, it is expected to increase system capacity by deploying RSs in a manner that enables more aggressive frequency reuse. Figure 14.1 illustrates the different scenarios in which relay mode could be used. However, introducing such RSs considerably alters the architecture of the network and raises many issues and questions. It is still unclear what system design is appropriate and can be realized at a low cost while still providing good coverage with an enhancement of the throughput.

The 802.16j task group's scope is to specify OFDMA PHY and MAC enhancement to the IEEE 802.16 standards for licensed bands. These specifications aim to enable the operation of fixed, nomadic, and mobile RSs by keeping the backward compatibility with SS/MS. In other words, the standard will define a new RS entity and modify the BS to support Mobile Multihop Relay (MMR) links and aggregation of traffic from multiple sources. An MMR link represents a radio link between an MMR-BS and an RS or between a pair of RSs. Such link can support fixed, portable, and mobile RSs and multihop communications between a BS and RSs on the path. An access link is a radio link that originates or terminates at an SS/MS. Table 14.1 illustrates the main scope of the project.

### 14.2.2   RELAY STATION CAPABILITIES

As the standard is still evolving, it is not clear what the final variant will look like. However, at present, it appears that two categories of RS will be defined: low capability RS (simple RS) and high capability RS (full function RS). The simple RS is used for low cost deployment, and operates on one OFDMA channel. It contains no control functionality (i.e., control functions are centralized in the MMR-BS) with one transceiver and optionally supports multiple input multiple output (MIMO). The full function RS can operate on multiple OFDMA channels, implement distributed control functions, and support MIMO. This type of RS has a further two variants: fixed/nomadic full function RS and mobile full function RS. Mobile RSs add support for handover and the ability to deal with a varying channel due to mobility. Table 14.2 summarizes the different RSs capabilities.

At present, it is considered that an MMR network could be composed of multiple usage models [7] including multiple RS types specifically deployed. But at present, there is only a little work about the heterogeneous functionalities of the RSs in different scenarios.

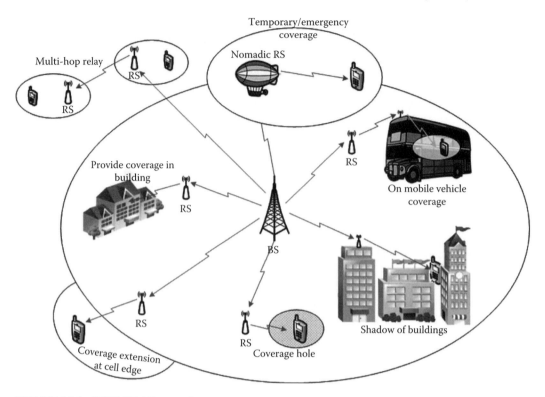

**FIGURE 14.1**   IEEE 802.16j example use cases.

**TABLE 14.1**
**IEEE 802.16j Project Scope**

|  |  | **Define New** |  |
| --- | --- | --- | --- |
| **No Change** | **Changes to BS** | **RS Entity** | **"802.16j Relay" Link Air Interface** |
| • To SS/MS<br>• To 802.16e OFDMA PMP link | • Add support for MMR links<br>• Add support for aggregation of traffic from multiple RSs | • Supports PMP links<br>• Supports MMR links<br>• Supports aggregation of traffic from multiple RSs | • Support fixed, portable, and mobile RSs<br>• Based on OFDMA PHY<br>• MAC to support multi-hop communication<br>• Security and management |

*Source*:   Extracted from Nohara, M., et al., IEEE 802 tutorial: 802.16 mobile multihop relay, March 2006.

**TABLE 14.2**
**RS Capabilities**

| | Simple RS | Full Function Fixed/Nomadic RS | Mobile RS |
|---|---|---|---|
| Number of OFDMA channels | 1 | $\geq 1$ | $\geq 1$ |
| Duplexing on MMR and access links | TDD | TDD or FDD | TDD or FDD |
| Frequency sharing between access and MMR links | Yes | Yes or No | Yes or No |
| Mobility | Centralized in MMR-BS | Centralized in MMR-BS or distributed in RSs | Centralized in MMR-BS or distributed in RSs |
| Antenna support | SISO or MIMO | MIMO | MIMO |

For example, an MS can move from the coverage provided inside a building by fixed/nomadic RS to a train where the coverage is provided by a mobile RS. Furthermore, there is no direct mapping between the usage models and the types of RS. An operator may deploy a variety of different RS types depending on traffic, mobility, topology (two hops or more) within the area of each RS location for a specific usage model.

In fact, the future standard will not answer all the issues raised by the RS incorporation to provide vendor differentiation. For instance, intelligent scheduling either at the BS (in a centralized approach) or at the BS and RSs (in a distributed approach) are required to minimize the interference that occurs at the RSs.

## 14.3  RADIO PROPAGATION MODELS

In any wireless network planning problem, the radio model is a key component. Because of the variety of the propagation environment, there is no universal propagation model. In general, radio models can be almost arbitrarily complex. However, working with such models can be very computationally intensive and it is important to find the model with the right balance of abstraction and complexity for the problem under study. For the WiMax network planning problems, two propagation models can be suitable and are described below.

### 14.3.1  FREE-SPACE MODEL

The free-space model [8] (originally published by H.T. Friis in 1946) is the simplest model that can only be applied in open area, i.e., no obstruction on the transmission line. This model is considered as a standard propagation model, a reference and benchmark of all other propagation models.

The path loss of the free-space model is

$$L_{fs}(f, d) = 32.44 + 20 \log_{10} f + 20 \log_{10} d \tag{14.1}$$

where
  $L_{fs}$ is the free space path loss in decibels
  $d$ is the distance between the transmitter and the receiver in kilometer
  $f$ is the frequency in MHz

In the free space model, many factors, such as reflection/multipath, shadowing, fading, atmosphere factors, etc., that may affect radio on its transmission path are omitted. This model, consequently, does not capture key transmission characteristics of radio, so it is not a very appropriate model for real world scenarios.

### 14.3.2 SUI Model

The Stanford University Interim (SUI) model was developed for design, development, and testing in the multipoint microwave distribution system frequency band [9] (2–3 GHz). It was recommended by the IEEE 802.16 standard body. The SUI model is valid for radio propagation within the 2–3 GHz range and has different parameter settings for urban, suburban, and rural scenarios. The maximum path loss (type A) is hilly terrain with moderated-to-heavy tree density. The minimum path loss (type C) is mostly flat terrain with light tree densities. The intermediate path loss condition is type B.

The SUI model is used for receiver's antenna height between 2 and 10 m. The path loss model is given by

$$L_{SUI}(d,f,h_m) = A + 10\delta \log_{10}\left(\frac{d}{d_0}\right) + X_f + X_h + s, \quad \text{for } d > d_0 \tag{14.2}$$

with the correction factors for the operating frequency and for the customer-premises equipment (CPE) antenna height of the model:

$$X_f = 6\log_{10}\left(\frac{f}{2000}\right) \tag{14.3}$$

$$X_h = -10.8\log_{10}\frac{h_m}{2}, \quad \text{for terrain type A and B} \tag{14.4}$$

$$X_h = -20\log_{10}\frac{h_m}{2}, \quad \text{for terrain type C} \tag{14.5}$$

where
$L_{SUI}$ is the SUI path loss in decibels
$d$ is the distance between the BS and the CPE antennas in meters, $d_0 = 100$ m
$h_m$ is the CPE height above ground
$s$ is a log normally distributed factor that is used to account for the shadow fading owing to trees and other clutter and has a value between 8.2 and 10.6 dB

The other parameters are defined as

$$A = 20\log_{10}\frac{4\pi d_0}{\lambda} \tag{14.6}$$

$$\delta = a - bh_b - c/h_b \tag{14.7}$$

where
$h_b$ is the base station height above the ground in meters and should be between 10 and 80 m parameters
$a, b, c$ are the constants dependent on the terrain type and are shown in Table 14.3.

The SUI model was chosen to be used in the following network planning models based on the following reasons: (1) the model was accepted by the IEEE 802.16 standard body; (2) it has a good compromise between simplicity and accuracy, i.e., it models the key characteristics of the radio frequency and it is simple, computationally with a relatively small number of parameters.

**TABLE 14.3**
**Constant Values for the SUI Model Parameters**

| Model Parameters | Terrain Type A | Terrain Type B | Terrain C |
|---|---|---|---|
| $a$ | 4.6 | 4.0 | 3.6 |
| $b$ | 0.0075 | 0.0065 | 0.005 |
| $c$ | 12.6 | 17.1 | 20 |

## 14.4 IEEE 802.16J NETWORK PLANNING PROBLEM FORMULATION

Here, a specific problem formulation for planning of multihop 802.16 networks is developed. The following inputs are assumed:

- A set of candidate BS and RS sites
- User demand, modeled by a set of discrete test points (TPs)
  - This approach has been widely used in previous work [9] and originally appeared in Ref. [11]
- A suitable propagation model
- A set of costs associated with BS and RS

The objective is to determine the set of BSs and RSs from the total set of candidate BS/RS sites that can accommodate the user demand at lowest cost.

The propagation model used here is the well-known SUI channel model. To use the model, it is necessary to define a number of parameters: terrain type, frequency of operation, antenna height, etc. In the experiments described below, a single set of parameters was used as in Table 14.4.

The height of BSs and RSs is the general height of a building, radio tower, or other constructions that can be mounted as BS/RS. The height of TP is average height of an adult. 2.5 GHz is the frequency recommended by WiMax forum for mobile WiMax systems.

### 14.4.1 INTEGER PROGRAMMING MODEL

The following problem inputs are defined:

- $S = \{1, \ldots, m\}$: Candidate site for BSs
- $R = \{1, \ldots, n\}$: Candidate site for RSs
- $T = \{1, \ldots, t\}$: TPs
- $c_j^b (j \in S)$: Cost of each BS
- $c_j^r (j \in R)$: Cost of each RS
- $u_i (i \in T)$: Traffic demand for each TP (number of connections)

**TABLE 14.4**
**Parameters Used in SUI Model**

| Parameter Name | Value |
|---|---|
| Height of BSs and RSs | Random value between 10 and 80 m |
| Height of TPs | 1.6 m |
| Frequency | 2.5 GHz |
| Terrain type | C, mostly flat terrain with light tree densities |

The set of BS sites differs from those of RS, as a BS is larger than an RS in size and has more functionalities. Also, the multihop concept is limited to nodes which are at most two hops from the BS: hence subscriber stations (SSs) can connect to an RS which is connected to the BS, or they can connect directly to the BS.

The gain matrices are determined based on the SUI model:

- $g_{ij}^b (0 < g_{ij}^b < 1, i \in T, j \in S)$: Propagation factor of the radio link between TP $i$ and candidate site of BS $j$
- $g_{ij}^r (0 < g_{ij}^r < 1, i \in T, j \in R)$: Propagation factor of the radio link between TP $i$ and candidate site of RS $j$
- $g_{ij} (0 < g_{ij} < 1, i \in R, j \in S)$: Propagation factor of the radio link between candidate site of RS $i$ and candidate site of BS $j$

The decision variables of the problem are a set of binary variables as follows:

$$y_j = \begin{cases} 1, \text{if a BS is installed in } j \\ 0, \text{otherwise} \end{cases} \quad \text{for } j \in S \tag{14.8}$$

$$z_j = \begin{cases} 1, \text{if an RS is installed in } j \\ 0, \text{otherwise} \end{cases} \quad \text{for } j \in R \tag{14.9}$$

$$x_{ij}^b = \begin{cases} 1, \text{TP } i \text{ is assigned to BS } j \\ 0, \text{otherwise} \end{cases} \quad \text{for } i \in T \text{ and } j \in S \tag{14.10}$$

$$x_{ij}^r = \begin{cases} 1, \text{TP } i \text{ is assigned to RS } j \\ 0, \text{otherwise} \end{cases} \quad \text{for } i \in T \text{ and } j \in R \tag{14.11}$$

$$r_{ij} = \begin{cases} 1, \text{RS } i \text{ is assigned to BS } j \\ 0, \text{otherwise} \end{cases} \quad \text{for } i \in R \text{ and } j \in S \tag{14.12}$$

It is now possible to write the objective function:

$$\min_{x,y,z,r} \left[ \left( \sum_{j=1}^m c_j^b y_j + \sum_{j=1}^m c_j^r z_j \right) + \lambda_1 \left( \sum_{i=1}^t \sum_{j=1}^m u_i \frac{1}{g_{ij}^b} x_{ij}^b \right) \right.$$

$$\left. + \lambda_2 \left( \sum_{i=1}^t \sum_{j=1}^n u_i \frac{1}{g_{ij}^r} x_{ij}^r \right) + \lambda_3 \left( \sum_{i=1}^n \sum_{j=1}^m \frac{1}{g_{ij}} r_{ij} \right) \right] \tag{14.13}$$

subject to the following constraints:

$$\sum_{j=1}^m x_{ij}^b + \sum_{j=1}^n x_{ij}^r = 1, \quad \forall i \in T \tag{14.14}$$

$$\sum_{j=1}^m r_{ij} = z_i, \quad \forall i \in R \tag{14.15}$$

$$x_{ij}^b \leq y_j, \quad \forall i \in T, \quad \forall j \in S \tag{14.16}$$

$$x_{ij}^r \leq z_j, \quad \forall i \in T, \quad \forall j \in R \tag{14.17}$$

$$r_{ij} \leq y_j, \quad \forall i \in R, \quad \forall j \in S \tag{14.18}$$

In the objective function (Equation 14.13), the first term constitute the cost of installing the BSs and RSs. The next two terms relate to the transmit power of the mobile stations—it is desirable to limit the required transmit power for the mobile devices since the mobile devices normally have less power and thus, the radio frequency can transmit shorter than that from a BS. The first of these terms relates to the transmit power of those devices that are communicating directly with the BS, while the second relates to devices that are using the relays. The final term ensures that RSs are associated with their closest BSs. The parameters $\lambda_1$, $\lambda_2$, and $\lambda_3$ are weight parameters which determine how much weight is given to each of these terms in the optimization process. Note that this formulation is somewhat independent of node transmit power—it simply tries to find a solution with lowest path loss between nodes.

The set of constraints are quite natural. Constraint 14.14 ensures every TP is assigned to either a BS or an RS. Constraint 14.15 ensures every RS assigned to only one BS; also if the RS is not installed, it cannot be assigned to a BS. Constraints 14.16 through 14.18 ensure that TPs are not assigned to BSs that are not present, TPs are not assigned to RSs that are not present and RSs are not assigned to BSs that are not present.

### 14.4.2 STATE SPACE REDUCTION

The above problem formulation is a 0-1 IP problem. Standard approaches can be used to solve this problem, such as the branch and bound algorithm. However, since it is an NP-hard problem, it can take huge amount of time to solve when the problem size scales up. To reduce the execution time, some more constraints could be added to reduce the problem state space. These extra constraints derive from understanding of the problem.

As each TP is connected to only a single RS or BS which is close to it, it is possible to add constraints which limit the set of RSs/BSs that any given TP can be associated with. In this way, the problem state space can be reduced considerably. There are two natural choices to determine whether it should be possible to allocate a TP to an RS/BS: the decision could be based on distance or path loss. In this work, the latter was chosen because the path loss reflects naturally the received signal quality. By means of close, a percentage parameter is set during the experiments and the first percentage number of TPs are considered close to the corresponding BS or RS (Refer to Section 14.5.2). The following matrices were then introduced:

$$f_{ij}^b = \begin{cases} 1, \text{TP } i \text{ is close to BS } j \\ 0, \text{otherwise} \end{cases} \quad \text{for } i \in T \text{ and } j \in S \qquad (14.19)$$

$$f_{ij}^r = \begin{cases} 1, \text{TP } i \text{ is close to RS } j \\ 0, \text{otherwise} \end{cases} \quad \text{for } i \in T \text{ and } j \in R \qquad (14.20)$$

$$f_{ij} = \begin{cases} 1, \text{RS } i \text{ is close to BS } j \\ 0, \text{otherwise} \end{cases} \quad \text{for } i \in R \text{ and } j \in S \qquad (14.21)$$

And the following constraints are added for this variant of the problem:

$$x_{ij}^b \leqslant f_{ij}^b, \quad \forall i \in T, \quad \forall j \in S \qquad (14.22)$$

$$x_{ij}^r \leqslant f_{ij}^r, \quad \forall i \in T, \quad \forall j \in R \qquad (14.23)$$

$$r_{ij} \leqslant f_{ij}, \quad \forall i \in R, \quad \forall j \in S \qquad (14.24)$$

Constraints 14.22 and 14.23 ensure that TPs can only be associated with nearby BSs or RSs, respectively. Constraint 14.24 ensures that RSs will only be associated with nearby BSs.

### 14.4.3 Clustering Approach

Clustering is a very standard technique for grouping entities together which are somehow related. In this case, these entities are related due to their geographical proximity. The main idea behind the clustering approach described here is to divide the larger problem into a number of smaller problems, each of which can be solved separately using the formulation above. The advantage of this is that the resulting time to find a solution is significantly lower. A clustering approach is used rather than more primitive methods of dividing the state space as it tends to result in fewer problems at the boundaries between the different clusters [12].

The clustering approach employed here comprises of three steps:

1. Use the standard $k$-means clustering based on a particular metric to divide the state space into $k$ separate clusters.
2. Use the problem formulation described above to solve the problem for each of the clusters independently.
3. Given the resulting set of RS and BS locations, perform a reallocation of TPs to RS/BS and RS to BS using another IP formulation.

The first two steps mentioned above are quite intuitive; the purpose of the third step is to address the problems that arise at the boundary. More specifically, it is possible that TPs or RSs at the boundary of a cluster are associated with an RS/BS within that cluster when there is a much closer RS/BS in a neighboring cluster: the third step above enables such points to be associated with nodes in other clusters.

The first step, generating clusters, was performed by generating clusters of nodes (TP, BS, RS) which have similar path losses to all other nodes (BS, RS) in the system. Note that this is a variant of an approach in which distance to other nodes is used as the metric.

A large gain matrix is first generated as input to the problem. This gain matrix comprises of all path losses between all nodes in the system as shown below:

$$M = \begin{bmatrix} A & B \\ C & D \\ E & F \end{bmatrix} \tag{14.25}$$

In which, $A$, $B$, $C$, $D$, $E$, and $F$ are all matrices as follow:

$$A = \begin{bmatrix} l_{1,1} & \cdots & l_{1,m} \\ \vdots & \ddots & \vdots \\ l_{t,1} & \cdots & l_{t,m} \end{bmatrix}; \quad B = \begin{bmatrix} l_{1,m+1} & \cdots & l_{1,m+n} \\ \vdots & \ddots & \vdots \\ l_{t,m+1} & \cdots & l_{t,m+n} \end{bmatrix};$$

$$C = \begin{bmatrix} l_{t+1,1} & \cdots & l_{t+1,m} \\ \vdots & \ddots & \vdots \\ l_{t+n,1} & \cdots & l_{t+n,m} \end{bmatrix}; \quad D = \begin{bmatrix} l_{t+1,m+1} & \cdots & l_{t+1,m+n} \\ \vdots & \ddots & \vdots \\ l_{t+n,m+1} & \cdots & l_{t+n,m+n} \end{bmatrix};$$

$$E = \begin{bmatrix} l_{t+n+1,1} & \cdots & l_{t+n+1,m} \\ \vdots & \ddots & \vdots \\ l_{t+n+m,1} & \cdots & l_{t+n+m,m} \end{bmatrix}; \quad F = \begin{bmatrix} l_{t+n+1,m+1} & \cdots & l_{t+n+1,m+n} \\ \vdots & \ddots & \vdots \\ l_{t+n+m,m+1} & \cdots & l_{t+n+m,m+n} \end{bmatrix} \tag{14.26}$$

where

matrix $A$ represents the gain matrix between TPs and BSs

matrix $B$ represents the gain matrix between TPs and RSs

matrix $C$ represents the gain matrix between RSs and BSs

matrix $D$ represents the gain matrix between RSs and RSs

matrix $E$ represents the gain matrix between BSs and BSs

matrix $F$ represents the gain matrix between BSs and RSs

Thus, the matrix $M$ has a dimension of $t + n + m$ by $m + n$.

The $k$-means clustering algorithm is then applied using the above gain matrix to obtain $k$ clusters: the nodes in each cluster are characterised by similar path loss to all RS and BS in the system. This results in $k$ distinct, nonoverlapping clusters, typically of comparable size for realistic node distributions.

In step 2, the standard branch and bound algorithm was used to obtain RS and BS locations for each cluster. This resulted in solutions for each cluster in which the TPs were allocated to RSs/BSs in that specific cluster.

A new problem formulation was developed for the final step to overcome the boundary issues. This problem differs from the original in that the RS and BS locations are now fixed, and the focus is on determining the relationships between TPs, RSs, and BSs: more specifically, which TPs should be allocated to which RS/BS and which BS each RS should be associated with.

The problem can be stated as follows:

$$\min_{x,r} \left[ \lambda_1 \left( \sum_{i=1}^{t} \sum_{j=1}^{m} u_i \frac{1}{g_{ij}^b} x_{ij}^b \right) + \lambda_2 \left( \sum_{i=1}^{t} \sum_{j=1}^{n} u_i \frac{1}{g_{ij}^r} x_{ij}^r \right) + \lambda_3 \left( \sum_{i=1}^{n} \sum_{j=1}^{m} \frac{1}{g_{ij}} r_{ij} \right) \right] \quad (14.27)$$

subject to

$$\sum_{j=1}^{m} x_{ij}^b + \sum_{j=1}^{n} x_{ij}^r = 1, \quad \forall i \in T \quad (14.28)$$

$$\sum_{j=1}^{m} r_{ij} = z_i, \quad \forall i \in R' \quad (14.29)$$

$$x_{ij}^b \leqslant y_j, \quad \forall i \in T, \quad \forall j \in S' \quad (14.30)$$

$$x_{ij}^r \leqslant z_j, \quad \forall i \in T, \quad \forall j \in R' \quad (14.31)$$

$$r_{ij} \leqslant y_j, \quad \forall i \in R', \quad \forall j \in S' \quad (14.32)$$

where $S'$ and $R'$ are the sets of BS and RS locations, respectively. As the formulation is very similar to that presented earlier, it is reasonably clear what the objective is and what the constraints represent.

## 14.5  RESULTS AND DISCUSSION

The objective of these tests can be divided into two parts. One is to obtain an understanding of the scalability of the problem formulation—the basic and the state space reduction model. More specifically, the objective was to understand if this problem formulation can be used to solve problems of realistic size. Given that it is, in principle, an NP-hard problem, it is important to understand the range of problems for which standard solution techniques are appropriate and the range of problems which require the development of heuristics which employ domain knowledge.

The second is to determine how the clustering approach compares with the more rudimentary approaches. The comparison was performed based on both the time taken to obtain a solution and the

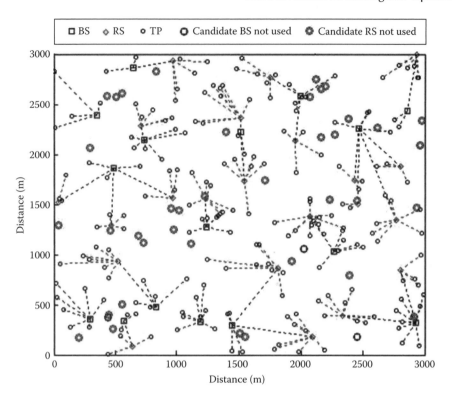

**FIGURE 14.2** A typical output of the planning tool.

quality of the resulting solution; naturally, the former relates directly to the scalability characteristics of the approach and its applicability for realistic scenarios.

A number of tests were performed in which the number of BSs, RSs, and TPs were varied. All tests were done using a standard desktop computer—Centrino Duo 2.0 GHz, 1 GB Memory, Windows Vista. Twelve tests were performed each time and the mean execution time taken. As there was some variation in the results, the minimum and maximum execution times were removed and the mean taken over the remaining ten results.

Problems were generated at random. The locations of each of the BSs, RSs, and TPs were chosen randomly from an area of size $3 \times 3$ km. The $(x, y)$ coordinates of each node were chosen by selecting two random variable from the distribution $U(0, 3000)$. For each of the problems the same set of weight parameters were used: $\lambda_1 = 8, \lambda_2 = 8$, and $\lambda_3 = 20$. However, it is worth noting that the values of these parameters have little impact on the time required to find solutions. In each of the problems, the BS cost was chosen at random and was three times the cost of the RS.

In all of the following tests, the branch and bound method found the optimal solution to the given problem. Figure 14.2 shows one possible result for planning a network with 20 candidate BSs, 60 candidate RSs, and 200 TPs. In the solution, 10 BSs are selected with 36 RSs.

## 14.5.1 INTEGER PROGRAMMING MODEL

Four sets of tests were performed with the basic variant of the problem to determine its sensitivity to different parameters.

In the first experiment, all three parameters were scaled—the number of candidate BSs, candidate RSs, and TPs. The number of BSs was varied and the numbers of RSs and TPs were three times and ten times this figure, respectively. Figure 14.3 shows how the time required finding a solution

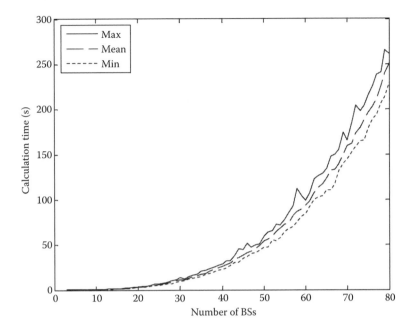

**FIGURE 14.3**  Calculation time when three parameters are scaled at the same time.

scales up. As it can be seen, the problem can be solved for up to 80 candidate BSs and 240 RSs with ease. Further, the results show that the problem complexity is scaling up quite rapidly. Indeed, further experiments were performed in which the number of candidate BSs was increased to 120 and the resulting execution mean time was under 30 min. The system is exhibiting scaling properties which are quite nonlinear, although some basic curve fitting has shown that for the available data set, the scaling is considerably less than exponential.

Figure 14.4 shows the calculation time when only the number of BSs is scaling. The number of RSs is set to 90 and the number of TPs is set to 300 in all tests.

A similar experiment was performed in which the number of RSs was scaled up and the number of BSs and TPs remained constant. Again it is clear that the system is scaling up linearly in this parameter (Figure 14.5). The number of BSs is set to 30 and the number of TPs is set 300 in all tests.

Finally, in this set of experiments, the sensitivity to the number of TPs was considered. The same characteristic is again observed: the system scales linearly as can be seen from Figure 14.6. The number of BSs is set to 30 and the number of RSs is set to 90 in all tests.

From the figures, it can be seen that this algorithm should suit small size network planning problems since the time cost is very short for small number of BSs. The time varies almost linearly if individual parameter is varying. For the problem sizes studied—which are typical for small metropolitan scenarios—the solution can be found quickly on typical desktop computers, e.g., under two minutes for problems with 50 candidate BS sites, and approximately ten minutes for problems with 100 candidate BS sites. The time cost for the planning could increase to one day long or a few days to plan a larger network, e.g., around 500 candidate sites, but it is still practicable.

### 14.5.2  EFFECT OF THE CONSTRAINTS REDUCING THE STATE SPACE

Based on the test results, the additional constraints do not affect the planning result obtained but shortened the execution time significantly.

Figure 14.7 shows the effect of the state space reduction in solving different problem sizes. The execution time of applying and not applying the state space reduction constraints are shown in the

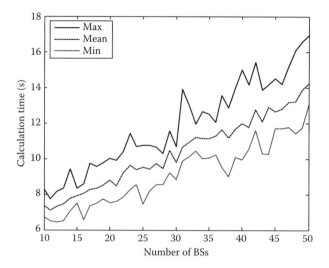

**FIGURE 14.4**   Calculation time when only the number of BS is scaled.

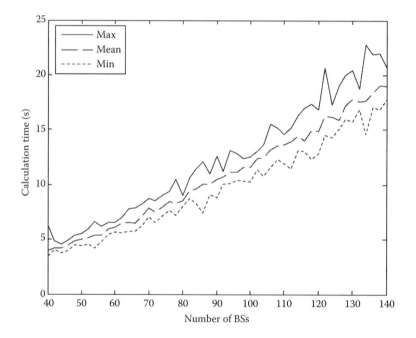

**FIGURE 14.5**   Calculation time when only the number of RS is scaled.

figure. The problem size varies depending on the number of candidate BS, which scaling between 30 and 75. The three parameters are scaling in the same manner as in Figure 14.3. It can be seen that it reduces half of the execution time.

Figure 14.8 shows the effect of the parameter which defines the term "close to" as in Section 14.4.2. The parameter defines the percentage of the state space is reduced. It vary between 0 and 75 percent which means the rest of the nodes, i.e., 100–25 percent of all the other nodes (TP/RS

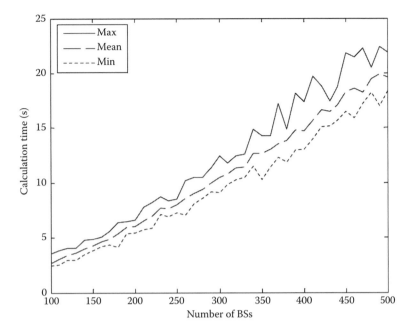

**FIGURE 14.6** Calculation time when only the number of TP is scaled.

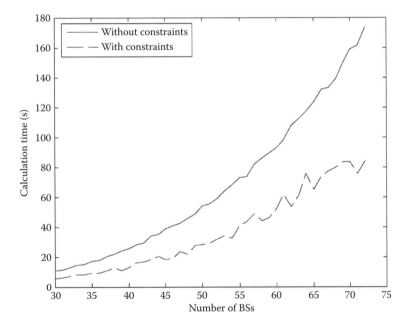

**FIGURE 14.7** Comparison of the calculation time, with and without the additional constraints.

that can connect to the node itself) are considered close to the node it self. In all tests, the number of candidate BSs is set to 80, the number of candidate RSs is set to 240, and the number of TPs is set to 800. From the figure, it is very obvious that the additional constraints could increase the performance and the more percentage of the state space is reduced, the more time can be saved.

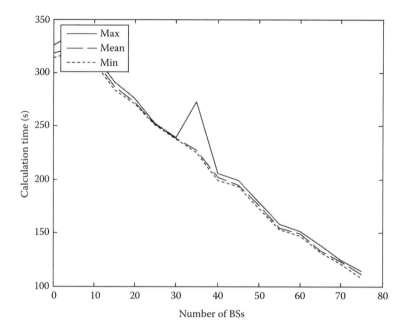

**FIGURE 14.8** Trend of the calculation time when different filtered percentage is applied.

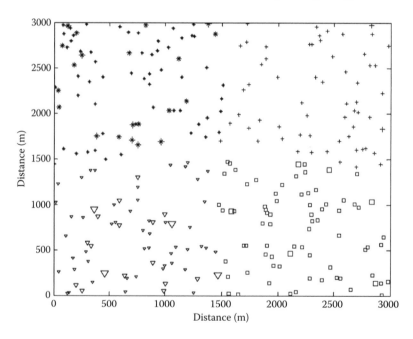

**FIGURE 14.9** A clustering output with 4 clusters in which different shapes represent different clusters.

### 14.5.3 USING THE CLUSTERING APPROACH

Before considering the final output of the clustering approach, it is interesting to consider how well the clustering algorithm works. Figure 14.9 shows a clustering output of a network with 20 candidate BSs, 60 candidate RSs, and 200 TPs. There, it can be seen that the clustering algorithm divides the nodes into four groups of approximately equal size.

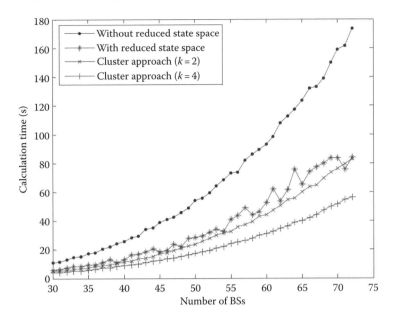

**FIGURE 14.10**  Comparison of the calculation time, with and without the state space reduction constraints and the clustering approach.

Once the clusters were obtained, the performance of the full approach was considered. Figure 14.10 compares the calculation time of using the clustering algorithm with that of using the basic model. Two variants of the clustering approach were considered: one in which the number of clusters was 2 and one in which 4 clusters were used.

The results show that the clustering approach results in significant improvements—the amount of time required obtaining the solution decreases by up to 60 percent. The results show that the 4-cluster solution operates significantly faster: this is to be expected, as it involves the solution of significantly smaller problems.

It is interesting to consider further the impact of the number of clusters. Figure 14.11 shows the calculation time of the whole clustering model with different $k$ values. All other parameters remain the same: 80 candidate BSs, 240 candidate RSs, and 800 TPs. When $k = 1$, the clustering model is the same as the basic model.

From the Figure 14.11, it can be seen that the calculation time drops as the number of clusters increases. As the number of clusters increases, the execution time drops to 20 percent of the nonclustered approach. It is worth noting, however, that the number of clusters reduces to have much impact above 4–6 clusters, for the problem size studied. This indicates that it is not necessary to have a large number of clusters to obtain significant savings: further, increasing the number of clusters does not result in further savings. It is anticipated, however, that larger problems could benefit from slightly larger numbers of clusters.

It is insufficient to consider execution time alone: it is necessary to consider the quality of the resulting solution. Figure 14.12 shows how the overall cost changes with $k$. The figure shows the variation normalized to the known optimal solution obtained via branch and bound. The scenarios are the same as those in Figure 14.3. While it is clear that $k$ does have an impact on the resulting overall cost—with increasing $k$ resulting in poorer solutions—the difference is so small as to be considered negligible. Further, there is a small improvement which results from the final step that reduces this difference.

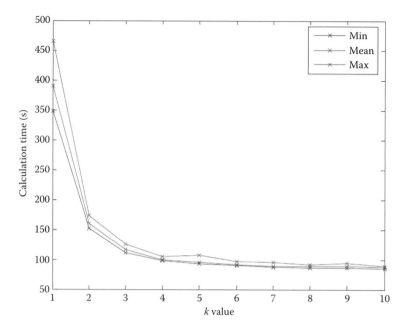

**FIGURE 14.11**   Calculation time for different $k$ values.

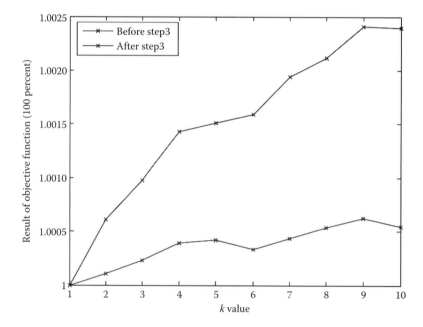

**FIGURE 14.12**   BS and RS costs for different $k$ values.

### 14.5.4   IMPACT OF THE RATIO OF THE COST OF BS TO RS ON SOLUTION

It is also worth to notice how the ratio of the cost of BS and RS affects the site selection. Intuitively, as the ratio raising, the RS becomes relatively cheaper, so it should tend to select more RS compared to the BS and connections from TP to RS should increase. Figure 14.13 shows the trend. It shows number of connections between TP and BS and between TP and RS as the cost ratio varying from

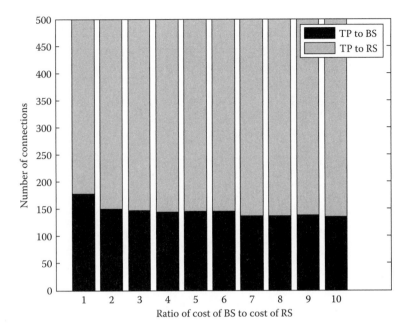

**FIGURE 14.13** Number of connections between TP and BS and between TP and RS as the ratio of the cost of BS and RS is varied.

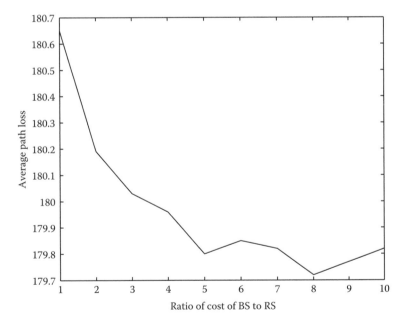

**FIGURE 14.14** Average path loss between each TP and its communicating node as the ratio of the cost of BS and RS is varied.

one to ten, i.e., from the cost of BS equals to the cost of RS to the cost of BS ten times the cost of RS. Figure 14.14 shows the corresponding average path loss between each TP and its communicating node. It can be seen that the path loss is decreasing which means the quality of the radio received becomes higher as the cost of RS becoming lower.

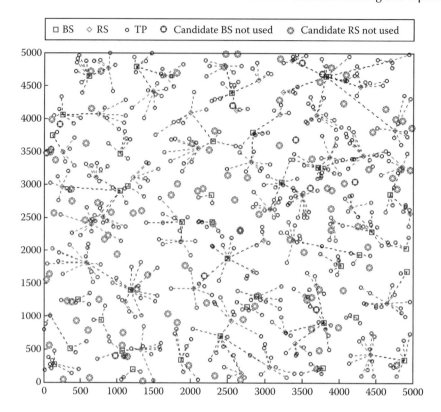

**FIGURE 14.15**  An output of the planning tool when the ratio of the cost of BS and RS is 1.

Figures 14.15 and 14.16 show two extreme cases. In both cases, the number of candidate BS sites is 50, the number of candidate RS sites is 150, and the number of TP is 500. Figure 14.15 shows the plan of the cost of BS equals to the cost of RS. In this case, there are 38 BSs and 50 RSs being selected; 177 connections between TP and BS; 320 connections between TP and RS. Figure 14.16 shows the plan of the cost of BS ten times to the cost of RS. In this case, there are 28 BSs and 75 RSs being selected; 133 connections between TP and BS; 367 connections between TP and RS.

## 14.6  CONCLUSION

In this chapter a model for planning 802.16-based relay networks is proposed. An IP formulation was developed and an investigation of the applicability of standard algorithms to this problem was performed. The results show that the standard branch and bound algorithm can find optimal solutions to problems of reasonable size on standard hardware. More specifically, these techniques can be used to solve planning problems for small metropolitan areas or areas of a city. Further, the results show that the time required to obtain solutions scales linearly with each of the individual parameters of the problem. However, when all parameters are scaled, the time complexity increases more quickly.

A simple state space reduction mechanism was also considered. While the performance of this can vary, it was found to reduce the computing time required by 50 percent for realistic cases. A clustering approach was also proposed and it was shown to deliver significant time improvements over the two previous approaches, finding solutions in 30 percent of the time required by the basic model with negligible impact on solution quality. The analysis found that the system has some sensitivity to the number of clusters used: for the size of problems studied, 4–6 clusters are optimal

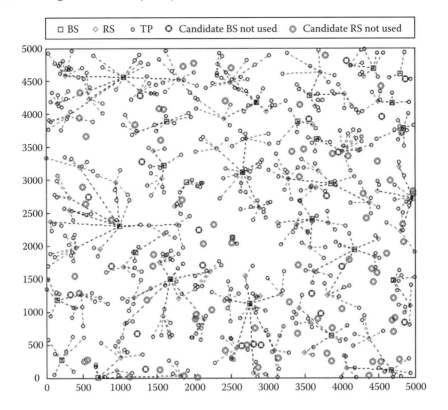

**FIGURE 14.16**   An output of the planning tool when the ratio of the cost of BS and RS is 10.

in terms of execution speed and quality of resulting solution. This new approach enables larger problems to be solved in realistic time on typical computing hardware.

Some analysis of the impact of the ratio of BS/RS cost was also performed. This analysis showed that as the cost of RS decreases (relative to that of the BS), the solutions comprise of more RSs. Further, the path loss between the mobile node and the infrastructure node is lower in the case that the RS cost is lower.

This initial work clearly leaves many questions unanswered. Future work will involve investigation of frequency reuse in this context, addition of QoS constraints to the model, further study of the impact of power constraints, and some investigation of the impact of the weighting parameters. Also heuristic techniques are necessary to be studied to significantly reduce computation complexity and present corresponding results.

## ACKNOWLEDGMENT

This work was supported in part by a UCD Ad Astra scholarship and an IRCSET postgraduate scholarship.

## REFERENCES

[1] IEEE, IEEE Std 802.16-2004, IEEE Standard for Local and Metropolitan Area Networks, Part 16: Air Interface for Fixed Broadband Wireless Access Systems, October 2004.
[2] IEEE, IEEE Std 802.16e-2005, IEEE Standard for Local and Metropolitan Area Networks, Part 16: Air Interface for Fixed Broadband Wireless Access Systems. Amendment for Physical and Medium Access Control Layers for Combined Fixed and Mobile Operation in Licensed Bands, February 2006.

[3] IEEE 802.16j MMR Work Group, http://www.ieee802.org/16/relay.

[4] Y. Yu, S. Murphy, and L. Murphy, Planning base station and relay station locations in IEEE 802.16j multi-hop relay networks, *Proc. 2nd IEEE Broadband Wireless Access Workshop, co-located with IEEE CCNC 2008*, Las Vegas, Nevada, January 2008.

[5] Y. Yu, S. Murphy, and L. Murphy, A cluster approach to planning base station and relay stations in IEEE 802.16j multi-hop relay networks, *Proc. IEEE International Conference on Communications* (*ICC 2008*), Beijing, China, 19–23, May 2008.

[6] M. Nohara, J. Puthenkulam, M. Hart, M. Asa, J. Cho, I.K. Fu, et al., IEEE 802 Tutorial: 802.16 mobile multihop relay, March 2006.

[7] J. Sydir, IEEE 802.16 broadband wireless access working group C harmonized contribution on 802.16j (mobile multihop relay) usage models, July 2006.

[8] V.S. Abhayawardhana, I.J. Wassell, D. Crosby, et al., Comparison of empirical propagation path loss models for fixed wireless access systems, IEEE VTC 2005.

[9] V. Erceg, K.V.S. Hari, et al., Channel models for fixed wireless applications, tech. ep., IEEE 802.16 Broadband Wireless Access Working Group, January 2001.

[10] E. Amaldi, A. Capone, and F. Malucelli, Planning UMTS base station location: Optimization models with power control and algorithms, *IEEE Transactions on Wireless Communication*, 2(5), 939–952, September 2003.

[11] K. Tutschku, Demand-based radio network planning of cellular mobile communication systems, *Proceedings IEEE INFOCOM'98*, vol. 3, San Francisco, CA, pp. 1054–1061, April 1998.

[12] H. Zhang, H. Gu, and Y. Xi, Planning algorithm for WCDMA base station location problem based on cluster decomposition, *Control and Decision*, 21(2), 213–216, February 2006.

# 15 Automatic Configuration and Optimization of WiMAX Networks

*Xuemin Huang and Jijun Luo*

## CONTENTS

Network planning meets its challenges along with the growing demand of data rates and the introduction of new air interfaces. The complexity of the optimized network planning increases due to a handful of constraints introduced by the system capacity, service quality, frequency features, mechanism for radio resource usage specified by the system and transceiving bandwidth, etc.

In this chapter, a method of the network design performance evaluation is introduced taking the characteristics of worldwide interoperability for microwave access (WiMAX) system into account. Features as AMC (Adaptive Modulation and Coding), packet switched radio connection, etc., have been considered to model the KPI (key performance indicator) for evaluation. The evaluation method is the kernel of several network engineering processes including the configuration and optimization subtasks.

Under the assumption of typical system configuration, optimal site selection and site configuration methods for WiMAX system are described. Using an advanced heuristic optimization algorithm, experimental results comparing different adjustable parameters are given that justify the feasibility and advantages of the introduced planning method.

## 15.1    INTRODUCTION

This section introduces the impacts and targets of network planning. By reviewing the most significant challenges such as traffic demand and quality of service (QoS) criteria, key WiMAX network planning process is explained.

### 15.1.1    Quality, Capacity, and Economic Issues of Network Design

The increasing demand for mobile communications leads mobile service providers to look for ways to improve the QoS and to support increasing numbers of users in their systems. Because the amount of frequency spectrum available for mobile communications is very limited, efficient use of the frequency resource is needed. Currently, cellular system design is challenged by the need for a better QoS and the need for serving an increased number of subscribers. Network planning is becoming a key issue in the current scenario, with exceedingly high growth rates in many countries which force operators to reconfigure their networks virtually on a monthly basis. Therefore, the search for intelligent techniques, which may considerably alleviates planning efforts (and associated costs), becomes extremely important for operators in a competitive market.

Cellular network planning is a very complex task, as many aspects must be taken into account, including the topography, morphology, traffic distribution, existing infrastructure, and so on. Things become more complicated because a handful of constraints are involved, such as the system capacity, service quality, frequency bandwidth, and coordination requirements. Nowadays, it is the network planner's task to manually place base stations (BSs) and to specify their parameters based on personal experience and intuition. These manual processes have to go through a number of iterations before achieving satisfactory performance and do not necessarily guarantee an optimum solution. It could work well when the demand for mobile services was low. However, the explosive growth in the service demand has led to a need for an increase in cell density. This in turn has resulted in greater network complexity, making it extremely difficult to design a high-quality network manually [1–3].

Furthermore, OFDM (orthogonal frequency division multiplexing) technology is emerging as an attractive solution for fast wireless access. It has been adopted for many future wireless networks, e.g., FLASH (fast low-latency access with seamless handoff) OFDM and WiMAX. The UMTS (Universal Mobile Telecommunication System) evolution will go into the direction of OFDM, e.g., LTE (Long Term Evolution). Similar to other technologies, the deployment of OFDM networks poses the problem to select antenna locations and configurations with respect to contradictory goals: low costs versus high performance. A key to successful planning is the fast and accurate assessment of network performance in terms of the coverage, capacity, and QoS [4]. This also makes the conventional design methods insufficient for planning mobile networks in the future. Thus, more advanced and intelligent network planning tools are required. A promising planning tool should be able to aid the human planner by automating the design processes.

### 15.1.2    Radio Planning Objective

The task of radio planning is to define a set of site locations and respective BTS (Base Transceiver Station) configurations with addressing the coverage and capacity figures derived from dimensioning. Dimensioning a new network/service is to determine the minimum capacity requirements that will still allow the GoS (Grade of Service) to be met. Site densities in each clutter type are one of the outputs. The site count, i.e., number of sites in a considered service area, derived in radio planning often differ from the site count derived from dimensioning since the actual site coverage may differ significantly from the assumed empirical model(s). There is always a risk that the planned site count may exceed the estimated site count from dimensioning. As a result several planning iterations are needed to reach a reliable figure.

One problem with radio planning deals with site density. Firstly, higher site density poses more difficulty in finding suitable candidates. This is true in all clutter types. In dense areas, most suitable sites are already overcrowded with 2G and 3G antennas. This will likely put the WiMAX antennas in less ideal positions. Secondly, there is a tendency that the candidate sites are not having comparable heights. This is a major drawback in radio planning because large differences in heights can distort the site dominance areas and cell ranges. The third problem is the bandwidth constraint which may require tighter frequency reuse. In this case, the radio plan must be as close as the ideal case.

Radio network planning normally follows the dimensioning exercise. Sometimes the dimensioning process includes a rough plan to justify the site count and coverage level using some commonly accepted propagation model and generic WiMAX system modules in the planning tool. In the actual planning phase, a number of inputs are needed to improve the quality and accuracy of the radio plan. Depending on the selected planning tool to use, a number of inputs maybe required to be fully utilized by the tool. For example, it is assumed that following items are already well considered:

- Propagation characteristics of various areas (propagation models tuned)
- Required inputs defined (clutter maps, terrain maps, building data, etc.)
- Traffic and demographic information, i.e., per clutter type
- WiMAX RF equipment parameters are defined (antennas, RF [radio frequency] features, etc.)
- Options for BTS configuration (sectorized, omni, PUSC [partial usage of subchannels], FUSC [full usage of subchannels])
- CPE (customer premises equipment) types and parameters defined (antenna types, mounting, diversity)

Two important decisions with regards to radio planning have to be considered prior to the actual planning exercise. Firstly, the level of accuracy when it comes to coverage and capacity needs to be considered and this highly depends on the accuracy of the propagation model in the planning tool. Secondly, the planner needs to decide how much RF optimization will be undertaken during the planning phase. This is only possible if the planning tool together with the planning parameters and equipments models are accurate enough. It is often the case where optimization is neglected during the planning process. Postplanning optimization exercise is often costly and produces only minor improvements. It is often limited to antenna adjustments (tilting and azimuth changes).

There are a number of features that are useful when selecting a planning tool such as

- Automatic frequency selection
- Optimal site selection—when existing or candidate sites are provided
- Support of mixed and multiple propagation models
- Support of model tuning and user defined models
- Support of OFDMA system including channel impairments
- Optimal downtilting
- Propagation parameters (or constants) for 2.5 and 3.5 GHz [5,6]

A number of commercial planning tools are available in the market. The major factor that determines the usability of the tool is the accuracy of the RF modeling such as propagation, BTS and CPE antenna models, interference prediction, frequency allocation, and channel models. Planning tools with OFDMA models for capacity planning are advantageous but not necessary since the capacity figures for each site of cluster can be estimated based on the signal quality outputs.

### 15.1.3 WiMAX Network Planning Process

WiMAX radio planning involves a number of steps ranging from tool setup to site survey. The process is similar to any wireless network. What differs between WiMAX and other technologies are the actual site configuration, KPIs, and the propagation environment as WiMAX may support mobile and fixed users where the latter may employ directional/rooftop antennas.

The final radio plan defines the site locations and their respective configuration. The configuration involves BTS height, number of sectors, assigned frequencies or major channel groups, types of antennas, azimuth and downtilt, equipment type, and RF power. The final plan will be tested against various KPI requirements mainly coverage criteria and capacity (or signal quality). Figure 15.1 can be used as a guide in developing a planning process. The planning process also largely depends on the planning tool used.

The planning process in Figure 15.1 includes measurements (i.e., drive test and verifications) after the site survey. This procedure is not mandatory for all sites if the site count is too large. Usually, site survey and the KPI analysis give an indication of which areas are expected to have poor RF quality and which sites are involved. This is usually done when the candidate site(s) are not located in ideal locations or if the site survey finds some discrepancies of the candidate(s).

The differences between WiMAX and 3G radio planning is discussed in Ref. [7]. WiMAX radio offers modest processing gain in a form of repetition coding and subchannelization. These features are only exploited when the signal quality demands more processing. To support high data rates, the radio plan must offer very good SINRs (signal to interference and noise ratio) even with very limited spectrum. For example, in the absence of subchannelization and repetition coding, the

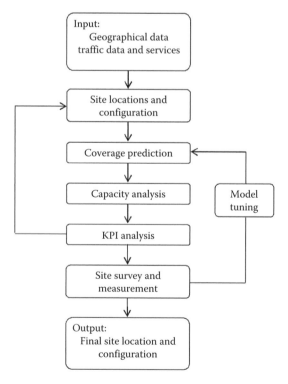

**FIGURE 15.1** WiMAX radio planning process.

required SINR for the lower MCS (modulation and coding scheme) is around 5 dB and this needs to be achieved even with very tight frequency reuse factor of 1/3 or 1/4 in the presence of shadowing, where the reuse factor is the reciprocal of the number the cells using different frequencies and the sum of the frequencies presents the whole spectrum resource allocated to the planned system. Another consideration in the case of WiMAX planning is the high SINR requirements to support high data rates. Although a site is expected to support high data rates for CPEs closer to it, SINR values $>30$ dB are only possible in the absence of interference. This requires accurate modeling of the propagation and RF equipments. For example, in 3G, high data rates are possible even with $C/(I+N)$ of $<10$ dB as the processing gain enables the receiver to tolerate some amount of interference. In WiMAX, this is not the case as the processing gain is only provided through channel coding and limited coding repetition.

## 15.2 WiMAX COVERAGE AND CAPACITY EVALUATION

The performance of a network design must be evaluated and the evaluation measures must be quantifiable and efficiently calculated during the optimization phase. Therefore, pointing out the desired KPI is the necessary step. We model the most significant KPIs according to WiMAX air interface characteristics with necessary performance evaluation.

In a WiMAX network, all measures of system load are time-averaged and scaled by the channel activity factor. The channel activity factor is defined as the ratio of the required user data rate and the maximum available throughput. As WiMAX supports AMC, the maximum available throughput is a function of the channel quality in terms of SINR. In Table 15.1, we give an example of names of some coding and modulation schemes together with the necessary SINR and achieved throughput. The values can be obtained by means of link level simulations. Interested readers are referred to Figure 6.3 in [13], Table 84 in [14], and Table 3 in [15] for more information.

### 15.2.1 Downlink Evaluation

#### 15.2.1.1 Downlink Coverage

In the downlink all links are transmitted at full power $P_{\alpha B}^{DL}$. The downlink is covered if the received power is sufficient.

$$P_{\alpha B}^{DL}/L_{Bk} \geq P_{\alpha B}^{\downarrow requiredRSS} \tag{15.1}$$

where $L_{Bk}$ is the pathloss between the best server, B and the terminal, $k$. $\alpha$ stands for one subcarrier under evaluation.

**TABLE 15.1**
**Resource Allocation Scheme**

| MCS | Net Throughput (Mbps) | SINR (dB) |
|---|---|---|
| QPSK 1/2 | 6.34 | 2.9 |
| QPSK 3/4 | 9.50 | 6.3 |
| 16QAM 1/2 | 12.67 | 8.6 |
| 16QAM 3/4 | 19.01 | 12.7 |
| 64QAM 2/3 | 25.34 | 16.9 |
| 64QAM 3/4 | 28.51 | 18.0 |
| 64QAM 5/6 | 31.68 | 19.9 |

The SINR requirement of the terminal is also checked.

$$(SINR)^{\downarrow}_{\alpha Bk} \geq (SINR)^{\downarrow \text{required}}_{\alpha B} \tag{15.2}$$

where $(SINR)^{\downarrow \text{required}}_{\alpha B}$ is the required SINR for DL bearer.

The DL SINR for best server, B is given by

$$(SINR)^{\downarrow}_{\alpha Bk} = \frac{P^{DL}_{\alpha B}/L_{Bk}}{N^{\text{thermal}}_{\alpha k} + \sum_{\beta} \sum_{J \neq B} A^{\downarrow}_{\alpha \beta} \bar{P}_{\beta J}/L_{Jk,B}} \tag{15.3}$$

where $N^{\text{thermal}}_{\alpha k} = k_o TW \eta_{\alpha k}$ is the thermal noise at the terminal, with $k_o$ the Boltzmann's constant, $T$ the temperature, $\eta_{\alpha k}$ the noise figure of the terminal $k$ at subcarrier $\alpha$ and $W$ the subcarrier bandwidth. And $L_{Jk,B}$ is the pathloss between a neighboring cell, $J$ and the terminal, $k$.

Let $A^{\downarrow}_{\alpha \beta}$ be adjacent carrier interference ratio. This is the power leakage from carrier $\beta$ to carrier $\alpha$. By definition, $A^{\downarrow}_{\alpha \alpha} = 1$. $\bar{P}_{\alpha J}$ is the time average TX power of the cell $J$.

$$\bar{P}_{\alpha J} = \phi^{\text{control}}_{\alpha J} P^{DL}_{\alpha J} + \sum_{k \in J} \phi^{\downarrow}_{k} P^{DL}_{\alpha J} \tag{15.4}$$

where $\phi$ is used to present the activity factor for both control signaling denoted by superscript "control" and user-plane traffic. It is defined as $\phi = \text{DataRate}/\text{MaxAvailableThroughput}$. $k \in J$ denotes terminal $k$ associated to cell $J$. This expression applies also for later equations.

The total DL time averaged transmit power is the sum of the time averaged control channel power and the time averaged traffic power transmitted by the cell. All links are transmitted at a power of $P^{DL}_{\alpha J}$. The use of the channel activity factor enables the modeling of an averaged power value in the time domain.

### 15.2.1.2 Downlink Capacity

The sum of the downlink channel activity factors of the served terminals within the best server, B is the downlink cell load, which should not exceed a certain load target $T_{DL}$.

$$\phi^{\text{control}}_{\alpha B} + \sum_{k \in B} \phi^{\downarrow}_{k} \leq T_{DL} \quad \text{and} \quad T_{DL} \leq 1 \tag{15.5}$$

The control signaling is modeled by specifying an activity factor, control for the period of time that the control channel is transmitting.

### 15.2.2 UPLINK EVALUATION

### 15.2.2.1 Uplink Coverage

In the uplink all terminals transmit at full power during the active period. Hence, the total UL TX power in the active period is $P^{UL}$. Whether or not a link can be established is determined by calculating the link budget in the UL and checking that the SINR obtained is greater than the required SINR defined for the bearer.

$$P^{UL}/L_{Bk} \geq P^{\uparrow \text{requiredRSS}}_{\alpha B} \tag{15.6}$$

$$(SINR)^{\uparrow}_{\alpha Bk} \geq (SINR)^{\uparrow \text{required}}_{\alpha B} \tag{15.7}$$

The UL SINR for the best server, B is given by

$$(\text{SINR})^{\uparrow}_{\alpha Bk} = \frac{P^{UL}_{\alpha k}/L_{Bk}}{I^{total}_{\alpha B}} \tag{15.8}$$

$$I^{total}_{\alpha B} = R^{out-cell}_{\alpha B} + N^{thermal}_{\alpha B} \tag{15.9}$$

$$R^{out-cell}_{\alpha B} = \sum_{\beta} \sum_{J \neq B} \sum_{k \in J} \frac{A^{\uparrow}_{\alpha\beta}\alpha^{\uparrow}_k P^{UL}_{\beta k}}{L_{Jk,B}} \tag{15.10}$$

$$N^{thermal}_{\alpha B} = k_o TW \eta_{\alpha B} \tag{15.11}$$

#### 15.2.2.2 Uplink Capacity

The sum of the uplink channel activity factors of the served terminals within the best server, B is the uplink cell load, which should not exceed a certain load target $T_{UL}$.

$$\sum_{k \in B} \phi^{\uparrow}_k \leq T_{UL} \quad \text{and} \quad T_{UL} \leq 1 \tag{15.12}$$

### 15.2.3 EVALUATION SUPPORTING FAST PLANNING

Similar to other technologies, the deployment of WiMAX networks poses the problem to select antenna locations and configurations with respect to contradictory goals: low costs versus high performance. A key to successful planning is the fast and accurate assessment of network performance in terms of the coverage, capacity, and QoS.

Snapshot analysis based on Monte-Carlo simulation is widely used for UMTS network performance evaluation. It can be adapted for a very detailed, flexible, and accurate analysis of WiMAX networks. However, it is not so efficient in terms of computational time and memory requirements, in particular for an application in search-based optimization techniques, which require a large number of different network configurations to be compared in a short-time frame.

In contrast to the coverage, which basically requires sufficient received signal strengths, the capacity of a network is much more difficult to assess: Due to the AMC mechanism used in OFDM systems, on one hand, the available throughput and implied channel activity of a user is determined by the channel quality that is represented by the SINR. On the other hand, the SINR is strongly impacted by the interference from neighboring cells and the strength of the interference power is mainly determined by the channel activities of interfering cells. This complicated interrelationship makes the load assessment of OFDM networks a very challenging task.

In Ref. [4] an analytical cell load calculation was proposed. A linear approximation is used to capture the behavior of AMC by adapting the effective data rate of a user to his channel quality. Thus, it gives a system of linear equations and conditions for the total channel activities and service probabilities of the cells, which characterize the cell loads and are not difficult to solve numerically.

## 15.3 SITE DEPLOYMENT AND CONFIGURATION

This section examines the site deployment problem with typical configuration assumptions for WiMAX network. Optimal site locations can be found in an automatic manner by applying site selection algorithms, in which a trade-off between coverage, capacity, and costs including CAPEX (Capital Expenditure) and OPEX (Operational Expenditure) is achieved. Note that the optimal site locations are found under limitations, therefore they are different to ideal locations.

## 15.3.1   SITE SELECTION

The site selection targets at optimizing mobile radio networks. It provides flexible methods to modify the configuration of a network in such a way that key performance measures are optimized and pivotal performance targets are met [8].

The working engine behind the site selection is a fast construction and solution of downlink or uplink cell equations for a given network configuration from which most performance measures can be deduced. Based on an analysis of the "permitted configurations," a preprocessing tree is constructed which contains all partial coefficients of the cell equations which might be relevant for the construction of the downlink/uplink/combined cell equations of an arbitrary permitted network configuration. This preprocessing tree then allows for an efficient accumulation of the coefficients of the equation system which in turn yields a fast calculation of the cell transmit/interference powers/activities from which most performance measures can be deduced. One can observe how many configurations are evaluated by looking at the number of evaluations in the status messages (which are issued about once in a second) when the algorithm is running after preprocessing.

Based on this fast evaluation, an optimization of the network is performed by evaluating all neighboring configurations of the current configuration and moving to the best of them if this improves the current network configuration, where the neighboring configuration deviates from the current configuration with short distance in the searching space. As long as an improvement can be obtained this optimization step is iterated. When no improvement can be found the last configuration is stored as the end design. On top of this basic procedure (construct preprocessing tree—perform descent to local minimum), there is the possibility for a repetition (since when many sites are switched off the coefficients in the preprocessing tree might become more and more irrelevant hence yield too conservative cell equations) and for a division of large problems into smaller ones, where one part of the network is optimized after another in a circular way.

A key role in the simple local descent iteration is played by the comparison operator between configurations. Here the algorithm allows choosing the relevant performance measures/constraints by assigning priorities to them. For example, one often assigns the monetary cost the highest priority and the coverage constraint the second highest (nonzero) priority.

The site selection can be used for optimizing cell parameters and selecting sites/cells. It can start from given configurations or empty/full networks. It can look through neighborhoods with selectable depths and iterate through these neighborhoods with a deterministic steepest or gradient descent, or a greedy random descent. However, this flexibility is at the price of a more complex user interaction. Through a large number of parameters the user has to tell the program what kind of optimization is intended. In addition, in large optimization problems, the preprocessing tree requires a huge amount of computer memory and if a large number of partial interferer coefficients are stored the evaluation can become slow. Partial coefficients which cannot be stored are accumulated in a so-called remainder coefficient. This remainder coefficient can accumulate a large number of small interferer and then makes sure that these interferers are not neglected in the cell equations. This saves computation time and computer memory, but might yield overconservative cell transmit powers/channel activities, when the number of stored coefficients becomes too small or when the interferer which belong to the stored coefficients are not active (e.g., is switched off). In these cases one might have to recalculate the preprocessing tree or allow for a larger number of coefficients within the nodes of the tree. Or one could divide a large problem into smaller ones which also reduces the preprocessing tree.

### 15.3.1.1   Candidate Sites and Permitted Configurations

The site selection algorithm starts from a description of the permitted network configurations. Such descriptions consist of a set of potential cells (configured cells which might appear in a network design) and subsets of these cells which we call selection sets and choice sets.

The meaning of a selection set is that all potential cells in this set have to appear together in a permitted network configuration or none of them. In a site selection, problem one can think of a selection set as the set of cells which belong to a site: One has to select all of them or none of them. Since selection sets can consist of a single cell, one can also define "cell selection" problems, where individual cells are selected by putting each potential cell into an individual selection set.

The meaning of a choice set is that exactly one of the potential cells in this set has to appear in a permitted network configuration. In a tilt choice problem such a set would consist of all potential cells which model the same cell but with different tilt settings. Defining such a choice set means to tell the optimization kit that it should choose exactly one of these cells for a solution design. Through a choice set which contains only one potential cell one can tell the program that this cell must appear in a solution design.

### 15.3.1.2  Optimization Algorithm

Given this definition of permitted configurations and the results of the preprocessing, the algorithm can walk through the permitted configurations and try to improve the performance measures of interest during this walk. The permitted configurations are the possible search choices for the optimization algorithm. The user can choose between several different manners of walking towards an optimized configuration by either "local steepest descent," "local gradient descent," "random greedy descent," or "local immediate descent."

The local steepest descent method evaluates all neighbor network configurations and then replaces the current with the best one found, where the comparison checks the selected performance measures exactly according to their priority.

The local gradient descent method evaluates all neighbor network configurations and then replaces the current one with the best one, where the comparison is based on gradients instead of the absolute values of the performance measures.

The random greedy descent method chooses randomly one new configuration in the neighborhood of the current one and moves to it if it performs better than the old one.

The local immediate descent is an exotic procedure where improvements are immediately accepted and the search continues from the improved configuration. This can speed up the algorithm, but at the cost of perhaps not giving as good results as the above more thorough search methods.

In all cases the walk moves from the current configuration to a configuration within a certain neighborhood of the current one. The neighborhood is defined as all configurations which can be reached by a given number of permitted elementary steps. An elementary step is either changing a given permitted configuration by switching one selection set on or off (adding its cells to the configuration or removing them) or modifying one cell in a choice set (replacing it by another one of the same choice set).

However, increasing the search depth to more than one, usually dramatically increases the computation time because a large number of configurations might have to be considered before accepting a move. This increase can be controlled through algorithm parameter.

### 15.3.1.3  Performance Evaluation and Comparison

There are two groups of performance measures: Ordinary performance measures and performance constraints, which additionally have a target value which should be reached. The ordinary performance measures are

- ActiveCells: The number of active cells in the evaluated network configuration. (Can be used instead of TotalFirstYearCost if detailed costs are irrelevant.) Small values are better than large ones.
- ActiveSites: The number of active sites. (Can be used instead of TotalFirstYearCost if the detailed costs of the base stations are irrelevant.) Small values are better than large ones.
- MaxOtherCoefficient: The maximum of the nondiagonal coefficients in the coupling matrix. Small values are better than large ones.

- MaxOwnCoefficient: The maximum diagonal coefficient in the coupling matrix. Small values are better than large ones.
- MaxSiteCostPerAccess: The maximum first year cost of one BS location (including hardware) divided by its sum of area and traffic access coverage in percent. Small values are better than large ones.
- MaxUserLoadInPercent: The maximum user load of a cell in the network. Small values are better than large ones. If this measure has high priority, the optimization seeks for a configuration in which the maximum user load of a cell is as small as possible.
- OverloadedCells: The number of overloaded cells. Small values are better than large ones.
- TotalFirstYearCost: The most important (nontechnical) performance measure in site selection, namely the total first year cost of a given network configuration. Small values are better than large ones.

The performance constraints are

- AreaCoverageInPercent: The percentage of the area which can be served (i.e., considered pixels which have a large enough receive signal strength from at least one of the active cells) by the network configuration. Its target value is the area coverage defined in the underlying optimization profile. Large values are better than small ones.
- AccessCoverageInPercent: The number of users on the served area of a configuration as a percentage of the number of users in the total considered area. The target value is the access coverage from the underlying optimization profile. Large values are better than small ones.
- MinCoverageGapInPercent: Here one looks at the differences of the area coverage in percent and its target value and the traffic coverage in percent and its target value. The smaller one of these two differences is the MinCoverageGapInPercent. Large values are better than small ones. The target value of this constraint is 0.
- TrafficCoverageInPercent: After solving the cell equations one can assess the number of customers which can be served by the network. The target value is the traffic coverage from the underlying optimization profile. Large values are better than small ones.

Only performance measures are calculated and displayed during the optimization for which the priority is strictly positive. The one with the largest positive value has highest priority, the one with the lowest positive value the lowest priority. The positive priorities should be different.

When two evaluations of network configurations are compared, first the number of violations (i.e., number of performance constraints, where the target is not met) is compared. The configuration with fewer violations is better than the other one.

Next, if both configurations have the same nonzero number of violations, the violation with the highest priority is considered. The configuration where the corresponding performance measure has a better value is considered as better than the other one. If both values agree, the performance measure with a violation and next highest priority is considered, and so on.

If there are no violations or the performance measures of all violated constraints agree, the performance measure with the highest priority is considered: The configuration which has the better value of this performance measure is considered to be better than the other one. If both values agree one looks at the performance measure with the second highest priority, and so on.

Since the steepest descent algorithm accepts only "better" permitted configurations, this evaluation first tries to reach the performance targets and then to optimize the high priority measures.

If the optimization method is set to local gradient descent, the decision whether a neighbor of the current configuration is better than another neighbor of the current configuration is slightly more complex: instead of the absolute values one considers gradients with respect to the current configuration.

### 15.3.2　Automatic Cell Configuration

During the optimization phase after the first rollout, the network is still subject to be further optimized. The optimization problem still remains nondeterministic polynomial-time (NP) hard [9]. The parameters can be finely tuned and targets are analyzed and modeled for suitable heuristic search algorithms. Multidimensional optimization process given by a number of configuration parameters is performed to configure the WiMAX system optimally.

#### 15.3.2.1　Optimization Parameters and Targets

The targets of the radio network optimization are mainly twofold. First target is to minimize the interference caused by the individual cells, while a sufficient coverage over the planning area is maintained. This is in general a trade-off and needs to be balanced, e.g., tilting down the antenna causes lower coverage, but also lower interference in neighboring cells and thus a potentially higher network capacity. Second target is the traffic distribution between cells. It is desirable to maintain similar cell loading of neighboring cells to minimize blocking probabilities and to maximize spare capacity for traffic fluctuations and a future traffic evolution.

The most effective parameter in network optimization is the antenna tilt. Antenna tilts need to be set such that the traffic within the "own" cell is served with maximum link gain, but at the same time the interference in neighboring cells is minimized. The possible tilt angles are typically restricted because of technical and civil engineering reasons. Especially in case of collocated sites with multiband antennas there might be strong restrictions on the possible tilt angles to be taken into account during optimization.

The transmitted pilot channel power and the other common channel powers, which are typically coupled by a fixed offset, are also vital parameters of network optimization. It needs to be assured that these channels are received with sufficient quality by all users in the serving cell. At the same time a minimization of the common channel powers yields significant capacity gains: Firstly, additional power becomes available for other (user traffic) channels, and secondly, the interference is reduced. The gains obtained from reducing the pilot power are often underestimated. It is important to note that in a capacity-limited WiMAX network (e.g., in urban areas) the reduction of pilot power levels by a certain factor also reduces the total transmit power of cells and as a consequence the cell loading by up to the same factor.

Optimization of azimuth angles of sectored sites is of great importance in particular in case of antennas with rather small horizontal beam-width (e.g., $65°$ vs. $90°$ in case of three-sectored sites). In this case the difference between antenna gains in direction of the main lobe and the half-angle between neighboring sectors is comparatively large, and cells of neighboring sites might need to be adjusted such that maximum coverage is achieved. It is observed that during optimization azimuth changes are in particular introduced to reduce coverage problems. For possible azimuth angles typically even stronger restrictions apply than for the tilt angles.

The antenna height is also often a degree of freedom for the optimization. Higher antennas can provide better coverage, but on the other hand also cause more interference in neighboring cells. Additional important parameters are the antenna type and the number of deployed sectors at a site. Both parameters are closely coupled, as a larger number of sectors also suggest the use of antenna pattern with smaller horizontal beam-width. The choice of sectorization is typically a trade-off between increased network capacity and higher monetary cost.

#### 15.3.2.2　Advanced Search Algorithm

The optimization method described in the previous section is well suited for an initial planning and in cases where the level of completeness or accuracy of input data is limited. For a detailed optimization that takes the full set of input data into account, a search approach is proposed [10]. That is, the space of possible configurations denoted as search space is explored to find the point in the search space that is optimal with respect to a certain criterion. This point is denoted as global optimum. Exhaustive search traverses the complete search space in a systematic manner. As all points in the search space are visited, with exhaustive search it is guaranteed to find the global optimum. The search space is very large for typical applications in network planning. For each site there can easily be several hundreds of possible configurations. Furthermore, configurations of different sites cannot be considered independently, so that the amount of possible network configurations grows expotentially with the number of sites. Hence targeting this area, an exhaustive search is too time-consuming and local search algorithms are commonly used for network optimization purposes.

Local search algorithms start at some point in the search space denoted as initial solution and subsequently move from the present to neighboring solutions, if they fulfill some criterion, i.e., appear to be better or more promising. Local search algorithms cannot guarantee to find the global optimum. The objective of the search algorithm—developed or tailored for a particular problem—is to find a solution that is at least close to the global optimum. Local search algorithms are thus often classified as heuristics [11].

The basic procedure of local search is independent of the actual search algorithm applied. Starting from the initial solution in each search step, first a search neighborhood is generated. The search neighborhood is a subset of points from the search space that are close to the current solution, i.e., that have some attributes in common with the current solution. The point that is most appropriate with respect to some criterion is selected from the search neighborhood and accepted as new initial solution for the next search step. If no appropriate new solution is found (or some other stop criterion is fulfilled) the search is terminated. The comparison of points from the search space is carried out by means of cost values associated with them. The cost values are generated from a cost function, in the literature often also referred to as objective function. The objective function maps a given point in the search space to a cost value. The cost value can be a scalar but could also be represented by a vector. In the latter case, an appropriate function to compare cost values needs to be defined.

Local search algorithms are very much application specific. However, several search paradigms have been developed in the last three decades. The simplest search paradigm is the descent method. This method always selects the solution from the neighborhood that has lowest cost. If this value is lower than the lowest value in the last search step, the solution is accepted as a new solution, otherwise the algorithm is terminated. The algorithm hence explores the search space by always moving in the direction of the greatest improvement, so that it typically gets trapped in a local minimum. A local minimum is a solution that is optimal with respect to the search neighborhood but which generally is worse than the global optimum. To escape from local minima, among several others, one widely applied approach is to carry out restarts, that is, the local search is restarted from a new solution that is selected from a different area of the search space. The new start solutions are often selected randomly. If restarts are applied, the algorithm strictly speaking is not a local search algorithm anymore.

Another option for escaping from local minima is to accept also the cost deteriorating neighbors under certain conditions. The most prominent local search paradigms that apply this strategy are simulated annealing and Tabu search [11,12].

Simulated annealing is based on an analogy with the physical annealing process. In simulated annealing improving points from the neighborhood are always selected when exploring the search neighborhood, nonimproving points are accepted as new solutions with a certain probability. The probability of acceptance is a function of the level of deterioration, but also gradually decreases during the algorithm execution. The reduction of the probability of acceptance is determined by the cooling scheme [12].

In contrast to the simulated annealing which comprises randomness, classical Tabu search is deterministic. The basic operation is equivalent to the descent method with the difference that the best point from the neighborhood is also accepted if it is worse than the current solution. In this way the search is directed away from local minima. To avoid a move back to already visited solutions, a Tabu list is introduced. The Tabu list typically contains sets of attributes of solutions that have already been visited. If a point from the neighborhood exhibits one of the sets of attributes stored in the Tabu list, the point is only accepted as new solution if its quality, i.e., cost, exceeds a certain aspiration level. The Tabu list is updated, keeping the individual entries only for a number of iterations. The size of the Tabu list is a very important design parameter of the Tabu search, it in particular needs to be chosen large enough to prevent cycling, but a too large list might introduce too many restrictions. Several enhancements to the basic operation of the Tabu search have been introduced most of which modify the handling of the Tabu list. These include intensification and diversification schemes [12].

### 15.3.2.3 Optimization Process

The basic structure of the local search algorithm that has been developed for an optimization of WiMAX networks is depicted in Figure 15.2.

The algorithm comprises the basic elements of a local search method that have been presented in the previous section. The local search starts from an initial solution, which can for example be the current configuration of the network to be optimized, a manually planned solution, or a solution suggested by the fast heuristics presented in the previous section.

At the beginning of each search step, the search neighborhood is generated. At first a cluster of cells is selected for which parameter changes are considered. Based on the selected cells the search neighborhood is generated and explored to yield a new solution.

The quality of a certain solution is assessed by a performance analysis. The choice of the method depends on the particular application and is a trade-off between accuracy and speed of the optimization process. From the results of the performance analysis, a cost value is generated by means of a cost function. The cost function is basically a linear combination of the evaluated quantities. In addition, penalty components are added if certain thresholds are exceeded for any of the different cost function components (e.g., coverage probability below some design target). The search process can be guided by appropriately setting weights for the different cost function components.

As a search paradigm either a descend method or a Tabu search can be applied. The Tabu list is maintained independent of the selected search paradigm. The search paradigm only influences the way in which the search is terminated. In case of the descend method, the search process is terminated if no improvement could be found. For the Tabu search, nonimproving moves are accepted to escape from local minima. The Tabu search is terminated once no improving moves were found for a certain number of search steps.

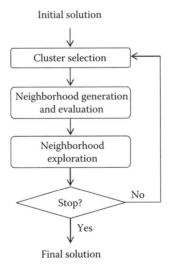

**FIGURE 15.2**    Local search algorithm for WiMAX network optimization.

The performance of the local search method strongly depends on the applied performance evaluation. The choice between the methods is a trade-off between accuracy and running time. The basic static method is the fastest but as was shown has some weaknesses in terms of accuracy of results. The presented statistical methods significantly outperform the latter method in accuracy of results but even if implemented efficiently are of higher computational complexity, especially if the experimental analysis is applied for evaluating quantities per pixel. The local search optimization presented in this section can be extended to yield a hybrid method which makes use of two methods for performance evaluation. The exploration of the neighborhood is split into two parts. In a first step, the neighborhood is explored by the use of a simple and fast basic performance evaluation method. As a result a list of candidate solution is generated. The list is sorted with respect to cost values. The next candidate solution is selected from this list using a more accurate but also more time consuming advanced performance evaluation. Either the first improving solution from the list or the best solution from a subset of most promising solutions ("short list") is selected.

### 15.3.3  EXPERIMENTAL RESULTS

We start with an experiment on site selection. There are 55 candidate sites as shown in the left of Figure 15.3. Each site has three cells. Without loss of generality, it can be assumed that all the cells have the same configuration as given in Table 15.2. The configuration of subscriber station is given in Table 15.3.

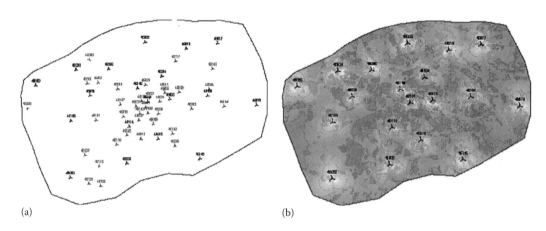

(a)                                                        (b)

**FIGURE 15.3**  Site deployment before and after optimization.

**TABLE 15.2**
**Cell Configuration**

| Parameter | Value |
| --- | --- |
| TX power (dBm) | 40 |
| Operating frequency | 2350 |
| Bandwidth (MHz) | 10 |
| Antenna gain (dBi) | 18 |
| Antenna 3 dB angle | 65° |
| Antenna height (m) | 40 |
| Receiver sensitivity (dBm) | −96 |
| Noise figure (dB) | 4 |

**TABLE 15.3**
**Configuration of Subscriber Station**

| Parameter | Value |
|---|---|
| TX power (dBm) | 23 |
| Antenna gain (dBi) | 0 |
| Antenna height (m) | 1.5 |
| Receiver sensitivity (dBm) | −100 |
| Noise figure (dB) | 7 |

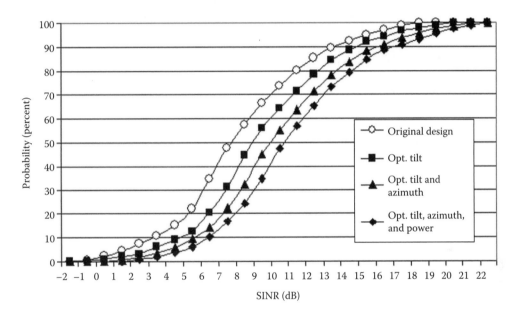

**FIGURE 15.4**   SINR CDF of different designs.

Inhomogeneous traffic distribution is assumed. Furthermore, we assume that all the candidate sites have equal cost in terms of CAPEX and OPEX. The target is to cover 95 percent of the planning area and 98 percent of the predicted traffic.

The site selection method described in Section 15.3.1 is applied in this experiment. The obtained result is displayed in the right hand side of Figure 15.3, which consists of 19 sites (57 cells). This design is (locally) optimal in the sense that one cannot remove a site without violating the coverage constraint, one cannot increase coverage by replacing a site by another site, and one cannot replace two sites by another site without violating the coverage constraint. This experiment shows that significant reduction of deployment cost can be achieved by using the site selection technology.

In the next step, this design with 19 sites (57 cells) is further optimized by applying the advanced search algorithm described in Section 15.3.2. The optimization is carried out by changing cell parameters including antenna tilt, azimuth, and TX power to improve the network performance. A main performance measure is SINR. Figure 15.4 shows the SINR CDF (Cumulative Density Function) of different optimization scenarios.

## 15.4 SUMMARY AND DISCUSSIONS

WiMAX network planning has to take a set of system parameters into account. The ray-tracing to be modeled into the pathloss model is one of the most important factors, that affect in addition the coverage and capacity analysis. Due to the frequency selective feature of the broadband communication, OFDM signal processing technology is of favor for physical transmissions. The time domain selectivity of the mobile communication system in addition requires an effective average of the SINR values over both the frequency and time domain. Furthermore, the features as Hybrid ARQ, AMC, and flexible channel allocation require the network planning not anymore straightforward as the classic network planning method. It has been analyzed in this chapter that the appropriate network engineering technology requires careful dimensioning methods including separating the planning phase, the performance evaluation phase, and optimization phase. To select a powerful optimization tool to search for the optimal reconfiguration in a mass combinations of possibilities is always placed as the first task before the commercial launch of the mobile network.

The network planning method described in this chapter is in principle applicable for other similar air interfaces. In fact, considerable commercial systems as IEEE 802.11x and 3G LTE use also similar technology components. There is a trend that network operator utilize a common platform embedded with multistandard multiband technologies. In that case, the proposed method can be installed in the common *Operation and Maintenance Center* for the automatic planning purpose for the multistandard system and becomes very useful.

- Optimization of antenna tilt
- Optimization of antenna tilt and azimuth
- Optimization of antenna tilt, azimuth, and TX power

Significant improvement of network performance can be observed when comparing the SINR CDF of optimized designs with that of the original design.

## REFERENCES

[1] Saleh Faruque, *Cellular Mobile Systems Engineering*, Artech House Mobile Communications Series, ISBN-0-89006-518-7, New York, 1997.
[2] X. Huang, U. Behr, and W. Wiesbeck, A new approach to automatic base station placement in mobile networks, in *Proceedings of International Zurich Seminar on Broadband Communication*, Zurich, Switzerland, 2000.
[3] X. Huang, Automatic cell planning for mobile network design: Optimization models and algorithms, Forschungsberichte aus dem Institut fuer Hoechstfrequenztechnik und Elektronik der Universitaet Karlsruhe (TH), Band 30, Karlsruhe, Germany, 2001.
[4] K. Majewski, U. Tuerke, X. Huang, and B. Bonk, Analytical cell load assessment in OFDM radio networks, in *Proceedings of the 18th Annual IEEE International Symposium on Personal, Indoor and Mobile Radio Communications (PIMRC'07)*, Athens, Greece, 2007.
[5] V. Erceg, L. J. Greenstein, S. Y. Tjandra, S. R. Parkoff, A. Gupta, B. Kulic, A. A. Julius, and R. Bianchi, An empirically based path loss model for wireless channels in suburban environments, *IEEE Journal on Selected Areas in Communications*, 17, 1205–1211, July 1999.
[6] V. S. Abhayawardhana, I. J. Wassell, D. Crosby, M. P. Sellars, and M. G. Brown, Comparison of empirical propagation path loss models for fixed wireless access systems, in *Proceedings of IEEE Vehicular Technology Conference (VTC2005-Spring)*, Stockholm, Sweden, 2005.
[7] B. Upase, M. Hunukumbure, and S. Vadgama, Radio network dimensioning and planning for WiMAX networks, *Journal of Fujitsu Science and Technology*, 43, 435–450, October 2007.
[8] K. Majewski, Network optimization kit, Technical Report, Siemens AG, Munich, Germany, 2006.
[9] I. Siomina and D. Yuan, Minimum pilot power for service coverage in WCDMA networks, *ACM/Kluwer Wireless Networks Journal (WINET)*, 14(3), 393–402, 2007.

[10] U. Tuerke, Efficient Methods for WCDMA Radio Network Planning and Optimization, Deutscher Universitaets-Verlag, Wiesbaden, Germany, 2007.

[11] C. R. Reeves, *Modern Heuristic Techniques for Combinatorial Problems*, McGraw-Hill, London, 1995.

[12] E. Aarts and J. K. Lenstra, *Local Search in Combinatorial Optimization*, John Wiley & Sons, New York, 1997.

[13] H. Liu and G. Li, OFDM-based broadband wireless networks, *Design and Optimization*, John Wiley and Sons, Hoboken, NJ, 2005.

[14] WiMAX Forum: Mobile system profile release 1.0, November 2006, approved specification.

[15] WiMAX Forum: Mobile WiMAX, Part 1: A technical overview and performance evaluation, August 2006.

# 16 Automatic and Optimized Cell-Mesh Planning in WiMAX

*Roberto Carlos Hincapié
and Roberto Bustamante Miller*

## CONTENTS

Fixed broadband wireless access is an alternative to provide coverage to remote rural users that cannot be covered by traditional wired solutions. We use automatic cell planning techniques to study different alternatives during the design process. We include packet transmission models, realistic topographic conditions, and point-to-multipoint (PMP) and multihop topologies. We consider features such as adaptive modulation and coding, which allows different link rates as a function of link conditions, but also adds additional complexities. Realistic topographic conditions make use of detailed cartography to include mountains and obstacles in the analysis. Finally, we include multihop topologies to show that under certain conditions, relay and mesh topologies can improve users coverage without an increase in the number of active sites.

## 16.1 INTRODUCTION

Wireless access is an alternative for remote users to connect to existent telecommunication networks. This concept is known as fixed broadband wireless access (FBWA), whose objective is to provide a high quality access to satisfy the requirements of users. The design process, known as *cell planning*, must search for the optimum configuration of the system given a set of requirements. We make use of *automatic cell planning* to automatically search for the best solution using an iterative search over different design alternatives. Automatic cell planing has been extensively studied for voice networks as it is described in Section 16.2. However, the additional characteristics of data networks make the design process even more difficult, but they provide additional alternatives for the solution.

We present an automatic cell planning model for wireless access networks, with realistic terrain and propagation conditions to give an insight into the rural wireless access problem. We analyze the wireless access system under different terrain and traffic conditions for point-to-multipoint (PMP) and multihop topologies. We show the conditions under which the number of connected users is increased with the use of multihop topologies as an extension to PMP topologies. We first present the related work and a detailed description of the problem we are trying to solve. After that, we discuss our model and the results we obtained.

## 16.2 RELATED WORK

The problem of rural wireless access has received a lot of attention from the research community. In Ref. [1] authors use WiMAX technology for community wireless access in Africa. Refs. [2,3] describe experiences to provide rural access in India using hybrid PMP and multihop topologies. The geographical distribution of users cause that a pure PMP solution is not optimal. In Ref. [4] a wireless mesh network (WMN) is suggested in a mountainous region where a PMP topology cannot provide full coverage. We divide the related work into two aspects: automatic cell planning is presented in Section 16.2.1 and fixed broadband access in Section 16.2.2.

### 16.2.1 AUTOMATIC CELL PLANNING

Wireless design has a high complexity because of the random characteristics and the shared nature of wireless medium. A cell planning example based on WiMAX standard can be found in Ref. [5], even though it is not based on automatic cell-based planning. A complete synthesis of automatic cell planning process is presented in Ref. [6]. The model uses a multiobjective function, built by a weighted sum of functions, each one representing signal coverage, capacity, system growth capabilities, and cost. The decision variables are channel assignment, sites location, and transmission power. Because of the nonlinear characteristics of this model, author uses genetic algorithms to solve it.

In Ref. [7], there is a general description of the optimization problem related to wireless network design. It presents a simplified model for global system for mobile communications (GSM) cell planning based on the activation or deactivation of a set of candidate base stations. In Refs. [8,9], there are other models based on multiobjective functions. Authors solve the problem by iteratively changing the transmission power used by base stations to guarantee signal reception and interference reduction. They use heuristic techniques and artificial intelligence algorithms in the solution process. Automatic cell planning based on artificial intelligence algorithms is also presented in Refs. [10–12]. Reference [13] makes a comparison among different techniques, showing a better performance of *tabu search* over the other techniques. In Ref. [14] authors allow the variation of the height of antennas and the transmission power by using genetic algorithms, but they do not consider capacity criteria. Particle swarm optimization is used in Ref. [15], with an optimization criteria similar to that of Ref. [6].

Two heuristic techniques to solve high complexity nonlinear optimization problems are tabu search [16,17] and simulated annealing [18]. In Ref. [19] authors use tabu search to solve an integer linear programming problem. In Ref. [20] authors describe a design process which is similar to ours. It uses simulated annealing to choose active base stations from a set of candidate base stations. A similar problem is solved in Ref. [21] using simulated annealing too.

Previous references are oriented to cellular networks to provide voice services. WiMAX [22] networks support different adaptive modulation and coding (AMC) schemas according to link quality. It also defines different types of connections ranging from a circuit-like access to a completely random access. In WiFi networks, there are different link conditions as in WiMAX, but there are not different types of flows. Most of the references for the design of WiFi networks, use the position of access points, their transmission power, and the channel assignment as the decision variables. In Ref. [7] there is a simple but illustrative description of the problems involved in wireless LAN design. In Ref. [23], authors present a genetic algorithm to solve channel assignment in Wireless LAN Networks. In Ref. [24] the algorithm modifies transmission power of fixed access points to react to changes in user traffic requirements. The model described in Ref. [25] uses a heuristic search model to provide coverage and a minimum data rate at test points. In Ref. [26], authors solve the joint problem of access points location and channel assignment. In Refs. [27,28], there is a good description of Wireless LAN Network planning. They use a penalty function to avoid placing access points near each other, to increase the probability of a posterior feasible channel assignment solution. The objective function is a weighted sum of a coverage variable, an interference mitigation variable, and a QoS variable. Authors use tabu search to solve the problem.

In Ref. [29], authors consider the problem of locating relay nodes to improve access point covered area. The decision variables are user nodes and relay nodes location. PMP and multihop topologies can be mixed as it is described in Ref. [30], where an IEEE 802.16 wireless mesh network interconnects a cell-based IEEE 802.11 access network. In more dynamic scenarios, the design process must prepare a feasible scenario for operation. In Ref. [31] authors discuss several issues for channel allocation and transmission scheduling and in Ref. [32], a connection admission control and a transmission power control for WiMAX networks is presented.

### 16.2.2  FIXED BROADBAND WIRELESS ACCESS AND WiMAX

A complete state of the art in FBWA can be found in Ref. [33]. We consider networks with characteristics similar to WiMAX and IEEE 802.16 [22] networks. Transmission is divided into frames of equal duration which are also divided into information units known as slots. A base station uses some overhead slots for operation and configuration and assigns the rest of them to schedule users' transmissions. It supports AMC, so different users will have different link rates. To achieve a certain data rate, users would need a number of time slots that is inversely proportional to their link rate.

In PMP topology, a base station is responsible for scheduling transmission periods to every user, according to their traffic requirements and contract guarantees. WiMAX allows different types of flows guarantees that range from a capacity/delay guarantee flow, known as unsolicited grant service (UGS) to a nonguarantee flow known as best effort (BE). Each flow type defines a set of requirements for different applications. Standard IEEE 802.16-2004 [22] allows operation in both PMP and mesh modes. Relay topologies are under study by the IEEE 802.16j study group [34]. Networks based on mesh topologies with fixed users are also known as WMN. WMN and relay networks use multihop paths to reach additional users. Relay topologies are a special case of mesh topologies in which paths to the base station can only have up to two hops.

Additional information on WiMAX standard and flows guarantees can be found in several references as [35–37]. Good references on WMN are Refs. [38–41].

## 16.3   PROBLEM DESCRIPTION

We look for the optimum conditions required by a fixed broadband wireless access system to cover remote rural users. We extend cellular automatic cell planning models to build an automatic cell planning tool. Our goal is to design a system to provide access to remote rural users under realistic conditions considering data networks. And also to find out how multihop topologies can improve over PMP. We describe these issues in the following.

### 16.3.1   SCENARIO DESCRIPTION

We suppose a set of potential users, which are placed on real villages and country houses. Users are not necessarily uniformly distributed. We suppose that all the users have the same traffic requirements, they are fixed and have an external energy source. Every region corresponds to a real place in Colombia.

- *High population density, flat terrain*: This scenario represents a city with uniformly distributed users, with shadowing caused by surrounding obstacles. One base station covers many users and usually operates saturated. There are also usually several base stations on the same site.
- *Medium population density, medium mountainous rural region*: This scenario represents a typical rural region, where some of the users are uniformly distributed and some of them are placed on small towns or near roads or trails. We suppose that some of those users cannot be easily covered because of nearby obstacles.
- *Medium population density, mountainous rural region*: We suppose a user distribution similar to the previous scenario. We suppose the existence of high mountains and rivers that cause deep canyons. There are several users with difficult coverage conditions, i.e., there are no privileged places that can cover a high percentage of the region.
- *Low population density, flat terrain*: We suppose users widely separated from others. This is common in regions dedicated to agriculture, pasture lands, and forestry. In this case the main problem is caused by the long distance links. We also suppose some places with higher population concentration over the region average such as small villages.

### 16.3.2   DATA MODELS

Data models differ from voice systems in many ways. There are different QoS requirements, they are based on packet multiplexing and there are different transmission schemas that depend on link quality. QoS requirements for data networks include several criteria such as delay, delay variation, and guaranteed data rates. Base stations make use of statistical multiplexing to increase system capacity. An analysis of different transmission flows and the resources assignment problem can be found in Ref. [42]. The base stations perform a process known as *packet scheduling* to assign transmission opportunities to packets. Some packets can have priority over others, to allow transmission of more urgent packets [43]. Schedulers and multiplexing models for data traffic are difficult to use in the design process. Data networks like WiFi and WiMAX support AMC. As users have different spectrum efficiency values, they might require different number of slots on transmission frames to achieve the same data rate.

### 16.3.3   DIFFERENT TOPOLOGIES TO SOLVE THE PROBLEM: PMP, MESH, RELAY

In PMP, a user connects to a single base station using a direct link. It chooses which base station to connect to from a set of available base stations, depending on link quality and available capacity. In multihop networks, information can go through several links until it reaches the base station. Packets transmitted through multiple hops have higher delay and require more capacity, i.e., multihop topologies extend coverage at the expense of more capacity consumption. Operation of multihop networks makes use of spatial reuse, controlled by a scheduler. Two different links on the same channel can transmit simultaneously if they do not interfere with each other. Our assumption is that there is only one active link among all links belonging to paths that end on the same base station [22], but links of users connected to different base stations can be active simultaneously even though they use the same channel.

In multihop topologies, users must decide not only which base station to connect to, but also the path that the packets should follow. The amount of resources required at every hop is not the same, as different links can have different modulation and coding schema. In our case, a certain node chooses the route that requires the lowest amount of resources. We limit our problem to routes up to two hops in relay topologies and up to five hops in mesh networks. A larger number of hops would be prohibitive in terms of delay and resources consumption.

## 16.4  AUTOMATIC AND OPTIMIZED CELL-MESH PLANNING MODEL

The model is composed of two submodels:

- *Coverage and capacity assignment submodel*: Given a fixed set of active base stations, with fixed parameters, we find the higher number of users that can connect to the system. This model requires a propagation model to check for link conditions, a user-base station assignment model to decide which base station every user should connect to, and a traffic model that finds out a feasible assignment under QoS guarantees. This model must support all the proposed topologies.
- *Optimization and automatic cell planning submodel*: In this model, we change the topology by the activation, deactivation, or modification of sites and base stations. The main idea is to find a configuration that optimizes network in terms of capacity, coverage, interference, and cost.

Both submodels interact with each other. The optimization submodel chooses a configuration that the coverage and capacity submodel problem evaluates. After that, the second one returns performance measures to the optimization algorithm to help it deciding on new configurations.

### 16.4.1  COVERAGE AND CAPACITY ASSIGNMENT SUBMODEL

The coverage process is based on some calculations made before algorithm execution. We use a digital elevation map of the chosen region and ICS Telecom propagation software. We estimate the propagation losses from every candidate site to every user. The propagation loss includes phenomena as diffraction or shadowing between transmitter and receiver. We fix sites towers to a height of 15 m and the users' towers to be 4 m high. We make this assumption to reduce problem complexity. We also fix a frequency value for the propagation model and consider that all available channels are near this central frequency so the propagation losses are independent of channel assignment. This information is used as data by the *coverage* subprocess.

#### 16.4.1.1  Coverage

We start from a fixed distribution of users with the same traffic requirement as described in Section 16.3.2 and a set of active base stations at each of the active sites. One site can have more than one base station with fixed parameters as the radiation pattern, orientation, and transmission power. The link quality depends on link losses, transmission power, antenna orientation, transmitter and receiver antenna gain, and radiation pattern. The power of the received signal is calculated supposing a perfect orientation of the user's antenna toward the chosen base station. Other base stations that use the same channel interfere with the desired signal. We calculate their received interference power in a similar way, but considering appropriate orientations. With direct signal from base station *i*, interference from other base stations different from *i* and noise power at the receiver of user *j*, we estimate the signal to noise plus interference ratio of the link from base station *i* to user *j* as (SNIR$_{i,j}$). This process is repeated from every user to every active base station, changing receiver antenna orientation. We finally obtain a matrix [SNIR]$_{M \times N}$ for *N* users and *M* active base stations. Then we check for the existence of a link between a user and a base station and its spectrum efficiency. In multihop topologies, we suppose every user has an omnidirectional antenna and we calculate the SNIR for links from every base station to every user and also between each pair of users.

After this, we run the routing process. In PMP topologies, this process is trivial as we can have only one-hop link from a user to any base station. In multihop topologies, we consider a minimum cost routing Dijkstra algorithm [44]. The cost of every link is proportional to the inverse of its spectral efficiency. If there is no link between two nodes, it is considered as an infinite cost link. Even though a direct links exists, one node could choose a multihop path if it had a better cumulate link efficiency. If there is no direct link, a multihop route would be only solution available. Coverage model returns a matrix with the cost of the optimal path from every user towards every base station. In PMP topologies, the path cost is the same direct link cost. In multihop topologies, the cost is the sum of the costs of links along the path. If there is no path between a user and a base station, the cost is set to infinite.

### 16.4.1.2 User-Base Station Assignment Model

This problem tries to maximize the number of users that can connect to the system. In each iteration, we search through the *path cost matrix* for the lowest cost path from an unconnected user to any base station. If there are multiple lowest cost paths, then we choose one of them randomly. If we can connect the user to the base station according to the *Capacity assignment model*, then we mark this user as connected. If not, we set the cost of this path to infinite. The algorithm ends when all users have been connected to one base station or when no additional user can be connected. This algorithm maximizes system spectral efficiency because lowest cost paths are the first ones to be assigned.

### 16.4.1.3 Capacity Assignment Model

This model decides if a specific base station can accept a new user without violating guarantees of its currently connected users. It needs a traffic model that considers different types of flows, system model, and QoS guarantees.

We consider three types of flows to represent typical applications, which can be used or not accordingly to users' needs.

- *Guaranteed data rate service*: For applications that require a fixed capacity assignment independently of the instantaneous transmission rate. An example of these applications include control channels, video surveillance, and on demand streaming of audio and video.
- *Guaranteed connections*: For applications that require capacity and delay guarantees but only during an active period or call. It is oriented to applications such as voice over Internet Protocol (VoIP), in which calls are established according to a statistical arrival and occupancy model. After the call finishes, the resources are released. We suppose that the data rate requirement is constant during the call.
- *Best effort*: For elastic traffic applications that do not require instantaneous guarantees. These applications require a minimum time average data rate.

Our capacity assignment model is based on this three types of flows and a new user can connect to a base station only when the required guarantees for these flows are satisfied. Our model is highly configurable, as we can choose all of the tree flows or just some of them. We could model a flow with a minimum sustainable rate and a time average rate by the combination of a guaranteed flow and a BE flow. We must emphasize that we consider in this work only the design process, which is independent from a later scheduler operation.

We divide users into $K$ groups according to the modulation and coding schema. A user belonging to a group $k$, has an equivalent *link transmission rate* $R_k$ [bps] when it transmits data. If the user has a direct link toward the base station, $R_k$ is the same link rate $r_e$. If the path goes through a set of links $Path_k$, then its equivalent link transmission rate is given by $R_k = 1/(\sum_{e \in Path_k} 1/r_e)$, where $r_e$ is the link rate of every link on the path. If the required instantaneous transmission rate of a user is $R$ [bps], then transmission requires $R \cdot T_{frame}/R_k$ seconds on the frame during its activity period to achieve it. $T_{frame}$ represents the duration in seconds of a transmission frame. $R$ is the required data rate of a flow and it is defined depending on the flow type. We define $R_{granted}$ [bps] to be the data rate needed by a granted flow, $R_{voice}$ [bps] as the peak data rate for a voice flow, and $R_{be}$ [bps] as the time average data rate needed by BE flows. According to our definition, we can find a small number $K$ of groups in PMP topologies, but several groups with multihop topologies. However, the number of groups is upper bounded by the number of nodes.

As we add more users to the base station, the transmission frame will have higher occupation. We define $m_k$ as the total number of users of the system connected to the base station that belong to group $k$. $n_k(t)$ is the number of voice calls established at any time of users belonging to group $k$ and connected to the base station, $n_k(t) \leq m_k$. With these definitions, we can find the total occupancy in seconds of the frame with Equation 16.1.

$$T_{occup}(t) = T_{frame} \left( R_{granted} \sum_{k=0}^{K-1} \frac{m_k}{R_k} + R_{voice} \sum_{k=0}^{K-1} \frac{n_k(t)}{R_k} + R_{be}(t) \sum_{k=0}^{K-1} \frac{m_k}{R_k} \right) \quad (16.1)$$

If we consider that the frame can only be filled up to a maximum value $T_{disp}$ reserved for data transmission, we can estimate the instantaneous data rate per BE flow. This quantity gives a deeper insight into the model. If an instantaneous $R_{be}(t)$ is greater than zero, it means that granted and voice flows do not fill the frame completely and there is available capacity for BE data. If this value is zero, it means that the system is blocked as there is no capacity left for other incoming calls or for BE flows. By controlling the number of connected users $m_k$, we can control the probability that an incoming call can be blocked and we can estimate the average data rate that BE traffic receives. This way we decide how many users are allowed to connect to a base station. The average value for the capacity assigned to a BE flow can be found in Equation 16.2

$$E[R_{be}] = \frac{1}{\sum_{k=0}^{K-1} \frac{m_k}{R_k}} \left( \frac{T_{disp}}{T_{frame}} - R_{granted} \sum_{k=0}^{K-1} \frac{m_k}{R_k} - R_{voice} \sum_{k=0}^{K-1} \frac{E[n_k]}{R_k} \right) \quad (16.2)$$

We have to estimate the time expectation of $n_k$, which is the average number of active voice calls of group $k$. We suppose an exponential model for inter arrival times with time average value $1/\lambda$ [s] and durations of voice calls with time average value $1/\mu$ [s]. We formulate the model as a birth–death process with a Markov chain [45]. If calls are never blocked (there is always left capacity for them), then it can be shown that $E[n_k] = \rho \cdot m_k/(1+\rho)$; with $\rho = \lambda/\mu$. Then, the acceptance criteria is that $T_{occup}$ is never higher than $T_{disp}$ and that the average BE data rate is equal to or higher than the required minimum value.

However, if calls can be blocked, then the model is a multidimensional loss system [45]. We represent the requirements of a call from every user type $k$ as a number of blocks with a fixed size in seconds. We choose a block size such that the requirements of every user type can be represented by an integer number of blocks. The requirements from any user are described by a probability mass function, which requires 0 blocks with a probability $P_{off} = \mu/(\lambda + \mu)$ and requires a defined number of blocks with probability $P_{on} = \lambda/(\lambda + \mu)$. The whole model is solved by a convolution method, in which, the whole system probability of using $q$ blocks is calculated by the convolution of individual calls' probability mass function. This model must be solved by numerical techniques. When the connection of a new user is analyzed, we calculate the convolution of the block requirements probability mass function of currently connected users with the probability mass function of the block requirements of the new user. With the duration of these blocks, we transform block requirements into frame occupancy. We find that the system is blocked if the number of blocks is over a threshold value. We must guarantee that the frame is never occupied beyond its available capacity, i.e., any occupancy beyond this value must be considered as a blocked state. So, we can find a blocking probability and the average number of established calls and complete the model. In this case we have two criteria, we must guarantee that blocking probability is under a threshold value and that average BE rate is above a minimum value.

In both cases, we suppose that a certain combination of $m_k$ users for each group $k$ has already been accepted. When we try to connect a new user of any type, then we increase the appropriate $m_k$ value and check if we have violated any of the requirements. If not, then we accept the new user.

## 16.4.2 OPTIMIZATION AND CELL PLANNING MODEL

The second problem is to find the set of active base stations, their orientation, and transmission power to achieve the optimum coverage and capacity assignment to users. The problem is separated into two parts.

### 16.4.2.1 *Transmission Towers* Construction

We define the concept of *Transmission tower* as a fixed set of active base stations placed at one active site. One site can have many *transmission towers* but only one of them can be active. We try to explore different alternatives for the number of antennas, their transmission power, orientation, and radiation pattern. After we build them, the problem reduces to choose one of them from every available active site.

The process begins with a set of candidate sites. We discard candidate sites with very low coverage. Also, if there are two sites with similar coverage, we discard the one with the lowest coverage. To build the *transmission towers* at one site, we begin placing one omni-directional antenna with the maximum transmission power. If this base station is not saturated, i.e., all covered users can connect to it, then we create several *transmission towers* with one single antenna and different transmission powers, chosen from a set of discrete values. We use also 120° and 180° sectorized antennas.

On the other case, if the first omni-directional antenna base station is saturated, i.e., not all covered users can connect because of capacity restrictions, then we build a set of *transmission towers* composed of several antennas with 120° and 180° sectors. We use all possible combinations of transmission power and sectors to build several options for the site. We solve coverage and capacity assignment by previously described algorithms to find the orientation of each set of active base stations. We finally remove redundant *transmission towers* from the set of available ones. We solve this for every site to get a set of *transmission towers* and a matrix that keeps a record of the users that connect to each one of them. This information is used in the optimization process.

### 16.4.2.2 Optimization Process

In this process, we try to find the set of active *transmission towers* to optimize coverage and connect the highest number of users. We fix the number of sites during one execution of the algorithm to find the best solution. After that, we increment the number of sites and run the algorithm once again. At the beginning, we start from an empty solution, then we try to improve it by activating, deactivating, or moving *transmission towers*. We do this every iteration using a probabilistic model to decide if a new solution is chosen or not over the current one. However, we keep track of the best solution that has been reached so far. The iterative process has two main components:

- *Building of a new solution*: In this process, we start from the current solution and try to improve it by a randomly chosen modification. Our modifications are based on those presented in Ref. [20]. We can deactivate any *active transmission* tower and activate any inactive *transmission tower*. We deactivate an active *transmission tower* randomly by assigning a deactivation probability inversely proportional to the number of users connected to it. We activate a new transmission tower randomly by assigning an activation probability proportional to the number of uncovered users that could connect to it. A new solution is analyzed for a feasible channel assignment by trying several combinations to reduce interference. We fix the number of available channels. We finally use the channel combination that has the lowest interference level, represented by the highest number of connected users.
- *Optimization process*: This algorithm iterates, comparing the new candidate solution and the current solution. This is the core process for simulated annealing, in which a new candidate solution replaces the current solution according to the improvement and the temperature of the system as described in Ref. [18]. If the candidate solution is better than the current solution, it is accepted. Otherwise, it has an acceptance probability that depends on how bad is the new solution with respect to the current solution and the temperature of the system. This process keeps track of the current solution and the best solution ever found.

The metric used to decide the performance of a solution is the percentage of users that could connect to any base station. It means that a better solution has more connected users than a previous one. We must recall, that other optimization criteria are included in the inner process.

- Interference is reduced during the building of a candidate solution by choosing the lowest interference channel assignment, i.e., each new candidate solution tries to increase the number of connected users with the minimal interference.
- For the iteration process we add and remove base transmission towers to the new candidate solution. If we have two solutions with similar number of connected users, we choose the one with the lowest number of base stations. This way we reduce the cost related to the number of base stations. We leave the site cost criterion for a later post process which is described in Section 16.4.3.
- The QoS guarantees are included in the User-Base station assignment model, where we try to connect users to base stations according to their spectrum efficiency. Also, in the Capacity assignment model, if we connect a user to a base station, we guarantee that the requirements are satisfied.

### 16.4.3  FINAL OPTIMIZATION CRITERION

We run the optimization problem for a certain scenario, by iteratively fixing the maximum number of active sites and solving the optimization problem. After we complete all the iterations, we have a set of solutions for different number of active sites. We identify every iteration by an index $u$. Then, we choose a solution based on the analysis of the results, by building a weighted sum of different functions that represent the number of users that can connect to the system, the efficiency of the solution, and the cost of sites and base stations. We define all the functions to have a normalized value between 0 and 1, being 1 the better. First we have to define some quantities. The total number of users in the system that need to access the network is $N$. We define the set of connected users as $\mathbf{C}$ and the number of connected users as $|\mathbf{C}|$. For the iteration $u$, the number of connected users would be $|\mathbf{C}_u|$. The set of active sites for any iteration is $\mathbf{S}_u$ and the number of active sites is $|\mathbf{S}_u|$. The set of active base stations is given by $\mathbf{B}_u$ and the number of active base stations is given by $|\mathbf{B}_u|$. Now we proceed to define the metrics used.

The percentage of connected users is given by $x_\alpha^k = |\mathbf{C}_u|/N$. This is just the ratio between the number of connected users and the total number of users. An efficiency function is defined as $x_\beta^u = (|\mathbf{C}_u|/|\mathbf{B}_u|)/\max_u\{|\mathbf{C}_u|/|\mathbf{B}_u|\}$, which is the ratio between the total number of users connected to any base station and the total number of base stations used in the solution. We normalize this value by using the highest value from all iterations. The third value is directly related to the cost of the number of base stations on the solution. We define $x_\gamma^u = (\max_u\{|\mathbf{B}_u|\} - |\mathbf{B}_u|)/\max_u\{|\mathbf{B}_u|\}$, as an index that has a value equals to 0 at the iteration with the highest number of base stations and the highest value on the iteration with the lowest number of base stations. In a similar way, we define $x_\theta^u = (\max_u\{|\mathbf{S}_u|\} - |\mathbf{S}_u|)/\max_u\{|\mathbf{S}_u|\}$, as a measure of the cost of the number of sites. This quantity has the highest value when the number of sites is the lowest.

With these definitions, we define our optimization function to be maximized in Equation 16.3, with $\alpha + \beta + \gamma + \theta = 1$.

$$\max_u\left\{z = \alpha \cdot x_\alpha^u + \beta \cdot x_\beta^u + \gamma \cdot x_\gamma^u + \theta \cdot x_\theta^u\right\} \tag{16.3}$$

## 16.5 RESULTS

Our goal is to define the best conditions to cover rural remote users by the usage of PMP and multihop topologies. We defined four scenarios with different population density and topographical conditions as described in Section 16.3.1. We also defined three different set of values for traffic requirements classified as low, medium, and high requirements, presented in Table 16.1. Finally, we included in the results PMP, relay, and mesh topologies.

We use the following metrics to deeply analyze the performance of the solutions found by the optimization algorithm and compare the different topologies considered.

- $x_\alpha^u$ as previously defined.
- The ratio between the number of connected users and the number of used base stations ($|\mathbf{C}_u|/|\mathbf{B}_u|$). This is the value $x_\beta^u$ previously defined before normalization.
- *Average frame occupancy percentage*: This variable determines how much of the frame is occupied on every base station. This variable is important to measure if a multi-point solution can improve a PMP solution.
- *Equivalent modulation and coding schema*: In WiMAX standard [22], there are several levels of transmission profiles defined, ranging from 1/2 BPSK to 3/4 64QAM. Every transmission profile is defined by the *coding factor* and the *modulation factor*. We define a quantity by the product of these two values, equivalent to the amount of bits that are sent within one QAM symbol into an OFDM symbol. The lowest value is 0.5 and the higher value is 4.5. We calculate the average value for all the users connected to every base station. An average value near 4.5 means that the solution has a high spectrum efficiency, equivalent to say that all users have the best link conditions possible.

**TABLE 16.1**

**Traffic Parameters for Different Requirements**

| Traffic Requirements | $R_{\text{granted}}$ | $R_{\text{voice}}$ | $R_{\text{be}}$ | $\rho$ |
|---|---|---|---|---|
| Low | 20,000 | 24,000 | 10,000 | 0.025 Erl |
| Medium | 60,000 | 24,000 | 40,000 | 0.025 Erl |
| High | 120,000 | 24,000 | 80,000 | 0.050 Erl |

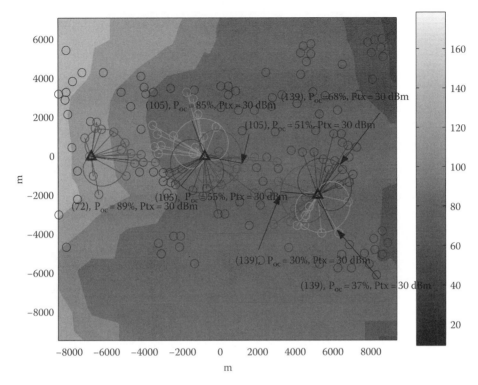

**FIGURE 16.1**    Example solution for a PMP topology.

In Figure 16.1, we present a solution for the low population density and flat terrain scenario with PMP topology. Dark triangles represent sites location. Each antenna in every site is represented by its radiation pattern with a black line indicating its orientation. The connection from users to base stations is represented by a solid line with the same color of the radiation pattern. Information text near the base station indicates the site index, the frame occupancy percentage, and the base station transmission power. In Figure 16.2 we present a mesh topology solution for the same terrain shown in Figure 16.1. Every site includes all the base stations that were used on the PMP solution. The information about each site is the site index and the average frame occupancy percentage of all base stations used. Gray curves represent terrain heights.

In the following we discuss the behavior of each one of the performance metrics with respect to the number of sites. We focus then on the number of users that can connect to the system as a function of traffic requirements and the terrain characteristics for all three topologies. We finally discuss the optimization objective function defined in Equation 16.3. We found interesting to split multihop solutions into relay and mesh, because when the mesh solution performs near the relay solution, it means that the relay solution as proposed in study group IEEE 802.16j [34] could be enough over a full mesh solution.

## 16.5.1   SYSTEM PERFORMANCE VARIABLES

We begin by presenting our metrics results in a comparative way, to show that they can give a good insight into the model performance. In Figure 16.3 we show the four previously defined performance metrics for the scenario with low population density and flat terrain and with low traffic requirements. We can see that the percentage of connected users $x_\alpha^u$ increases with the the number of sites. There are additional connected users with relay topology and even more with mesh topology. We can also see that the increase in the number of sites reduces the number of users per base station, i.e., reduces

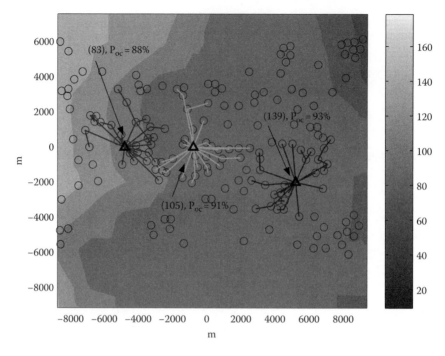

**FIGURE 16.2** Example solution for a mesh topology.

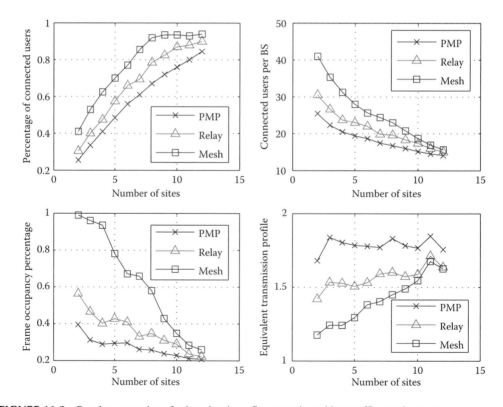

**FIGURE 16.3** Results comparison for low density—flat scenario and low traffic requirements.

efficiency. The percentage of frame occupancy shows that as the number of sites increases, every base station has a lower number of users connected to it and this causes a lower frame occupancy. With many base stations, the frame occupancy is really low. We can see that a mesh solution has a higher occupancy up to four active sites. This is because it increases the number of connected users but consumes more resources because of multihop links.

The equivalent transmission profile clearly shows that a PMP topology has a higher spectrum efficiency than a mesh or relay solution. This is because multihop topologies require more capacity due to multiple hops. In PMP topology this parameter remains almost constant, which means that the addition of new base stations allows the inclusion of more users but does not improve link quality. We analyze that this is caused by the reduction of the transmission power of some base stations due to interference restrictions. The increase of link efficiency with the number of sites for mesh and relay topologies, means that users reduce the number of hops to reach the nearest base station. The similarity between mesh and relay topology means that with many active sites, there are paths with at most two hops for some users. As the number of sites is higher, there is a low use of multihop links.

### 16.5.2 System Performance as a Function of Traffic Requirements

We present in Figure 16.4 the relative gain of multihop topologies with respect to PMP topologies as a function of traffic requirements and the number of active sites for low population density and flat terrain scenario. We define the relative gain by Equation 16.4. This definition is the same for mesh topologies and it indicates how better relay and mesh solutions are relative to PMP in terms of the number of connected users.

$$g_{u,\text{Relay}} = \frac{|\mathbf{C}_{u,\text{Relay}}| - |\mathbf{C}_{u,\text{PMP}}|}{|\mathbf{C}_{u,\text{PMP}}|} \tag{16.4}$$

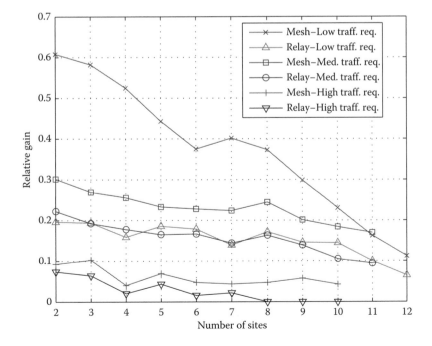

**FIGURE 16.4** Relative gain as a function of traffic requirements for low user density and flat terrain.

We can see in all cases that a mesh solution performs better than a relay solution according to this criterion. The higher gain is obtained with the lower traffic requirement. This means that frame occupancy is lower and more users can connect to the base station. As traffic requirements increase, relative gain decreases. As an extreme case, when traffic requirement is high, there is almost no improvement with mesh or relay solutions.

### 16.5.3  SYSTEM PERFORMANCE AS A FUNCTION OF THE SCENARIO

In Figures 16.5 and 16.6 we present the relative gain (Equation 16.4) for relay and mesh topologies with low and medium traffic requirements, respectively, as a function of the scenario. We do not discuss high traffic requirements case as we have shown that multihop topologies have almost no improvement over PMP. In this case the higher relative gain is obtained in flat, low population density scenario. This scenario has long distance links and multihop topologies help to extend coverage.

The medium population density and a mountainous terrain scenario is the second in performance improvement by the use of multihop technologies. We can see that there is not much gain with mesh relative to relay solution. The problem seems to be avoiding obstacles by using a relay node, as there are not many paths with more than two hops. When there are only two sites there is a higher gain with mesh solution, due to low connectivity. But the inclusion of a third site reduces the need for high number of hops. After three sites, the improvement can be just achieved by two hops topologies. A similar behavior is seen in a medium population density with a medium mountainous terrain. There seems to be higher connectivity, so the multihop topologies can be limited to relay. Finally, in a high population density and medium size city, there is virtually no gain with multihop topologies. City scenarios have high connectivity and shorter links. These cases are well covered by a PMP topology or another wired technology. As we can see, PMP topologies perform better in high density scenarios where a base station is saturated. When the problem is caused by obstacles, a two-hop (relay) solution can be enough to reach the base station. When the problem is related to distance, where coverage

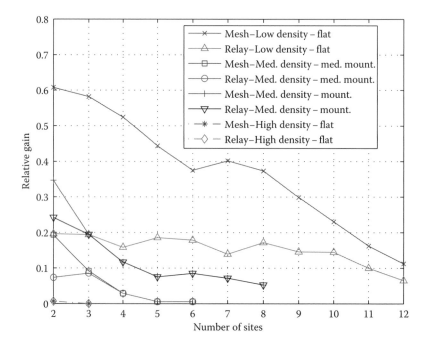

**FIGURE 16.5**  Relative gain as a function of terrain with low traffic requirements.

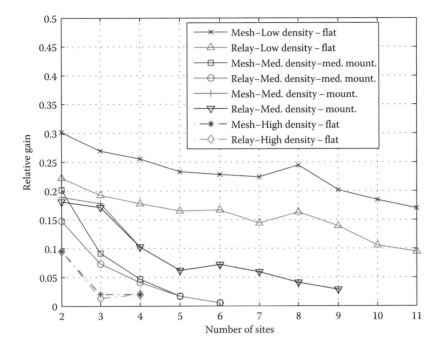

**FIGURE 16.6** Relative gain as a function of terrain with medium traffic requirements.

is limited by long distance links, a mesh solution with more than two hops can certainly improve system coverage.

We have then several behaviors clearly identified.

- An increase in the number of sites increases system coverage and the number of connected users. After a high number of active sites, there is a notorious reduction on frame occupancy and efficiency. Base stations become subutilized and this reduces network efficiency.
- When there is capacity left in transmission frames, coverage can be extended with multihop topologies. This causes an extra occupancy of the transmission frame that reduces spectral efficiency but increases network coverage.
- When the main problem is the presence of obstacles that block the line of sight, the use of two hops link helps the extension of coverage to those obstructed users. When the low coverage problem is caused by long distance links, then a mesh solution can improve system coverage by dividing a long link into shorter link. In both cases, the spectral efficiency is reduced.
- To increase system coverage and capacity assignment, we can add sites or we can use multihop topologies. In the first case there is an increase in the network cost but there is no reduction in spectral efficiency. The use of multihop topologies reduces spectral efficiency but does not increase the number of base stations.

### 16.5.4 OPTIMIZATION CRITERION

We used objective function defined in Equation 16.3 to find the optimal solution value. We used the following set of parameters: $\alpha = 0.4$, $\beta = 0.2$, and $\gamma = 0.2$ and $\theta = 0.2$ for Figure 16.7 and $\alpha = 0.6$, $\beta = 0.4/3$, and $\gamma = 0.4/3$ and $\theta = 0.4/3$ for Figure 16.8. The first case gives a lower importance to the percentage of connected users than the second one. We show in Figure 16.7, the $z$ value as a function of the number of sites and for different traffic requirements.

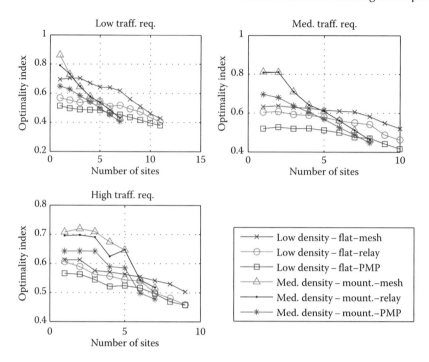

**FIGURE 16.7** Objective function for $\alpha = 0.4$, $\beta = 0.2$, $\gamma = 0.2$, $\theta = 0.2$.

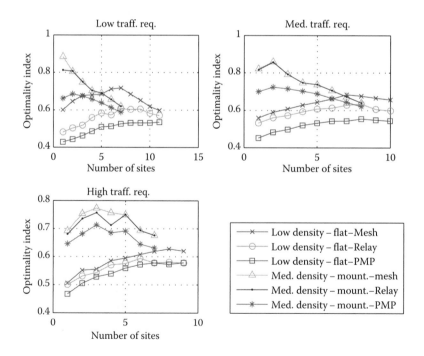

**FIGURE 16.8** Objective function for $\alpha = 0.6$, $\beta = 0.4/3$, $\gamma = 0.4/3$, $\theta = 0.4/3$.

For Figure 16.7 there is a lower priority for the percentage of connected users and a higher priority for the cost and the efficiency. We see that with low traffic requirement, the optimal solution is found with a low number of base stations for both scenarios. With higher traffic requirements, there is an increase in the required number of base stations and sites. This model reduces considerably the number of base stations to reduce cost.

In Figure 16.8, there is a clear increase in the number sites at the optimal solution as we give more priority to the percentage of connected users. For the low density, flat terrain scenario, there is a need for more sites as the traffic requirements increase. We can see that at low traffic requirements, the mesh solution needs less sites compared to the PMP. The optimal solution is reached at 7 sites for mesh solution, but for relay and PMP solutions the optimal point is reached with more sites.

We can see that the criteria for choosing the optimal solution depends on the terrain characteristics, the user distribution, and the traffic requirements. There are different conditions under which it is better to increase the number of sites or to use multihop topologies as was previously described. This confirms out previous assumptions.

## 16.6  OPEN ISSUES AND FUTURE WORK

Automatic cell planning for data networks is still an open research subject. Its unique characteristics justify the efforts to extend current voice models to obtain new models. We have found few models oriented to the design of multiuser wireless data networks based on WiMAX like technologies. A complete solution that considers traffic models, link channel quality, sectorized antennas, channel allocation, and variable transmission profile, as far as we know, has not been found.

Right now we are considering a medium complex traffic model. There are several things to do to complete this model. Even though granted services require a fixed data rate, there might be some multiplexing gain based on statistical usage of this flows by end users. In voice model, there is an extension based on silence suppression which reduces capacity requirements for the system. In other types of flows, there are more realistic traffic models based on variable bit rate sources, which are harder to include. There is a need to include models that are nearer the operation scheduler in the design process. Finally, in BE traffic, there is a serious problem with the uplink scheduling. This model is more suitable for downlink transmissions, where the base station assigns transmission opportunities to contending flows. In the uplink, the backoff process defined in IEEE 802.16 standard [22] requires a low usage of the available capacity for it to be effective. It is known that many users contending for the channel simultaneously would cause a low equivalent data rate per user. This model is not included in this work. We need to keep an equilibrium between precision and complexity. As this is a design process, there would be a prohibitive computational cost if the traffic model is far more complicated.

Another issue to be extended, is that users could dynamically change the base station which they connect to. Temporarily, during congestion periods of a base station, a node could connect through another base station which is not optimal in terms of propagation but that has available capacity. Our model only considers static assignment of users to base stations.

Another extension of our model is based on uplink channels estimation. We do not consider this in our model because of its complexity. Usually, uplink data rate requirements are lower than downlink ones, which gives a margin for uplink link budget. We consider symmetric channels for propagation between a user and a base station. Under this consideration, uplink channel quality should be similar to downlink quality. The issue we find difficult to include is the interference. The estimation of uplink received power can be easily done. When we consider interference signals at the base station, we have to consider transmission power of other base stations and other users. The interference from other base stations can be easily considered, but we have no method to estimate the interference caused by other users because we do not know a priori the base station they are connected to, their transmission channel, nor their antennas orientation. We are working on better ways to improve this part of the model.

We are working right now on simplifications of this model to account mathematically tractable solutions. We are considering an optimization problem based on activation and deactivation of candidate base stations, but also to include power control and data traffic models. Variable transmission profile based on adaptive modulation and coding is a really complex part, because of discontinuity. We are looking forward to find the simplifications needed to write it as a mathematical model and compare it with this solution.

## 16.7  CONCLUSIONS

We studied wireless solutions to provide coverage and connect remote users to an existing telecommunication network. We used multihop topologies such as mesh and relay, rather than just PMP. We found that an increase in the number of sites, increases the number of connected users. However, an excessive increase in the number of sites, reduces the efficiency of the solution. Multihop solutions increase the number of connected users without an increase in the number of base stations, but reduce spectral efficiency. The necessary condition to allow a multihop solution is to have available capacity in transmission frames.

Another condition to consider is the terrain characteristics. Relay topologies seem to perform better in mountainous terrains, in which the main problem is to avoid obstacles to reach the base station. Mesh topologies seem to have better performance for low coverage regions like mountainous terrain with obstacles or when the coverage problem is caused by long distance links. In this case, there is not enough coverage gain by two hops links and there is a need to extend it further with more hops. This solution can perform better if there is a low traffic requirement. We finally found that for high population density and high traffic requirements, a PMP solution is the best choice. There is no need to extend coverage by multihop topologies.

## ACKNOWLEDGMENT

We would like to thank TES AMERICA for the cartography information used during this study.

## REFERENCES

[1] M.T. Mandioma, G.S.V.R.K. Rao, A. Terzoli, H. Muyingi, Deployment of WiMAX for telecommunication and Internet access in Dwesa-Cwebe rural areas, in *The 9th International Conference on Advanced Communication Technology*, 3(1): 2141–2143, February 12–14, 2007.

[2] B. Raman, K. Chebrolu, Experiences in using WiFi for rural internet in India, *IEEE Communications Magazine*, 45(1): 104–110, January 2007.

[3] K. Paul, A. Varghese, S. Iyer, B. Ramamurthi, A. Kumar, WiFiRe: Rural area broadband access using the WiFi PHY and a multisector TDD MAC, *IEEE Communications Magazine*, 45(1): 111–119, January 2007.

[4] R. Hincapie, J. Sierra, R. Bustamante, Remote locations coverage analysis with wireless mesh networks based on IEEE 802.16 Standard, *IEEE Communications Magazine*, 45(1): 120–127, January 2007.

[5] J. Garcia-Fragoso, G.M. Galvan-Tejada, Cell planning based on the WiMax standard for home access: A practical case, in *2nd International Conference on Electrical and Electronics Engineering*, 1(1): 89–92, 7–9 September 2005.

[6] X. Huang, Automatic cell planning for mobile network design: Optimization models and algorithms, PhD thesis. Institut für Höchstfrequenztechnik und Elektronik (IHE), Großchönau, Germany, May 2001.

[7] A. Eisenblatter, H. Geerd, Wireless network design: Solution-oriented modeling and mathematical optimization, *IEEE Wireless Communications*, 13(6): 8–14, December 2006.

[8] I.H. Cavdar, O. Akcay, The optimization of cell sizes and base stations power level in cell planning, in *IEEE 53rd Vehicular Technology Conference*, 4(1): 2344–2348, 2001.

[9] S.M. Allen, S. Hurley, R.M. Whitaker, Spectrally efficient cell planning in mobile wireless networks, in *IEEE 53rd Vehicular Technology Conference*, 2(1): 931–935, 2001.

[10] X. Huang, U. Behr, W. Wiesbeck, Automatic cell planning for a low-cost and spectrum efficient wireless network, in *IEEE Global Telecommunications Conference GLOBECOM 2000*, 1(1): 276–282, 2000.

[11] H.P. Lin, R.T. Juang, S.S. Jeng, C.W. Tsung, A cell planning scheme for WCDMA systems using genetic algorithm and performance simulation platform, in *15th IEEE International Symposium on Personal, Indoor and Mobile Radio Communications PIMRC 2004*, 4(1): 2560–2564, September 5–8, 2004.

[12] X. Huang, U. Behr, W. Wiesbeck, Automatic base station placement and dimensioning for mobile network planning, in *IEEE 52nd Vehicular Technology Conference*, 4(1): 1544–1549, 2000.

[13] R. Subrata, A.Y. Zomaya, A comparison of three artificial life techniques for reporting cell planning in mobile computing, *IEEE Transactions on Parallel and Distributed Systems*, 14(2): 142–153, February 2003.

[14] X. Wang, T. Long, Y.H. Lee, Automated cell planning based on propagation loss, in *Proceedings of the 2003 Joint Conference of the Fourth International Conference on Information, Communications and Signal Processing, 2003* and the *Fourth Pacific Rim Conference on Multimedia*, 1(1): 134–138, 15–18 December 2003.

[15] H.M. Elkamchouchi, H.M. Elragal, M.A. Makar, Cellular radio network planning using particle swarm optimization, in *National Radio Science Conference NRSC 2007*, 1(1): 1–8, 13–15 March 2007.

[16] F. Glover, Tabu Search—Part I, *ORSA Journal on Computing*, 1(3): 190–206, 1989.

[17] F. Glover, Tabu Search—Part II, *ORSA Journal on Computing*, 2(1): 4–32, 1990.

[18] S. Kirkpatrick, C.D. Gelatt, M.P. Vecchi, Optimization by simulated annealing, *Science*, 20(4598): 671–680, May 1983.

[19] C.Y. Lee, H.G. Kang, Cell planning with capacity expansion in mobile communications: A tabu search approach, *IEEE Transactions on Vehicular Technology*, 49(5): 1678–1691, September 2000.

[20] S. Hurley, Planning effective cellular mobile radio networks, *IEEE Transactions on Vehicular Technology*, 51(2): 243–253, March 2002.

[21] I. Siomina, P. Varbrand, D. Yuan, Automated optimization of service coverage and base station antenna configuration in UMTS networks, *IEEE Wireless Communications*, 13(6): 16–25, December 2006.

[22] IEEE standard for local and metropolitan area networks Part 16: Air Interface for fixed broadband wireless access systems, IEEE std 802.16-2004, pp. 1–857, 2004.

[23] T. Vanhatupa, M. Hännikäinen, T.D. Hämäläinen, Evaluation of throughput estimation models and algorithms for WLAN frequency planning, *Computer Networks*, 51: 3110–3124, August 2007.

[24] P. Bahl, M.T. Hajiaghayi, K. Jain, S.V. Mirrokni, L. Qiu, A. Saberi, Cell breathing in wireless LANs: Algorithms and evaluation, *IEEE Transactions on Mobile Computing*, 6(2): 164–178, February 2007.

[25] C. Prommak, J. Kabara, D. Tipper, Demand-based network planning for large scale wireless local area networks, *Workshop on Broadband Wireless Services and Applications Broadnets 2004*, California, October 2004.

[26] X. Ling, K.L. Yeung, Joint access point placement and channel assignment for 802.11 wireless LANs, *IEEE Transactions on Wireless Communications*, 5(10): 2705–2711, October 2006.

[27] K. Jaffrès-Runser, J.M. Gorce, S. Ubéda, Mono—and multiobjective formulations for the indoor wireless LAN planning problem, *Computers and Operations Research*, 35(12): 3885–3901, 2008.

[28] K. Jaffrès-Runser, J.M. Gorce, S. Ubéda, QoS constrained wireless LAN optimization within a multiobjective framework, *IEEE Wireless Communications*, 13(6): 26–33, 2006.

[29] A. So, B. Liang, Enhancing WLAN capacity by strategic placement of tetherless relay points, *IEEE Transactions on Mobile Computing*, 6(5): 474–487, May 2007.

[30] D. Niyato, E. Hossain, Integration of IEEE 802.11 WLANs with IEEE 802.16-based multi-hop infra-structure mesh/relay networks: A game-theoretic approach to radio resource management, *IEEE Network*, 21(3): 6–14, May–June 2007.

[31] S.H. Ali, K.D. Lee, V.C.M. Leung, Dynamic resource allocation in OFDMA wireless metropolitan area networks [Radio Resource Management and Protocol Engineering for IEEE 802.16], *IEEE Wireless Communications*, 14(1): 6–13, February 2007.

[32] B. Rong; Y. Qian; H.H. Chen, Adaptive power allocation and call admission control in multiservice WiMAX access networks [Radio Resource Management and Protocol Engineering for IEEE 802.16], *IEEE Wireless Communications*, 14(1): 14–19, February 2007.

[33] H. Bolcskel, A.J. Paulraj, K.V.S. Hari, R.U. Nabar, W.W. Lu, Fixed broadband wireless access: State of the art, challenges, and future directions, *IEEE Communications Magazine*, 39(1): 100–108, January 2001.

[34] http://ieee802.org/16/relay/

[35] D. Sweeney, *WiMax Operator's Manual: Building 802.16 Wireless Networks*, APress, New York, 2004.

[36] F. Ohrtman, *WiMAX Handbook*, McGraw-Hill Communications, New York, 2005.

[37] J. Andrews, A. Ghosh, R. Muhamed, *Fundamentals of WiMAX: Understanding Broadband Wireless Networking*, Prentice Hall Communications Engineering and Emerging Technologies Series, Upper Saddle River, NJ, 2007.

[38] I.F. Akyildiz, X. Wang, A survey on wireless mesh networks, *IEEE Communications Magazine*, 43(9): S23–S30, September 2005.

[39] I.F. Akyildiz, X. Wang, W. Wang, Wireless mesh networks: A survey, *Computer Networks*, 47: 445–487, March 2005.

[40] Y. Zhang, J. Luo, H. Hu, *Wireless Mesh Networking: Architectures, Protocols and Standards*, Auerbach Publications, Taylor & Francis Group, New York, 2006.

[41] R. Bruno, M. Conti, E. Gregori, Mesh networks: Commodity multihop ad hoc networks, *IEEE Communications Magazine, IEEE*, 43(3): 123–131, March 2005.

[42] C. Huang, H.H. Juan, M.S. Lin, C.J. Chang, Radio resource management of heterogeneous services in mobile WiMAX systems [Radio Resource Management and Protocol Engineering for IEEE 802.16], *IEEE Wireless Communications*, 14(1): 20–26, February 2007.

[43] A. Kumar, D. Manjunath, J. Kuri, *Communication Networking: An Analytical Approach*, Elsevier/Morgan Kaufmann, Morgan Kaufmann series in networking, Amsterdam, 2004.

[44] E.W. Dijkstra, A note on two problems in connexion with graphs, *Numerische Mathematik*, 1: 269–271, 1959.

[45] ITC in cooperation with ITU-D SG2, *Teletraffic Engineering Handbook*, ITU-D, Available online: http://www.tele.dtu.dk/teletraffic/, Geneva, 2005.

# 17 Capacity Planning and Design of WiMAX Access Networks

*Shekhar Srivastava, André Girard,
and Brunilde Sansò*

## CONTENTS

In this chapter, we study the problem of capacity planning for worldwide interoperability for microwave access (WiMAX) backhaul network based on IP transport. With backhaul networks forming the significant part of buildout cost, it is crucial to evaluate dimensioning approaches that facilitate cost-effective deployment of WiMAX networks. In this reference, we have considered models guaranteeing absolute and average delay requirements. We consider three kinds of source models: Poisson based, on–off based and leaky bucket (LB) filter based, and present explicit expressions capturing the required capacity. We compare the required capacity for average delay for these models with the required capacity for absolute delay and comment on the differential. Further, using simulation results, we study the average and variance of the observed delay for voice and video sources and compare the three models. We found that on–off and LB models are very powerful and ensure that the actual delays are less than the required delay and that the variance remains acceptable. These results seem to indicate that it may be possible to dimension systems based on average quality of service (QoS) requirements and still get adequate performance for other requirements such as jitter. The results serve to highlight that wireless carriers need to take a closer look at average delay models in the face of high bandwidth cost, particularly in the WiMAX backhaul segment.

## 17.1   INTRODUCTION

Wireless users across the world are constantly exploring newer and richer applications for their wireless handsets. Multimedia video, wireless voice over IP, Gaming applications, location-based services, etc., are recent set of applications that have moved to wireless-based devices such as cell phones, pdas, laptops, etc. These applications are putting the pressure on currently available wireless networks based on technologies such as EVDO, UMTS, etc. Service providers are getting challenged to provide faster data rates and support for new applications and devices, even in the face of constant or decreasing average revenue per customer (ARPU). This has forced service providers to look at alternate technologies, which are cheaper to deploy and can provide faster downloads. WiMAX is being touted as the appropriate technology for the next generation of wireless networks, sometimes referred to as "4G."

Initially, WiMAX evolved as a standard for fixed microwave communication, referred to as IEEE 802.16-2004 [10] or fixed WiMAX. Later amendments incorporated mobility and led to the creation of IEEE 802.16e [11], mobile WiMAX, or just "WiMAX." While fixed WiMAX has interesting problems and challenges, this chapter will focus on the mobile version and the associated design and planning issues and approaches. By design, WiMAX [17] standard attempts to bring together disparate pieces of access networks, to form a convergent platform which provides uninterrupted connectivity and cost-effective services to mobile end-points. Therefore, one of the fundamental under-pinnings driving the WiMAX adoption is the potential for fixed mobile convergence (FMC). Additional salient features include high-data rates, QoS support, scalability, security, and mobility.

High-data rates are achieved by the physical layer based on orthogonal frequency division multiplexing (OFDM) along with multiple input multiple output (MIMO) antenna techniques. Coupled with flexible subchannelization schemes, advanced coding and modulation schemes, it could deliver peak downlink rates of 50–60 Mbps per sector in a 10 MHz channel. This translates into a peak downlink rate of 150–180 Mbps per cell.

High-speed air link, asymmetric downlink/uplink capability, fine resource granularity, and a flexible resource allocation mechanism ensure that WiMAX can meet the resource requirements of wide range of existing and future applications. WiMAX medium access control (MAC) provides controlled performance to each service flow over the air based on the QoS category of the session. Subsequently, the session traverses the radio access network (RAN) transport which is based on IP. For this part of the network, differentiated services (DiffServ) architecture is deployed to provide controlled QoS. The defined service flows of the MAC are subsequently mapped to DiffServ code points of the IP transport. The IP transport connects the base station (BS) to the access gateway (AG), together referred to as access service network (ASN), which further connects to the connectivity service network (CSN). It is critical to ensure that the QoS requirements are honored across the ASN and in the CSN. In this chapter, we focus on the capacity planning issues of the IP transport which connects the BS to the AG and then to the CSN.

The above-mentioned IP transport network is sometimes referred to as "backhaul network." In the traditional 2G networks, it forms a small part of total cost of ownership (TCO), particularly a part of operational expenditure (OpEx). These backhaul networks were mostly built on capacity leased from local exchange carriers (LECs). Consider a typical global system for mobile communications (GSM) BS with three sectors, each serving one carrier, which costs about USD 35,000. Such a BS would require one T-1 (about 1.5 Mbps) to connect to the BS controller (colocated with mobile switching center). While the leased line cost of the T-1 would depend on the location of

BS and the BSC, average cost in an urban environment could be USD 2500–3500 per year.* Observe that for a GSM BS, it takes about ten years for the backhaul lease cost to compare to the one time capital expenditure (CapEx).

Consider a similar TCO study for a WiMAX network. Current estimates of the BS cost are about USD 40,000–50,000 for a three sector cell, each processing 10 MHz of carrier. As discussed above, such a BS would generate 150–180 Mbps of data. Additional compression has also been supported to decrease the required capacity, some available ratios are 2:1, leading to approximate data capacity of 75–90 Mbps. Disregarding any considerations of delay and loss budgets, it would at least require two DS-3s (45 Mbps each) for backhaul. Again, cost of DS-3 depends upon the location of BS and the BSC, on an average could be USD 20,000–28,000 per year. This means that the lease OpEx of backhaul will be higher than the BS CapEx after only two years of service. This has led to a paradigm shift in terms of design of wireless networks.

Wireless service providers are increasingly less worried about the CapEx for the BS and more concerned about the OpEx for the backhaul. In other words, OpEx for backhaul as a fraction of TCO is significantly higher for WiMAX vis-a-vis GSM. Therefore, one of the major impediments in the rollout of WiMAX networks is the exorbitant backhaul cost. Although, comparison presented above for backhaul and BS cost numbers are based on North American carriers, their ratios are representative of other parts of the World as well.

In this chapter, we present a comparative study of various mechanisms available for dimensioning of the backhaul IP transport for WiMAX networks. The mechanisms are compared in terms of required capacity and in terms of their capability to guarantee the QoS to individual applications. We base our study on voice and video applications and consider different values of required delay bounds. In Section 17.2, we present various components of a WiMAX network and discuss their relationship and requirements such as IP transport, DiffServ model, etc. In Section 17.3, we discuss the absolute and average delay bound requirements and discuss the amount of capacity required for given multimedia applications. In Section 17.4, we take specific examples of voice and video and determine the amount of capacity required to guarantee absolute and average delay. In Section 17.5, we simulate an actual IP link based on DiffServ architecture and present results in terms of encountered delay for the capacities computed in the previous section. In Section 17.6, we summarize the chapter and discuss possible future directions and extensions.

## 17.2 NETWORK ARCHITECTURE

In this section, we introduce the reader to various components of a WiMAX network and discuss their relationship and requirements such as BS, GateWay (ASN-GW), ASN, CSN, etc. We also discuss the architecture of the IP interface at the BS and delve into the details capturing the configuration and settings required to guarantee the delay requirements of different applications.

### 17.2.1 WiMAX NETWORK ARCHITECTURE

In Figure 17.1 (from Ref. [17]), we present the overall network architecture of a WiMAX network. The network can be logically partitioned into three components, user terminals, ASN, and CSN. User terminals capture the data origination points, could be using the fixed, mobile, or portable WiMAX technology. All the three variations can be supported using a common air interface. ASN spans the BS and the ASN-GW. BS receives the transmitted signal, processes it, and converts into an IP packet and sends to the GW on the outgoing IP transport link. GW receives and upon processing determines the destination on the network side and sends the packet. BS and GW are connected to each other using an IP transport. Typical implementations would have BS located in the field/coverage area and the GW will be centrally located in the switch centers. Therefore, the IP link between BS and

---

* Efforts by competitive local exchange carriers has led to sharp decrease in the T-1 pricing. Previously, average price for T-1 would be USD 6000–7000 per year.

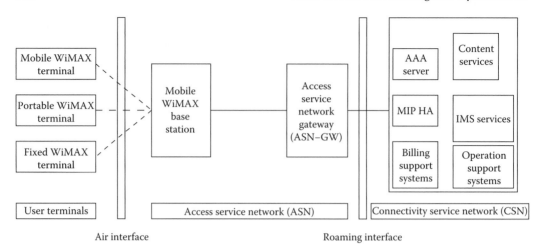

**FIGURE 17.1**   Logical network architecture of a WiMAX network.

GW forms the transport backhaul network. CSN contains many different commercial off-the shelf (COTS) components, which provide connectivity services to the WiMAX subscribers. Addressing, authentication, and availability (AAA) servers, mobile IP home agent (MIP HA), IP multimedia services (IMS), content services, etc. provide support for seamless services to subscribers. AAA servers ensure that a user is uniquely identified and authenticated as legitimate customer. MIP HA ensures that roaming across IP networks is handled and accurate routing of data packets is ensured. Call processing related services are provided by IMS entity. Billing and operational support systems help in managing the overall network.

In Figure 17.2, we present typical implementation of a WiMAX network in a market. For example, say a carrier plans to lay down WiMAX network in Washington D.C. market. Typically, we would have more than 100 BSs connecting to a GW location, based on the anticipated traffic, each GW location might require a cluster of servers providing the functions of the GW. Each IP transport link would be leased from the local carrier and provisioned. Based upon the cost points and required capacity, the carrier can choose to directly lease a TDM segment, Ethernet link, fiber connectivity, etc. Components of the CSN located at each switch center might also be implemented using clusters and would have enough capacity to support the entire market. Switch centers could be connected to each other using a high speed IP network running on an OC-192 (or higher) SONET ring leased from local exchange carrier. Actual network would also include connectivity to the other markets, trunking with public switched telephone network (PSTN) via the end office (EO), tandem connections with other wireless carriers, etc.

For most WiMAX networks, it is unlikely that the carriers would provision the IP transport based on the capacity of the WiMAX air interface. According to WiMAX forum [17], air interface built on 10 MHz channel with $2 \times 2$ MIMO can support peak downlink rate of 63 Mbps and peak uplink rate of 28 Mbps per sector. Assuming three sectors per BS, this would translate into close to 200 Mbps of backhaul transport for each BS. When we share the symbols 3:1 between DL and UL, it could provide data rates of 46 Mbps DL and 8 Mbps UL per sector. Even then it would require about 150 Mbps of capacity between BS and GW. Such a requirement would lead to an unmanageable backhaul cost, which might become a road block in the large-scale adoption of the WiMAX technology.

Our contention is that the service providers will only provision based on the anticipated demand. For example, they might provision just enough capacity for $N$ voice calls, $M$ video calls, and few more Mbps for best effort. This would ensure that the initial cost of building the network is manageable, and as the users grow, more backhaul can be added to ensure acceptable QoS for the subscribers.

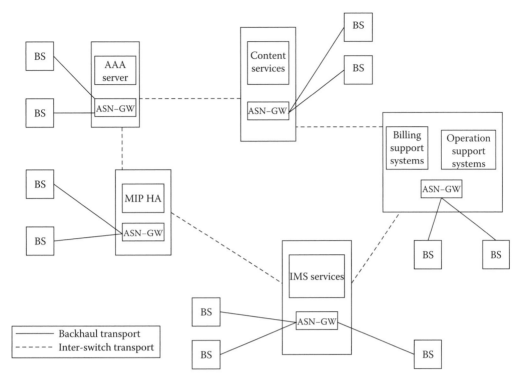

**FIGURE 17.2**   Physical network architecture of a WiMAX network.

In this chapter, we systematically pursue various techniques for mapping the voice and video calls with QoS requirements to required capacity.

### 17.2.2   IP TRANSPORT ARCHITECTURE

Next, we present the architectural details for the IP transport link between the BS and the GW. Presence of multiple service classes with different QoS requirements running on a native IP link with constrained capacity necessitates prioritization and scheduling among the arriving packets. Internet Engineering Task Force (IETF) has standardized the differentiated services architecture (DiffServ) for large-scale deployment of IP networks with QoS support [14]. They have provided three types of service to packets: expedited forwarding (EF) [12], assured forwarding (AF) [9], and best effort. The applications requiring absolute delay bound are mapped to EF class. For providing the average delay bound, we use the AF class. We propose to use the proportional delay differentiation (PDD) model of Dovrolis et al. [4,6] for providing different delays to the subclasses within the AF class. The PDD-based approach is unique in its simplicity and tractability. Recently, many real-time applications have been successfully mapped to delay and loss differentiation parameters of the PDD subclasses [16,18,19].

Next, we discuss the architectural details of a forwarding interface of the BS. For this purpose and toward discussions in later sections, consider multiple sources which want to send their traffic from BS (A) to GW (B) connected by a direct link $\ell$.

Consider now the BS (A) and GW (B) routers. Assume that the concerned forwarding interface on BS A to GW B has been configured for $n$ EF subclasses and $m$ AF subclasses. At the interface, each source is mapped to EF or AF class based on whether the class requires absolute or average delay. The mapping to subclass (such as $i$) within the class (EF or AF) is based on the source application running at the source (voice, video, etc.). The IP link (A-B) has to support a set of sources $s \in \mathcal{S}$.

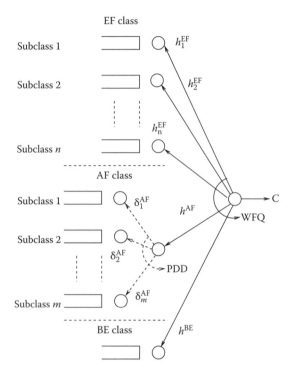

**FIGURE 17.3** Forwarding interface of BS A.

In Figure 17.3, we present the architecture of the forwarding interface of BS A, supporting DiffServ. Let the capacity of the direct link connecting BS (A) to GW (B) be $c$. We assume that the bandwidth is distributed among the $n$ EF subclasses, AF class and BE class using a weighted fair queuing scheduler (WFQ) where the vector $\mathbf{h}$ determines the weights used in scheduling. This ensures that each EF subclass on the link gets no less bandwidth than $c_i^{\text{EF}}$, where

$$c_i^{\text{EF}} = \frac{h_i^{\text{EF}}\, c}{h^{\text{BE}} + h^{\text{AF}} + \sum_{i=1}^n h_i^{\text{EF}}}. \tag{17.1a}$$

Here, $c_i^{\text{EF}}$ is the minimum bandwidth required for subclass $i$ to provide the target delay $D_i^{\text{EF}}$ to the sources belonging to the class. Similarly, $h^{\text{AF}}$ captures the weight for the AF class which translates into minimal bandwidth of

$$c^{\text{AF}} = \frac{h^{\text{AF}}\, c}{h^{\text{BE}} + h^{\text{AF}} + \sum_{i=1}^n h_i^{\text{EF}}}. \tag{17.1b}$$

$c^{\text{AF}}$ is the total bandwidth available to the AF class such that can be shared between the $m$ subclasses. The bandwidth $c^{\text{BE}}$ available to the BE class can be computed as

$$c^{\text{BE}} = \frac{h^{\text{BE}}\, c}{h^{\text{BE}} + h^{\text{AF}} + \sum_{i=1}^n h_i^{\text{EF}}} \tag{17.1c}$$

Multiple sources belonging to the same subclass (EF or AF) at the BS A are placed in the queue for the subclass on a first-come-first-serve basis. Let the set of sources belonging to the $i$th EF subclass be $\mathcal{S}_i^{\text{EF}}$, then every source $s \in \mathcal{S}_i^{\text{EF}}$ has a absolute delay requirement $D_s \geq D_i^{\text{EF}}$. Similarly, for the $i$th AF subclass, sources $s \in \mathcal{S}_i^{\text{AF}}$ have an average delay requirement such that $D_i^{\text{AF}} \leq D_s$.

Source $s$ belonging to AF or EF class has an average arrival rate of $r_s$. When generated by an on–off source model, it has a peak rate of $R_s$ and the on period of average length $I_s$. Such a source can be effectively shaped by an LB filter of parameter $(\sigma_s, \rho_s)$, where $\rho_s$ is the average arrival rate and $\sigma_s$ is the maximum allowed burst length of the LB filter. To ensure low losses, it is advisable to have $\rho_s > 1.1r_s$ and high value of $\sigma_s$.

Observe that sometimes the bandwidth allocated to the AF class needs to be shared between the $m$ subclasses such that each subclass meets its target delay requirement. This is done by using PDD scheduling [6] between the subclasses where the value of parameter $\delta_i^{AF}$ determines the extent of differentiation as discussed in Section 17.3.2. Furthermore, each AF subclass can have an end-to-end delay requirement for the concerned hop. Providing hop-by-hop delay allows greater flexibility and options of better mapping the sources to subclasses. This is outside the scope of current chapter and is currently under study.

## 17.3 DELAY MODELS FOR LINK DIMENSIONING

In this chapter, we focus on delay as the main performance measure. We claim that the other important measure, the delay jitter, could be controlled using playout buffers assuming that the delay encountered is small enough. For nonelastic sources such as UDP-based interactive services having small delays, losses might be totally avoidable. Observe that only when the delays are high and queues in the routers build up, do we have losses. Since the sources are nonelastic, losses are not coupled with throughput.

### 17.3.1 PROVIDING ABSOLUTE DELAY BOUND

In this section, we discuss the amount of capacity required to provide an absolute delay guarantee to sources belonging to the EF class. The capacity is a function of the characteristics of the sources and their delay requirement. Such a problem is sometimes referred to as the equivalent capacity problem (see Ref. [13], and references therein). Absolute bounds on delays can be obtained for sources which are shaped by a LB filter. Behavior of such shaped sources has been extensively studied in the literature. Cruz studied such shaped sources in isolation, when multiplexed and on an end-to-end basis using fluid flow models [1–3]. As discussed before, we consider sources whose traffic is shaped by a LB of parameter $(\sigma_s, \rho_s)$. The value of the maximum delay $D_s$ that can be incurred by any packet can be determined based on the nature of the application connected to the source $s$. First, we consider a subclass $i$ and we define

$$\rho_i^{EF} = \sum_{s \in S_i^{EF}} \rho_s, \quad \sigma_i^{EF} = \sum_{s \in S_i^{EF}} \sigma_s.$$

Then the maximum backlogged traffic from A to B for EF subclass $i$ [15] will be

$$Q_i^{EF} \leq \sigma_i^{EF}.$$

Observe that direct addition of burst lengths could be a conservative approach. However, it is necessary to guarantee deterministic delays to each individual source. Furthermore, we have

$$c_i^{EF} = \max \left\{ \rho_i^{EF}, \frac{\sigma_i^{EF}}{D_i^{EF}} \right\}. \tag{17.2}$$

The required minimal capacity can be ensured to the subclass $i$ by adjusting the weights based on Equation 17.1a and b. The above presented capacity requirement is only used for comparison with the required capacity for AF classes which provide average delay guarantee.

### 17.3.2 PROVIDING AVERAGE DELAY BOUND

In this section, we determine the minimal bandwidth $c^{AF}$ required by the AF class to ensure that the average delay for each AF subclass meets or exceeds its required average delay $D_i^{AF}$. Recall that AF class only provides average delay guarantee to its sources.

Such a problem was originally considered by Dovrolis et al. in Ref. [5]. They accounted for the average arrival rates of the sources but did not account for their burstiness and used simulation to arrive at the required capacity. The burstiness of the sources impacted the derived capacity in an indirect way. The approach is simple yet effective. For each AF subclass, define the target average delay as $d_i^{AF}$. Then, we know that

$$D_i^{AF} \geq d_i^{AF}. \tag{17.3}$$

To ascertain this, consider an imaginary queue which is being fed by the sources belonging to the AF class. The packets are serviced at the rate $c^{AF}$ on a first-come-first-serve (FCFS) basis. Let $q^{AF}$ denote the average length of such a queue. The aggregate arrival rate will be

$$r^{AF} = \sum_{i=1}^{m} \sum_{s \in \mathcal{S}_i^{AF}} r_s. $$

Recall that $r_s$ is the average arrival rate for source $s \in \mathcal{S}$, and here we assume that $r^{AF} < c^{AF}$. Then, the required capacity $c^{AF}$ is such that the imaginary queue has queue length,

$$q^{AF} \leq \sum_{i=1}^{m} \sum_{s \in \mathcal{S}_i^{AF}} r_s d_i^{AF}. \tag{17.4}$$

Alternately, the average waiting time for the imaginary queue should be

$$d^{AF} \leq \frac{1}{r^{AF}} \sum_{i=1}^{m} \sum_{s \in \mathcal{S}_i^{AF}} r_s d_i^{AF}. \tag{17.5}$$

It will ensure that each source belonging to subclass $i$ will have an average delay of $d_i^{AF}$ or less.

The approach argues that if the number of packets in the imaginary queue conforms to the condition (Equation 17.4) then in the real queue, PDD scheduler can distribute the available capacity among the contending AF subclasses such that each one of them conforms to the desired average delay on short as well as long-time scales. The PDD scheduler requires the parameter $\mathbf{d}^{AF}$ which can be computed as follows [5]. Without loss of generality, we assume that subclass $m$ has the maximum delay requirement, then

$$\delta_i^{AF} = \frac{d_i^{AF}}{d_m^{AF}}, \quad i = 1, 2, \ldots, m-1, \quad \text{and} \quad \delta_m^{AF} = 1. \tag{17.6}$$

We then have to determine the capacity $c^{AF}$ required to achieve an average queue length of $q^{AF}$. For a given value of average queuelength, characteristics of sources impact the amount of capacity required. The capacity was determined using simulations in Ref. [5]. The approach provides good estimates for required capacity but has limited utility towards network dimensioning. Due to the use of simulation, it would be hard to incorporate in an overall network design problem. This would greatly curtail its usefulness toward the goal of designing QoS-based IP networks.

In this chapter, we refine on the approach and use it towards the goal of determining the value of $c^{AF}$. In some cases, we do have computational models for the imaginary queue. We can then use them in the Dovrolis framework and this gives us a fast computational technique for estimating delays. We apply this approach to Poisson, on–off, and LB-controlled sources. For each of the three scenarios, we develop the expressions for average queue length and use it to compute the required capacity.

### 17.3.2.1 Poisson Sources

We first consider that each source generates packets with an exponential inter-arrival time. Such systems are fairly well studied in the literature. We know that each source $s$ has the average arrival rate of $r_s$, i.e., its inter-arrival times are exponential with a mean of $1/r_s$. The required capacity is referred to as $c_P^{AF}$. Using the $M/M/1$ queuelength formula, we have

$$d^{AF} = \frac{\left(r^{AF}/c_P^{AF}\right)^2}{r^{AF}\left(1 - r^{AF}/c_P^{AF}\right)}.$$

The equation can be rearranged to get

$$\left(c_P^{AF}\right)^2 - r^{AF}c_P^{AF} - \frac{r^{AF}}{d^{AF}} = 0.$$

Then the required capacity will be

$$c_P^{AF} = \frac{r^{AF} + \sqrt{(r^{AF})^2 - 4r^{AF}/d^{AF}}}{2}. \tag{17.7}$$

The Poisson-based model was also considered in Ref. [5] and is presented here for comparison.

### 17.3.2.2 On–Off Sources

Next we consider sources which have a two-state, on–off behavior. Such models are sometimes used to characterize voice or video sources. As mentioned before, each source has an average rate of $r_s$, peak rate of $R_s$, and average on-period of $I_s$. We develop upon the work presented in Ref. [7]. The required capacity, referred to as $c_{OO}^{AF}$ and the average delay $d^{AF}$, are related by

$$r^{AF}d^{AF} = \frac{1}{\left(c_{OO}^{AF} - r^{AF}\right)} \sum_{i=1}^{m} \sum_{s \in \mathcal{S}_i^{AF}} \left[(R_s - r_s)\left(R_s - c_{OO}^{AF} + r^{AF} - r_s\right)\frac{r_s I_s}{R_s}\right].$$

Upon solving, we get

$$c_{OO}^{AF} = \frac{\left(r^{AF}\right)^2 d^{AF} + \sum_{i=1}^{m}\sum_{s \in \mathcal{S}_i^{AF}}(R_s - r_s)(R_s - r_i + r^{AF})\frac{r_s I_s}{R_s}}{r^{AF}d^{AF} + \sum_{i=1}^{m}\sum_{s \in \mathcal{S}_i^{AF}}(R_s - r_s)\frac{r_s I_s}{R_s}}. \tag{17.8}$$

### 17.3.2.3  Shaped Sources

Now consider the scenario where each source is policed by a LB with parameters $\rho_s$ and $\sigma_s$. For this we consider the results presented in Ref. [8]. They have shown the following result.

---

**THEOREM 1**

*The average delay for a queue serving at rate c to multiplexed stream of sources ($s = 1, 2, \ldots, S$) policed by leaky bucket ($\rho_s$, $\sigma_s$) is*

$$\bar{d} = \frac{\sum_{s=1}^{S} \sigma_s \rho_s}{2c \left( c - \sum_{s=1}^{S} \rho_s \right)}. \tag{17.9}$$

For ease of presentation, define the average regulated arrival rate for the AF class as

$$\rho^{AF} = \sum_{i=1}^{m} \sum_{s \in S_i^{AF}} \rho_s,$$

and the regulated arrival rate weighted burstiness for the AF class as

$$\gamma^{AF} = \sum_{i=1}^{m} \sum_{s \in S_i^{AF}} \rho_s \sigma_s.$$

Then the value of capacity, referred to as $c_{LB}^{AF}$, should satisfy the relation

$$\left( c_{LB}^{AF} \right)^2 - c_{LB}^{AF} \rho^{AF} - \frac{\gamma^{AF}}{2\, d^{AF}} = 0, \tag{17.10}$$

where $d^{AF}$ was expressed in Equation 17.5. Solving the quadratic equation and discarding the value which is smaller than $\rho^{AF}$, we get

$$c_{LB}^{AF} = \frac{1}{2} \left( \rho^{AF} + \sqrt{(\rho^{AF})^2 + \frac{4\, \gamma^{AF}}{2\, d^{AF}}} \right). \tag{17.11}$$

Observe that the required capacity increases with burstiness weighted with average arrival rate and decreases with increasing target delay.

## 17.4  NUMERICAL RESULTS

In this section, we present numerical results comparing the required capacity for providing average delay guarantee for Poisson, on–off, and shaped sources. We also present for comparison the results for absolute delay guarantees. We consider two types of on-off sources, voice and video.

### 17.4.1 Voice Application

We model a voice source as a two state on–off source where it generates packets with a deterministic inter-arrival time of 15 ms in the on-state. On-periods are exponential with rate 2.5 and off-period are also exponential with a rate 1.67. This leads to an average rate $r = 25.632$ kbps, $R = 64$ kbps, and $I = 2.5$. Each source can be policed by an LB with parameters $\rho = 28$ kb and $\sigma = 192$ kb which incurs losses of less than 0.1 percent.

### 17.4.2 Video Application

The video source is also modeled as a two state on–off source. During each burst, the source generates 184.4 packets per second, each packet of size 1000 bytes for an packet inter-arrival time of 5.4 ms during the active period. The length of the active period is exponentially distributed with an average of 0.23 s. This produces an average rate $r = 1.08$ Mbps and peak rate $R = 1.475$ Mbps. Each application can be policed by an LB with parameter $\rho = 1200$ kb and $\sigma = 360$ kb which causes fairly low losses.

Now, we compare the values of $c$ for absolute delay requirement on one hand, and average delay for Poisson, on–off, and LB-shaped sources on the other. To normalize the voice and video sources, we compute the ratio of capacity to the average arrival rate ($c/r$), referred to as overprovisioning. The amount of overprovisioning is dependent upon the source models and the delay requirements. To ensure that the link is not overloaded and that the queues do not build up excessively, we impose the condition that the value of capacity is always greater than 1.05 time average arrival rate. Furthermore, we consider two variations, application-based partitioning and application-based sharing.

### 17.4.3 Application-Based Partitioning

Here, we reserve separate bandwidth for each subclass. In other words, different applications such as voice and video do not share bandwidth. However, sources belonging to each subclass share the capacity. The scenario could be useful for service providers who wish to guard applications from each other by isolation. As an outcome, benefits of multiplexing between applications can not be yielded. Here, $c_1/r$ and $c_2/r$ represent the overprovisioning required to guarantee absolute delay for voice and video sources. $c_{LB}/r$, $c_{OO}/r$ and $c_P/r$ refer to overprovisioning required when dimensioning for average delays using LB-based model, using on-off based model and Poisson-based model, respectively. In this regard, we present the results for voice sources in Figure 17.4 and that of video sources in Figure 17.5. We see that the requirement of absolute delay causes an

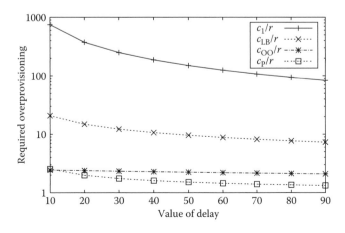

**FIGURE 17.4** Required overprovisioning for single voice source.

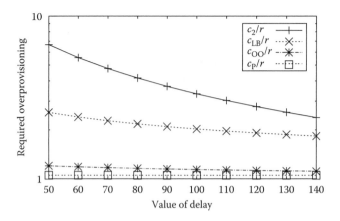

**FIGURE 17.5** Required overprovisioning for single video source.

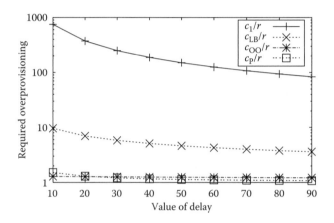

**FIGURE 17.6** Required overprovisioning for five voice sources.

overprovisioning as high as 1000 for voice traffic and more than 8 for video traffic. Guaranteeing average delay reduces the overprovisioning to the range 2–11 in voice sources and less than 1.2 in video sources. Particularly, design with Poisson models requires less than twice the arrival rate for target delays in the range 10–100 ms. For video, minor overprovisioning is sufficient to ensure the average delay of 50–100 ms.

In Figures 17.6 through 17.9, we present the required overprovisioning for five and ten voice and video sources. Observe that the benefits of multiplexing between the sources of the same subclass further help in decreasing the overprovisioning for the average delay scenario. For absolute delay, the values are same. Interestingly, for five and ten voice sources requiring 10 ms of average delay (Figures 17.6 and 17.8), the on–off-based model requires marginally less capacity than the Poisson based model. This can be ascribed to the fact that for on–off-based model, the packets are generated at regular deterministic intervals during an active period and hence multiplexed on–off sources can lead to a smoother traffic than Poisson.

### 17.4.4 APPLICATION-BASED SHARING

The previous scenario did not require scheduling between the voice and video sources since they are not sharing the capacity. Now, we consider such a sharing. For the absolute delay requirement, sharing is still not possible although WFQ ensures that capacity unused by other classes is made available

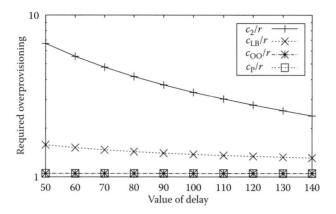

**FIGURE 17.7** Required overprovisioning for five video sources.

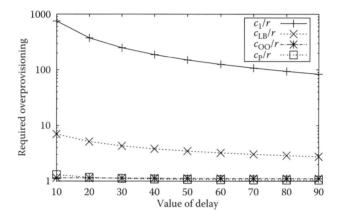

**FIGURE 17.8** Required overprovisioning for ten voice sources.

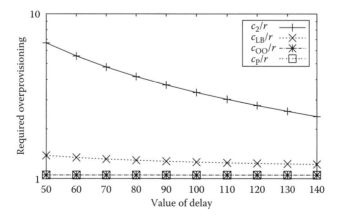

**FIGURE 17.9** Required overprovisioning for ten video sources.

to active classes, but no guarantee can be provided. Therefore, we do not present results for absolute delay requirement. When considering average delay requirements, in the previous sections we have discussed that the bandwidth could be shared between the subclasses using the PDD scheduling based

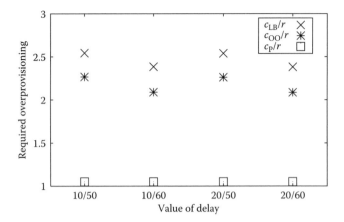

**FIGURE 17.10**   Required overprovisioning for single sources.

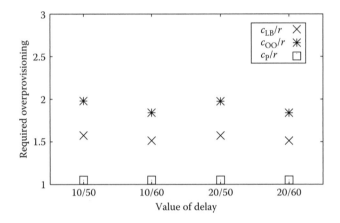

**FIGURE 17.11**   Required overprovisioning for five sources.

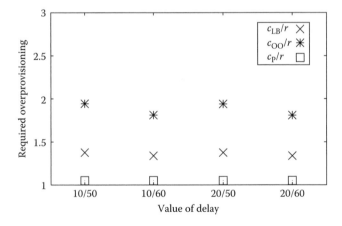

**FIGURE 17.12**   Required overprovisioning for ten sources.

on the parameter $\mathbf{d}^{AF}$ (see Equation 17.6). We now consider that the voice sources are allocated to AF subclass 1 and video sources are mapped to AF subclass 2. In Figures 17.10 through 17.12, we present the overprovisioning required to support both the AF subclasses 1 and 2, each having one, five, and ten sources. Here also $c_{LB}/r$, $c_{OO}/r$, and $c_P/r$ refer to overprovisioning required when dimensioning for average delays using LB-based model, using on–off-based model and Poisson-based model, respectively.

Note that for single voice and video source, design using LB filter requires more capacity than the on–off source-based design model whereas for five and ten voice and video sources, on–off-based model requires more capacity. This can be attributed to the way these two models derive the benefit of multiplexing. The LB-based model better accounts for the multiplexing gain as compared to the on–off-based model. These interactions are subjects for further work.

So far we have investigated the extent of overprovisioning required to guarantee average delay vis-a-vis absolute delay for an IP transport link for a WiMAX backhaul network. Using analytical results, we have shown that 100–1000 times more bandwidth is required for voice and five to ten times more for video sessions for guaranteeing absolute delays to these applications. Such high numbers might be financially untenable for the backhaul network, particularly in the face of high bandwidth costs. As previously discussed in the introduction, WiMAX requires 150–180 Mbps of data rate (including voice, video, and best effort) and even forcing that the capacity be five to ten times the data rate, would require more than 1 Gbps. Such a high amount would be prohibitively expensive if not unavailable in most parts of North America and the World.

In this backdrop, we stress that wireless carriers need to take a closer look at the financial pressures of WiMAX buildout and the techniques used to dimension their backhaul. We contend that average delay-based techniques might be the only financially viable approach and hence deserves a closer look. In the next section, we further investigate average delay-based models using simulations.

## 17.5 SIMULATION RESULTS

The average-based design models lead us to much smaller estimates of required capacity. Therefore, they run the risk of not being able to guarantee acceptable performance for many real-time applications such as voice or video where jitter must also be taken into account. We currently do not have design models that can take jitter into account so we need to evaluate whether the jitter remains acceptable in a system designed with an average delay method.

In this section, we present simulation results to study the delays encountered by the individual voice and video sources under various provisioning scenarios and compare them with the required delays for voice and video, respectively. We used ns-2 to conduct simulations. In this section, we only consider AF subclasses where multiple sources send packets to each subclass, and packets of each subclass are served in the order of their arrival while sharing bandwidth between the subclasses using PDD scheduling. The parameter for the PDD scheduling are determined based on the discussions in Section 17.2. The simulation model for AF class is shown in Figure 17.13. We simulate a voice source using a two state on–off model where it generates packets with a deterministic inter-arrival time of 15 ms in the on-state. On-periods are exponential with rate 2.5 and off-periods are also exponential with a rate 1.67. Each packet is of size 120 bytes. The video source is modeled using deterministic batch arrivals with batch inter-arrival time of 33 ms. The number of packets in a batch are geometrically distributed with an average of five packets. In each burst, the last packet has size distributed as uniform (0,1000) bytes. All other packets have 1000 bytes.

Here also $c_{LB}/r$, $c_{OO}/r$, and $c_P/r$ refer to overprovisioning required when, dimensioning for average delays using LB based model, using on–off-based model and Poisson-based model, respectively. Observe that the capacity computed using these models along with PDD-based scheduling were presented in Figures 17.10 through 17.12 for single, five, and ten voice and video sources. Now we use that capacity for the simulation and compare in Figures 17.14 through 17.19 the delays for single, five, and ten voice and video sources. We have plotted the observed mean delay and error bars corresponding to twice the sample standard deviation for voice and video sources. We also present a horizontal line showing the required average delay for each source.

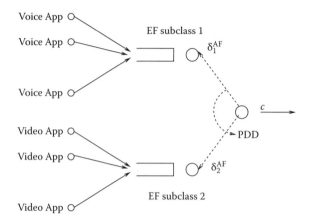

**FIGURE 17.13** Simulation model for AF class.

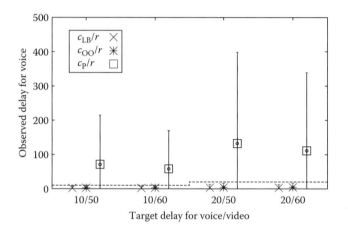

**FIGURE 17.14** Delay for single voice source.

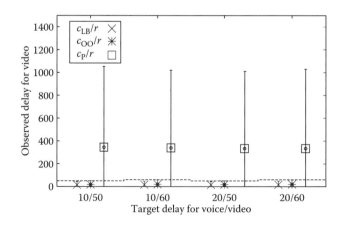

**FIGURE 17.15** Delay for single video source.

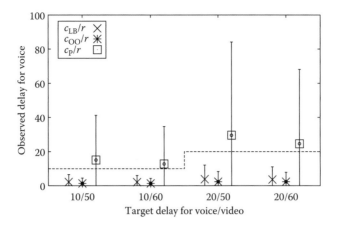

**FIGURE 17.16**  Delay for five voice source.

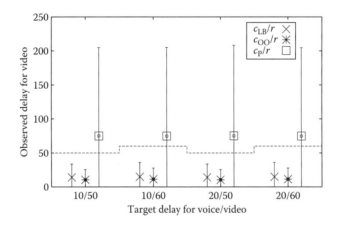

**FIGURE 17.17**  Delay for five video source.

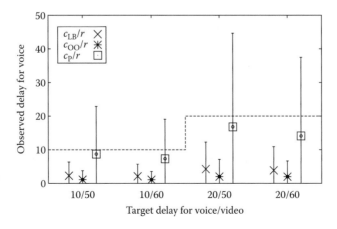

**FIGURE 17.18**  Delay for ten voice source.

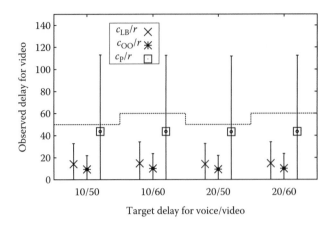

**FIGURE 17.19**    Delay for ten video source.

Note that for the Poisson-based capacity model with single sources, the actual mean delay is many times the target delay, both for voice and video. Moreover, some voice packets can have a delay as high as 400 ms and will be useless at the receiver. For video also, packets can have delays as much as 1 s. Such a capacity planning is not very useful and could lead to unsatisfied customers. When we multiplex five or ten voice and video sources, the average delays get closer to the target delays and for ten sources, they are even acceptable for both voice and video. However, there is still a large variance in the observed delays and voice packets could still have as high as 40 ms and video as high as 100 ms. Note that such high delays could be tolerable if they affect only a small number of packets.

Next, we consider on–off and LB-based design models. Observe that both the approaches provide acceptable delays, average as well as average along with two times standard deviation. The values are smaller than the required delays and hence a significant fraction of packets belonging to voice and video sources will encounter less than required delays. These models remain consistent for single, five, or ten sources and provide acceptable, performance to individual sources. Note that the LB-based model provides delays which are less than the target for both voice and video, although it requires lesser capacity than the on–off-based models. Observe that not only the delays are acceptable but also the variance is quite small.

Based on these results, it can be argued that LB-based model could be used to determine required capacity for a source requesting an average delay QoS. When allocating capacity for a small number of sources, it can achieve the multiplexing gain and provides minimal capacity to meet the required delays.

## 17.6    SUMMARY AND FUTURE WORK

In this chapter, we consider capacity requirements for IP transport for backhaul of WiMAX networks. We contend that the high cost of bandwidth in the backhaul space challenges the carriers to evaluate average delay-based models for providing QoS of different applications. For average delay guarantee, we consider AF service class of the differentiated services based architecture for QoS aware IP networks. Three kinds of models were accounted for: Poisson, on–off, and LB based. We present closed form expressions to determine the capacity required to ensure average delay to each service class. Using numerical results, we compare the required capacity for the three models with the capacity required to guarantee absolute delays for voice and video applications. It was observed that absolute delays require many orders of more capacity than the average delay models. We then use these capacity values to simulate a typical link and present results demonstrating the delays encountered by voice and video sources for these capacity models. It was also found that LB-based model is suitable for classes with few sources. However, for networks with high number of individual sources, Poisson-based models can also be used successfully.

We are in the process of incorporating other source structures such as three state, long range, dependent, etc. into the design models. We are also in the process of extending the analysis to multilink or end-to-end network-based models.

## ACKNOWLEDGMENT

We would like to thank C. Dovrolis from Georgia Tech. for providing the simulation code for PDD scheduling.

## REFERENCES

[1] R. L. Cruz. A calculus for network delay, Part I: Network elements in isolation. *IEEE Transactions on Information Theory*, 37(1):114–131, 1991.

[2] R. L. Cruz. A calculus for network delay, Part II: Network analysis. *IEEE Transactions on Information Theory*, 37(1):132–141, 1991.

[3] R. L. Cruz and H. Liu. End-to-end queueing delay in ATM networks. *Journal of High Speed Networks*, 3(4):413–427, 1994.

[4] C. Dovrolis and P. Ramanathan. A case for relative differentiated services and the proportional differentiation model. *IEEE Network*, 13(5):26–34, September/October 1999.

[5] C. Dovrolis and P. Ramanathan. Dynamic class selection and class provisioning in proportional differentiated services. *Computer Communications*, 26:204–221, 2003.

[6] C. Dovrolis, D. Stiliadis, and P. Ramanathan. Proportional differentiated services: Delay differentiation and packet scheduling. *IEEE/ACM Transactions on Networking*, 10(1):12–26, 2002.

[7] H. Dupuis and B. Hajek. Simple formulas for multiplexing delay for independent regenerative sources. In *Proceedings of INFOCOM*, pp. 28–35. IEEE, San Francisco, CA, 1993.

[8] F. M. Guillemin, N. Likhanov, R. R. Mazumdar, and C. Rosenberg. Extremal traffic and bounds for the mean delay of multiplexed regulated traffic streams. In *Proceedings of INFOCOM*, pp. 985–993. IEEE, New York, 2002.

[9] J. Heinanen, F. Baker, W. Weiss, and J. Wroclawski. Assured forwarding PHB group. In *Request For Comments, RFC 2497*. Internet Engineering Task Force, June 1999.

[10] IEEE. *Air Interface for Fixed Broadband Wireless Access Systems*, STD 802.16-2004, October 2004.

[11] IEEE. *Air Interface for Fixed and Mobile Broadband Wireless Access Systems*, STD 802.16e/D12, February 2005.

[12] V. Jacobson, K. Nichols, and K. Poduri. An expedited forwarding PHB group. In *Request For Comments, RFC 2498*. Internet Engineering Task Force, June 1999.

[13] F. P. Kelly. Notes on effective bandwidths. In F. P. Kelly, S. Zachary, and I. Ziedins, editors, *Stochastic Networks: Theory and Applications*, volume 4 of *Royal Statistical Society Lecture Notes Series*, pp. 141–168. Oxford University Press, 1996.

[14] K. Nichols, S. Blake, F. Baker, and D. Black. Definition of the differentiated services field (DS Field) in the IPv4 and IPv6 headers. In *Request For Comments, RFC 2474*. Internet Engineering Task Force, December 1998.

[15] A. K. Parekh and R. G. Gallager. A generalized processor sharing approach to flow control in integrated services networks: The multiple node case. *IEEE/ACM Transactions on Networking*, 2(2):137–150, 1994.

[16] W. Tan and A. Zakhor. Packet classification schemes for streaming MPEG video delay and loss differentiated networks. In *Proceedings of Packet Video Workshop*, Kyongju, Korea, April 2001.

[17] WiMAX Forum. *Mobile WiMAX – Part I: A Technical Overview and Performance Evaluation*, August 2006.

[18] L. Zhao, J. Kim, and C.-C. J. Kuo. Rate adaptation and error control for scalable video streaming over Diffserv networks. In *Proceedings of Symposium on Electronic Imaging*, pp. 20–25, San Jose, CA, 2002.

[19] L. Zhao, J. Kim, and C.-C. J. Kuo. Constant quality rate control for streaming MPEG-4 FGS video, *Proceedings of International Symposium on Circuits and Systems (ISCAS)*, pp. 544–547, Scottsdale, AZ, IEEE 2002.

# 18 An Optimization Model for WiMAX Network Planning

*Fabio D'Andreagiovanni and Carlo Mannino*

## CONTENTS

Mixed-integer linear programming (MILP) is a well-established technique to model and solve a large variety of network planning problems. In this chapter, we show how to derive an MILP formulation for worldwide interoperability for microwave access (WiMAX) network planning. The model includes as decision variables the major physical and radio-electrical parameters, namely, base stations locations, emission powers, transmission frequencies, burst profiles, and service areas. An optimum solution to this model assigns suitable values to all parameters to maximize the expected overall profit.

## 18.1 INTRODUCTION

The extraordinary growth of demand for wireless connections resulted in highly congested frequency spectrums and in a drastic reduction of available sites to accommodate transmitting antennas. This stimulated the development of optimization models and algorithms to support planning decisions. In this chapter, we describe an optimization model for the WiMAX network planning problem (WPP). To do this, the physical and radio-electrical parameters relevant to the model are first identified and described. Such parameters are then associated with binary and semicontinuous decision variables. Logical relations, coverage, and capacity requirements are represented by linear inequalities in the decision variables.

## 18.2    OPTIMIZATION IN WIRELESS NETWORK DESIGN

A standard model, suitable for planning purposes, identifies a wireless network with a set of transmitting and receiving antennas scattered over a territory. Such antennas are characterized by a position (geographical coordinates and elevation) and by a number of radio-electrical parameters. The network design process consists in establishing locations and suitable radio-electrical parameters of the antennas. The resulting network is evaluated by means of two basic performance indicators: (1) network coverage, that is the quality of the wanted signals perceived in the target region and (2) network capacity, that is the ability of the network to meet traffic demand. On the basis of quality requirements and projected demand patterns, suitable target thresholds are established for both indicators. In principle, coverage and capacity targets should be pursued simultaneously, as they both depend on the network configuration. However, to handle large real-life instances, conventional network planning resorts to a natural decomposition approach, which consists in performing coverage and capacity planning at different stages (see Ref. [28]). In particular, the network is designed by first placing and configuring the antennas to ensure the coverage of a target area, and then by assigning a suitable number of frequencies to meet (projected) capacity requirements. The final outcome can be simulated and evaluated by an expert, and the whole process can be repeated until a satisfactory result is obtained (Figure 18.1). Future change in demand patterns can be met by increasing sectorization (i.e., mounting additional antennas in a same site), by selecting new sites, and by assigning additional transmission frequencies (see Ref. [25]).

The network planning process requires an adequate representation of the territory. In the past years, the standard approach was to subdivide the territory into equally sized hexagons (see Ref. [27]) and basic propagation laws were implemented to calculate field strengths. By straightforward analytical computations, these simplified models could provide the (theoretical) position of the antennas and their transmission frequencies. Unfortunately, the approximations introduced by this approach were in most cases unacceptable for practical planning, as the model does not take into account

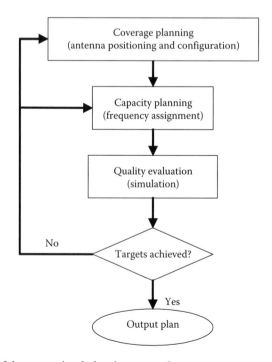

**FIGURE 18.1**    Phases of the conventional planning approach.

several fundamental factors (e.g., orography of target territories, equipment configurations, actual availability of frequencies and of geographical sites to accommodate antennas, etc.). Furthermore, the extraordinary increase of wireless communication quickly resulted in extremely large networks and congested frequency spectrum, and asked for a better exploitation of the available band. It was soon apparent that effective automatic design algorithms were necessary to handle large instances of complex planning problems, and to improve the exploitation of the scarce radio resources. These algorithms were provided by mathematical optimization. Indeed, already in the early 1980s, it was recognized that the frequency assignment performed at the second stage of the planning process is equivalent to the Graph Coloring Problem (or to its generalizations). The graph coloring problem consists in assigning a color (= frequency) to each vertex (= antenna) of a graph so that adjacent vertices receive different colors and the number of colors is minimum. The graph $G = (V, E)$ associated with the frequency assignments of a wireless network is called interference graph, since edge $uv \in E$ represents interference between nodes $u \in V$ and $v \in V$ and implies that $u$ and $v$ cannot be assigned the same frequency (see Ref. [7]). The graph coloring problem is one of the most known and well studied topics in combinatorial optimization. A remarkable number of exact and heuristic algorithms have been proposed over the years to obtain optimal or suboptimal colorings. Some of these methods were immediately at hand to solve the frequency assignment problem.

The development of mathematical optimization methods triggered the introduction of more accurate representations of the target territories. In particular, also inspired by standard Quality-of-Service (QoS) evaluation methodologies, the coarse hexagonal cells were replaced with (the union of) more handy geometrical entities, namely the demand nodes introduced by Tutschku [28], and with the now universally adopted testpoints (TP). In the TP model, a grid of approximately squared cells is overlapped to the target area. Antennas are supposed to be located in the center of testpoints: all information about customers and QoS in a TP, such as traffic demand and received signals quality, are aggregated into single coefficients. The TP model allows for smarter representations of the territory, of the actual antennas position, of the signal strengths, and of the demand distributions. This in turn permits a better evaluation of the QoS and, most important, makes it possible to construct more realistic interference graphs, thus leading to improved frequency assignments. Indeed, by means of effective coloring algorithms, it was possible to improve the design of large real-life mobile networks (see the FAP-Web [13]), and also of analogue [18] and digital broadcasting networks [19].

Finally, basing on the TP model, it was also possible to develop accurate models and effective optimization algorithms to accomplish the first stage of the planning process, namely the coverage phase, to establish suitable positions and radio-electrical parameters for the antennas of a wireless network [9].

In recent years, thanks to the development of more effective optimization techniques and to the increase of computational power, a number of models integrating coverage and capacity planning have been developed and applied to the design of global system for mobile (GSM) [26], universal mobile telecommunication system (UMTS) [10,12], Analog and Digital Video Broadcasting [19] networks.

The optimization models above mentioned are defined by associating suitable decision variables $x \in R^n$ with the physical and radio-electrical antenna parameters (i.e., candidate locations, power values, activation statuses, transmission frequencies, service, and coverage requirements, etc.). Such variables must satisfy a number of constraints, which are represented by inequalities of the form $g(x) \leq b$, where $g : R^n \to R$ and $b \in R$. The set of the feasible values is defined as $X = \{x \in R^n : g_i(x) \leq b_i, i = 1, \ldots, m\}$. Finally, the general optimization problem can be written as

$$\min_{x \in X} f(x) \tag{18.1}$$

where $f : R^n \to R$ is the objective function and may represent, for example, unsatisfied traffic demand or installation costs to be minimized. When $f$ and $g_i$, for $i = 1, \ldots, m$, are linear functions, Problem (18.1) is a Linear Program (LP) [22]. If, in addition, some of the variables can only assume integer values, Problem 18.1 becomes MILP. We will show that the WiMAX network planning problem can be reduced to the solution of a suitable MILP.

Several optimization models (most often MILPs) have been proposed for wireless network planning, in broadcasting, mobile, military, and civil communications. Such models can be solved by ad hoc algorithms or by commercial solvers. However, the large size of most instances of practical relevance and the notorious difficulty of the corresponding optimization problems, makes it often necessary to resort to heuristic procedures which typically yield suboptimal solutions.

In the following sections, we formalize the WiMAX network planning problem and formulate it as a MILP. In Section 18.3, we describe the technological features which will be modeled by our decision variables. The overall MILP model is then shown in Section 18.4. The model has been tested on several realistic instances, both in urban and rural environment. Computational results are presented in Section 18.5.

## 18.3 SYSTEMS ELEMENTS

In this section, we introduce the technological elements and the modeling assumptions which provide the basis of the optimization model presented in Section 18.4. We consider the design of a Fixed WiMAX Network [1]. It consists of a set of installations: the base stations (BS)—distributed over a number of sites to provide connectivity to a set of customers' equipments; the subscriber stations (SS)—located in a portion of territory called target area (Figure 18.2).

### 18.3.1 REPRESENTATION OF TERRITORY

The definition of an appropriate territory representation is an essential requirement for effective wireless network planning. The coverage is in fact evaluated on the basis of predicted propagation conditions and these are in turn calculated by a propagation model which takes into account the characteristics of the area. In recent years, the so-called TP model has been universally adopted for quality evaluation and planning purposes. A grid of equally sized squares is overlapped to the target area. By choosing a suitable dimension of the basic square, the TP, the signal strength in the center of the corresponding area can be considered as an acceptable approximation for the whole square. In the sequel, the set of all testpoints is denoted by $T$. Every SS located in a TP originates a demand for WiMAX services that turns into a bandwidth request. The traffic demand $d_t$ of testpoint $t$ is equal to the traffic generated by all the SSs in $t$. A parameter $r_t$ is also introduced to represent the potential revenue associated with the customers in $t$.

### 18.3.2 BSS AND TRANSCEIVERS

The backbone of every cellular network is constituted by the BSs, typically consisting of a pylon accommodating a number of transceivers (TRX). In the following, the set of all the TRXs that can be deployed in the potential sites is denoted by $B$. Every TRX is characterized by a position

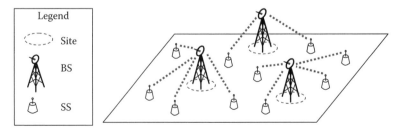

**FIGURE 18.2**   Scheme of a fixed WiMAX network.

represented by geographical coordinates, height, and orientation (azimuth and tilt). Candidate sites are conventionally identified with the center of (a subset of) TP. A TRX is also characterized by a set of radio-electrical parameters. Typically, only a subset of such parameters is tackled by the optimization model. Specifically, we consider: (1) frequency channel, which belongs to a finite set of available channels $F$, each having a constant bandwidth $D$; (2) *emitted power* $P_b^f$ at which TRX $b \in B$ transmits on frequency $f \in F$ (with $P_b^f \in [P_b^{min}, P_b^{max}]$). In every site and for every frequency channel, it is possible to install either a single TRX, mounting an omnidirectional antenna, or several TRXs, each with a directive antenna. Note that the activation of a TRX may prevent the simultaneous activation of other potential TRXs in the same site and operating on the same frequency. This is the case, for example, of TRXs with similar azimuths. A family $\mathcal{G} = \{G_1, \ldots, G_{|\mathcal{G}|}\}$ is therefore introduced, where $G_i \subseteq B$, $i = 1, \ldots, |\mathcal{G}|$, is a set of mutually exclusive TRXs. Finally, a parameter $c_b$ is introduced to represent the overall cost of installation and activation of TRX $b$.

## 18.3.3 PROPAGATION MODELS

The planning process requires the adoption of a propagation model to predict signal propagation conditions and to calculate the overall strength attenuation $a_{tb} \in [0, 1]$ from the center of the TP accommodating transmitter $b$ to the center of each TP $t \in T$. The signal power $P_b^f(t)$ received in TP $t$ from TRX $b$ on frequency $f$ is calculated as

$$P_b^f(t) = a_{tb} \cdot P_b^f \tag{18.2}$$

## 18.3.4 SERVICE COVERAGE

A TP is said to be covered by the network if the wanted signal is received with suitable quality. Coverage depends not only on the wanted signal strength, but also on the strength of other unwanted (interfering) signals. Specifically, the quality of the received signal is measured by means of the signal to interference ratio (SIR), which is defined as the ratio between the wanted signal power and the sum of the interfering signals powers, also including thermal noise (see Ref. [24]). A TP is regarded as covered if the SIR value is above a given threshold $\delta$. If we denote by $\beta$ the TRX serving $t$ on a frequency $f \in F$, from Constraint 18.2 the above requirement can be expressed by the following inequality:

$$\frac{a_{t\beta} \cdot P_\beta^f}{\sum_{b \in B \setminus \{\beta\}} a_{tb} \cdot P_b^f + N} \geq \delta \tag{18.3}$$

where

$N$ is the thermal noise

$\sum_{b \in B \setminus \{\beta\}} a_{tb} \cdot P_b^f$ is the cumulative interference generated by all other TRXs

## 18.3.5 BURST PROFILES

By adopting adaptive modulation and coding (AMC), WiMAX supports a number of combinations of modulation schemes and forward error correction (FEC) coding schemes [11]. These combinations are referred to as burst profiles. By defining different burst profiles, different spectral efficiencies can be reached [29]. To provide higher capacity using limited bandwidth, it would be desirable to select more efficient schemes. However, these schemes are more vulnerable to interference and thus require a higher signal-to-noise ratio to ensure a fixed bit error ratio. A set $H$ is thus introduced to represent the available burst profiles. For every profile $h \in H$ two parameters are also introduced, namely $\delta_h$ representing the SIR threshold that must be reached to ensure service coverage according to Inequality 18.3, and $s_h$ representing the spectral efficiency associated with the burst profile. In Table 18.1 we resume the notation introduced so far.

**TABLE 18.1**
**Summary of Notation**

| | |
|---|---|
| $T$ | Set of testpoints (TP) |
| $B$ | Set of transceivers (TRX) |
| $F$ | Set of frequency channels |
| $H$ | Set of burst profiles |
| $D$ | Channel bandwidth |
| $d_t$ | Traffic demand of testpoint $t$ |
| $r_t$ | Revenue from service supply of testpoint $t$ |
| $\mathcal{G}$ | Family of sets of mutually exclusive TRXs |
| $c_b$ | Cost of installation and activation of TRX $b$ |
| $P_b^{\min}$ | Minimum emitted power of TRX $b$ |
| $P_b^{\max}$ | Maximum emitted power of TRX $b$ |
| $a_{tb}$ | Total attenuation of signal power transmitted by TRX $b$ to TP $t$ |
| $N$ | Thermal noise |
| $\delta_h$ | SIR threshold of burst profile $h$ |
| $s_h$ | Spectral efficiency of burst profile $h$ |

We are finally able to state the (WPP):

**PROBLEM 18.1   (WPP)**

Given sets $B$, $T$, $F$, $H$, $\mathcal{G}$, the attenuation matrix $[\mathbf{a}]_{t \in T, b \in B}$, the minimum and maximum power vectors $\mathbf{P}^{\min}, \mathbf{P}^{\max} \in R^{|B|}$ the demand vector $\mathbf{d} \in R^{|T|}$, the capacity vector $\mathbf{D} \in R^{|B|}$, the testpoint revenue vector $\mathbf{r} \in R^{|T|}$ and the TRXs installation costs $\mathbf{c} \in R^{|T|}$, and all fixed radio-electrical parameters, the *WiMAX Network Planning Problem* is the one of finding positions and emission powers of the TRXs to maximize the overall profit.

In Ref. [21], a hierarchy of major wireless network planning problems is identified. The classification takes into account a wide literature and is based on the decision parameters included in the optimization model. The root of the hierarchy is a general planning problem, namely the *power and frequency assignment problem* (PFAP), which asks for establishing suitable emission powers and transmission frequencies of the antennas to maximize the total profit. Interestingly, by considering capacity constraints and multiple burst profiles, WPP generates a new class of problems, which further generalizes PFAP.

## 18.4   FORMULATION

In this section, we describe a MILP formulation for WPP along with reduction techniques to improve the quality of the coefficient matrix. We concentrate on downlink transmission. Uplink can be modeled similarly and easily included. We use the notation introduced in the previous section and summarized in Table 18.1.

We introduce three sets of Boolean variables:

$$z_t = \begin{cases} 1 \text{ if testpoint } t \in T \text{ is served} \\ 0 \text{ otherwise} \end{cases}$$

$$y_b^f = \begin{cases} 1 \text{ if TRX } b \in B \text{ transmitting on channel } f \in F \text{ is} \\ \quad \text{activated} \\ 0 \text{ otherwise} \end{cases}$$

$$x_{tb}^{fh} = \begin{cases} 1 \text{ if testpoint } t \in T \text{ is served by TRX } b \in B \text{ on} \\ \quad \text{channel } f \in F \text{ with burst profile } h \in H \\ 0 \text{ otherwise} \end{cases}$$

We introduce also a set of semicontinuous variables:

$$p_b^f \in \{0, [P_b^{\min}, P_b^{\max}]\}$$

representing the power emitted by TRX $b \in B$ on channel $f \in F$.

The aim is to maximize total profit, which corresponds to the difference between total revenue from served TPs and total TRXs activation costs. This can be expressed by the following objective function:

$$\max \sum_{t \in T} r_t \cdot z_t - \sum_{b \in B} \sum_{f \in F} c_b \cdot y_b^f \tag{18.4}$$

A testpoint $t \in T$ can be served only if there exists at least one TRX $b$ serving $t$ on a frequency $f$ with burst profile $h$, that is at least one variable $x_{tb}^{fh}$ is equal to 1, for $b \in B, f \in F, h \in H$. This can be expressed by the following linear constraints:

$$z_t \leq \sum_{b \in B} \sum_{f \in F} \sum_{h \in H} x_{tb}^{fh} \quad t \in T \tag{18.5}$$

In fact, if $x_{tb}^{fh} = 0$ for all $b \in B, f \in F, h \in H$, then $z_t = 0$ and $t$ is not served.

If $x_{tb}^{fh} = 1$, for some $b \in B, f \in F, h \in H$, then TRX $b$ must be activated on frequency $f$. This is ensured by the following variable upper bound constraints:

$$x_{tb}^{fh} \leq y_b^f \quad t \in T, b \in B, f \in F, h \in H \tag{18.6}$$

In fact, $y_b^f = 0$ implies $x_{tb}^{fh} = 0$.

Observe that if $x_{t\beta}^{fh} = 1$, for some $t \in T, \beta \in B, f \in F, h \in H$, then $t$ is served by $\beta$ on frequency $f$ with profile $h$ and the corresponding SIR Inequality 18.3 must be satisfied. We can easily rewrite Inequality 18.3 in the following linear form:

$$a_{t\beta} \cdot p_\beta^f - \delta_h \sum_{b \in B \setminus \{\beta\}} a_{tb} \cdot p_b^f \geq \delta_h \cdot N \tag{18.7}$$

Clearly, when $x_{t\beta}^{fh} = 0$ the above constraint does not hold. This can be modeled by introducing a large constant $M$ and by including in the MILP model the following constraint for all $t \in T, b \in B, f \in F, h \in H$:

$$a_{t\beta} \cdot p_\beta^f - \delta_h \sum_{b \in B \setminus \{\beta\}} a_{tb} \cdot p_b^f + M \cdot (1 - x_{t\beta}^{fh}) \geq \delta_h \cdot N \tag{18.8}$$

Note that when $x_{t\beta}^{fh} = 1$, then Constraint 18.8 reduces to the SIR Inequality 18.7. On the other hand, when $x_{tb}^{fh} = 0$, Constraint 18.8 reduces to the following inequality:

$$a_{t\beta} \cdot p_\beta^f - \delta_h \sum_{b \in B \setminus \{\beta\}} a_{tb} \cdot p_b^f \geq \delta_h \cdot N - M \tag{18.9}$$

which, for $M \geq \delta_h \sum_{b \in B \setminus \{\beta\}} a_{tb} \cdot P_b^{\max} + \delta_h \cdot N$, is satisfied by every feasible power vector **p** and is therefore redundant.

When a $x_{tb}^{fh} = 1$, testpoint $t$ consumes a portion $d_t \cdot \frac{1}{s_h}$ of the bandwidth of channel $f$. The sum of the bandwidth consumed by all testpoints served by $b$ on $f$ must therefore not exceed the total bandwidth $D$ of the channel. This is expressed by the following set of constraints:

$$\sum_{t \in T} \sum_{h \in H} d_t \cdot \frac{1}{s_h} \cdot x_{tb}^{fh} \leq D \quad b \in B, f \in F \tag{18.10}$$

To prevent the activation of mutually exclusive TRXs, we introduce the following family of constraints:

$$\sum_{b \in G} y_b^f \leq 1, \quad f \in F, G \in \mathcal{G} \tag{18.11}$$

Finally, observe that if TRX $b$ is not activated on frequency $f$ then $p_b^f = 0$, otherwise $P_b^{\min} \leq p_b^f \leq P_b^{\max}$. This can be expressed by

$$p_b^f \leq y_b^f \cdot P_b^{\max} \quad b \in B, f \in F \tag{18.12}$$

and

$$p_b^f \geq y_b^f \cdot P_b^{\min} \quad b \in B, f \in F \tag{18.13}$$

We are finally able to summarize the overall MILP formulation for (WPP):

$$\max \quad \sum_{t \in T} r_t \cdot z_t - \sum_{b \in B} \sum_{f \in F} c_b \cdot y_b^f \quad (WPP - LP)$$

$$\text{s.t.} \quad a_{t\beta} \cdot p_\beta^f - \delta_h \sum_{b \in B \setminus \{\beta\}} a_{tb} \cdot p_b^f + M \cdot (1 - x_{tb}^{fh}) \geq \delta_h \cdot N$$

$$\quad t \in T, \beta \in B, f \in F, h \in H \tag{18.14}$$

$$z_t \leq \sum_{b \in B} \sum_{f \in F} \sum_{h \in H} x_{tb}^{fh} \quad t \in T \tag{18.15}$$

$$\sum_{t \in T} \sum_{h \in H} d_t \cdot \frac{1}{s_h} \cdot x_{tb}^{fh} \leq D \quad b \in B, f \in F \tag{18.16}$$

$$\sum_{b \in G} y_b^f \leq 1 \quad f \in F, G \in \mathcal{G} \tag{18.17}$$

$$x_{tb}^{fh} \leq y_b^f \quad t \in T, b \in B, f \in F, h \in H \tag{18.18}$$

$$p_b^f \leq y_b^f \cdot P_b^{\max} \quad b \in B, f \in F \tag{18.19}$$

$$p_b^f \geq y_b^f \cdot P_b^{\min} \quad b \in B, f \in F \tag{18.20}$$

$$y_b^f \in \{0, 1\} \quad b \in B, f \in F \tag{18.21}$$

$$x_{tb}^{fh} \in \{0, 1\} \quad t \in T, b \in B, f \in F, h \in H \tag{18.22}$$

$$z_t \in \{0, 1\} \quad t \in T \tag{18.23}$$

$$p_b^f \in \{0, [P_b^{\min}, P_b^{\max}]\} \quad b \in B, f \in F \tag{18.24}$$

## 18.4.1 STRENGTHENING

It is common experience that the above formulation cannot be solved to optimality when applied to large real-life instances. In some cases, even finding feasible solutions can represent a difficult task, also for effective commercial MILP solvers such as CPLEX [30]. These difficulties are mainly determined by the following reasons: (1) the general wireless network planning problem belongs to the class of NP-hard problems [19], and no polynomial time algorithm is known to the solution of (WPP-LP). This implies that the solution time can grow very fast as the number of (binary) variables grows. (2) The presence of the notorious "big-M" coefficient $M$ makes (WPP-LP) a weak formulation, that is the solution to its linear programming relaxation, obtained by removing the integrality stipulation on the variables, yields poor quality upper bounds. This in turn drives standard MILP solution algorithms to generate larger search trees. (3) The coefficient matrix of WPP-LP is (very) ill conditioned, because of the large range of feasible power values and attenuation coefficients of most real-life instances. Indeed, the ratio between the largest and the smallest coefficient in a SIR Constraints 18.14 can be up to $10^{12}$.

To overcome these difficulties, two re-formulation techniques have been recently investigated for DVB and UMTS network planning. In particular, the Dantzig-Wolfe decomposition can be applied to generate a set packing reformulation of (WPP-LP) with exponentially many columns [20] while an analogous of Benders' decomposition yields a set covering formulation with exponentially many rows [21]. Such noncompact reformulations have (0,1) coefficient matrices and must be solved by applying appropriate row and column generation mechanisms.

Another promising coefficient reduction approach has been proposed in Ref. [20], allowing to improve the quality of the coefficient matrices and reduce the value of $M$. It is based on the simple observation that the activation of an interfering TRX at its minimum emission power may be sufficient to prevent service in a testpoint. Analogously, the activation of a serving TRX at its minimum emission power may be sufficient, in special cases, to ensure coverage for any possible emission value of the interfering TRXs. We show now how to extend these ideas to reduce WPP-LP.

---

### DEFINITION 1

*Let $SIR(t, \beta, f, h)$ be a constraint of type (18.7). Then $\gamma \in B - \{\beta\}$ is a* superinterferer *for $SIR(t, \beta, f, h)$ if $\delta_h \cdot a_{t\gamma} \cdot P_\gamma^{\min} > a_{t\beta} \cdot P_\beta^{\max} - \delta_h \cdot N$.*

It is worth noting that a superinterferer $\gamma$ is typically associated with a large coefficient $\delta_h \cdot a_{t\gamma}$ of Constraint 18.7. If a superinterferer $\gamma$ for a constraint $SIR(t, \beta, f, h)$ is activated on frequency $f$ (i.e., $y_\gamma^f = 1$), then the corresponding SIR Inequality 18.7 cannot be satisfied. This implies that $x_{t\beta}^{fh} = 0$, and the corresponding Constraint 18.14 in WPP-LP becomes redundant. On the other hand, if $\gamma$ is not activated on frequency $f$ (i.e., $y_\gamma^f = 0$), then $p_\gamma^f = 0$ and the term $a_{t\gamma} \cdot p_\gamma^f$ can be removed from Constraint 18.14. In other words, the Constraint 18.14 associated with $SIR(t, \beta, f, h)$ can be replaced with the pair of inequalities:

$$y_\gamma^f \leq 1 - x_{t\beta}^{fh} \tag{18.25}$$

and

$$a_{t\beta} \cdot p_\beta^f - \delta_h \sum_{b \in B \setminus \{\beta, \gamma\}} a_{tb} \cdot p_b^f + \bar{M} \cdot (1 - x_{t\beta}^{fh}) \geq \delta_h \cdot N \tag{18.26}$$

where $\bar{M} = \delta_h \sum_{b \in B \setminus \{\beta, \gamma\}} a_{tb} \cdot P_b^{\max} + \delta_h \cdot N$.

If fact, when $y_\gamma^f = 1$, then Inequality 18.25 implies $x_{t\beta}^{fh} = 0$ and Inequality 18.26 becomes redundant. When $y_\gamma^f = 0$ then $p_\gamma^f = 0$ and Inequality 18.14 reduces to Inequality 18.26.

It is easy to see that the two new inequalities have "nicer" coefficients than the original one. In particular, Inequality 18.25 is a standard packing constraint (all nonzero coefficients are 1). The large coefficient $a_{t\gamma}$, which appears in Constraint 18.14, does not appear in Inequality 18.26. Finally, we also have $\bar{M} \leq M - a_{t\gamma}$.

If the set of superinterferers of $\text{SIR}(t, \beta, f, h)$ is denoted by $\text{Super}(t, \beta, f, h)$, then the corresponding Constraint 18.14 can be replaced with the following family:

$$y_\gamma^f \leq 1 - x_{t\beta}^{fh} \quad \gamma \in \text{Super}(t, \beta, f, h) \tag{18.27}$$

and

$$a_{t\beta} \cdot p_\beta^f - \delta_h \sum_{b \in \hat{B}} a_{tb} \cdot p_b^f + \hat{M} \cdot (1 - x_{t\beta}^{fh}) \geq \delta_h \cdot N \tag{18.28}$$

where $\hat{B} = B \backslash \text{Super}(t, \beta, f, h) \cup \{\beta\}$ and $\hat{M} = \delta_h \sum_{b \in \hat{B}} a_{tb} \cdot P_b^{\max} + \delta_h \cdot N$.

Now, suppose $a_{t\beta} \cdot P_\beta^{\min} \geq \delta_h \sum_{b \in \hat{B}} a_{tb} \cdot P_b^{\max} + \delta_h \cdot N$, i.e., $\beta$ is a superserver. This implies that if all superinterferers are not activated on frequency $f$ while the superserver $\beta$ is active, then the SIR inequality is satisfied. Then Constraint 18.28 can be replaced by the following logic constraint:

$$y_\beta^f - \sum_{\gamma \in \text{Super}(t, \beta, f, h)} y_\gamma^f \leq x_{t\beta}^{fh} \tag{18.29}$$

In fact, when $y_\gamma^f = 0$ for all $\gamma \in \text{Super}(t, \beta, f, h)$ and $y_\beta^f = 1$, then $x_{t\beta}^{fh} = 1$. Note that in the above constraint, all coefficients are in the set $\{-1, 0, 1\}$.

## 18.5 COMPUTATIONAL EXPERIENCE

The MILP formulation presented in Section 18.4 has been applied to several instances of WiMAX network planning referred to two distinct scenarios: an urban one corresponding to a district of the city of Rome and a rural one corresponding to an area in Central Italy near the town of Avezzano. Besides assessing the capability of the model to address realistic instances, the experiments were also designed to evaluate the behavior of two major network performance indicators, namely coverage and missed traffic, as functions of traffic demand.

All of the MILPs were solved by means of the commercial solver *ILOG CPLEX 10.0* on an *Intel Core 2 Duo* 1.80 Ghz machine with 2 Gb RAM.

The urban scenario refers to Montesacro district, in the Northeastern part of Rome. Originally a residential district, Montesacro has been recently pervaded by a network of small and medium enterprises. For this reason, it appears to be suitable for the penetration of WiMAX services and has been selected to build up 23 distinct instances. The performance is measured through two indices: (1) coverage, which corresponds to the percentage of covered testpoints (i.e., satisfying SIR Constraint 18.7); (2) missed traffic, which corresponds to the percentage of unsatisfied traffic demand.

The bandwidth demand $d_t$ of each testpoint $t \in T$ is estimated according to the methodology described in Ref. [4]. In particular, the projected demands are obtained by considering the extent and composition of the population, the offered range of services and the expected market penetration rate. The district of Montesacro is classified as an urban environment in which two groups of customers, namely residential and small and medium enterprises (SME), generate an average demand of 22 Mbps/km$^2$. This demand is then distributed over the testpoints taking into account the particular area demographics. For each customer class, two types of services are supposed to be available: a slower and cheaper basic service and a faster and more expensive premium service. The revenue $r_t$ of a testpoint is computed by assuming a particular distribution (see Ref. [4]) of the demand $d_t$ among the four possible combinations of customers and service classes.

The notification of public auction for WiMAX licenses of the *Italian Communications Regulatory Authority (Agcom)* provides for the assignment of three licenses in the $(3.4 \div 3.6)$ GHz band [5]. The channels set $F$ coincides with the three 7 Mhz frequency channels that a license grants in downlink. In every site and for every frequency channel, it is possible to install either a single TRX, equipped

**TABLE 18.2**

**Omnidirectional Antenna**

**Parameters**

| | |
|---|---|
| Gain | 8.5 dBi |
| Power | [10,45] dBm |

**TABLE 18.3**

**Directive Antenna Parameters**

| | |
|---|---|
| Gain | 15 dBi |
| Power | [10,45] dBm |
| Azimuth | [0°, 360°] 10° step |

with an omnidirectional antenna (see Table 18.2 for the radio-electrical parameters), or up to three TRXs, each equipped with a 120° directive antenna (Table 18.3).

The azimuth of a directive antenna may vary in the range [0°, 360°] with a step of 10° thus allowing 36 distinct orientations. For every orientation, a TRX $b$ is added to set $B$. In these tests, we did not take into account tilt. However, a discrete number of tilts, say $k$, can be easily considered in the model by including $k$ copies of each $b \in B$. Denoting by $S$ the set of candidate sites, then for each $s \in S$, 37 potential TRXs are introduced

- One TRX corresponding to the omnidirectional antenna and denoted by $(s, 0)$
- Thirty-six directive TRXs, each corresponding to one specific orientation and denoted by $(s, i)$, for $i = 1, 2, \ldots, 36$

The activation in a same site and on a same frequency of overlapping directive (120°) transceivers is not allowed: in other words, the activation of TRX $(s, i)$ prevents the activation of TRX $(s, i - k)$, for $k = 1, \ldots, 11$, and of TRXs $(s, i + k)$, for $k = 1, \ldots, 11$, associated with the preceding and the following 11 orientations. Analogously, the activation of the omnidirectional antenna $(s, 0)$ prevents the activation of any other TRX $(s, i)$, for $i = 1, \ldots, 36$. In other words, for each site $s \in S$, exactly $2 \times 36$ mutually exclusive sets of TRXs are identified, namely the sets $G_{s,i}^- = \{(s, 0), (s, i - 11), \ldots, (s, i)\}$ and $G_{s,i}^+ = \{(s, 0), (s, i), \ldots, (s, i + 11)\}$, for $i = 1, \ldots, 36$. Thus, the family of mutually exclusive TRXs is given by $\mathcal{G} = \{G_{s,i}^+ : s \in S, i = 1, \ldots, 36\} \cup \{G_{s,i}^- : s \in S, i = 1, \ldots, 36\}$.

We define positive transceivers activation costs $c_b$ for each TRX, according to the WiMAX Forum estimations [3,4]. The cost is obtained by summing up a fixed and a variable component: the fixed component includes all the elements that precede the installation of any WiMAX equipment (acquisition of sites and pylons positions, backhaul interface equipment, etc.), while the variable component is strictly related to the mechanical and radio-electrical specifications of the TRX.

To predict propagation conditions, the *COST-231 Hata Model* [6] is adopted. However, it must be noted that the optimization model is independent of the adopted propagation model as it only affects the coefficients of the input attenuation matrix. COST-231 Hata is a path loss model developed as an extension of the *Hata-Okomura* model [16,23]: it is designed for transmissions on the band from 500 to 2000 MHz and includes a correction term to take into account the area type (urban, suburban, or rural). It is widely used and taken as reference model for predictions in WiMAX networks [2]. Though out of its frequency range of definition, it is also used in WiMAX networks operating in the 3.5 GHz band [8,15]. In the case of Montesacro, the correction factors for urban environment are applied. Elevation data required by the model are provided by a digital terrain database of the area.

**TABLE 18.4**

**Supported Burst Profiles**

| Modulation Scheme | Code Rate |
| --- | --- |
| BPSK | 1/2 |
| QPSK | 1/2 |
| 16-QAM | 1/2 |
| 16-QAM | 3/4 |
| 64-QAM | 1/2 |
| 64-QAM | 3/4 |

**FIGURE 18.3**   Satellite image of Montesacro district.

In most WiMAX implementations, the number of offered burst profiles is typically a small fraction of all the possible [11]. For the Montesacro network a set of six profiles involving the four possible modulation schemes are provided (Table 18.4).

The basic instance involves a target area that is approximately 3.6 × 3.6 km wide (Figure 18.3) and is decomposed into a grid of 1600 TP. On the basis of territory configuration and demand distribution, seven potential sites are selected to accommodate candidate TRXs (Figure 18.4). One of the sites is located in the court of an industrial service center: a 15 m high pylon is supposed to be directly installed on the ground. The other sites correspond to rooftops of buildings placed in high locations, and transceivers are mounted on 10 m high pylons.

**FIGURE 18.4**   Target area: testpoints and sites.

It is worth noting that, due to the small size of the target area, every omnidirectional or three 120° directional antennas BS, located in any candidate site, is able to cover all testpoints. Moreover, every BS is able to serve at least 97 percent of the testpoints by means of the best available burst profile. In particular, BSs located in nearly baricentric sites can ensure a full 100 percent 64-QAM based coverage of the entire area.

In the first instance, we suppose that the revenue of a testpoint is proportional to the demanded traffic (i.e., $r_t = k \cdot d_t$, with $k > 0$, $t \in T$), and that TRXs activation costs are negligible (i.e., $c_b = 0$, $b \in B$). The objective function (Equation 18.4) thus reduces to $k \cdot \max \sum_{t \in T} d_t z_t$. The solution returned by CPLEX was able to cover 99 percent of the area, while missing only 2 percent of traffic demand, by activating five complete trisectorial TRXs in five distinct sites. Only 1 percent of the testpoints was not covered due to interference.

The second instance is derived from the previous one by introducing TRXs activation costs. The output plan is shown in Figure 18.5. The grey squares are associated with zero demand and zero revenue testpoints, which in most cases correspond to unbuilt areas. Four out of the seven potential installations are activated (darker circles in the figure) allowing a 98.5 percent coverage. The testpoints served by each installation are represented by different colors. The white testpoints are unserved. Observe that some of the white testpoints are completely surrounded by served areas: they correspond to residential areas with low revenue customers. In this case, the limited bandwidth capacity is allocated to more profitable testpoints that generate business premium traffic. Each BS transmits on all available channels and thanks to quite fair propagation conditions is able to use, in most cases, 64-QAM-based burst profiles. The results show a three percent missed traffic, which represents a three percentage points increase with respect to the first instance. This is a consequence

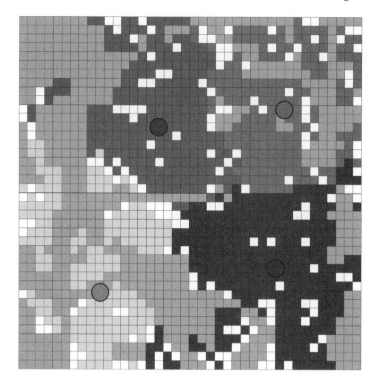

**FIGURE 18.5**   TP assignment and activated BSs.

**TABLE 18.5**

**Results for Increasing Average Traffic Density (3 × 7 Mhz Channels)**

| Traffic Density (Mbps/km²) | Coverage (percent) | Service by QAM64 (percent) | Missed Traffic (percent) | BS Spacing (km) |
|---|---|---|---|---|
| 10 | 100.0 | 100.0 | 0.0 | 3.25 |
| 12 | 100.0 | 100.0 | 0.2 | 3.10 |
| 14 | 100.0 | 100.0 | 0.4 | 3.00 |
| 16 | 100.0 | 100.0 | 0.5 | 2.85 |
| 18 | 100.0 | 98.0 | 1.0 | 2.65 |
| 20 | 100.0 | 98.0 | 2.2 | 2.50 |
| 22 | 99.5 | 97.0 | 3.0 | 2.20 |
| 24 | 99.0 | 96.0 | 3.5 | 2.05 |
| 26 | 98.5 | 95.5 | 4.9 | 2.00 |
| 28 | 97.5 | 93.0 | 5.5 | 1.80 |
| 30 | 97.0 | 92.5 | 6.5 | 1.75 |
| 32 | 96.5 | 92.0 | 7.4 | 1.70 |
| 34 | 95.5 | 90.5 | 7.8 | 1.65 |
| 36 | 94.0 | 89.0 | 8.5 | 1.55 |
| 38 | 93.5 | 87.5 | 9.1 | 1.30 |
| 40 | 93.0 | 85.0 | 9.7 | 1.15 |
| 42 | 91.0 | 85.0 | 10.5 | 1.10 |
| 44 | 89.0 | 84.5 | 11.6 | 1.00 |
| 46 | 89.0 | 84.0 | 14.5 | 1.00 |
| 48 | 87.5 | 84.0 | 19.4 | 1.00 |
| 50 | 86.5 | 83.5 | 23.3 | 1.00 |

of the reduction in the number of activated TRXs, which in turn depends on the trade-off between activation costs and testpoints revenues.

To investigate the behavior of coverage and missed traffic as functions of traffic demand, we tested the formulation on a set of 21 instances with increasing traffic density. Every instance is based on an ideal $10 \times 10$ km target area where traffic is uniformly distributed and 169 candidate sites are regularly distributed over the area. The results obtained by CPLEX are showed in Table 18.5. Both coverage and served traffic reduce as traffic density increases. This is a consequence of the increasing number of needed BSs: to fulfill higher traffic demands, a larger number of BSs has to be activated thus reducing the average spacing, which in turn results in stronger interference. It is worth noting that, once the minimum spacing is reached, further increases in traffic density correspond to higher rates of missed traffic. All testpoints are served through QAM modulations; the percentage of served testpoints reached by 64-QAM is shown in the table.

The previous results are based on a $3 \times 7$ MHz channelization. We then investigated the effect of an alternative channelization using $6 \times 3.5$ MHz channels on the set of instance of Table 18.5. The results are shown in Table 18.6. Thanks to the larger number of channels, interference is mitigated and thus coverage and service degradation is slower. In turn, as channels becomes narrower, a larger number of BSs is needed to serve a given traffic demand. Thus, spacing decreases at a higher rate.

The previous computational experience confirms that, in an urban Fixed WiMAX deployment, service is mainly limited by channel capacity and, secondarily, by interference and signal path attenuation. In contrast, in rural scenarios, due to sparse customers, the service in a testpoint is typically limited by its distance from the BSs. In our experiment, we referred to a territory located in Central Italy near the town of Avezzano: the land is mostly flat and under cultivation, and is characterized by the presence of five main settlements. The target area is $10 \times 10$ km wide, with

**TABLE 18.6**

**Results for Increasing Traffic Density ($6 \times 3.5$ Mhz Channels)**

| Traffic Density (Mbps/km²) | Coverage (percent) | Service by QAM64 (percent) | Missed Traffic (percent) | BS Spacing (km) |
|---|---|---|---|---|
| 10 | 100.0 | 100.0 | 0.0 | 2.95 |
| 12 | 100.0 | 100.0 | 0.0 | 2.95 |
| 14 | 100.0 | 100.0 | 0.0 | 2.85 |
| 16 | 100.0 | 100.0 | 0.2 | 2.65 |
| 18 | 100.0 | 100.0 | 0.7 | 2.40 |
| 20 | 100.0 | 100.0 | 1.2 | 2.35 |
| 22 | 100.0 | 100.0 | 1.6 | 2.10 |
| 24 | 99.5 | 99.0 | 2.6 | 1.90 |
| 26 | 99.5 | 98.5 | 3.1 | 1.85 |
| 28 | 99.0 | 98.5 | 3.7 | 1.70 |
| 30 | 98.5 | 98.0 | 4.2 | 1.60 |
| 32 | 98.5 | 97.0 | 4.8 | 1.50 |
| 34 | 98.0 | 95.5 | 5.5 | 1.35 |
| 36 | 96.0 | 95.0 | 6.1 | 1.20 |
| 38 | 96.0 | 95.0 | 6.7 | 1.05 |
| 40 | 95.5 | 94.0 | 8.1 | 1.00 |
| 42 | 95.0 | 92.5 | 8.4 | 1.00 |
| 44 | 94.5 | 92.0 | 8.9 | 1.00 |
| 46 | 93.5 | 91.5 | 9.5 | 1.00 |
| 48 | 93.0 | 89.5 | 10.3 | 1.00 |
| 50 | 93.0 | 87.0 | 10.7 | 1.00 |

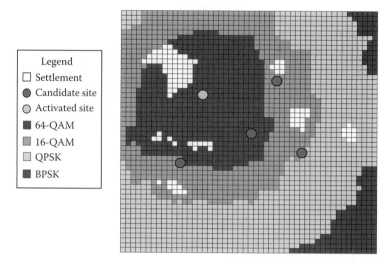

**FIGURE 18.6**    Service results for the rural instance.

**TABLE 18.7**

**Incidence of Each Modulation Scheme
on Service**

| Modulation | Percentage of Served Area |
|---|---|
| 64-QAM | 27 |
| 16-QAM | 28 |
| QPSK | 39 |
| BPSK | 6 |

very low traffic density (0.4 Mbps/km$^2$ on average). Five candidate sites are available for BSs deployment (Figure 18.6) to represent the objective of a public authority, which is interested in ensuring broadband services to the maximum number of people and to the largest possible area, we consider a modified objective function: we let $c_b = 0$, for all $b \in B$ and $r_t = 2$ if $t \in T$ is a village testpoint, $r_t = 1$ otherwise. Observe that the constraints of the model imply that, to ensure coverage in a testpoint, it suffices that the corresponding SIR level is above the lowest burst profile threshold. However, by including suitable penalty terms in the objective functions, solutions with a larger portion of testpoints served by higher burst profiles are preferred. This corresponds to a robust planning principle, typically adopted by practitioners. The optimum solution shows that a single omnidirectional BS suffices to serve the entire area. In particular, the activated BS is located in the site which allows a complete 64-QAM-based coverage of the largest settlement in the area. Not surprisingly, all modulation schemes are exploited as reported in Table 18.7.

## 18.6   CONCLUDING REMARKS

In this chapter, we presented an optimization model for the WiMAX network planning problem. The major physical and radio-electrical parameters are identified and represented by the decision variables of a suitable MILP. The capability of the model to cope with realistic instances has been assessed through a number of experiments.

Similarly to other relevant wireless planning problems, the WiMAX network planning problem poses complex modeling and computational challenges. The experiments show that the running times are quite sensitive to the number of testpoints and candidate antennas and exhibit an even more drastic dependency on the number of frequencies and burst profiles. Besides the fact that we are dealing with large integer programs (which are inherently difficult), other specific factors contribute to such a bad behavior:

1. The coefficients of the fading matrix may differ by several orders of magnitude (up to $10^{12}$), which implies severe numerical restrictions and increasing solution difficulties.
2. The presence of the notorious "big-M" coefficients in Constraints 18.9 yields poor relaxations and consequently a larger number of nodes in the search trees.
3. Finally, the availability of multiple interchangeable channels results in a very large number of equivalent solutions; consequently, most computing time is spent in visiting barren regions of the solution space.

The three points listed above must be addressed to solve larger real-life instances, which may include thousands of testpoints, candidate sites, and several alternative hardware configurations. In particular, (1) and (2) can be tackled by suitable decomposition schemes (such as Dantzig-Wolfe or Benders decomposition, see Ref. [21]) and related reformulations. As to point (3), it is worth noting that a very recent line of research is devoted to the development of symmetry breaking methodologies (see, for example, Ref. [17]), which could be effectively applied to reduce the number of visited solutions.

## ACKNOWLEDGMENT

We wish to thank Maria Missiroli from Fondazione "Ugo Bordoni" for her precious comments and suggestions. This work was partially supported by MIUR, Project Ref. No. 2878, within the European framework COST—Action 2100.

## REFERENCES

[1] IEEE Std. 802.16-2004, IEEE standard for local and metropolitan area networks, Part 16: Air interface for fixed broadband wireless access system, 2004.

[2] IEEE 802.16 broadband wireless access working group, Channel models for fixed wireless apllications system, 2003.

[3] Business case models for fixed broadband wireless access based on WiMAX technology and the 802.16 standard, WiMAX Forum, White Papers, October 2004.

[4] WiMAX deployment considerations for fixed wireless access in the 2.5 GHz and 3.5 GHz licensed bands, WiMAX Forum, White Papers, June 2005.

[5] Procedure per l'assegnazione di diritti d'uso di frequenze per sistemi Broadband Wireless Access (BWA) nella banda a 3.5 GHz—Delibera n. 209/07/CONS, Italian Communications Regulatory Authority, 2007.

[6] COST Action 231, Digital mobile radio towards future generation systems—COST 231 Final Report, EURO-COST, 1999.

[7] K. Aardal, S.P.M. van Hoesel, A.M.C.A. Koster, C. Mannino, and A. Sassano, Models and solution techniques for frequency assignment problems, *Annals of Operations Research* 153(1), 79–129, 2007.

[8] V.S. Abhayawardhana, I.J. Wassell, D. Crosby, M.P. Sellars, and M.G. Brown, Comparison of empirical propagation path loss models for fixed wireless access systems, *Proceedings of the 61st IEEE Vehicular Technology Conference*, Stockholm, Sweden, May 2005.

[9] E. Amaldi, A. Capone, F. Malucelli, and C. Mannino, Optimization problems and models for planning cellular networks, *Handbook of Optimization in Telecommunications*, Springer Science, New York, 2006.

[10] E. Amaldi, A. Capone, and F. Malucelli, Radio planning and coverage optimization of 3G cellular networks, *ACM Wireless Networks*, published online, January 2007.

[11] J.G. Andrews, A. Ghosh, and R. Muhamed, *Fundamentals of WiMAX*, Prentice Hall, Upper Saddle River, NJ, 2007.

[12] A. Eisenblätter, A. Fügenschuh, T. Koch, A. Koster, A. Martin, T. Pfender, O. Wegel, and R. Wessaly, Modelling feasible network configurations for UMTS, *Zentrum für Informationstechnik*, Berlin, Germany, Report 02-16, 2002.

[13] *FAP web – A web site about Frequenct Assignment Problems*, http://fap.zib.de/

[14] A. Gamst, Homogeneous distribution of frequencies in a regular hexagonal cell system, *IEEE Transactions on Vehicular Technology*, 31(3), 132–144, 1982.

[15] P. Grønsund, P.E. Engelstad, T. Johnsen, and T. Skeie, The physical performance and path loss in a fixed WiMAX deployment, *Proceedings of the 2007 International Conference on Wireless Communications and Mobile Computing* , Honolulu, HI, August 2007.

[16] M. Hata, Empirical formula for propagation loss in land mobile radio services, *IEEE Transactions on Vehicular Technology*, 29(3), 317–325, 1980.

[17] V. Kaibel and M. Pfetsch, Packing and partitioning orbitopes, *Mathematical Programming*, 114(1), 1–36, 2008.

[18] C. Mannino and A. Sassano, An enumerative algorithm for the frequency assignment problem, *Discrete Applied Mathematics*, 129(1), 155–169, 2003.

[19] C. Mannino, F. Rossi, and S. Smriglio, The network packing problem in terrestrial broadcasting, *Operations Research*, 54(4), 611–626, 2006.

[20] C. Mannino, F. Marinelli, F. Rossi, and S. Smriglio, A tight reformulation of the power-and-frequency assignment problem in wireless networks, *Technical Report* TRCS 003/2007, Dip. di Informatica, Università L'Aquila, 2007.

[21] C. Mannino, F. Rossi, and S. Smriglio, A unified view in planning broadcasting networks, Dipartimento di Informatica e Sistemistica, Universitá di Roma, Italy, Report 08-07, 2007.

[22] G.L. Nemhauser and L.A. Wolsey, *Integer and Combinatorial Optimization*, John Wiley & Sons, New York, 1988.

[23] Y. Okumura, E. Ohmori, T. Kawano, and K. Fukuda, "Field strength and its variability in VHF and UHF land-mobile service", *Review of the Electrical Communications Laboratory*, 16, September/October 1968.

[24] T.S. Rappaport, *Wireless Communications: Principles and Practice*, 2nd edn., Prentice Hall, Upper Saddle River, NJ, 2001.

[25] W. Stallings, *Wireless Communications & Networks*, 2nd edn., Prentice Hall, Upper Saddle River, NJ, 2004.

[26] R. Mathar and M. Schmeinck, Optimisation models for GSM Radio, *International Journal of Mobile Network Design and Innovation*, 1(1), 70–75, 2005.

[27] R. Steele, Introduction to digital cellular radio, in *Mobile Radio Communications*, edited by R. Steele, John Wiley and Sons, New York, 1992.

[28] K. Tutschku, Demand-based radio network planning of cellular mobile communication systems, Institute of Computer Science, University of Würzburg, Germany, Report No. 177, 1997.

[29] Intel in Communications Adaptive Modulation (QPSK, QAM), Intel White Papers on WiMAX, http://www.intel.com/technologies/wimax, 2004.

[30] CPLEX, commercial optimizer for Mathematical Programming, http://www.ilog.com/products/cplex

# 19 Adaptations for Optimized Performance in WiMAX Networks

*Hossam S. Hassanein, Ahmed Iyanda Sulyman, and Mohamed Ibnkahla*

## CONTENTS

Adaptive signal processing plays a pivotal role in the emerging next-generation wireless networks. The IEEE 802.16 standard-based worldwide interoperability for microwave access (WiMAX) system has deployed various adaptive processing techniques both at the physical (PHY) and the Medium Access Control (MAC)-layer. In this chapter, we discuss the role of adaptations at the various layers in the WiMAX protocol stack, in enhancing resource utilizations and system performance. We take a close look at the WiMAX wireless network standards and discuss key enhancements provided by the use of adaptation techniques specified in the standard. Current protocols and algorithms employed in the system are reviewed, and future research and deployment trends are also discussed.

## 19.1   INTRODUCTION

WiMAX is an emerging telecommunications technology aimed at providing wireless access based on IEEE 802.16 standard. IEEE 802.16 is an evolving standard for broadband metropolitan area networks (MAN). The early version of the standard approved in June 2004, IEEE 802.16-2004 [1], details the PHY and MAC protocols for fixed broadband wireless access (BWA) in a metro area network. In December 2005, following compelling interests emerging for mobile and personal broadband markets, the IEEE 802.16 working group won approval for the IEEE 802.16e extension of the standard, addressing the PHY and MAC layer changes necessary to support mobility. The IEEE 802.16e-2005 standard was published in February 2006 [2], and is referred to as the mobile WiMAX extension of the WiMAX technology. Devices manufactured to this standard will support both fixed and mobile BWA services. WiMAX certified products based on 802.16-2004 for fixed applications have been available in the market since January 2006. Some of these products are still in the form of trial WiMAX network deployments [3], while others have matured into full commercial products. Deployments of products for mobile applications started in June 2006, when Samsung launched the first commercial mobile WiMAX services in Korea, with the name "WiBro" (wireless broadband). Products for WiBro services operate in the licensed 2.3 GHz frequency band with an 8.75 MHz channel bandwidth and are fully compliant with IEEE 802.16e-2005 standard.

Adaptation plays an important role in WiMAX networks, both fixed and mobile applications. The use of adaptive processing techniques provide vital enhancements in system performance and resource utilizations. Adaptive processing techniques currently deployed in the system, both at the PHY and the MAC layers include adaptive antenna and beamforming technologies, adaptive modulation and coding techniques, adaptive forward error correcting codes (FEC) and rate adaptation systems, as well as adaptive resource allocation and quality-of-service (QoS) scheduling. In this chapter, we take a close look at these adaptive processing techniques and discuss the enhancements provided by their use at the various layers in the WiMAX protocol stack. Current protocols and algorithms employed in the standard are reviewed, and future research and deployment trends are also discussed.

## 19.2   WiMAX PHYSICAL AND MAC LAYERS

WiMAX PHY is responsible for the transmission of data over the air interface (physical medium). The PHY receives MAC layer data packets through its interface with the lowest MAC sublayer, and transmits them according to the MAC layer QoS scheduling. WiMAX MAC layer comprises of three sublayers, which interact through service access points (SAP) to provide the MAC layer services, as shown in Figure 19.1. The convergence sublayer (CS) interfaces the WiMAX network with other networks by mapping external network data (from ATM, Ethernet, IP, etc.) to the WiMAX system. MAC common part sublayer (MAC CPS) provides majority of the MAC layer services. The MAC CPS receives data from the CS as MAC service data unit (MAC SDU) and efficiently packs them on to the payload of the MAC packet data unit (MAC PDU) through the process of fragmentation and aggregations. Fragmented parts of MAC SDU are used to fill (aggregate) remnant portions of MAC PDU payloads that cannot accommodate full MAC SDU during package. As WiMAX provides connection-oriented service, MAC CPS is also responsible for bandwidth request/reservation for a requested connection, connection establishment, and maintenance. In the WiMAX standard, bandwidth request/reservation is an adaptive process that takes place on a frame-by-frame basis. This allows more efficient resource utilization and optimized performance. Thus the MAC CPS

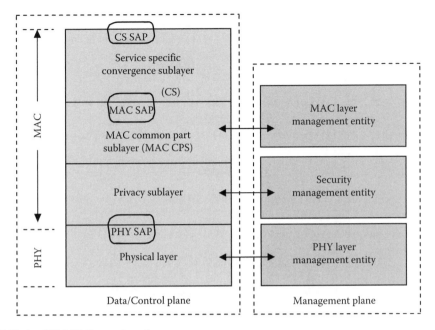

**FIGURE 19.1** WiMAX Protocol stack.

is required to provide up-to-date data on bandwidth request/reservation for each connection, on a frame-by-frame basis. The MAC CPS also provides connection ID for each established connection and marks all MAC PDUs traversing the MAC interface to the PHY with the respective connection ID. This sublayer also performs QoS scheduling by deciding the orders of packet transmissions on the PHY, based on the service flow decided during connection establishments. Privacy sublayer provides authentication to prevent theft of services, and encryption to provide security of services.

  The ensemble of the activities of the three sublayers of the WiMAX MAC layer constitutes the MAC layer services. MAC layer services can broadly be categorized into two: periodic and aperiodic activities. Periodic activities are fast- or delay-sensitive types of activities and are carried out to support ongoing communications, thus they must be completed in one frame duration. Examples include QoS scheduling, packing, and fragmentation. Aperiodic activities are slow- or delay-insensitive types of activities. They are executed when and as required by the system, and are not bounded by frame durations. Examples include ranging and authentications for network entry.

## 19.3 ADAPTATIONS AT THE PHYSICAL LAYER

### 19.3.1 PHY Layer Specifications in the Standard

Operations in high frequencies ranging between 10 and 66 GHz were initially specified in the earlier versions of the 802.16 standard for fixed access. With this specification, only line-of-sight (LOS) signal propagation, with unobstructed path from the transmitter to the receiver, is feasible. Though high-frequency operations have the advantage of less interference, however most wireless technologies prefer lower frequencies because RF signals penetrate structures much better at low frequencies, enabling non-LOS propagation techniques. In non-LOS or multipath propagation modes, the transmitted signals are scattered, reflected, and diffracted by objects in the propagation paths between the transmitter and the receiver as shown in Figure 19.2. Thus, the receiver receives multiple copies of the transmitted signal, each arriving with different amplitude and phase or delay. These multipath signals may combine destructively at the receiver resulting in severe signal fades. To accommodate

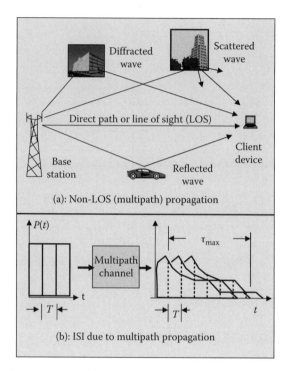

**FIGURE 19.2**   Non-LOS propagation and intersymbol interference (ISI).

services in non-LOS conditions in the WiMAX system, 802.16-2004 standard subsequently speci-fies operations at lower frequencies, between 2 and 11 GHz. Single-carrier transmission, known as wirelessMAN-SC, as well as two multicarrier transmissions, wirelessMAN-OFDM (orthogonal frequency division multiplexing) and wirelessMAN-OFDMA (orthogonal frequency division multiple access) are also specified. The WiMAX system also specified a number of advanced PHY layer and antenna technologies, both fixed and adaptive, to combat the severe fading effect of the multipath propagation channel, to enhance system performance.

### 19.3.2   Antenna Technologies in WiMAX

Advanced antenna technologies specified in the WiMAX system to mitigate the non-LOS propaga-tion problems and ensure high quality signal receptions include [4–6] Diversity and multiple-input multiple-output (MIMO) systems, adaptive antenna systems (AAS), as well as beamforming systems.

#### 19.3.2.1   Diversity Systems

Diversity technique provides the receiver with multiple copies of the transmitted signal, each of them received over independently fading wireless channel. The notion of diversity relies on the fact that with $M$ independently fading replicas of the transmitted signals available at the receiver, the probability of an error detection is improved to $p^M$, where $p$ is the probability that each signal will fade below a usable level. The link error probability is therefore improved without increasing the transmitted power [10,12–14]. Recently, the use of diversity technique at the transmitter side also gained wide attentions, and has resulted in the consideration of the more general case of multiple transmit–multiple receiving antennas or MIMO systems.

### 19.3.2.2 MIMO Systems

The two options for MIMO transmissions in the WiMAX standard are space-time codes and multi-plexing. For space-time codes, both space-time trellis codes and Alamouti space-time block codes are specified. However, it is the Alamouti space-time block codes that has yet been implemented by vendors due to its reduced complexity (eventhough space-time trellis code has better link performance improvements). In the Alamouti scheme designed for two transmitting antennas, a pair of symbol is transmitted at a time instant, and a transformed version of the symbols are transmitted in the next time instant. At the receiver, the decoder detects the four symbols transmitted over two time slots and processes them to obtain 2-branch diversity gain. Thus the Alamouti scheme achieves full diversity, with a rate-1 code. For the multiplexing option, the multiple antennas are used for capacity increase. In this option, original high-rate stream is partitioned into $N$ low-rate substreams and each substream is transmitted in parallel over the same channel, using different antennas. If there are enough scatterers between the transmitter and the receiver, adequate MIMO detection algorithms like zero-forcing, minimum mean-square error (MMSE), or vertical Bell labs Layered Architecture for space-time codes (V-BLAST), etc., can be designed to separate the substreams. Thus the link capacity (theoretic upper-bound on the throughput) is increased linearly with $\min(N,M)$, where $N$ is the number of transmit and $M$ is the number of receiving antennas [7].

### 19.3.2.3 MIMO Systems with Antenna Selection

For MIMO systems to be deployed on mobile WiMAX devices, the concept of antenna selection is very essential. Because RF chain dominates the link budget in wireless systems, mobile devices are unable to implement large numbers of RF chains to incorporate high order MIMO systems. For such systems therefore, a reduced numbers of RF chains are implemented and antennas with the best received energies are adaptively selected and switched on to the implemented RF chains for MIMO signal processings, as illustrated in Figure 19.3. The performance of such system has been studied quite elaborately in the literature [9–16], in comparison to the full complexity system that utilizes all available antennas. It was shown that the diversity gain performance is maintained in the reduced-complexity system despite the use of antenna selection, while the coding gain deteriorates proportional to the ratio of the selected antennas to the total available antennas [15,16].

### 19.3.2.4 MIMO Technologies in IEEE 802.16m Standard

The IEEE 802.16 standards committee has recently initiated the process of extending the existing IEEE 802.16e standard (mobile WiMAX) for high capacity, high-QoS mobile application. The new standard was dubbed IEEE 802.16m at the IEEE January session in London, 2008. The working group tasked with the responsibility of producing the working documents for the new standard was named task group m (TGm). The group hopes to complete the specification for the new standard by the end of 2009. When completed, the standard will be backward compatible with IEEE 802.16e,

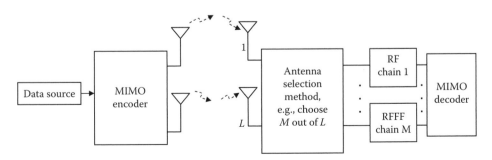

**FIGURE 19.3**   MIMO subset antenna selection.

and interoperable with 4G cellular standards supporting the IMT-advanced technologies. Although the details of the IEEE 802.16m standard is not available at the moment, the most important features being touted for the standard include

- Target downstream speed of 100 Mbps in highly mobile mode, and upto 1 Gbps in normadic mode (upstream rate are not yet known, but would be at least at par with 802.16e).
- Channel sizes upto 40 MHz (802.16e currently supports upto 20 MHz channel size).
- Use of TDD and FDD.
- Backward-compatibility with 802.16e.
- OFDMA radio (same as in 802.16e).
- Mandatory MIMO antenna technology of size $4 \times 4$ (four transmitting, and four receiving antennas).

In contrast to the IEEE 802.16e, which supports mandatory MIMO antenna technology of size $2 \times 2$ (two transmitting, and two receiving antennas), the use of mandatory higher-capacity MIMO technology in 802.16m will provide extra capacity to support the targeted high-speed in the downstream. Since downstream has been the bottleneck in wireless services, this improvement will provide significant boost in system capacity, to enable the system support wide range of multimedia services expected in 4-G compatible technologies.

### 19.3.3    Adaptive Antenna and Beamforming Systems in WiMAX

#### 19.3.3.1    Beamforming

Beamforming takes advantage of interference to change the directionality of an antenna array system. A beamformer controls the amplitude and phase of the signal at each transmitting antenna element, to create a pattern of constructive interference (beamspots) and destructive interference (null) in the wavefront. To create a beamspot, the beamformer uses an array of closely spaced antennas, often enclosed in a single enclosure as illustrated in Figure 19.4. $\lambda/2$ antenna spacing between the antenna elements is commonly used (where $\lambda$ is the wavelength of the transmitted signals, given by $\lambda = c/f$; $f$ is the frequency of the transmitted signals, $c$ is the speed of light). By varying the amplitude and phase of each antenna element, the beamformer is able to focus electromagnetic energy (beam) in the desired directions. The beams are directed to intended users, while nulls are focused on other

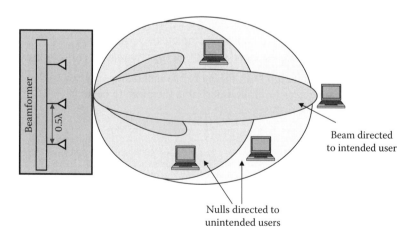

**FIGURE 19.4**    Beamforming technique.

unintended users reducing interference to the unintended users while increasing received SNR for the intended user. This provides a stronger link to the intended user and improves reach and capacity.

### 19.3.3.2 Adaptive Antenna System

AAS is one of the advanced antenna technologies specified in the WiMAX standard to improve performance and coverage. In the AAS system, the transmitter (base station, BS) adaptively tracks a mobile receiver as it moves around the coverage area of the transmitter (BS), and steers the focus of the beam (beam spot) on the receiver unit as it moves. The beam steering method can either be mechanical or electronic. Thus AAS creates narrow beams to communicate with desired user device, which helps to reduce interferences to unintended user devices and improves carrier-to-interference ($C/I$) and frequency reuse, giving rise to high spectral efficiency. Thus, through the use of adaptive processing (beam steering), AAS improves performance and coverage of the system significantly. In WiMAX networks, AAS will find wide applications both in the point-to-multipoint (PMP) as well as mesh network deployments. In the PMP mode, AAS operates in similar way as in the current 3G cellular system, and is used for enhancing coverage and performance. In the mesh mode, AAS is used to form physical or directed mesh links. Physical or directed mesh is a form of mesh where substantially directional antennas are used to create physical links between neighboring devices. Mesh nodes adaptively steer antennas towards other nodes in their neighborhood and direct the focus of the electromagnetic radiations accordingly, to create the physical link with the intended neighboring device. One of the main drawback of AAS however is the high complexity involved in designing antenna systems capable of adaptively switching (steering) antenna directionality toward users who may be highly mobile. The use of AAS technology in mobile network deployment is therefore very challenging from a complexity perspective. Another drawback of AAS technology is that in an urban environment, with rich scatterer, the beams get blurred at the receiver and are not focused as expected, due to the reflections of waves as it propagates from the transmitter to the receiver. This effect is known as angle spread and it impacts significantly the performance of AAS in urban areas with cluttered structures. The gains achieved using AAS in such places thus reduce considerably from the theoretical expectations. For example, an AAS system using an eight-column array would have an ideal gain of 6.9 dB but angle spread would reduce this to only 3.2 dB in an urban environment and 4.7 dB in a suburban environment. There are some techniques however to mitigate the effects of angle spread. Active research works are ongoing in this area [17,18].

## 19.4 ADAPTATIONS AT THE MAC LAYER

### 19.4.1 ADAPTIVE MODULATIONS, POWER, AND CODING RATE CONTROL

In wired networks the channel impairments tend to be constant or at least very slowly varying. Wireless networks in contrast are well known for rapidly fluctuating channel conditions even when the transmitter and receiver are stationary. Broadly speaking, the lower the modulation and coding rate, and the higher the transmitted power, the more channel fading a system can tolerate and still maintain a link at a constant error level. It is desirable therefore to be able to dynamically change the transmitted power, modulation, and data rate to best match the channel conditions at the moment to continually support the highest capacity channel possible. WiMAX systems support adaptive modulation and coding on both the downlink (DL) and uplink (UL) and adaptive power control on the UL. Adaptive modulation allows the WiMAX MAC layer to adjust the signal modulation rate depending on the channel or radio link quality [8]. When the channel quality is good, the MAC layer chooses the highest modulation rate, e.g., 64QAM, giving the system the highest throughput. When the channel quality degrades, the MAC layer reduces the modulation rate, e.g., 16QAM, reducing the throughput. In practice, adaptive modulation and coding rate control are used in conjunction with power control. In the PMP network deployments with multiple users in a cell serviced by a BS, when a link degradation arises for a user, the BS first increases the transmitted power of the user to provide

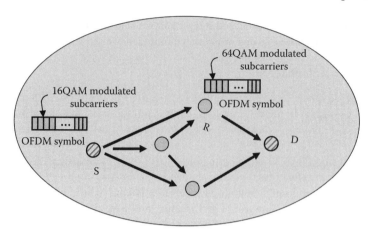

**FIGURE 19.5** Adaptive modulation at WiMAX mesh nodes.

extra link budget gain, until it reaches the maximum permitted. If the received signal quality does not improve, then the coding rate is reduced. Extra redundancy is added to provide more coding gain for better error correction performance. If the received signal quality still does not improve, then the modulation rate is reduced as a last resort (as this significantly affects the throughput than others). Similar (reverse) process is also followed when link quality appreciates. For WiMAX mesh networks using the amplify-and-forward relaying option, mesh relaying cannot exploit adaptive modulation technology because relaying nodes are not able to decode the contents of the received OFDM symbols to retrieve the modulated data and remodulate them at higher or lower rate, to increase or reduce the transmission rate (or throughput) of the mesh streams in response to link quality condition. However for mesh networks using the decode-and-forward relay option, adaptive modulation and coding rate control can benefit the mesh relaying operation as mesh nodes can decode the mesh data streams and adjust the coding and modulation rate, depending on the forwarding link quality. For example in Figure 19.5 a relay node $R$ decodes a data stream originally transmitted from the source node $S$ using 16QAM modulation, and remodulates the data stream using 64QAM as it has good channel quality to the destination node $D$ that can support this modulation rate. This results in fast and efficient use of the mesh links. Power control is applicable in WiMAX networks (PMP and mesh) in two ways: One, when nodes are transmitting data, they are regulated to transmit only the minimum power required to achieve successful reception at the receiver. Two, when mobile nodes do not have data (mesh relay or access service data) to transmit or receive, they go on sleep modes to save battery life.

### 19.4.2 Forward Error Correction and Rate Adaptations

FEC allows the WiMAX MAC layer to detect errors introduced during the transmissions of frames over the air link. There are three methods of FEC specified in the WiMAX system; Reed-Solomon concatenated with convolutional code (RS-CC), block turbo code (BTC), and convolutional turbo code (CTC). RS-CC is mandatory, while BTC and CTC are made optional due to their complexity, eventhough they provide 2–3 dB better coding gain than RS-CC. For the 802.16e, a hybrid ARQ (H-ARQ) has been included as an optional feature. There are three types of H-ARQ, classified based on the manner in which they handle the retransmissions. Type I H-ARQ retransmits lost or unacknowledged blocks using chase combining in which the old erroneous block is stored at the receiver and compared with the retransmitted copy. This helps to increase the probability of successful decoding at the FEC block during the retransmission attempts. Type II/III H-ARQ uses incremental coding rate to ensure successful decoding at the FEC block during the retransmission attempts. Rate adaptation works hand-in-hand with the FEC block in the WiMAX system. When a user experiences good channel condition, it is desirable to exploit these peaks in the channel gain to increase throughput. This is achieved by having the SS increase the coding rate, e.g., from rate 1/2 code to rate 1/4 code, so that more information bits can be transmitted per channel use while still keeping to the target bit error rate (BER). When the channel degrades, the rate is reduced back to the next minimum to ensure that the target BER is met. This dynamic process is carried out on a frame-by-frame basis in the WiMAX system, using the flexibility provided by MAP signaling to adaptively adjust the UL/DL rates.

### 19.4.3 QoS Scheduling and Adaptive Resource Allocations

IEEE 802.16 supports fixed-length frame, with flexible (adaptive) DL/UL resource usage ratios. The BS adaptively adjusts DL and UL subframe lengths on a frame-by-frame basis depending on the DL/UL traffics and channel conditions. Typically, the DL:UL resources can be varied from 3:1 to 1:1 in a PMP WiMAX network. Figure 19.6 illustrates the fixed-length frame in the PMP WiMAX network, and the flexible DL/UL subframes. The figure also depicts the network entry process for subscriber stations (SS) and the scheduling periods for assigning transmission opportunities to SS already initiated into the network. For access (PMP) mode, new SS detects preamble and frame control header (FCH), and identifies the number of DL burst transmissions from the DL MAP in the FCH. At the end of the last DL burst (Figure 19.6), new SS uses a contention period to exchange network entry request signal with the BS. If successful, the BS process the request and sends entry instruction (assigned DL/UL transmission opportunities, power, etc.) in the DL/UL MAPS of the next frame, and the SS gets initiated into the network. For the mesh mode, new SS waits for network entry signal broadcast at the beginning of a frame, to which they can respond within a specified

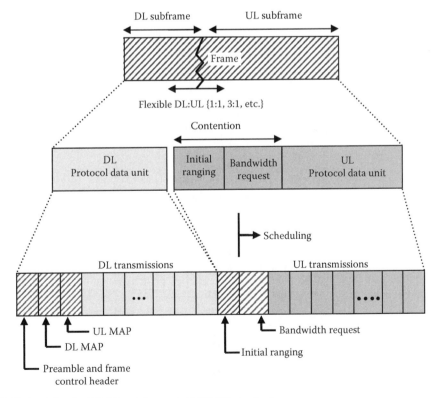

**FIGURE 19.6**  Adaptive DL/UL subframes in WiMAX standard.

period. Scheduling process is used for initiating new SS into the network. SS transmits on the scheduled slots.

In the WiMAX standard (802.16e), UL and DL assignments are based on time division multiple access (TDMA). In each frame, the BS scheduler assigns UL and DL transmission opportunities to SS until their negotiated data periods expire. The resources given to an SS for its data transmission are both in the frequency and time domain. WiMAX MAC thus supports frequency-time resource allocation in both DL and UL on a per-frame basis. The resource allocation is delivered in media access protocol (MAP) messages at the beginning of each frame. Therefore, the resource allocation can be dynamically changed frame-by-frame in response to traffic and channel conditions. Additionally, the amount of resource in each allocation can range from one slot to the entire frame in the time domain, and from one subchannel to the entire subchannels in an OFDM symbol, in frequency domain. Also WiMAX employs fast scheduling both in the DL and UL to respond to fast variations in channel conditions. This fast and fine granular resource allocation allows superior QoS for data traffic in a bursty traffic and rapidly changing channel condition. The fundamental premise of the IEEE 802.16 MAC architecture is QoS. It defines services flows which can map to Diffserve code points or MPLS flow labels that enable end-to-end IP-based QoS. Additionally, subchannelization and MAP-based signaling schemes provide a flexible mechanism for optimal scheduling of space, frequency, and time resources over the air interface on a frame-by-frame basis. This flexible scheduling allows QoS to be better enforced and enable support for guaranteed service levels including committed and peak information rates, latency, and jitter for various types of traffic on a customer-by-customer basis.

### 19.4.3.1 Scheduling Algorithms

In the WiMAX standard (802.16), four scheduling classes have been defined: Unsolicited grant service (UGS), real-time polling service (rtPS), nonreal-time polling service (nrtPS), and best effort (BE). As illustrated in Figure 19.7, each traffic connection is associated with one of the four scheduling services, and the SS scheduler selects packets to be transmitted from each queue based on the scheduling policy employed. Usually, the scheduler selects packets to be transmitted from the

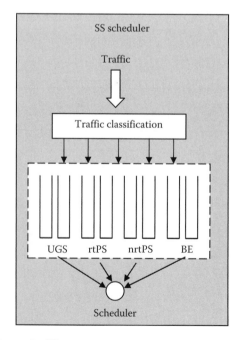

**FIGURE 19.7** UL scheduling at the SS.

highest priority queue that is not empty. Transmission of packets from lower priority queues are postponed until there is no packet available to send from a higher priority queue. Since UL traffic is generated at SS, the SS scheduler is able to arrange the transmission based on the up-to-date information on the current numbers/status of UL connections, which help to improve QoS performance. In the following, we review the various scheduling algorithms provided in the standard for handling the transmissions of packets belonging to these various services [26–34].

### 19.4.3.2 Unsolicited Grant Service

The UGS algorithm is designed to support real-time service flows, such as Voice over Internet Protocol (VoIP), that generate fixed size data packets periodically. BS periodically assigns fixed-size grants to voice users. These grants are sufficient to send voice data packets generated by the maximum data rate of enhanced variable rate "voice" codec (EVRC). The grant period are negotiated during the initialization process of the connection. Thus, MAC overhead and UL access delay caused by bandwidth request process are minimized. The drawback of the UGS algorithm is the following. Generally, voice users do not always have voice data packets to send throughout the duration of a connection, because voice users have frequent silence periods. A typical voice codec switches intermittently between "on" and "off" states as illustrated in Figure 19.8. While in "on" state, popular voice codecs like the EVRC also have variable data rates. For example, the EVRC operates at 1/8 of the full data rate during the off state, while the device has three different rates during the on state (rates 1, 1/2, and 1/4). Therefore for a UGS algorithm that reserves a flat amount of resources capable of sending data at the maximum rate of EVRC periodically, a significant amount of UL resources is wasted when the codec is in silence (or off) mode as well as when the codec is on but not operating at the full rate. This is illustrated in Figure 19.9. A number of other algorithms have thus been designed to adaptively determine the actual UL needs of each connection during frame periods, so as to minimize these resources wastage.

### 19.4.3.3 Real-Time Polling Service

The rtPS algorithm is designed to support real-time service flows, such as MPEG video or tele-conference, that generate variable size data packets periodically. In this algorithm, the BS assigns UL resources that are sufficient for unicast bandwidth request to the voice users. This is called the polling process. The duration for which the BS continues to poll an SS with rtPS connection is nego-tiated in the initialization process of the connection. The SSs utilize the assigned polling resources to send their bandwidth requests, reporting the exact bandwidth need for their rtPS connection.

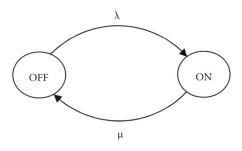

$1/\lambda$ = Mean on time
$1/\mu$ = Mean off time

**FIGURE 19.8** Voice codec status.

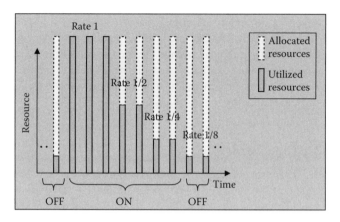

**FIGURE 19.9**   UL resource allocation using UGS algorithm.

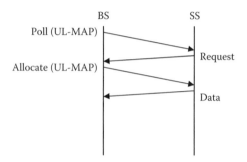

**FIGURE 19.10**   Polling process in rtPS.

The BS in response then allocates the exact bandwidth requested to the SS for transmission of the data. Figure 19.10 illustrates this dynamic polling process.

Because rtPS always carry out polling process, it is able to adaptively determine suitable resource allocation from frame to frame. This adaptive request-grant process goes on until the connection is terminated. Because of the dynamic request-grant process, the algorithm has more optimum data transport efficiency than the UGS algorithm. The algorithm is able to dynamically follow all data rate of the voice codec without any resource wastage as illustrated in Figure 19.11 (allocated and utilized resources are equal). This is a major advantage over the UGS algorithm. The drawback of the rtPS algorithm however is that the dynamic polling process causes MAC overhead and access delay. Hence rtPS has more MAC overhead and larger access delay than the UGS.

### 19.4.3.4   Unsolicited Grant Service with Activity Detection

The UGS-AD algorithm is designed to support real-time service flows that generate fixed-size data packets on a semi-periodic basis (e.g., VoIP using on–off voice codec). It incorporates activity detection, which makes it suitable for use with on/off voice codecs. The algorithm uses combined features of UGS and rtPS. UGS-AD has two scheduling modes: UGS and rtPS, and can switch between these modes depending on the status of the voice users (on or off). On initialization of VoIP services, this algorithm starts with the rtPS mode. While in rtPS mode, if the voice user requests bandwidth size of zero bytes, the BS maintains this (rtPS) mode. However if the user requests bandwidth size greater than zero, the BS switches its mode to UGS. While in UGS mode, if the voice user requests bandwidth size = 0, the BS switches to rtPS, and if the user requests

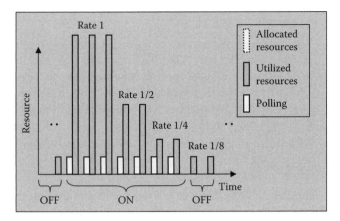

**FIGURE 19.11** UL resource allocation using rtPS algorithm.

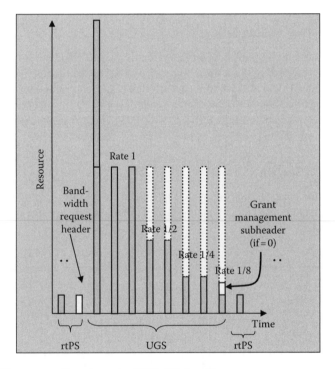

**FIGURE 19.12** UL resource allocation using UGS-AD algorithm.

bandwidth greater than zero, the BS stays in UGS. By switching between rtPS and UGS modes, the UGS-AD algorithm significantly addresses the problem of UL resources wastage in the UGS algorithm, and the MAC over head and access delay in the rtPS. This is however only for the case where the voice user uses voice codecs with only two data rates (on–off). Where voice codecs with variable data rates like EVRC is used, resources wastage still occur in the UGS-AD algorithm. In this case, the wastage occurs during the on duration, when full resources is assigned eventhough the variable data rate of voice codecs means that it will not operate at full rate for all of the time the resources is allocated. The operation of the UGS-AD algorithm is illustrated in Figure 19.12.

### 19.4.3.5   Extended Real-Time Polling Service

The ertPS algorithm is designed for real-time service flows that generate variable-sized data packets on a periodic basis (e.g., VoIP using EVRC). The ertPS algorithm was recently added in the IEEE 802.16e version of the WiMAX standard. In the ertPS, the BS and SS use grant management subheader and bandwidth request header to exchange information. Users make new bandwidth request (data rate increment) using bandwidth request header, and inform the BS of changes to existing data rate (data rate decrement) using the grant management subheader.

*Data rate increment:* When voice codec switches to on (or data rate increases), the user informs the BS using bandwidth request header. The BS changes the grant size to the exact amount of bandwidth requested and keeps this size until the user sends another request.

*Data rate decrement:* Voice user informs BS of its new (reduced) data rate using the grant management subheader (the extended piggyback request bits of the grant management subheader is set to zero), then the BS reduces grant size accordingly. ertPS can follow all voice codec data rates, offering better data transport efficiency compared to UGS, rtPS, and UGS-AD. Figure 19.13 illustrates the ertPS algorithm.

### 19.4.3.6   Nonreal-Time Polling Service

The nrtPS service is designed for non-real-time service flows. The BS ensures that certain minimum bandwidth guarantees are provided for these services. Additional bandwidth can be provided above the minimum guaranteed if extra resources are available and commitments with higher priority services have been met. No delay deadline is required for the nrtPS, as this service is used only for nonreal-time applications such as bandwidth intensive FTP. Therefore the service is just better than BE service.

### 19.4.3.7   Best Effort Service

BE service is used for data streams for which no minimum service level (or QoS guarantee) is required. Examples include HTTP or E-mail services. Therefore BE service is handled only if the resources are available.

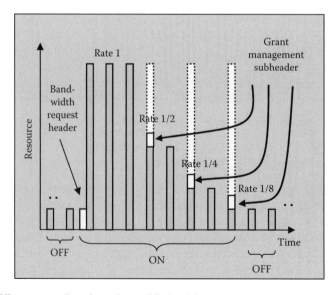

**FIGURE 19.13**   UL resource allocation using ertPS algorithm.

### 19.4.3.8  Downlink Scheduling and Adaptive Resource Allocations

To support broadband applications under limited radio resources and harsh wireless channel conditions, WiMAX system employs dynamic scheduling and resource allocation in the DL transmission from the BS to the users, in a PMP network architecture. This enables the system to achieve both higher spectral efficiency and better QoS. A key principle exploited in adaptive resource allocation is the inherent system diversities in various domains [19–22]. Inherent diversities typically result from the time variation and frequency selectivity of the wireless channels, independent fading of multiple users, parallel channels available in space and frequency domains, random traffic arrival, user mobility, as well as the interaction between different layers of the network protocol stack. Adaptive management of bandwidth allocation, multiple access, scheduling, rate, and power adaptations are therefore used to exploit the extra degree of freedom provided by these system diversities for optimized performance.

### 19.4.3.9  Multiuser Diversity-Based Scheduling

The dynamics in wireless communication systems can be exploited in a multiuser environment to enhance the utilization of scarce radio resources and optimize system performance. The variation in wireless channel gains has traditionally been regarded as a limitation in the performance of wireless systems because the BER performance is typically dominated by the worst-case (weakest) channel condition. With adaptive resource allocation, however, the time- and frequency-varying characteristics of wireless channels can be exploited and indeed considered an advantage. This is the principle of multiuser diversity-based scheduling. In the WiMAX system, multiuser diversity-based scheduling is employed in the DL. This choice is predicated on the fact that in OFDM system, subcarriers in deep fade for a given user may not be faded for other users. Different users experience mutually independent attenuations on the subcarriers. Therefore, dynamically allocating the subcarriers to users according to their channel conditions ensures that each subcarrier is allocated to the user with high channel gains in those subcarrier frequencies. Users are allocated subcarriers that are not in deep fade for them, and this improves spectrum utilization (maximize data rate per channel use). This is commonly referred to as frequency-selective scheduling. The haul-mark of this scheduling method is the dynamic frequency (subcarrier) allocation that takes place on a frame-by-frame basis. Figure 19.14 illustrates this system.

Because all available subcarriers may be deeply faded for a user at a particular time (or due to the use of fixed multi-access scheme like TDMA or FDMA), a user can still be allocated deep-faded subcarrier. Such subcarriers will not be efficiently utilized as they can only carry a few information bits. For example, in the case of M-QAM modulation, the number of bits transmitted on subchannel $n$ in an OFDM system can be approximated as $r_i = \log_2(1 + \gamma_n P_n)$, where $\gamma_n$ is the channel gain-to-noise ratio on subchannel $n$ and $P_n$ is the transmit power on subchannel $n$. Therefore for same power allocation, supportable rates are less for deeply faded subchannels. Since frequency-selective fading is random, it is expected that these deep-faded subcarriers will be fairly evenly distributed among users in the long run. Thus frequency-selective scheduling somewhat achieves rate allocation fairness in the long run. In general however, there is the need to provide in the scheduling discipline, measures to incorporate higher-layer QoS issues such as throughput, delay, and fairness jointly with the PHY-layer issue such as fading and modulation. A number of other scheduling disciplines exploring these QoS issues have thus been presented for DL scheduling in the WiMAX system. These include the opportunistic fair scheduling [23–25] where priority is given to user with temporarily better channels subject to maximizing a utility function, which incorporates rate fairness and throughput indices.

### 19.4.3.10  Uplink Scheduling and Adaptive Resource Allocations

In the WiMAX system, BS scheduler adaptively makes UL bandwidth allocation based on the requests from SSs. Because an SS may have multiple connections at the same time, the bandwidth

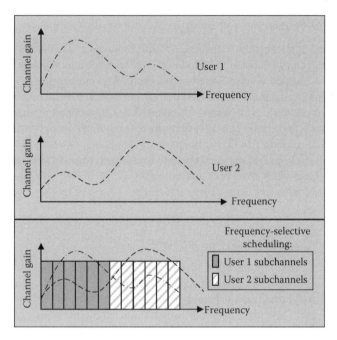

**FIGURE 19.14**    Adaptive frequency-selective scheduling using multiuser diversity principle.

request message of each SS reports the bandwidth requirement of each connection in the SS. A BS scheduler processes these messages and determines transmission opportunities. After processing all the requests, the BS sends in the DL subframe of the next frame, an UL MAP message with transmission opportunities assigned to all reserved SSs. WiMAX network deployments may employ grant per connection (GPC), in which the BS assigns transmission slots to the individual connection reported by the SS during the bandwidth requests, or grant per SS (GPSS), in which the BS estimates the total need of the SS and provides a lump sum bandwidth allocation to the SS for all its needs. For GPC, no extra scheduling is required at the SS. The SS simply transmits the flows from each connection on the assigned bandwidth to the connection. If GPSS is used, the SS has the flexibility of making scheduling at the SS unit to redistribute the assigned transmission opportunities among its connections. This allows adaptive scheduling based on most updated status of the connections in the SS, and ensures that the best QoS is achieved. The haul-mark of all the resource allocation processes described above is the use of adaptive techniques to enhance efficiency both in resource allocations and utilizations. Active research works are still ongoing to design new algorithms that improve on the performance of existing ones. Deployments of specific algorithms by the vendors have been left open in the standard, to provide quality differentiations among products from different vendors. It is therefore expected that future research works will continue to focus on this area.

## 19.5   CONCLUSIONS AND FUTURE TRENDS

This chapter reviews adaptive processing techniques in WiMAX networks. We take a close look at the WiMAX system and discuss key enhancements provided by the use of adaptation techniques specified in the standard. Current protocols and algorithms employed in the system are reviewed, and future research and deployment trends are also discussed. As the WiMAX technology matures, adaptive processings will continue to play more prominent roles. At the PHY layer, a number of vendors have extensively explored the use of AAS in their WiMAX products. Therefore more active research/deployment activities related to adaptive antenna technology is expected in the future WiMAX systems. For MIMO systems, the role of adaptation is expected to be in the area of antenna selection. Mobile WiMAX system is especially expected to use adaptive antenna switching system to benefit from the gains of MIMO technology using the concept of antenna selection. At the MAC layer, adaptive modulation and coding rate control is expected to be the bedrock of QoS in future broadband systems in general, and WiMAX system in particular. Therefore more active research/deployment activities are also expected on this techniques in the future. Furthermore, for QoS guarantees and efficient use of system resources, a number of adaptive resource allocations and scheduling methods have been specified in the WiMAX system. Future WiMAX system will also see more prominent roles for adaptations in this area.

## REFERENCES

[1] IEEE Standard for Local and Metropolitan Area Networks—Parts 16: Air Interface for Fixed Broadband Wireless Access System, October 2004.

[2] IEEE Standard for Local and Metropolitan Area Networks—Parts 16: Air Interface for Fixed and Mobile Broadband Wireless Access System Amendment2: Physical and Medium Access Control Layers for Combined Fixed and Mobile Operation in Licensed Bands, February 2006.

[3] F. Behmann, Impact of wireless (Wi-Fi, WiMAX) on 3G and next generation—An initial assessment, *Proceedings of IEEE International Conference On Electro Information Technology*, pp. 1–6, Lincoln, NE, May 2005.

[4] C. Eklund et al., *WirelessMAN: Inside the IEEE 802.16 Standard for Wireless Metropolitan Networks*, IEEE Press, New York, 2006.

[5] Z. Abichar, Y. Peng, and J. M. Chang, WiMAX: The emergence of wireless broadband, *IEEE IT Professional*, 8(4): 44–48, August 2006.

[6] C. Eklund, R. B. Marks, K. L. Stanwood, and S. Wang, IEEE Standard 802.16: A technical overview of the wirelessMAN air interface for broadband wireless access, *IEEE Communications Magazine*, 40(6): 98–107, June 2002.

[7] G. J. Foschini, G. D. Golden, R. A. Valenzuela, and P. W. Wolniansky, Simplified processing for high spectral efficiency wireless communication employing multi-element arrays, *IEEE Journal on Selected Areas in Communications*, 17(11): 1841–1852, November 1999.

[8] M. Settembre, M. Puleri, S. Garritano, P. Testa, R. Albanese, M. Mancini, and V. L. Curto, Performance analysis of an efficient packet-based IEEE 802.16 MAC supporting adaptive modulation and coding, *Proceedings of IEEE International Symposium on Computer Networks*, 11–16, Istanbul, Turkey, June 2006.

[9] S. Sanayei and A. Nosratinia, Antenna selection in MIMO systems, *IEEE Communications Magazine*, 42(10): 68–73, October 2004.

[10] M. K. Simon and M. -S. Alouini, Performance analysis of generalized selection combining with threshold test per branch (TGSC), *IEEE Transactions on Vehicular Technology*, 51(5): 1018–1029, September 2002.

[11] A. I. Sulyman and M. Ibnkahla, Performance of MIMO systems with antenna selection over nonlinear fading channels, *Proceedings of IEEE-SPAWC'06*, France, July 2006.

[12] A. I. Sulyman and M. Kousa, Bit Error rate performance of a generalized selection diversity combining scheme in Nakagami fading channels, *Proceedings of IEEE-WCNC 2000*, 3: 1080–1085, September 2000.

[13] A. I. Sulyman and M. Kousa, Bit Error rate performance analysis of a threshold-based generalized selection combining scheme in Nakagami fading channels, *EURASIP Journal on Wireless Communications and Networking*, 2005(2): 242–248, April 2005.

[14] N. Kong and L. B. Milstein, Combined average SNR of a generalized diversity selection combining scheme, *IEEE Communications Letters*, 3(3): 57–59, March 1999.

[15] A. Ghrayeb and T. M. Duman, Performance analysis of MIMO systems with antenna selection over quasi-static fading channels, *IEEE Transactions on Vehicular Technology*, 52(2): 281–287, March 2003.

[16] D. A. Gore and A. J. Paulraj, MIMO antenna subset selection with space-time coding, IEEE Transactions On Signal Processing, 50(10): 2581–2588, October 2002.

[17] D. Gerlach, Transmit antenna beamforming for the advanced mobile phone system, *Proceedings of 29th Asilomar Conference on Signals, Systems, and Computers* (2-Volume Set), p. 1162, Pacific Grove, CA, 1995.

[18] L. Ping'an and H. Yi, Effects of angle spread on beamforming in wireless communications, *Proceedings of IEEE International Conference on Wireless Communication, Networks, and Mobile Computing*, 1: 615–618, September 2005.

[19] K. B. Letaief and Y. J. Zhang, Dynamic multiuser resource allocation and adaptation for wireless systems, *IEEE Wireless Communications*, 13(4): 38–47, August 2006.

[20] Y. J. Zhang and K. B. Letaief, An efficient resource allocation scheme for spatial multiuser access in MIMO/OFDM systems, *IEEE Transactions on Communication*, 53(1): 107–116, January 2005.

[21] C. Y. Wong et al., Multiuser OFDM with adaptive subcarrier, bit and power allocation, *IEEE Journal on Selected Areas in Communications*, 17(10): 1747–1758, October 1999.

[22] Y. J. Zhang and K. B. Letaief, Multiuser adaptive subcarrier-and-bit allocation with adaptive cell selection for OFDM systems, *IEEE Transactions on Wireless Communication*, 3(5): 1566–1575, September 2004.

[23] M. Mehrjoo, M. Dianati, X. Shen, and K. Naik, Opportunistic fair scheduling for the downlink of IEEE 802.16 wireless metropolitan area networks, *Proceedings of 3rd ACM International Conference on Quality of Service in Heterogeneous Wired/Wireless Networks*, Waterloo, Canada, ACM Press, New York, August 2006.

[24] X. Liu, E. K. Chong, and N. B. Shroff, Opportunistic transmission scheduling with resource-sharing constraints in wireless networks, *IEEE Journal on Selected Areas in Communication*, 19(10): 2053–2064, 2001.

[25] P. Viswanath, D. Tse, and R. Laroia, Opportunistic beamforming using dumb antennas, *IEEE Transactions on Information Theory*, 48(6): 1277–1294, June 2002.

[26] L. F. Moraes and P. D. Maciel Jr., Analysis and evaluation of a new MAC protocol for broadband wireless access, *Proceedings of IEEE International Conference on Wireless Networks, Communication and Mobile Computing*, 1: 107–112, June 2005.

[27] H. Lee, T. Kwon, D.-H. Cho, G. Lim, and Y. Chang, Performance analysis of scheduling algorithms for VoIP services in IEEE 802.16e systems, *Proceedings of IEEE-VTC 2006*, 3: 1231–1235, 2006.

[28] H. Lee, T. Kwon, and D.-H. Cho, Extended-rtPS algorithm for VoIP services in IEEE 802.16 systems, *Proceedings of IEEE-ICC 2006*, 5: 2060–2065, June 2006.

[29] H. J. Zhu and R. H. Hafez, Novel scheduling algorithms for multimedia service in OFDM broadband wireless systems, *Proceedings of IEEE-ICC 2006*, 2: 772–777, June 2006.

[30] Z. Sun, Y. Zhou, M. Peng, and W. Wang, Dynamic resource allocation with guaranteed diverse QoS for WiMAX system, *Proceedings of IEEE-ICCCAS 2006*, 2: 1347–1351, June 2006.

[31] H. Zhang, Y. Li, S. Feng, and W. Wu, A new extended rtPS scheduling mechanism based on multipolling for VoIP service in IEEE 802.16e system, *Proceedings of IEEE-ICCT'06*, pp. 1–4, Guilin, China, November 2006.

[32] H. Lee, T. Kwon, and D.-H. Cho, An enhanced uplink scheduling algorithm based on voice activity for VoIP services in IEEE 802.16d/e system, *IEEE Communication Letters*, 9(8): 691–693, August 2005.

[33] H. Lee, T. Kwon, and D.-H. Cho, An efficient uplink scheduling algorithm for VoIP services in IEEE 802.16 BWA systems, *Proceedings of IEEE-VTC 2004*, 5: 3070–3074, September 2004.

[34] R. Iyengar, P. Iyer, and B. Sikdar, Delay analysis of 802.16 based last mile wireless networks, *Proceedings of IEEE-Globecom 2005*, pp. 3123–3127, St. Louis, MO, 2005.

# 20 Performance Evaluation and Dimensioning of WiMAX

## Georges Nogueira, Bruno Baynat, Masood Maqbool, and Marceau Coupechoux

## CONTENTS

This chapter tackles the challenging task of performance evaluation and dimensioning of WiMAX networks. It provides a simple analytical model which is able to take into account the effects of elastic traffic, radio channel variations, and scheduling policy. Compared to packet-level simulation-based evaluations, our model instantaneously delivers the dimensioning parameters necessary for the deployment of a WiMAX network. Compared to existing analytical solutions, we derive closed-form expressions for all performance metrics. We compare the results obtained through analytical model with those of simulations. We show that our analytical model is not only accurate but also robust with respect to the modeling assumptions. Finally, the quick results produced through our analytical tool allow to carry out dimensioning analyses that otherwise require several thousands of evaluations, which would not be tractable with any simulation tool.

## 20.1 INTRODUCTION

In recent years, the demand for broadband access has increased substantially. To date, most of the deployed broadband networks are wired ones. The evolution of last-mile infrastructure for wired networks faces acute implications such as difficult terrain and high cost-to-serve ratio. Latest developments in the wireless domain not only could address these issues but could also complement the existing framework. One such highly anticipated technology is WiMAX (worldwide interoperability for microwave access) based on the standard IEEE 802.16. The first operative version of IEEE 802.16 is 802.16-2004 (fixed/nomadic WiMAX) [2]. It was followed by a ratification of amendment IEEE 802.16e (mobile WiMAX) in 2005 [3]. A new standard, IEEE 802.16m, is currently under definition for providing even higher efficiency. Besides, the consortium WiMAX forum was founded to specify profiles (technology options are chosen among those proposed by the IEEE standard), define an end-to-end architecture (IEEE does not go beyond physical and MAC layer), and certificate products (through interoperability tests). Some WiMAX networks are already deployed but most operators are still under trial phases. As deployment is approaching, the need arises for manufacturers and operators to have fast and efficient tools for network design and performance evaluation. In this chapter, we develop a simple and accurate analytical model that allows to rapidly derive the capacity parameters such as throughput per user, channel utilization, or mean number of active users for different scheduling policies.

## 20.2 WiMAX PERFORMANCE EVALUATION

Literature on WiMAX performance evaluation is mainly constituted of two sets of papers. One set discusses detailed packet-level simulations that precisely implement system details and scheduling schemes while the other one focuses on analytical models and optimizations.

In the former set, Lee et al. [18] have presented a simulation-based performance analysis for three different classes of services proposed in IEEE 802.16e: UGS (unsolicited grant service), rtPS (real-time polling service), and ertPS (extended real-time polling service). The application in context was VoIP (Voice-over IP). The authors concluded that ertPS could accommodate more voice calls while satisfying the constraint of minimum packet delay. The performance analysis was focused on MAC layer. Cicconetti et al. [8] analyzed the performance of IEEE 802.16 system in providing multiple services (i.e., Web and VoIP). The authors have investigated quality of service (QoS) support mechanisms of the standard in conjunction with classical scheduling schemes like DRR (deficit round robin) or WRR (weighted round robin). See, for example, also Refs. [15,23,25] on the same subject.

Among the second set of papers, authors of Ref. [26] propose an analytical model for studying the random access scheme of IEEE 802.16d. Authors consider a perfect radio channel, which is certainly unrealistic in WiMAX networks. This model finally allows to configure an extension of exponential backoff algorithm applied to IEEE 802.16d. For the same standard, Singh and Sharma [24] consider UGS users and present a scheduling algorithm to minimize the global unsatisfaction of the circuit-switched class of users. Linear programming is used to formalize the problem, and heuristic algorithms are proposed. Finally, Niyato and Hossain [20] formulate the bandwidth allocation of multiple services with different QoS requirements by using linear programming. They also propose performance analysis, first at connection level, and then at packet level. In the former case, variations of the radio channel are, however, not taken into account. In the latter case, the computation of performance measures relies on a multidimensional Markovian model that requires numerical resolution.

Not specific to WiMAX systems, generic analytical models for performance evaluation of cellular networks with varying channel conditions have been proposed in Refs. [12,13,19]. The models presented in these articles are mostly based on multiclass processor-sharing queues with each class corresponding to the user that has similar radio conditions and subsequently equal data rates. The variability of radio channel conditions at flow level is taken into account by integrating a propagation model, classical mobility models, or, in some cases, a spatial distribution of users in a cell. For example, Refs. [12,13] consider a spatial distribution of users in a cell made of constant capacity rings obtained through a classical Rayleigh-fading distribution. To use classical PS-queues results, these papers consider implicitly that users can only switch class between two successive data transfers. However, as highlighted in the Section 20.4, in WiMAX systems, radio conditions and thus data rates of a particular user can change frequently during a data transfer. In addition, capacity of a WiMAX cell may vary as a result of varying radio conditions of users. As a consequence, any PS, discriminatory PS (DPS), or even generalized PS (GPS) queue is not appropriate for modeling these channel variations.

In this chapter, we develop a novel and generic analytical model that takes into account frame structure, precise slot-sharing-based scheduling, and channel quality variation of WiMAX systems. Unlike existing models [12,13,19], our model is adapted to WiMAX systems' assumptions and is generic enough to integrate any appropriate scheduling policy. Here, we consider three classical policies: slot sharing fairness, instantaneous throughput fairness, and opportunistic. For each of them, we develop closed-form expressions for all performance metrics. Moreover, our approach makes it possible to take into account the so-called outage situation. A given user experiences an outage, if at a given time its radio conditions are so bad that it cannot transfer any data and is thus not scheduled. Once again, classical PS-like queues are not appropriate to model this feature.

## 20.3  WiMAX SYSTEM DESCRIPTION

In this section, we briefly present the WiMAX system details needed to understand the proposed analytical model. Although the analysis is also valid for fixed WiMAX, we focus on mobile WiMAX, which is based on standard IEEE 802.16e and scalable orthogonal frequency division multiple access (SOFDMA) physical layer. In particular, the WiMAX frame structure, the notion of radio resource (slot), the access technique, and the different modulation and coding scheme (MCS) are presented.

The PHY layer of WiMAX is based on OFDMA. OFDM splits the available spectrum into a number of parallel orthogonal narrowband subcarriers, grouped into multiple subchannels. Radio resources are thus available in terms of OFDM symbols (time domain) and subchannels (frequency domain) providing a time–frequency multiple access technique [16]. In IEEE 802.16e, possible system bandwidths are 20, 10, 5, and 1.25 MHz with associated fast Fourier transform (FFT) sizes of 2048, 1024, 512, and 128, respectively [1]. The total number of subchannels depends on the subcarrier permutation, i.e., the way subcarriers are grouped together. Two main methods mentioned in Ref. [1] are distributed and adjacent subcarrier permutations. Full usage of subchannels (FUSC) and partial usage of subchannels (PUSC) are examples of distributed subcarrier permutations; they take advantage of channel diversity among subchannels. Adaptive modulation and coding (AMC) is a type of adjacent subcarrier permutation, which allows an opportunistic use of the channel.

IEEE 802.16e has specified time division duplex (TDD) as duplexing technique. The ratio of downlink (DL) to uplink (UL) has been left open in the standard. WiMAX forum has specified a duration of TDD frame of 5 ms. An example of a WiMAX TDD frame is shown in Figure 20.1. It has a two-directional structure with horizontal and vertical axes showing the time and frequency domain, respectively. A slot is the smallest unit of resource in a frame, which occupies space both in time and frequency domains. A burst is a set of slots using the same MCS. The total number of slots in the frame depends on the subcarrier permutation method. For numerical applications, we focus on PUSC, though our model is valid for any permutation scheme. In fact, a slot always carries 48 subcarriers whatever the type of subcarrier permutation used. In the DL subframe, the first part contains preamble, frame control header (FCH), UL_MAP, and DL_MAP. Preamble is used for synchronization. FCH provides length and encoding of two MAP messages and information about usable subchannels. The data mapping for users resides in the MAP messages. Their size depends on the number of scheduled users in the frame.

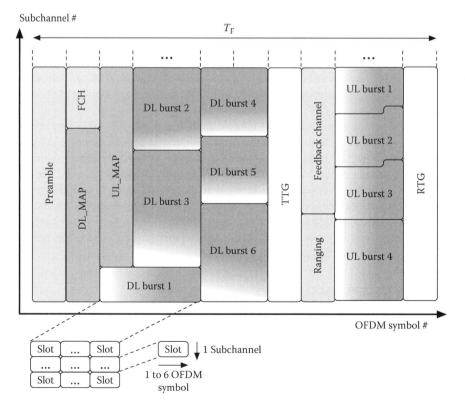

**FIGURE 20.1**   TDD frame structure.

One of the important features of IEEE 802.16e is link adaptation: different MCS allows a dynamic adaptation of the transmission to the radio conditions. As the number of data subcarriers per slot is the same for all permutation schemes, the number of bits carried by a slot for a given MCS is constant. The choice of the right MCS is made according to the signal to interference plus noise ratio (SINR). In case of outage, i.e., if the SINR is too low, no data can be transmitted without error.

The scheduling algorithm is responsible for allocating radio resources of a frame (or of a group of frames) to active users. In wireless networks, scheduling may take into account their radio link quality. In this work, we have considered three traditional schemes. The slot fairness scheduling allocates the same number of slots to all active users. The throughput fairness scheduling ensures that all active users have the same instantaneous throughput. The opportunistic scheduling gives all resources to active users with the best channel.

Let us now define the notations concerning the WiMAX system needed in this chapter:

- $N_S$ is the total number of slots available for data transmission in the DL part of the TDD frame. As mentioned before, $N_S$ depends on the system bandwidth, the frame duration, the DL/UL ratio, the permutation scheme, and the overhead.
- $T_F$ is the TDD frame duration: $T_F = 5$ ms.
- Radio channel states are denoted $MCS_k$, $1 \le k \le K$, where $K$ is the number of MCS. By extension, we denote $MCS_0$ the outage state.
- $m_k$ is the number of bits transmitted per slot by an MS using $MCS_k$. Recall that the number of bits transmitted per slot is independent of the permutation method and is thus constant for a given MCS. For the particular case of outage, $m_0 = 0$.

## 20.4  WiMAX ANALYTICAL MODELING

This section provides the development of our generic analytical model for WiMAX networks. We consider a single WiMAX cell handling the data traffic. This study targets the analysis of bottleneck, i.e., the radio link, and focuses on the DL part, which is assumed to be a critical resource in asymmetric data traffic.

### 20.4.1  SYSTEM MODELING

The development of our analytical model is based on several assumptions related either to the system or to the traffic. All of them will be discussed in Section 20.4.4, and, as developed in that section, most of them can be relaxed, if necessary, by slightly modifying the basic model.

**System assumptions**

1. The size of the DL_MAP and UL_MAP parts of the TDD frame is assumed to be constant and independent of the number of concurrent active mobiles. As a consequence, the total number of slots available for data transmission in the DL part is constant and equals $N_S$.
2. We assume that the number of simultaneous mobiles that can be multiplexed in one TDD frame is not limited. As a consequence, any connection demand will be accepted and no blocking can occur.
3. At any given time, if there is only one active user, we assume that the scheduler can allocate all the available slots for its transfer.

**Channel assumptions**

4. The coding scheme used by a given mobile can change very often because of the high variability of the radio link quality. We assume that each mobile sends feedback channel estimation on a frame-by-frame basis, and thus the base station (BS) can change its coding

scheme every frame. Because we do not make any distinction between users and consider all mobiles as statistically identical, we associate a probability $p_k$ with each coding scheme $\text{MCS}_k$, and assume that, at each time-step $T_F$, any mobile has a probability $p_k$ to use $\text{MCS}_k$ (including outage).

**Traffic assumptions**

5. All the users have the same traffic characteristics. In addition, we do not consider any QoS differentiation.
6. We do not take handover into account.
7. We assume that there is a fixed number $N$ of mobile stations (MSs) that are sharing the available bandwidth of the cell.
8. Each of the $N$ mobiles is assumed to generate an infinite length ON/OFF elastic traffic. An ON period corresponds to the download of an element (e.g., a web page including all the embedded objects). As the downloading duration depends on the system load and the radio link quality, ON periods must be characterized by their size. An OFF period corresponds to the reading time of the last downloaded element, and is independent of the system load. Unlike ON periods, OFF periods must then be characterized by their duration.
9. We assume that both ON sizes and OFF durations are exponentially distributed. We denote the average size of ON data volumes (in bits) by $\bar{x}_{\text{on}}$ and the average duration of OFF periods (in seconds) by $\bar{t}_{\text{off}}$.

## 20.4.2   ANALYTICAL MODEL

To develop our WiMAX analytical model, we first consider a system with a single coding scheme (i.e., $K = 1$) and no outage. We denote the number of bits transferred by any slot by $m\ (= m_1)$, and define $\mu$, the average departure rate, as

$$\mu = \frac{m\,N_\text{S}}{\bar{x}_{\text{on}}\,T_\text{F}}. \tag{20.1}$$

We also define $\lambda$, the inverse of the average reading time, as

$$\lambda = \frac{1}{\bar{t}_{\text{off}}}. \tag{20.2}$$

With all the assumptions presented in Sections 20.4.1, this basic system can be modeled by a simple continuous time markov chain (CTMC) made up of $N + 1$ states. A state $n$ of this chain ($0 \leq n \leq N$) corresponds to the total number of concurrent active mobiles, i.e., mobiles that are in ON period.

- Transition out of a generic state $n$ to state $n + 1$ occurs when a mobile in OFF period starts its transfer. This "arrival" transition is performed with a rate $(N - n)\lambda$. It corresponds to one mobile among the $(N - n)$ in OFF period, ending its reading.
- Transition out of a generic state $n$ to state $n - 1$ occurs when a mobile in ON period completes its transfer. This "departure" transition is always performed with a rate $\mu$ corresponding to the total departure rate of the frame.

It turns out that this basic but unrealistic model is equivalent to the classical Engset model [9].

We now go back to the real system including several MCSs ($K > 1$). Because of coding scheme diversity, the average departure rate is no longer constant. It actually depends both on the active mobile population and on the scheduling policy integrated into the system. The analytical model we

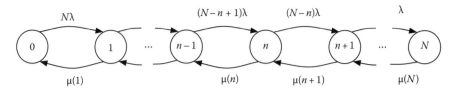

**FIGURE 20.2** General CTMC with variable departure rates.

propose keeps the same birth-and-death structure but integrates departure rates $\mu(n)$ that depend on the current state $n$ as shown in Figure 20.2.

The main difficulty consists in estimating accurately the average departure rates $\mu(n)$ of this model. To do so, we first express $\mu(n)$ as follows:

$$\mu(n) = \frac{\bar{m}(n)\, N_{\mathrm{S}}}{\bar{x}_{\mathrm{on}}\, T_{\mathrm{F}}}, \tag{20.3}$$

where $\bar{m}(n)$ is the average number of bits per slot when there are $n$ concurrent active transfers. Obviously, $\bar{m}(n)$ depends on $K$ the number of MCSs, and $p_k$, $0 \leq k \leq K$, the MCS vector probability. Also, $\bar{m}(n)$ strongly depends on $n$, because the average number of bits per slot must be estimated by considering all possible distributions of the $n$ mobiles between the $K+1$ possible coding schemes (including outage). Finally, the average number of bits per slot $\bar{m}(n)$ also depend on the scheduling policy. More precisely, for each possible mobile distribution, the scheduling policy defines the quantity of slots given to each of the $n$ mobiles that corresponds to the coding scheme they use.

At this step, our analytical model can represent any WiMAX system provided the average number of bits per slot $\bar{m}(n)$ can be estimated. In Section 20.4.3, we develop a generic analytical expression of these rates, whereas in Section 20.5 we present their detailed expressions depending on three specific scheduling policies.

### 20.4.2.1 Performance Parameters

The steady-state probabilities $\pi(n)$ can easily be derived from the birth-and-death structure of the Markov chain (depicted in Figure 20.2):

$$\pi(n) = \frac{N!}{(N-n)!} \left(\frac{T_{\mathrm{F}}}{N_{\mathrm{S}}}\right)^n \frac{\rho^n}{\prod_{i=1}^{n} \bar{m}(i)}\, \pi(0), \quad \text{with } \rho = \frac{\bar{x}_{\mathrm{on}}}{t_{\mathrm{off}}}, \tag{20.4}$$

and $\pi(0)$ is obtained by normalization.

The performance parameters of this system can be derived from the steady-state probabilities as follows. The average utilization $\bar{U}$ of the TDD frame is given by

$$\bar{U} = \sum_{n=1}^{N} \pi(n) \min\left(n\frac{\bar{x}_{\mathrm{on}}}{N_{\mathrm{S}}\, \bar{m}(n)}, 1\right). \tag{20.5}$$

The average number of active users $\bar{Q}$ is expressed as

$$\bar{Q} = \sum_{n=1}^{N} n\pi(n). \tag{20.6}$$

$X_d$, the mean number of departures (mobiles completing their transfer) per unit of time, is obtained as

$$\bar{X}_d = \sum_{n=1}^{N} \pi(n)\,\mu(n).$$ (20.7)

From Little's law, we can thus derive the average duration $\bar{t}_{on}$ of an ON period (duration of an active transfer):

$$\bar{t}_{on} = \frac{\bar{Q}}{\bar{X}_d}.$$ (20.8)

We finally compute the average throughput $\bar{X}$ obtained by each mobile in active transfer as

$$\bar{X} = \frac{\bar{x}_{on}}{\bar{t}_{on}}.$$ (20.9)

### 20.4.3 GENERIC AVERAGE NUMBER OF BITS PER SLOT

We now develop generic expressions of the average number of bits per slot $\bar{m}(n)$ without and with outage.

#### 20.4.3.1 Without Outage

We first consider a system without outage. To illustrate the derivation of the average number of bits per slot, we first consider a situation with two active mobiles (denoted as MS1 and MS2) in a system with two MCSs ($K = 2$), and develop the expression of $\bar{m}(2)$. $MCS_1$ is used with a probability $p_1$ and allows to transfer $m_1$ bits per slot. $MCS_2$ is used with a probability $p_2$ and allows to transfer $m_2$ bits per slot. We denote the average number of bits per slot in the TDD frame for one particular configuration having $j_1$ mobiles using $MCS_1$ and $j_2$ mobiles using $MCS_2$ ($j_1 + j_2 = 2$) by $\bar{m}(j_1, j_2)$. There are three possible configurations:

- $MCS_1 = 2$ MS and $MCS_2 = 0$ MS. This configuration occurs with a probability $p_1 p_1$. Whatever the scheduling policy, the corresponding average number of bits per slot $\bar{m}(2, 0)$ is obviously given as

$$\bar{m}(2, 0) = m_1$$ (20.10)

- $MCS_1 = 0$ MS and $MCS_2 = 2$ MS. Similarly, with a probability $p_2 p_2$, we have

$$\bar{m}(0, 2) = m_2$$ (20.11)

- $MCS_1 = 1$ MS and $MCS_2 = 1$ MS. This configuration can correspond to two different distributions of the two mobiles: MS1 = $MCS_1$ and MS2 = $MCS_2$, or MS1 = $MCS_2$ and MS2 = $MCS_1$. The associated probability is $2 p_1 p_2$, as both distributions have equal probabilities. The corresponding average number of bits per slot $\bar{m}(1, 1)$ can be expressed as

$$\bar{m}(1, 1) = m_1 x_1(1, 1) + m_2 x_2(1, 1),$$ (20.12)

- where $x_k(1, 1)$ is the proportion of the resource that is associated with mobiles using $MCS_k$, which strongly depends on the scheduling policy.

We finally express the average number of bits per slot when there are two active mobiles in the system as

$$\bar{m}(2) = \sum_{j_1=0}^{2} \bar{m}(j_1, 2 - j_1) \binom{2}{j_1} p_1^{j_1} p_2^{2-j_1},$$ (20.13)

where $\binom{2}{j_1}$ is a binomial coefficient that gives the number of distributions corresponding to the configuration of $j_1$ mobiles using $\text{MCS}_1$ and $2 - j_1$ mobiles using $\text{MCS}_2$. As a generalization, one can convince oneself easily that the average number of bits per slot, $\bar{m}(n)$, when there are $n$ active users, can be expressed as follows:

$$
\bar{m}(n) = \sum_{\substack{(j_1,...,j_K) = (0,...,0)| \\ j_1 + \cdots + j_K = n}}^{(n,...,n)} \bar{m}(j_1,...,j_K) \binom{n}{j_1,...,j_K} \left( \prod_{k=1}^{K} p_k^{j_k} \right)
$$

$$
= \sum_{\substack{(j_1,...,j_K) = (0,...,0)| \\ j_1 + \cdots + j_K = n}}^{(n,...,n)} \sum_{k=1}^{K} m_k j_k x_k(j_1,...,j_K) \binom{n}{j_1,...,j_K} \left( \prod_{k=1}^{K} p_k^{j_k} \right), \qquad (20.14)
$$

where

$\binom{n}{j_1,...,j_K}$ is the multinomial coefficient

$x_k(j_1,...,j_K)$ is the proportion of resource given to MS using $\text{MCS}_k$, when the current distribution of the $n$ mobiles among the $K$ coding schemes is $(j_1,...,j_K)$

Let us emphasize that this expression has a $\mathcal{O}(n^K)$ complexity, where $K$, the number of different coding schemes, is usually low. Section 20.5.4 will show that this complexity can be drastically reduced without any significant impact on the accuracy of $\bar{m}(n)$ values.

### 20.4.3.2   With Outage

We now come back to a system with possible outage ($\text{MCS}_0$ used with a probability $p_0$). Relation 20.14 can be extended straightforwardly, as the $j_0$ mobiles in outage of a given distribution do not contribute to the sharing of resource:

$$
\bar{m}(n) = \sum_{\substack{(j_0,j_1,...,j_K) = (0,0,...,0)| \\ j_0 + j_1 + \cdots + j_K = n \\ j_0 \neq n}}^{(n,n,...,n)} \left( \sum_{k=1}^{K} m_k j_k x_k(j_1,...,j_K) \right) \binom{n}{j_0,j_1,...,j_K} \left( \prod_{k=0}^{K} p_k^{j_k} \right). \qquad (20.15)
$$

### 20.4.4   DISCUSSION OF THE MODELING ASSUMPTIONS

Markovian model is based on several system and traffic assumptions presented in Section 20.4.1. We now discuss these assumptions one by one (item numbers are related to the corresponding assumptions), evaluate their accuracy, and provide, if necessary and possible, extensions and generalization propositions.

1. As described in Section 20.4, DL_MAP and UL_MAP are located in the DL part of the TDD frame. They contain the information elements that allow mobiles to identify the slots to be used. The size of these MAPs and as a consequence the number $N_S$ of available slots for DL data transmissions depend on the number of mobiles scheduled in the TDD frame. To relax assumption 1, we can express the number of data slots $N_S(n)$ as a function of $n$, the number of active users. This dependency can be easily integrated in the model by replacing $N_S^n$ by $\prod_{i=1}^{n} N_S(n)$ in Relation 20.4, and $N_S$ by $N_S(n)$ in Relations 20.3 and 20.5.

2. A limit $n_{\max}$ on the total number of mobiles that can simultaneously be multiplexed on the TDD frame can easily be introduced in the model, if required. The corresponding Markov chain shown in Figure 20.2, indeed, has just to be truncated to this limiting state (i.e., the last state becomes $\min(n_{\max}, N)$). As a result, a blocking can now occur when a new transfer

demand arrives and the limit is reached. The blocking probability can easily be derived from the Markov chain [5].

3. In some cellular networks (e.g., (E)GPRS), MSs have limited transmission capabilities because of hardware considerations. This constraint defines the maximum throughput the network interface can reach or the maximum number of resource units that can be used by the mobiles. Such limitations add a slight complexity to the development of model, as one single mobile may not be able to use all the available slots. This characteristic has been introduced in the case of (E)GPRS networks [5,21] and can be applied to WiMAX networks by simply modifying the departure rates of the first states of the Markov chain (i.e., replacing $N_S$ by $\min(n\,d, N_S)$ in Relation 20.3, where $d$ is the maximum number of slots a mobile can use in DL).

4. Radio channel may be highly variable (i.e., conditions change from one frame to another) or it may vary with some memory (i.e., conditions are maintained during a number of frames). Our analytical model only depends upon stationary probabilities of different coding schemes whatever be the radio channel dynamics. This approach is authenticated through simulations in Section 20.6.

5. All mobiles in the considered system have statistically the same traffic characteristics. More complex systems with multiple-traffic and/or differentiation between users would naturally result in more complex models that are not addressed here.

6. As our main concern is dimensioning, we do not take handover into account and consider the fixed mobile population in a stationary manner. However, mobility effects are taken into account in the channel model by means of radio conditions variation.

7. Poisson processes are currently used in the case of a large population of users, assuming independence between the arrivals and the current population of the system. As we focus on the performance of a single cell system, the potential population of users is relatively small. The higher the number of ongoing data connections, the less likely the arrival of new ones. Poisson processes are thus a nonrelevant choice for our models. In addition, the finite population assumption is used typically for network planning when geomarketing data allows the prediction of the active mobile population that will be served by the cell (for a network in service, traffic statistics can also provide estimates of this population). Note, however, if the Poisson assumption has to be made for connection demand arrivals, one can directly modify the arrival rates of the Markov chain (i.e., replace the state-dependent rates $(N - n)\lambda$ by some constant value, and limit the number of states of the Markov chain as explained above in point 2).

8. Each mobile is supposed to generate infinite length ON/OFF session traffic. In the context of (E)GPRS networks [4,6], we have studied an extension to finite length sessions, where each mobile generates ON/OFF traffic during a session and does not generate any traffic during an intersession. We show in these studies that a very simple transformation of traffic characteristics, which increases OFF periods by a portion of the intersession period, enables to derive the average performance from the infinite length session model. The accuracy of this transformation is related to the insensibility of the average performance parameters with regards to the traffic distributions (see next point). An equivalent transformation can be applied to our WiMAX model, even if it is no longer processor-sharing. Until any theoretical result can be proven, the resulting transformation remains a good approximation.

9. Memoryless traffic distributions are strong assumptions that are validated by several theoretical results. Several studies on insensitivity (see, e.g., Refs. [7,11,14]) have shown (for processor-sharing systems) that the average performance parameters are insensitive to the distribution of ON and OFF periods. As we are not able to formally demonstrate here that this result also holds for our WiMAX model, we present in Section 20.6 a comparison of the system performance obtained by simulation for several traffic distributions (exponential

and Pareto) and our analytical model. These results tend to prove that insensibility still holds or is at least a good approximation. Thus, memoryless distributions are the most convenient choice to model traffic.

## 20.5 SCHEDULING POLICY MODELING

We now present the analytical model adaptation to different scheduling policies. For each of them, we provide closed-form expressions for the average number of bits per slot $\bar{m}(n)$.

### 20.5.1 SLOT SHARING FAIRNESS

We study a scheduling policy providing fairness in slot sharing. Each time-step, the scheduler equally shares the $N_S$ slots among the active users that are not in outage.

#### 20.5.1.1 Slot Sharing Fairness without Outage

First we do not take into account the outage (only coding schemes $\text{MCS}_k$, $1 \leq k \leq K$, are used). If at a given time-step, there are $n$ active mobiles, then each of them receives a portion $N_S/n$ of the whole resource. As a consequence, the proportion of the resource that is associated with mobiles using $\text{MCS}_k$ is constant for any $k$ and for any possible distribution $(j_1, ..., j_K)$ of the $n$ mobiles among the $K$ coding schemes, and is thus given by

$$x_k(j_1, ..., j_K) = \frac{1}{n}. \tag{20.16}$$

By replacing these equal proportions in generic expression (Equation 20.14), the average number of bits per slot, $\bar{m}(n)$, when there are $n$ active users, becomes

$$\bar{m}(n) = \sum_{\substack{(j_1, ..., j_K) = (0, ..., 0)| \\ j_1 + \cdots + j_K = n}}^{(n, ..., n)} \left( \sum_{k=1}^{K} \frac{m_k j_k}{n} \right) \binom{n}{j_1, ..., j_K} \left( \prod_{k=1}^{K} p_k^{j_k} \right). \tag{20.17}$$

After a few simplifications, we obtain

$$\bar{m}(n) = \sum_{\substack{(j_1, ..., j_K) = (0, ..., 0)| \\ j_1 + \cdots + j_K = n}}^{(n, ..., n)} \sum_{\substack{k=1| \\ j_k \neq 0}}^{K} m_k p_k \binom{n-1}{j_1, ..., j_k - 1, ..., j_K} \left( p_1^{j_1} \cdots p_k^{j_k - 1} \cdots p_K^{j_K} \right). \tag{20.18}$$

By rearranging carefully the terms of the summations, we can show that this expression can drastically be simplified as

$$\bar{m}(n) = \sum_{k=1}^{K} m_k p_k = \bar{m}. \tag{20.19}$$

This nice and very simple expression shows us that, when there is no outage, the average number of bits per slot $\bar{m}(n)$ associated with the slot sharing fairness policy are constant and can be simply seen as an average number of bits per slot $\bar{m}$.

#### 20.5.1.2  Slot Sharing Fairness with Outage

If we now consider outage ($MCS_0$ is used with a probability $p_0$), the expression of the average number of bits per slot becomes

$$
\bar{m}(n) = \sum_{\substack{(j_0,j_1,\ldots,j_K) = (0,0,\ldots,0)| \\ j_0 + j_1 + \cdots + j_K = n \\ j_0 \neq n}}^{(n,n,\ldots,n)} \left( \sum_{k=1}^{K} \frac{m_k j_k}{n - j_0} \right) \binom{n}{j_0, j_1, \ldots, j_K} \left( \prod_{k=0}^{K} p_k^{j_k} \right)
$$

$$
= \sum_{\substack{(j_0,j_1,\ldots,j_K) = (0,0,\ldots,0)| \\ j_0 + j_1 + \cdots + j_K = n \\ j_0 \neq n}}^{(n,n,\ldots,n)} \frac{n!}{n - j_0} \left( \sum_{k=1}^{K} m_k j_k \right) \left( \prod_{k=0}^{K} \frac{p_k^{j_k}}{j_k!} \right),
\tag{20.20}
$$

as the resource has now to be shared only among the $n - j_0$ users that are not in outage. It is important to note that the expression of the average number of bits per slot cannot be further simplified, and is no longer constant (see Section 20.5.4).

### 20.5.2  Instantaneous Throughput Fairness

We now consider a scheduling policy that shares the resource to provide the same instantaneous throughput to all active users. Hence, at a given time-step, mobiles using an MCS with a low bit rate per slot will obtain proportionally more slots than the mobiles using an MCS with a high bit rate per slot.

#### 20.5.2.1  Throughput Fairness without Outage

Let us first consider that there is no possible outage. Recall that $x_k(j_1, \ldots, j_K)$ is the proportion of the resource that is associated by the scheduler with mobiles using coding scheme $MCS_k$, when the current distribution of the mobiles is $(j_1, \ldots, j_K)$. To respect instantaneous throughput fairness between active users, the $x_k(j_1, \ldots, j_K)$ must be such that

$$
m_k x_k(j_1, \ldots, j_K) = C \quad \text{for any } k,
\tag{20.21}
$$

where $C$ is a constant such that $\sum_{k=1}^{K} j_k x_k(j_1, \ldots, j_K) = 1$, thus

$$
C = \frac{1}{\sum_{k=1}^{K} \frac{j_k}{m_k}}.
\tag{20.22}
$$

By replacing the proportions $x_k(j_1, \ldots, j_K)$ in generic expression (Equation 20.14), the average number of bits per slot $\bar{m}(n)$, when there are $n$ active users, becomes

$$
\bar{m}(n) = \sum_{\substack{(j_1,\ldots,j_K) = (0,\ldots,0)| \\ j_1 + \cdots + j_K = n}}^{(n,\ldots,n)} \frac{n}{\sum_{k=1}^{K} \frac{j_k}{m_k}} \binom{n}{j_1, \ldots, j_K} \left( \prod_{k=1}^{K} p_k^{j_k} \right) = \sum_{\substack{(j_1,\ldots,j_K) = (0,\ldots,0)| \\ j_1 + \cdots + j_K = n}}^{(n,\ldots,n)} \frac{n \, n! \prod_{k=1}^{K} \frac{p_k^{j_k}}{j_k!}}{\sum_{k=1}^{K} \frac{j_k}{m_k}}.
\tag{20.23}
$$

### 20.5.2.2 Throughput Fairness with Outage

If we now consider outage, the expression of the average number of bits per slot simply becomes

$$\bar{m}(n) = \sum_{\substack{(j_0,j_1,...,j_K) = (0,0,...,0)| \\ j_0+j_1+\cdots+j_K = n \\ j_0 \neq n}}^{(n,n,...,n)} \frac{(n-j_0)\, n! \prod_{k=0}^{K} \frac{p_k^{j_k}}{j_k!}}{\sum_{k=1}^{K} \frac{j_k}{m_k}}. \tag{20.24}$$

## 20.5.3 Opportunistic Scheduling

We finally study an opportunistic scheduling policy where all the resources are given to the users who have the highest transmission bit rate, i.e., the better radio conditions and hence the better MCS. Without loss of generality, we assume, in this section, that the coding schemes are classified in increasing order: $m_1 < m_2 < \cdots < m_K$. And even if it is still possible to derive the average number of bits per slot from generic expressions (Equations 20.14 and 20.15) (without and with outage), we prefer to give here a more intuitive and strictly equivalent derivation.

### 20.5.3.1 Opportunistic without Outage

We consider a system with $n$ current active mobiles. We denote by $\alpha_i(n)$ the probability of having at least one active user (among $n$) using $MCS_i$ and none using an MCS that gives higher transmission rates (i.e., $MCS_j$ with $j > i$). As a matter of fact, $\alpha_i(n)$ corresponds to the probability that the scheduler gives at a given time-step all the resource to mobiles that use $MCS_i$. As a consequence, we can express the average number of bits per slot when there are $n$ active users as

$$\bar{m}(n) = \sum_{i=1}^{K} \alpha_i(n) m_i. \tag{20.25}$$

To calculate the $\alpha_i(n)$, we first express $p_{\leq i}(n)$, the probability that there is no mobile using an MCS higher than $MCS_i$:

$$p_{\leq i}(n) = \left(1 - \sum_{j=i+1}^{K} p_j\right)^n. \tag{20.26}$$

Then, we calculate $p_{=i}(n)$, the probability that there is at least one mobile using $MCS_i$ conditioned by the fact that there is no mobile using a better MCS:

$$p_{=i}(n) = 1 - \left(1 - \frac{p_i}{\sum_{j=1}^{i} p_j}\right)^n. \tag{20.27}$$

$\alpha_i(n)$ can thus be expressed as

$$\alpha_i(n) = p_{=i}(n)\, p_{\leq i}(n). \tag{20.28}$$

### 20.5.3.2 Opportunistic with Outage

It is easy to show that the previous development remains the same when some mobiles can be in outage. Indeed, as soon as there is at least one mobile using a real coding scheme $MCS_k$ ($k \neq 0$), the scheduler gives no resource to mobiles using a "lower" MCS (including mobiles in outage). As a result, the average number of bits per slot have the same expression (with a single modification in Relation 20.27, where index $j$ of the sum must vary from 0 to $i$).

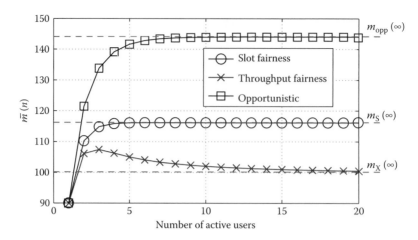

**FIGURE 20.3** $\bar{m}(n)$ asymptotic behavior.

### 20.5.4 ANALYTICAL ASYMPTOTIC STUDY

As a side study in the modeling of the average number of bits per slot $\bar{m}(n)$, we can observe asymptotic behavior of $\bar{m}(n)$ functions. Figure 20.3 shows the evolution of $\bar{m}(n)$, when $n$ increases for the three studied-scheduling policies, in the general case where some mobiles can be in outage. We can notice that the three resulting functions $\bar{m}(n)$ rapidly tend to an asymptote, as the number of active users $n$ increases. We thus derive, in Sections 20.5.4.1 through 20.5.4.3, the analytical expressions of these asymptotes for each scheduling policy. Note that one can benefit from this quick asymptotical behavior to avoid the calculation of the $\bar{m}(n)$ for large values of $n$ (by replacing, after a threshold, the exact value by the corresponding asymptote value).

#### 20.5.4.1 Slot Fairness Asymptote

In the case of slot fairness scheduling, as the number of active users grows, the proportion of mobiles using $MCS_k$ tends to $p_k$. If we denote the number of such mobiles by $J_k$, when $n \to \infty$, we have $J_k \sim p_k n$. As the resources are equally shared among $n - J_0$ mobiles that are not in outage, the limiting value of the average number of bits per slot is given by

$$m_S(\infty) = \lim_{n \to \infty} \bar{m}(n) = \lim_{n \to \infty} \sum_{k=1}^{K} m_k \frac{J_k}{n - J_0} = \frac{\sum_{k=1}^{K} m_k p_k}{1 - p_0}. \tag{20.29}$$

#### 20.5.4.2 Throughput Fairness Asymptote

We now detail the asymptote corresponding to the instantaneous throughput fairness policy. Again, the number of mobiles using $MCS_k$, when $n \to \infty$, is $J_k \sim p_k n$. Every such mobile obtains a proportion $x_k$ of the resource such that $\sum_{k=1}^{K} J_k x_k = 1$. To respect the fairness of the scheduling policy, these proportions must satisfy the following relation:

$$m_k x_k = C = \frac{1}{\sum_{k=1}^{K} \frac{J_k}{m_k}} \quad \text{for any } k \neq 0. \tag{20.30}$$

Note that mobiles in outage do not use any resource (and thus, $x_0 = 0$). By combining these relations, we obtain the expression of the asymptote value:

$$m_X(\infty) = \lim_{n\to\infty} \bar{m}(n) = \lim_{n\to\infty} \sum_{k=1}^{k} m_k J_k x_k = \frac{1 - p_0}{\sum_{k=1}^{K} \dfrac{p_k}{m_k}}. \tag{20.31}$$

### 20.5.4.3 Opportunistic Scheduling Asymptote

The asymptote value of $\bar{m}(n)$ for opportunistic scheduling simply corresponds to the highest bit rate per slot (obtained with the best coding scheme). Actually, as the number of active users grows, the probability of having at least one mobile using the best MCS tends to 1. Thus, we have

$$m_{\text{opp}}(\infty) = \lim_{n\to\infty} \bar{m}(n) = m_K. \tag{20.32}$$

## 20.6 VALIDATION

In this section, we discuss the validation of our analytical model through extensive simulations. We also show its robustness when traffic and channel models are complexified. For this purpose, a simulator has been developed that implements an ON/OFF traffic generator and a wireless channel for each user and a centralized scheduler that allocates radio resources, i.e., slots, to active users on a frame-by-frame basis. In the first phase, we validate our analytical model through simulations. In this validation study, the modeling assumptions (related to scheduling, traffic, and channel models) are reproduced in the simulator. The assumptions are related to scheduling, traffic, and channel models. This phase shows that describing the state of the system by the aggregation of all active users (whatever be the distribution of their coding schemes) is a very good modeling approximation. It also validates the analytical expression of the average number of bits per slot $\bar{m}(n)$. In the second phase, the robustness study, we relax the assumptions made for the analytical model by considering more realistic models for traffic and radio channel variations. By carrying out a comparison with simulation results, we thus show how robustly the analytical model reacts toward these relaxations.

### 20.6.1 SIMULATION MODELS

We now detail the simulation models before presenting the simulation results for the validation and robustness studies.

#### 20.6.1.1 System Parameters

As in Sections 20.2 and 20.3, we consider a single WiMAX cell and study the DL. Radio resources that are thus made of time–frequency slots in the DL TDD subframe. The number of slots depends on the system bandwidth, the frame duration, the DL/UL ratio, the subcarrier permutation (PUSC, FUSC, AMC), and the protocol overhead (preamble, FCH, MAPs). System bandwidth is assumed to be 10 MHz. The duration of one TDD frame of WiMAX is 5 ms and the DL/UL ratio is considered to be 2/3. Although a slot is made of the same number of data subcarriers whatever the subcarrier permutation is, their total number varies. For the purpose of simulation, PUSC has been kept as a reference. We assume for the sake of simplicity that the protocol overhead is of fixed length (2 symbols) although, in reality, it is a function of the number of scheduled users. These parameters lead to a number of data slots (excluding overhead) per TDD DL subframe of $N_S = 450$.

#### 20.6.1.2 Traffic Parameters

In our analytical model, we consider an elastic ON/OFF traffic. Mean values of ON data volume (main page and embedded objects) and OFF period (reading time) are 3 Mb and 3 s, respectively.

**TABLE 20.1**
**Traffic Parameters**

| Parameter | Value |
|---|---|
| Number of MS in the system $N$ | up to 50 |
| Mean ON data volume $\bar{x}_{on}$ | 3 Mb |
| Mean OFF duration $\bar{t}_{off}$ | 3 s |
| Pareto parameter $\alpha$ | 1.2 |
| Pareto low cutoff $q$ | 300 Mb |
| Pareto high cutoff $q$ | 3,000 Mb |
| Pareto parameter $b$ for low cutoff | 712,926 b |
| Pareto parameter $b$ for high cutoff | 611,822 b |

In the first phase (validation study), we assume that the ON data volume is exponentially distributed, as it is the case in the analytical model assumptions. Although well adapted to Markov theory-based analysis, exponential law does not always fit the reality for data traffic. This is the reason why we consider truncated Pareto distributions in the second phase (the robustness study). Recall that the mean value of the truncated Pareto distribution is given by

$$\bar{x}_{on} = \frac{\alpha b}{\alpha - 1} \left[ 1 - (b/q)^{\alpha - 1} \right], \tag{20.33}$$

where
   $\alpha$ is the shape parameter
   $b$ is the minimum value of Pareto variable
   $q$ is the cutoff value of truncated Pareto distribution

Two values of $q$ are considered: low and high. These have been taken as hundred times and thousand times the mean value, respectively. The mean value in both cases (high and low cutoff) is 3 Mb for the sake of comparison with the exponential model. The value of $\alpha = 1.2$ has been adopted from Ref. [10]. The corresponding values of parameter $b$ for high and low cutoff are calculated using Relation 20.33. Traffic parameters are summarized in Table 20.1.

### 20.6.1.3 Channel Models

The number of bits per slot, an MS is likely to receive, depends on the chosen MCS, which in turn depends on its radio channel conditions. The choice of an MCS is based on SINR measurements and SINR thresholds. A generic method for describing the channel between the BS and an MS is to model the transitions between MCSs by a finite state Markov chain (FSMC). The chain is discrete time, and transitions occur every $L$ frames, with $L\,T_F < \bar{t}_{coh}$, where $\bar{t}_{coh}$ is the coherence time of the channel. In our case, for the sake of simplicity, $L = 1$. Such an FSMC is fully characterized by its transition matrix $P_T = (p_{ij})_{0 \le i,j \le K}$. Note that an additional state (state 0) is introduced to take into account outage (when SINR is below the minimum radio quality threshold). Stationary probabilities $p_k$ provide the long-term probabilities for an MS to receive data with MCS $k$.

In our analytical study, channel model is assumed to be memoryless, i.e., MCS are independently drawn from frame to frame for each user, and the discrete distribution is given by $(p_i)_{0 \le i,j \le K}$. This corresponds to the case, where $p_{ij} = p_j$ for all $i$. This simple approach, referred as the memoryless channel model, is considered in the validation study, which exactly reproduces the assumptions of the analysis. Let $P_T(0)$ be the transition matrix associated with the memoryless model.

In the robustness study, we introduce two additional channel models with memory. In these models, the MCS observed for a given MS in a frame depends on the MCS observed in the previous frame according to the FSMC presented above. This transition matrix is derived from the following equation:

$$P_{\mathrm{T}}(a) = aI + (1-a)P_{\mathrm{T}}(0) \quad 0 \le a \le 1,$$

where

$I$ is the identity matrix

parameter $a$ is a measure of the channel memory

A mobile actually maintains its MCS for a certain duration with mean $\bar{t}_{\mathrm{coh}} = 1/(1-a)$. With $a = 0$, the transition process becomes memoryless. In the other extreme, with $a = 1$, the transition process will have infinite memory and mobiles will never change their MCS. For simulations, we have taken $a$ equal to be 0.5, so that the channel is constant in average two frames. This value is consistent with the coherence time given in Ref. [22] for a 45 km/h speed mobility in a 2.5 GHz bandwidth system. We call the case where all MSs have the same channel model with memory ($a = 0.5$) the average channel model. Note that the stationary probabilities of the average channel model are the same as those of the memoryless model.

As the channel depends on the BS–MS link, it is possible to refine the previous approach by considering part of the MS to be in a "bad" state, and the rest in a "good" state. Bad and good states are characterized by different stationary probabilities but have the same coherence time. In the so-called combined channel model, half of the MS is in a good state, the rest in a bad state, and $a$ is kept to 0.5 for both populations. For the sake of comparison, the overall MCS probabilities in the combined model are the same as those of the memoryless and average models. Three models are thus considered: the memoryless, the average, and the combined channel models. In Table 20.2, considered MCSs (including outage) are given, and for each of them, the number of bits transmitted per slot is also listed.

Channel stationary probabilities are given in Table 20.3. The respective MCS stationary probabilities for good and bad channel types can be obtained, for example, by performing system-level Monte Carlo simulations and recording channel statistics close (good state) or far (bad state) from the BS. Stationary probabilities for the combined model are obtained by averaging corresponding values of good and bad model stationary probabilities.

### 20.6.1.4 Scheduling Simulation

The simulator implements the three scheduling schemes considered in this study, i.e., opportunist, fair in throughput, and fair in slots. On a frame-by-frame basis, the scheduler allocates the DL slots to active users according to their radio conditions (their MCS) and the scheduling policy. As already mentioned, the scheduler does not allocate resources to active users in outage. The computation of number of slots to be allocated to each user is detailed hereafter. In a given frame, the number of slots allocated to active users should satisfy the following condition:

**TABLE 20.2**

**Channel Parameters**

| Channel State $\{0,...,K\}$ | MCS and Outage | Bits per Slot $m_k$ |
|---|---|---|
| 0 | Outage | $m_0 = 0$ |
| 1 | QPSK-1/2 | $m_1 = 48$ |
| 2 | QPSK-3/4 | $m_2 = 72$ |
| 3 | 16QAM-1/2 | $m_3 = 96$ |
| 4 | 16QAM-3/4 | $m_4 = 144$ |

**TABLE 20.3**

**Stationary Probabilities for Three Channel Models**

| Channel Model | Memoryless | Average | Combined Good 50 percent MS | Bad 50 percent MS |
|---|---|---|---|---|
| $a$ | 0 | 0.5 | 0.5 | 0.5 |
| $p_0$ | 0.225 | 0.225 | 0.020 | 0.430 |
| $p_1$ | 0.110 | 0.110 | 0.040 | 0.180 |
| $p_2$ | 0.070 | 0.070 | 0.050 | 0.090 |
| $p_3$ | 0.125 | 0.125 | 0.140 | 0.110 |
| $p_4$ | 0.470 | 0.470 | 0.750 | 0.190 |

$$N_S = \sum_{k=0}^{K} N_S^{(k)} n^{(k)},$$

where

$N_S^{(k)}$ is the number of slots allocated by the scheduler to an MS using MCS $k$

$n^{(k)}$ is the number of active mobiles using MCS $k$

Note that $N_S^{(k)}$ depends on the scheduling scheme and that the number of active users verifies $n = \sum_{k=0}^{K} n^{(k)}$. The way $N_S^{(k)}$ is chosen by the scheduler is detailed below in the scheduling pseudocode.

---

Scheduling pseudocode

---

Let $\mathbf{K_F} \subset [0 \cdots K]$ be the set of MCSs used by active MS in the considered frame.

▷ Opportunist

  **find** $k_{\max} = \max(\mathbf{K_F})$

   $N_S^{(k)} = \frac{N_S}{n^{(k_{\max})}}$    for $k = k_{\max}$

   $N_S^{(k)} = 0$     for all $k \neq k_{\max}$

▷ Fairness in slot

   $N_S^{(k)} = 0$    for $k = 0$

   $N_S^{(k)} = \frac{N_S}{\sum_{k=0}^{K} n^{(k)}}$    for $k \neq 0$

▷ Fairness in throughput

   **if**   $(k = 0)$    **then**   $N_S^{(k)} = 0$

   **else**   $N_S^{(k)} = \frac{N_S/m_k}{\sum_{k=1}^{K} \frac{n^{(k)}}{m_k}}$

---

The value of $N_S^{(k)}$ determined by the scheduling process may or may not be an integer. In case it is not, it is rounded down to the closest integer. It results in some spare number of slots that are allocated to active users (not in outage) in a round robin fashion.

## 20.6.2 SIMULATION RESULTS

In this section, we first present a comparison between the results obtained through our analytical model and scheduling simulator. The output parameters in consideration are $\bar{U}$, $\bar{X}$, and $\pi(n)$ (see Section 20.4.2).

### 20.6.2.1 Validation Study

In this study, simulations take into account the same traffic and channel assumptions as those of the analytical model. However, in simulator, MCS of users is determined on per frame basis and scheduling is carried out in real time, based on MCS at that instant. The analytical model, on the other hand, considers stationary probabilities of MCS only. Distributions of ON data volume and OFF period are exponential and the memoryless channel model is considered.

Figure 20.4a and b shows, respectively, the average channel utilization ($\bar{U}$) and the average instantaneous throughput per user ($\bar{X}$) for the three scheduling schemes. It is clear that simulation and analytical results show a good agreement: for both utilization and throughput, the maximum

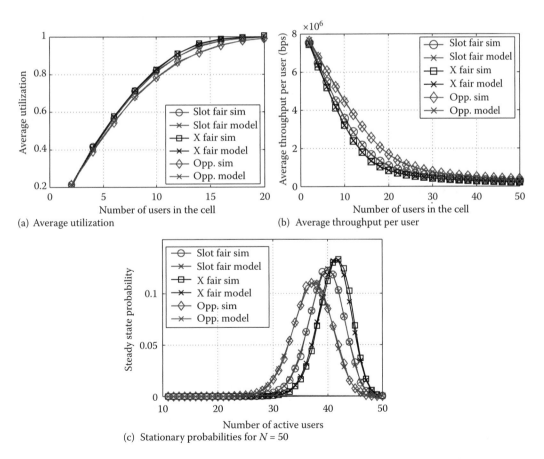

(a) Average utilization

(b) Average throughput per user

(c) Stationary probabilities for $N = 50$

**FIGURE 20.4** Performance validation for the three scheduling policies with $\bar{x}_{on} = 3$ Mb and $\bar{t}_{off} = 3$ s.

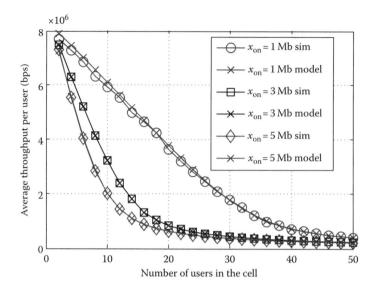

**FIGURE 20.5**   Average throughput per user for different loads.

relative error stays below 6 percent and the average relative error is less than 1 percent. Note that the analytical results have been obtained instantaneously whereas simulations have run for several days.

Figure 20.4c further proves that our analytical model is a very good description of the system: stationary probabilities $\pi(n)$ obtained by either simulations or analysis are compared for a given total number $N = 50$ of MSs. Again, results show a perfect match with an average relative error always below 9 percent. This means that not only average values of the output parameters but also higher moments can be derived from stationary probabilities with a high accuracy.

Finally, Figure 20.5 shows the validation for three different loads (1, 3, and 5 Mbps). Our model shows a comparable accuracy for all three load conditions with a maximum relative error of about 5 percent.

### 20.6.2.2   Robustness Study

We now move to the robustness study, where assumptions concerning traffic and channel models made by the analysis are relaxed in simulations. Note that we have run extensive simulations corresponding to various traffic and channel models that give very similar results.

To check the robustness of analytical model toward distribution of ON data volumes, simulations are carried out for exponential and truncated Pareto (with low and high cutoff). The results for this analysis are shown in Figure 20.6. The average relative error between analytical results and simulations stays below 10 percent for all sets. It is clear that considering a truncated Pareto distribution has little influence on the design parameters. This is mainly due to the fact that the distribution is truncated and is thus not heavily tailed. But even with a high cutoff value, the exponential distribution provides a very good approximation.

Until now we have always considered the memoryless channel model. To check the robustness of our analytical model with respect to the channel memory, we now compare the analytical results with simulation for the three precited channel models: memoryless, average, and combined (with stationary probabilities given in Table 20.3). If we look at the plot of Figure 20.7, we can say that even for a complex channel, our analytical model shows considerable robustness with an average relative error below 7 percent. We can thus deduce that for designing a WiMAX network, channel information is almost completely included in the stationary probabilities of the MCS.

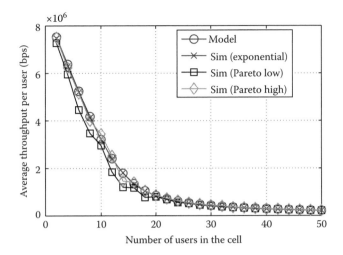

**FIGURE 20.6** Average throughput per user for different traffic distributions.

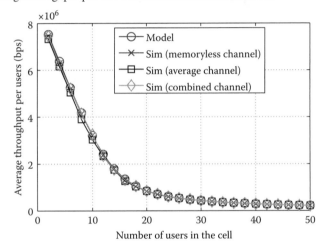

**FIGURE 20.7** Average throughput per user for different channel models.

## 20.7 PERFORMANCE ANALYSIS

In this section, we use the analytical models developed in Sections 20.4 and 20.5 to derive performance curves. We give a first set of general conclusions regarding system behaviors. Note, however, that this study does not intend to decide which scheduling policy is to be selected, as this would require a wider analysis. We consider a TDD frame with $N_S = 450$ DL data slots. The system contains $N = 50$ mobiles with the following traffic characteristics: the average size of a downloaded elements is $\bar{x}_{on} = 1$ Mb and the average reading time is $\bar{t}_{off} = 20$ s. The available coding schemes are the ones presented in Table 20.2 with the same corresponding probabilities $p_k$ of memoryless channel model given in Table 20.3. The influence of the main input parameters is studied on the following performance parameters: the average resource utilization $\bar{U}$ and the average throughput per user $\bar{X}$.

### 20.7.1 INFLUENCE OF THE NUMBER OF RESOURCES

First, we consider the influence of $N_S$, the number of available slots in the DL subframe. Figure 20.8a shows that, for a given number of users and a given traffic load, the average resource utilization $\bar{U}$

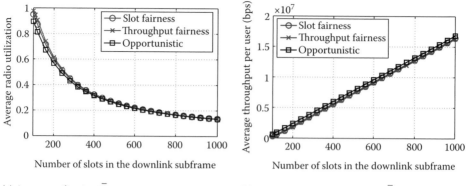

(a) Average utilization $\bar{U}$                           (b) Average throughput per user $\bar{X}$

**FIGURE 20.8**   Influence of the number of resources.

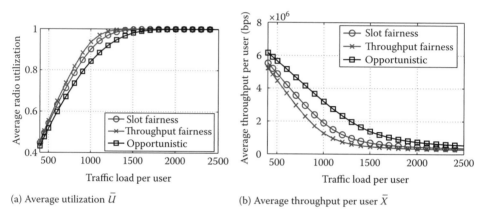

(a) Average utilization $\bar{U}$                           (b) Average throughput per user $\bar{X}$

**FIGURE 20.9**   Influence of the traffic load.

decreases as $N_S$ increases. As a direct consequence, Figure 20.8b shows that the average throughput per user linearly increases with $N_S$. When there are more resources, more slots are given to each active user (depending on the scheduling policy), which results in a better throughput. As a consequence, average transfer duration decreases and so does the average utilization of the cell. Obviously, increasing $N_S$ is beneficial as it enables an increase in the average throughput per user for all scheduling policies. Note, however, that in a real-world dimensioning exercise, increasing $N_S$ has a cost, because bandwidth requirements increases for raising the number of slots. This cost must then be balanced against the corresponding benefits.

### 20.7.2   INFLUENCE OF THE TRAFFIC LOAD

We now study the effect of changing the traffic load (by varying the mean size of downloaded elements). We vary the mean size $\bar{x}_{on}$ from 2 to 10 Mb, corresponding to $\rho \in [500; 2500]$ (see Relation 20.4). Performance curves on Figure 20.9a and b show a similar behavior than the one observed in Section 20.6 when the number $N$ of users in the cell varies. This common behavior reveals that traffic load is characterized by the combination of $N$, the population of the cell, and the traffic parameters expressed by $\rho = \bar{x}_{on}/\bar{t}_{off}$. Both have similar impact on performance. It is important to notice that, as the performance parameters depend on the ratio $\rho = \bar{x}_{on}/\bar{t}_{off}$, any couple of traffic characteristics that give the same value of $\rho$ provide the same performance results.

The large range of values used for this study allows us to observe two regimes on the performance curves corresponding to nonsaturated and saturated systems. In the former, the average utilization increases linearly with the traffic load (with a slope that depends on the scheduling policy), and throughput decreases accordingly. As a matter of fact, transfer durations increase linearly with the size of the downloaded elements, and thus with the traffic load. In the latter, resources are fully used and the system is under saturation. The utilization is close to 1 and the average throughput tends to attain very low values corresponding to infinite resource sharing. As shown in Figure 20.9a and b, the curve slope of the nonsaturated regime and thus the starting point of the saturated regime strongly depend on the scheduling policy. These curves show that opportunistic scheduling postpones the transition between the two regimes to higher traffic loads.

## 20.8 NETWORK DESIGN

We now provide and explain the use of some examples of graphs that can be instantaneously obtained with our analytical solution, and that cannot even be thought of with simulators because of their prohibitive computation time. All these graphs correspond to the throughput fairness scheduling but can be drawn as easily for the two other scheduling policies (as well as for any alternative policy provided the average number of bits per slot $\bar{m}(n)$ can be evaluated).

### 20.8.1 PERFORMANCE GRAPHS

We first draw three-dimensional surfaces where performance parameters are function of, for example, $N$, the number of users in the cell, and $\rho$, the combination of traffic parameters (see Relation 20.4). For each performance parameter, the surface is cut out into level lines and the resulting two-dimensional projections are drawn. The step between level lines can be arbitrarily chosen as a function of the required precision.

The average radio resource utilization of the WiMAX cell $\bar{U}$ and the average throughput per user $\bar{X}$ for any mobile in the system are presented in Figure 20.10a and b (corresponding to the radio link characteristics presented in Section 20.6). Each graph is the result of several thousands of evaluation points (corresponding to varying input parameters). Obviously, any simulation tool or even any multidimensional Markov chain requiring numerical resolution would have precluded the drawing of such graphs. These graphs allow to directly derive the corresponding performance parameter knowing the traffic load profile, i.e., the couple $(N, \rho)$. Measures on real systems can

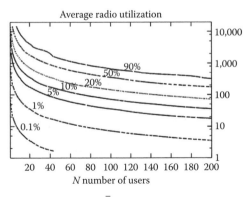

(a) Average utilization $\bar{U}$

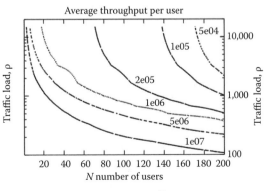

(b) Average throughput per user $\bar{X}$

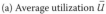

**FIGURE 20.10** Performance graphs.

provide such parameters as they depend on the corresponding traffic profiles of the used applications (FTP, Web, mail, etc.). These performance graphs can be used as easily and efficiently as the classical Erlang graphs employed for the dimensioning of telephone networks.

### 20.8.2 DIMENSIONING STUDY

In this section, we show how our model can be advantageously applied for dimensioning issues. Once again, our modeling framework allows very fast computations, which in turn allows complex iterative dimensioning analyses. Each point of the following graphs now corresponds to multiple iterations of our model resolution, to find the optimal value of an input parameter (e.g., the number of users in the system) that respects a given QoS criterion.

For example, Figure 20.10a gives minimum values $N_{min}$ of mobiles in the cell, to guarantee that the average radio utilization is over 50 percent. This kind of criterion allows operators to maximize the utilization of network resource in comparison with the traffic load of their customers. For a given traffic load profile and a given set of system parameters, the point of coordinates $(N_S, \rho)$ in the graph is located between two level lines, and the level line with the higher value gives the optimal value of $N_{min}$.

Figure 20.10b gives another example of compact and efficient dimensioning graphs, corresponding to the throughput offered to users. The QoS criterion requires that each mobile obtains a minimum average throughput for its transfers. Operators dimensioning issues usually require to find out an optimal value of $N$ that satisfies a maximum instantaneous rate (MIR) for users during their transfer, where the MIR simply corresponds to our average throughput per user $\bar{X}$. As explained above, the average throughput per user is a decreasing function of $N$. The higher the number of users, the higher the traffic load and the lower the throughput per user. We then have to find the maximum value $N_{max}$ of users in the cell to guarantee the minimum throughput threshold. In Figure 20.10b, a given point $(N_S, \rho)$ is located between two level lines. The line with the lower value gives $N_{max}$.

Note finally that these several dimensioning graphs can be used together for guaranteeing multiple QoS criteria. For example, if we have a WiMAX cell configured to have $N_S = 450$ slots and a traffic profile given by $\rho = 300$ (e.g., $x_{on} = 1.2$ Mb and $t_{off} = 20$ s), Figure 20.11a gives $N_{min} = 55$, and Figure 20.11b gives $N_{max} = 200$. As a consequence, the combination of these two graphs recommend to have a number of users $N \in [55; 200]$ to guarantee a reasonable utilization of the cell and to offer a minimum throughput to users.

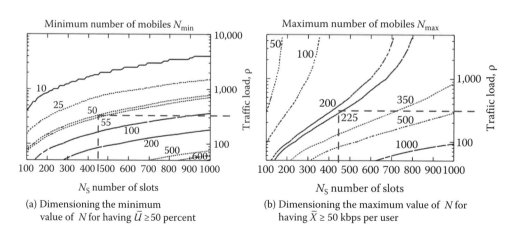

(a) Dimensioning the minimum value of $N$ for having $\bar{U} \geq 50$ percent

(b) Dimensioning the maximum value of $N$ for having $\bar{X} \geq 50$ kbps per user

**FIGURE 20.11**   Dimensioning graphs.

## 20.9    CONCLUSION

As deployment of WiMAX networks is underway, need arises for operators and manufacturers to develop dimensioning tools. In this chapter, we have presented novel analytical models for WiMAX networks and elastic ON/OFF traffic. The models are able to derive Erlang-like performance parameters such as throughput per user or channel utilization. Based on a one-dimensional Markov chain and the derivation of average number of bits per slot, whose expressions are given for three main scheduling policies (slot fairness, throughput fairness, and opportunistic scheduling), our model is remarkably simple. The resolution of model provides closed-form expressions for all the required performance parameters at a click speed. Therefore it will enable efficient and advanced dimensioning studies. The generic nature of model makes it flexible to be customized to scenario-specific requirements. For example, the Markov chain can be adapted to any other scheduling policy because a general expression for the average number of bits per slot is also given. Extensive simulations have validated the model's assumptions. The accuracy of the model is illustrated by the fact that, for all simulation results, maximum relative errors do not exceed 10 percent. Even if the traffic and channel assumptions are relaxed, analytical results still match very well with simulations that show the robust nature of our model.

## REFERENCES

[1] IEEE 802.16e: IEEE 802.16e Task Group (Mobile WirelessMAN) http://www.ieee802.org/16/tge/.

[2] IEEE Std. 802.16: IEEE Standard for local and metropolitan area networks—Part 16: Air Interface for Fixed Broadband Wireless Access Systems, 2004.

[3] IEEE Std. 802.16e: Draft IEEE std 802.16e/D9. IEEE Standard for local and metropolitan area networks—Part 16: Air Interface for Fixed Broadband Wireless Access Systems. Amendment 2: Physical and Medium Access Control Layers for Combined Fixed and Mobile Operation in Licensed Bands, 2005.

[4] B. Baynat, K. Boussetta, P. Eisenmann, and N. Ben Rached, Towards an Erlang-like formula for the performance evaluation of GPRS/EDGE networks with finite-length sessions, 3rd IFIP-TC6 Networking Conference, May 2004.

[5] B. Baynat and P. Eisenmann, Towards an Erlang-like formula for GPRS/EDGE network engineering, *IEEE International Conference on Communications (ICC)*, June 2004.

[6] B. Baynat, K. Boussetta, P. Eisenmann, and N. Ben Rached, A discrete-time Markovian model for GPRS/EDGE radio engineering with finite-length sessions traffic, *International Symposium on Performance Evaluation of Computer and Telecommunication Systems* (SPECTS04), July 2004.

[7] A. Berger and Y. Kogan, Dimensioning bandwidth for elastic traffic in high-speed data networks, *IEEE/ACM Transactions on Networking*, 8, 643–654, October 2000.

[8] C. Cicconetti, L. Lenzini, E. Mingozzi, and C. Eklund, Quality of service support in IEEE 802.16 networks, *IEEE Network*, 8(2), March 2006.

[9] T. O. Engset, On the calculation of switches in an automatic telephone system, Tore Olaus Engset: The man behind the formula, First published in Norwegian (1915), 1998.

[10] A. Feldmann, A. C. Gilbert, P. Huang, and W. Willinger, Dynamics of IP traffic: A study of the role of variability and the impact of control, *ACM SIGCOMM Computer Communication Review*, 29(4), October 1999.

[11] S. Ben Fredj, T. Bonald, A. Proutiere, G. Regnie, and J. Roberts, Statistical bandwidth sharing: A study of congestion at flow level, *ACM Special Interest Group on Data Communications (Sigcomm)*, August 2001.

[12] T. Bonald and A. Proutiere, Wireless downlink channels: User performance and cell dimensioning, *ACM Mobicom*, 2003.

[13] S. Borst, User-level performance of channel-aware scheduling algorithms in wireless data networks, *IEEE Infocom*, April 2003.

[14] D. Heyman, T. Lakshman, and A. Neidhardt, New method for analyzing feedback protocols with applications to engineering web traffic over the internet, *ACM Sigmetrics*, June 1997.

[15] C. Y. Huang, H.-H. Juan, M.-S. Lin, and C.-J. Chang, Radio resource management of heterogeneous services in mobile WiMAX systems, *IEEE WCNC*, February 2007.

[16] G. Kulkarni, S. Adlakha, and M. Srivastava, Subcarrier allocation and bit loading algorithms for OFDMA-based wireless networks, *IEEE Transactions on Mobile Computing*, 4(6), December 2005.

[17] T. Kwon, H. Lee, S. Choi, J. Kim, D.-H. Cho, S. Cho, S. Yun, W.-H. Park, and K. Kim, Design and implementation of a simulator based on a cross-layer protocol between MAC and PHY layers in a WiBro compatible IEEE 802.16e OFDMA system, *IEEE Communication Magazine*, 43(12), December 2005.

[18] H. Lee, T. Kwon, D. H. Cho, G. Lim, and Y. Chang, Performance analysis of scheduling algorithms for VoIP services in IEEE 802.16e systems, IEEE VTC Spring, 2006.

[19] S. Liu and J. Virtamo, Performance analysis of wireless data systems with a finite population of mobile users, 19th International Teletraffic Congress, 2005.

[20] D. Niyato and E. Hossain, A queuing-theoretic and optimization-based model for radio resource management in IEEE 802.16 broadband networks, *IEEE Transactions on Computers*, 55, 1473–1488, 2006.

[21] G. Nogueira, Methodes analytiques pour le dimensionnement des reseaux cellulaires, PhD thesis at UMPC – Paris 6 (2007), http://wwwrp.lip6.fr/~nogueira/pdf/theseGN.eps.zip.

[22] K. Ramadas and R. Jain, WiMAX System Evaluation Methodology (Wimax Forum), January 2007.

[23] M. Settembre, M. Puleri, S. Garritano, P. Testa, R. Albanese, M. Mancini, and V. Lo Curto, Performance analysis of an efficient packet-based IEEE 802.16 MAC supporting adaptive modulation and coding, 7th IEEE ISCN, 2006.

[24] V. Singh and V. Sharma, Efficient and fair scheduling of uplink and downlink in IEEE 802.16 OFDMA Networks, *IEEE WCNC*, April 2006.

[25] J. Sun, Y. Yanling, and H. Zhu, Quality of service scheduling for 802.16 broadband wireless access systems, *IEEE VTC Spring*, May 2006.

[26] A. Vinel, Y. Zhang, M. Lott, and A. Tiurlikov, Performance analysis of the random access in IEEE 802.16, *IEEE PIMRC*, September 2005.

# Index